Universitext

Editors

J.H. Ewing
F.W. Gehring
P.R. Halmos

Universitext

Editors: J.H. Ewing, F.W. Gehring, and P.R. Halmos

Booss/Bleecker: Topology and Analysis
Charlap: Bieberbach Groups and Flat Manifolds
Chern: Complex Manifolds Without Potential Theory
Chorin/Marsden: A Mathematical Introduction to Fluid Mechanics
Cohn: A Classical Invitation to Algebraic Numbers and Class Fields, 2nd ed.
Curtis: Matrix Groups, 2nd ed.
van Dalen: Logic and Structure
Devlin: Fundamentals of Contemporary Set Theory
Edwards: A Formal Background to Mathematics I a/b
Edwards: A Formal Background to Mathematics II a/b
Endler: Valuation Theory
Frauenthal: Mathematical Modeling in Epidemiology
Gardiner: A First Course in Group Theory
Gårding/Tambour: Algebra for Computer Science
Godbillon: Dynamical Systems on Surfaces
Greub: Multilinear Algebra
Hermes: Introduction to Mathematical Logic
Humi/Miller: Second Order Ordinary Differential Equations
Hurwitz/Kritikos: Lectures on Number Theory
Kelly/Matthews: The Non-Euclidean, The Hyperbolic Plane
Kostrikin: Introduction to Algebra
Luecking/Rubel: Complex Analysis: A Functional Analysis Approach
Lu: Singularity Theory and an Introduction to Catastrophe Theory
Marcus: Number Fields
McCarthy: Introduction to Arithmetical Functions
Mines/Richman/Ruitenburg: A Course in Constructive Algebra
Meyer: Essential Mathematics for Applied Fields
Moise: Introductory Problem Course in Analysis and Topology
Øksendal: Stochastic Differential Equations
Porter/Woods: Extensions of Hausdorff Spaces
Rees: Notes on Geometry
Reisel: Elementary Theory of Metric Spaces
Rey: Introduction to Robust and Quasi-Robust Statistical Methods
Rickart: Natural Function Algebras
Smith: Power Series From a Computational Point of View
Smoryński: Self-Reference and Modal Logic
Stanisić: The Mathematical Theory of Turbulence
Stroock: An Introduction to the Theory of Large Deviations
Sunder: An Invitation to von Neumann Algebras
Tolle: Optimization Methods
Tondeur: Foliations on Riemannian Manifolds

B. Booss
D. D. Bleecker

Topology and Analysis

The Atiyah-Singer Index Formula and Gauge-Theoretic Physics

Translated by D. D. Bleecker and A. Mader

With 75 Illustrations

Springer-Verlag
New York Berlin Heidelberg Tokyo

B. Booss
IMFUFA
Roskilde Universitetscenter
Postbox 260
4000 Roskilde
Denmark

D. D. Bleecker *(co-author and translator)*
A. Mader *(co-translator)*
University of Hawaii
Department of Mathematics
2565 The Mall
Honolulu, Hawaii 96822
U.S.A.

AMS Classification: Primary: 58-02, 58G05, 58G10, 58G15
Secondary: 00A25, 01A65, 14F05, 14H05, 15A60, 16A42, 22E10, 26A57, 30A52, 30A78, 30A88, 30A98, 31A25, 34B25, 35A30, 35J45, 35J50, 35S05, 35S15, 42A68, 45E10, 45K05, 46C05, 46E35, 47A05, 47A55, 47B05, 47B30, 47B35, 47D15, 55F45, 55F50, 58A05, 58C25, 58F10, 60J50

Library of Congress Cataloging in Publication Data
Booss, Bernhelm
Topology and analysis.
(Universitext)
Translation of: Topologie und Analysis.
Bibliography: p.
Includes index.
1. Operator theory. 2. Manifolds (Mathematics)
3. Index theorems. 4. Gauge theories (Physics)
I. Bleecker, David. II. Title: III. Title: Atiyah-Singer index formula and gauge-theoretic physics.
QA329.B6613 1985 515′.72 84-26684

Title of the original German edition: *Topologie und Analysis: Eine Einführung in die Atiyah-Singer-Indexformel,* Springer-Verlag Berlin Heidelberg New York, © 1977.

Printed and bound by R.R. Donnelley & Sons, Harrisonburg, Virginia.
Printed in the United States of America.

9 8 7 6 5 4 3 2

ISBN 0-387-96112-7 Springer-Verlag New York Berlin Heidelberg Tokyo
ISBN 3-540-96112-7 Springer-Verlag Berlin Heidelberg New York Tokyo

To Friedrich Hirzebruch, born October 17, 1927, for his 50th birthday.

Preface

The Motivation. With intensified use of mathematical ideas, the methods and techniques of the various sciences and those for the solution of practical problems demand of the mathematician not only greater readiness for extra-mathematical applications but also more comprehensive orientations within mathematics. In applications, it is frequently less important to draw the most far-reaching conclusions from a *single* mathematical idea than to cover a subject or problem area tentatively by a proper "variety" of mathematical theories. To do this the mathematician must be familiar with the shared as well as specific features of different mathematical approaches, and must have experience with their interconnections. The Atiyah-Singer Index Formula, "one of the deepest and hardest results in mathematics", "probably has wider ramifications in topology and analysis than any other single result" (F. Hirzebruch) and offers perhaps a particularly fitting example for such an introduction to "Mathematics": In spite of its difficulty and immensely rich interrelations, the realm of the Index Formula can be delimited, and thus its ideas and methods can be made accessible to students in their middle semesters.*

In fact, the Atiyah-Singer Index Formula has become progressively "easier" and "more transparent" over the years. The discovery of deeper and more comprehensive applications (see Chapter III.4) brought with it, not only a vigorous exploration of its methods particularly in the many-facetted and always new presentations of the material by M. F. Atiyah and I. M. Singer themselves, but also in the works of L. Hörmander which

*In the U.S., first or second year graduate students (Transl.)

carry independent weight. The sources and parts of the Index Formula have moved closer together. In comparison to the "classical" introduction to the Index Formula by [Palais 1965], we have among others the following advantages:

1. In addition to the original "cobordism proof" two further proofs are available today of which we will choose the "*imbedding proof*". It is the conceptually easiest proof of the Index Formula particularly because of its direct link to the Bott Periodicity Theorem (see Chapter III.3).

2. The structure of the space F of Fredholm operators has become much clearer, due to the K-theoretic notion of "index bundle" of a continuous family of Fredholm operators (see particularly the Theorem of Atiyah-Jänich, Chapter I.7, which is based on the Theorem of Kuiper on the contractibility of $B^{\times}$, Chapter I.6) and also due to the supply of practical examples furnished by the Wiener-Hopf operators (see Chapter I.9).

3. The calculus of "integral differential operators", which is essential for the analysis of elliptic operators, has been radically clarified in the theory of *pseudo-differential operators* (and Fourier integral operators; see the Theorem of Kuranishi, Chapter II.3).

4. The *Bott Periodicity Theorem*, which is the fundamental topological theorem of index theory, could be formulated much more simply in the "K-theory with compact support". Hereby its function analytic core became more visible, namely the index theorem of Gohberg-Krein for Wiener-Hopf operators.

The Plan. What is the relation between the topology of the general linear group $GL(N,\mathbb{C})$, the geometry of differentiable manifolds, and the behavior of the solutions of elliptic differential equations? How can the "index" which is defined analytically be expressed in terms of topological invariants? The sketch of the solution to this problem - which for better comprehension proceeds neither in greatest generality nor in geodetic brevity - focuses on presenting the contrasting methods which interact in the conquest of the index problem. I tried to exhibit at a focal point of the development of the mathematics in our century, the profound reasons and the promises of the union of topology and analysis, two separate main areas of mathematics. In order to cope with the large amount of material and in view of the modest background which can be assumed as experience teaches, I chose a mixed strategy. Most of the necessary material is proven *explicitly and in all detail*.

In addition there are *expositions in the form of exercises* with precise formulations, hints and references. In widely disseminated *comments,* I tried to summarize the material and provide the background needed.

On the Use. This book is intended for readers with a minimal background of knowledge and experience. It intends to make very clear a definite theme, the index formula, and to confer a certain awareness of methods. It is a textbook for independent study and is suitable for a one or two semester course on this theme and related topics, but most of all, it was written as a text for seminars for students in the middle of the studies. Each chapter is reasonably self-contained and separate and is suitable as a study assignment for a group of students, whereby the weightier chapters should be subdivided. For the presentation of each chapter in seminars, one has to allot one to four talks of 60 to 90 minutes duration. Experience tells that two semesters are enough to cover the material (up to the index formula) with a judicious choice of topics. On the other hand, it is hardly possible to cover adequately the full expanse of applications of the Atiyah-Singer theory in the course of a normal term of study. The weight of each individual area of application is too great. It is therefore advisable to concentrate on a single area in working with students - depending on their background and research interests - and to work out, with greater depth and detail than in our survey in Chapter III.4, its relationship to the index formula.

The Prerequisites. We assume the following background knowledge:

1. Basic notions of *linear algebra* (vector spaces, linear transformations, matrix rings, determinants) and of *real analysis* (open, closed, compact sets, continuous and differentiable functions, integration and, at least in Chapters I.8/9, the basics of Lebesgue integration).

2. Notion of a *complex separable Hilbert space* (scalar products, Schwarz Inequality, norm, completeness, orthogonality, orthogonal basis, orthogonal projection onto closed subspaces and identification of a Hilbert space with its dual space by means of the inner product). The precise definitions and theorems with complete proofs are readily available.

3. The *open mapping principle* according to which every continuous map on a Hilbert space *onto* another is open, i.e., maps open sets to open sets. (Included here is, as a corollary, the "Theorem of the Inverse Operator" which says that the inverse of a bijective continuous linear map

on a Hilbert space to another is again continuous.)

4. A few standard results of the *elementary theory of ordinary differential equations*. Essentially, one need only be familiar with "linear homogeneous systems of differential equations with constant coefficients" and linear differential equations of n-th order with constant coefficients. In individual cases, further references will be given.

5. Notion of a *differentiable manifold*. We will assemble, in a crash course, the definitions and theorems most important for us and give extensive references to the literature. However, it is better if the reader - for example in the context of quadratic surfaces (conic sections) in linear algebra - acquires a certain spacial imagination and facility with computations (change of coordinates).

Roskilde and Bielefeld
July 1977

B. Booss

Preface to English Edition

In the six years since the publication of the German edition, the main results and their applications have been extended in several directions. Among these directions, we have the following:

- Elliptic boundary-value problems. (e.g., see Rempel, S. & B.-W. Schulze, Index Theory of Elliptic Boundary Problems, Akademie-Verlag, Berlin, 1982.)
- C*-algebras and foliated manifolds. (Cf. Connes, A., Non-commutative differential geometry, IHES, Paris [to appear].)
- Applications to field theories in physics. (e.g., see [Eguchi-Gilkey-Hanson, 1980], [Madore, J., 1981], and Part IV of this book.)
- Topology of smooth 4-manifolds. (i.e., S. K. Donaldson's Theorem; cf. [H. B. Lawson 1983], [Freed-Freedman-Uhlenbeck, 1983], and Section IV.3.E of this book.)

Moreover, the machinery used has developed considerably and has become standardized in a number of monographs. This is especially true for the theory of pseudo-differential operators. We refer to the following:

Kumano-Go, H., Pseudo-differential Operators, MIT Press, Cambridge, Mass., 1981.

Shubin, M., Pseudo-differential Operators and Spectral Theory, Nauka, Moscow 1978 (in Russian).

Taylor, M. E., Pseudo-differential Operators, Princeton Univ. Press, Princeton, 1981.

Treves, F., Introduction to Pseudo-differential and Fourier Integral Operators I & II, Plenum Press, New York, 1980.

and for K-theory

Karoubi, M., K-theory - An Introduction, Springer-Verlag, New York, 1978.

However, the core of index theory is of course the Atiyah-Singer Index theorem in its now classical form, as presented in the German edition. Hence, only minor adjustments were made to the main text (Parts I, II & III). The new Part IV (The Index Formula in Gauge-Theoretical Physics) will hopefully serve a number of purposes. Chapter IV.1, provides some insight as to how the concepts surrounding the index theorem (e.g., manifolds, vector bundles, connections, pseudo-differential operators, etc.) have necessarily entered into our description of the fundamental laws of physics. Chapter IV.2 covers the essential preliminary geometrical notions which are needed to understand the application of the index formula to instanton parameter counting in gauge theory. In IV.3, the index of the relevant operator is computed, and the methods of non-linear functional analysis are used to show that the moduli space of irreducible self-dual connections (i.e., instantons) is naturally a manifold under suitable hypotheses. All of this is presented within the grasp of the reader who has only a basic familiarity with manifolds, differential forms, and the index formula. The final section IV.3.E is devoted to a sketch of the proof of Donaldson's Theorem.

The author B. Booss thanks D. Bleecker and A. Mader for their efficient work in translating and editing the new edition, and especially D. Bleecker for writing the new Part IV. He also would like to thank the many colleagues who have commented on the first edition; this helped a great deal in getting rid of errors. D. Bleecker thanks B. Booss for inviting him to write Part IV, and he thanks A. Mader for translating the more verbal passages which would have been impossibly difficult otherwise.

Honolulu and Roskilde
December 1983

B. Booss
D. Bleecker
A. Mader

Contents

PART II. ANALYSIS ON MANIFOLDS

PART III. THE ATIYAH-SINGER INDEX FORMULA

PART IV. THE INDEX FORMULA AND GAUGE-THEORETICAL PHYSICS

Part I. Operators with Index

"If we do not succeed in solving a mathematical problem, the reason frequently consists in our failure to recognize the more general standpoint from which the problem before us appears only as a single link in a chain of related problems."
(D. Hilbert, 1900)

1. FREDHOLM OPERATORS

A. Hierarchy of Mathematical Objects

"In the hierarchy of branches of mathematics, certain points are recognizable where there is a definite transition from one level of abstraction to a higher level. The first level of mathematical abstraction leads us to the concept of the individual numbers, as indicated for example by the Arabic numerals, without as yet any undetermined symbol representing some unspecified number. This is the stage of elementary arithmetic; in algebra we use undetermined literal symbols, but consider only individual specified combinations of these symbols. The next stage is that of analysis, and its fundamental notion is that of the arbitrary dependence of one number on another or of several others - the function. Still more sophisticated is that branch of mathematics in which the elementary concept is that of the transformation of one function into another, or, as it is also known, the operator."

Thus Norbert Wiener characterized the "hierarchy" of mathematical objects [Wiener 1933, 1]. Very roughly we can say: Classical questions of analysis are aimed mainly at investigations within the third or fourth level. This is true for real and complex analysis, as well as the functional analysis of differential operators with its focus on existence and uniqueness theorems, regularity of solutions, asymptotic or boundary behavior which are of particular interest here. Thereby research progresses naturally to operators of more complex composition and greater generality without usually changing the concerns in principle; the work remains directed mainly towards qualitative results.

In contrast it was topologists, as Michael Atiyah variously noted, who turned systematically towards quantitative questions in their topological investigations of algebraic manifolds, their determination of quantitative measures of qualitative behavior, the definition of global topological invariants, the computation of intersection numbers and dimensions. In this way, they again broadly broke through the rigid separation of the "hierarchical levels" and specifically investigated relations between these levels, mainly of the second and third level (algebraic surface - set of zeros of an algebraic function) with the first, but also of the fourth level (Laplace operators on Riemannian manifolds, Cauchy-Riemann operators, Hodge theory) with the first.

This last direction, starting with the work of William V. D. Hodge, continuing with Kunihiko Kodaira and Donald Spencer, with Henri Cartan and Jean Pierre Serre, with Friedrich Hirzebruch, Michael Atiyah and others, can perhaps be best described with the key word "differential topology" or "analysis on manifolds". Both its relation with and distinction from analysis proper is that

> "Roughly speaking we might say that the analysts were dealing with complicated operators and simple spaces (or were only asking simple questions), while the algebraic geometers and topologists were only dealing with simple operators but were studying rather general manifolds and asking more refined questions." [Atiyah 1968, 57]

We can read, e.g. in [Brieskorn 1974, 278-283] and the literature given there, to what degree the contrast between quantitative and qualitative questions and methods must be considered a driving force in the development of mathematics beyond the realm sketched above.

Actually, in the 1920's already, mathematicians such as Fritz Noether and Torsten Carlemann had developed the purely function analytical concept of the "index" of an operator in connection with integral equations, and had determined its essential properties. But "although its" (the theory of Fredholm operators) "construction did not require the development of significantly different means, it developed very slowly and required the efforts of very many mathematicians" [Gohberg-Krein 1957, 185]. And although Soviet mathematicians such as Ilja V. Vekua had hit upon the index of elliptic differential equations at the beginning of the 1950's, we find no reference to these applications in the quoted principal work on Fredholm operators. In 1959 Israel M. Gelfand published a programmatic article asking for a systematic study of elliptic differential equations from this quantitative point of view. He took as a starting point the theory of Fredholm operators with its theorem on the homotopy invariance of the index (see below). Only after the subsequent work of Michail S. Agranovich, Alexander S. Dynin, Aisik I. Volpert, and finally of Michael Atiyah, Raoul Bott, Klaus Jänich and Isadore M. Singer, did it become clear that the theory of Fredholm operators is indeed fundamental for numerous quantitative computations, and a genuine link connecting the higher "hierarchical levels" with the lowest one, the numbers.

B. The Concept of Fredholm Operator

Let H be a (separable) complex Hilbert space, and let $\mathcal{B}$ be the Banach algebra (e.g., [Schechter 1971, 208]) of bounded linear operators $T: H \to H$ with the operator norm

$$||T|| := \sup\{|Tu| : |u| \leq 1\} < \infty$$

where $|\cdot\cdot|$ is the norm in H induced by the inner product $<..,..>$.

An operator $T \in \mathcal{B}$ is called a *Fredholm operator*, if

$$\text{Ker } T := \{u \in H : Tu = 0\} \quad \text{and} \quad \text{Coker } T := H/\text{Im}(T)$$

are finite dimensional. This means that the homogeneous equation $Tu = 0$ has only finitely many linearly independent solutions, and to solve $Tu = v$, it is sufficient that v satisfy a finite number of linear conditions (e.g., see Exercise 3.1b below). We write $T \in \mathcal{F}$ and define the *index* of T by

$$\text{index } T := \dim \text{Ker } T - \dim \text{Coker } T.$$

The dimension of Coker T is called the *codimension* of $\text{Im}(T) = T(H) = \{Tu : u \in H\}$ or the *defect* of T. □

Remark. We can analogously define Fredholm operators $T : H \to H'$, where H and H' are Hilbert spaces, Banach spaces, or general topological vector spaces. In this case, we use the notation $\mathcal{B}(H,H')$ and $\mathcal{F}(H,H')$, corresponding to $\mathcal{B}$ and $\mathcal{F}$ above. However, in order to counteract a proliferation of notation and symbols in this section, we will deal with a single Hilbert space H and its operators as far as possible. The general case $H \neq H'$ does not require new arguments anywhere.

All results are valid for Banach spaces and a large part for Fréchet spaces also. For details see for example [Przevorska-Rolevicz, 182-318]. We will not use any of these but will be able to restrict ourselves entirely to the theory of Hilbert space whose treatment is in parts far simpler.

Motivated by analysis on symmetric spaces - with transformation group G - operators have been studied whose "index" is not a number but an element of the representation ring $R(G)$ generated by the characters of finite-dimensional representations of G [Atiyah-Singer 1968a, 519 ff] or, still more generally, is a distribution on G [Atiyah 1974, 9-17]. We will not treat this largely analogous theory, nor the generalization of the Fredholm theory to the discrete situation of von-Neumann algebras as it has been carried out - with real valued index - in [Breuer 1968/1969].

<u>Exercise 1</u>. Let $L^2(\mathbb{Z}_+)$ be the space of sequences $c = (c_0,c_1,c_2,\dots)$ of complex numbers with square-summable absolute values; i.e.,

$$\sum_{n=0}^{\infty} |c_n|^2 < \infty.$$

$L^2(\mathbb{Z}_+)$ is a Hilbert space in a natural way (see Ch. 8 below). Show that the forward shift

$$\text{shift}^+ : c \longmapsto (0,c_0,c_1,c_2,\dots)$$

and the backward shift

$$\text{shift}^- : c \longmapsto (c_1,c_2,c_3,\dots)$$

are Fredholm operators with $\text{index}(\text{shift}^+) = -1$ and $\text{index}(\text{shift}^-) = +1$.

<u>Warning</u>: Just as we can regard $L^2(\mathbb{Z}_+)$ as the limit of the finite dimensional vector spaces $\mathbb{C}^m$ (as $m \to \infty$), we can approximate shift^+ by endomorphisms of $\mathbb{C}^m$ given, relative to the standard basis, by the matrix

$$\begin{pmatrix} 0 & & & 0 \\ 1 & \ddots & & \\ & \ddots & \ddots & \\ 0 & & 1 & 0 \end{pmatrix}$$

Note that the kernel and cokernel of this endomorphism are one dimensional, whence the index is zero. (See Exercise 2.1 below.)

We have yet another situation, when we consider the Hilbert space $L^2(\mathbb{Z})$ of sequences $c = (\dots c_{-2},c_{-1},c_0,c_1,c_2,\dots)$ with

$$|c_0|^2 + \sum_{n=1}^{\infty} (|c_n|^2 + |c_{-n}|^2) < \infty.$$

The corresponding shift operators are now bijective, and hence have index zero. □

2. ALGEBRAIC PROPERTIES. OPERATORS OF FINITE RANK

Exercise 1. For finite-dimensional vector spaces the notion of Fredholm operator is empty, since then every linear map is a Fredholm operator. Moreover, the index no longer depends on the explicit form of the map, but only on the dimensions of the vector spaces between which it operates. More precisely, show that every linear map $T : H \to H'$ where H and H' are finite-dimensional vector spaces has index given by

$$\text{index } T = \dim H - \dim H'.$$

Hint: One first recalls the vector-space isomorphism $H/\text{Ker}(T) \cong \text{Im}(T)$ and then (since H, H' are finite-dimensional) obtains the well-known identity from linear algebra

$$\dim H - \dim \text{Ker } T = \dim H' - \dim \text{Coker } T.$$

Warning: If we let the dimensions of H and H' go to ∞, we obtain only the formula index $T = \infty - \infty$. Thus, we need an additional theory, to give this difference a particular value. □

Exercise 2. For two Fredholm operators $F : H \to H$ and $G : H' \to H'$, consider the direct sum

$$F \oplus G : H \oplus H' \to H \oplus H'.$$

Show that $F \oplus G$ is a Fredholm operator with

$$\text{index}(F \oplus G) = \text{index } F + \text{index } G$$

Hint: First verify that $\text{Ker}(F \oplus G) = \text{Ker } F \oplus \text{Ker } G$, and the corresponding fact for $\text{Im}(F \oplus G)$. Then show that

$$H \oplus H'/(\text{Im } F \oplus \text{Im } G) \cong H/\text{Im } F \oplus H'/\text{Im } G.$$

Warning: When $H = H'$, we can consider the sum (not direct)

$$F + G : H \to H,$$

but in general this is not a Fredholm operator; e.g., we could set $G := -F$. □

Exercise 3. Show by algebraic means, that the composition $G \circ F$ of two Fredholm operators $F : H \to H'$ and $G : H' \to H''$ is again a Fredholm operator.

Hint: Which inequality holds between

$\dim \operatorname{Ker} G \circ F$ and $\dim \operatorname{Ker} F + \dim \operatorname{Ker} G$

and between

$\dim \operatorname{Coker} G \circ F$ and $\dim \operatorname{Coker} F + \dim \operatorname{Coker} G$?

Warning: Why are these in general not equalities? Nevertheless, the chain rule $\operatorname{index} G \circ F = \operatorname{index} F + \operatorname{index} G$ can be proved, since the inequalities cancel out when we form the difference. Sometimes in mathematics a result of interest appears as an offshoot of the proof of a rather dull general statement. Such is the case when we obtain the chain rule for differentiable functions by showing that the composite of differentiable maps is again differentiable. Here, however, the proof of the index formula requires extra work, and we either must utilize the topological structure by function analytic means (see Exercise 3.2 below) or refine the algebraic arguments. The latter will be done next. □

A. The Snake Lemma. We recall a basic idea of "diagram chasing": $H_1, H_2, \ldots$ is a system of vector spaces and $T_k : H_k \to H_{k+1}$ is a linear map for each $k = 1,2,\ldots$. We call

$$H_1 \xrightarrow{T_1} H_2 \xrightarrow{T_2} H_3 \xrightarrow{T_3} \cdots$$

an *exact sequence*, if for all k

$$\operatorname{Im} T_k = \operatorname{Ker} T_{k+1}.$$

In particular, the following is a list of equivalences (where 0 denotes the 0-dimensional vector space)

$0 \longrightarrow H_1 \xrightarrow{T_1} H_2$ exact : $\operatorname{Ker} T_1 = 0$, T_1 injective

$H_1 \xrightarrow{T_1} H_2 \longrightarrow 0$ exact : $\operatorname{Im} T_1 = H_2$, T_1 surjective

$0 \longrightarrow H_1 \longrightarrow H_2 \longrightarrow 0$ exact : $H_1 \cong H_2$

$0 \longrightarrow H_1 \longrightarrow H_2 \longrightarrow H_3 \longrightarrow 0$ exact : $H_3 \cong H_2/H_1$ and $H_2 \cong H_1 \oplus H_3$.

<u>Theorem 1</u> (<u>Snake Lemma</u>): Assume that the following diagram of vector spaces and linear maps is commutative (i.e., $jF = F'i$ and $qF' = F''p$) with exact vertical sequences and Fredholm operators for horizontal maps. Then we have

$$\text{index } F - \text{index } F' + \text{index } F'' = 0.$$

$$\begin{array}{ccc} 0 & & 0 \\ \downarrow & & \downarrow \\ H_1 & \xrightarrow{F} & H_2 \\ i\downarrow & & \downarrow j \\ H_1' & \xrightarrow{F'} & H_2' \\ p\downarrow & & \downarrow q \\ H_1'' & \xrightarrow{F''} & H_2'' \\ \downarrow & & \downarrow \\ 0 & & 0 \end{array}$$

<u>Proof</u>: (See also the standard algebra texts; e.g., [MacLane, §II.5]). We do the proof in two parts.

1. Here we show that the following sequence is exact:

$$\begin{aligned} &0 \to \text{Ker } F \to \text{Ker } F' \to \text{Ker } F'' \to \\ &\to \text{Coker } F \to \text{Coker } F' \to \text{Coker } F'' \to 0. \end{aligned} \qquad (*)$$

For this, we first explain how the individual maps are defined. By the commutativity of the diagram, the maps $\text{Ker } F \to \text{Ker } F'$ and $\text{Ker } F' \to \text{Ker } F''$ are given by i and p; and the maps $\text{Coker } F \to \text{Coker } F'$ and $\text{Coker } F' \to \text{Coker } F''$ are induced by j and q in the natural way (please check).

Also $\text{Ker } F'' \to \text{Coker } F$ is well defined: Let $u'' \in \text{Ker } F''$; i.e., $u'' \in H_1''$ and $F''u'' = 0$. Since p is surjective, we can choose $u' \in H_1'$ with $pu' = u''$. Then $F'u' \in \text{Ker } q$, since $qF'u' = F''pu' = F''u'' = 0$. By exactness, we have $\text{Ker } q = \text{Im } j$ and a unique (by the injectivity of j) element $u \in H_2$ with $ju = F'u'$. We map $u'' \in \text{Ker } F''$ to the class of u in $H_2/\text{Im } F = \text{Coker } F$. It remains to show that we get the same class for another choice of u'. Thus, let $\tilde{u}' \in H_1'$ be such that $p\tilde{u}' = u''$ (p is in general not injective; hence, possibly $\tilde{u}' \neq u'$). As above, we have a $\tilde{u} \in H_2$ with $j\tilde{u} = F'\tilde{u}'$. We must now find $u_0 \in H_1$ with $Fu_0 = u - \tilde{u}$. Then we are done. For this, note that $j(u-\tilde{u}) = ju - j\tilde{u} = F'u' - F'\tilde{u}' = F'(u'-\tilde{u}')$. Since $pu' = p\tilde{u}' = u''$, we have $u'-\tilde{u}' \in \text{Ker } p$. By exactness, we have $u_0 \in H_1$ with $iu_0 = u'-\tilde{u}'$, whence $F'iu_0 = F'(u'-\tilde{u}')$. The left side is jFu_0 and the right side is $j(u-\tilde{u})$ from above. Hence, $Fu_0 = u-\tilde{u}$ by the injectivity of j, as desired.

We introduced the map $\text{Ker } F'' \to \text{Coker } F$ in such great detail in order to demonstrate what is typical for diagram chasing: it is straightforward, largely independent of tricks and ideas, readily reproduced, and hence somewhat monotonous. Therefore we will forgo showing the exactness of (*) except at $\text{Ker } F'$, and leave the rest as an exercise.

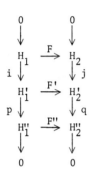

The exactness of $\text{Ker } F \xrightarrow{\tilde{i}} \text{Ker } F' \xrightarrow{\tilde{p}} \text{Ker } F''$, where $\tilde{i}$ and $\tilde{p}$ are the restrictions of i and p, means $\text{Im } \tilde{i} = \text{Ker } \tilde{p}$. Thus, we have two inclusions to show:

$\subset$: This is clear, since $p \circ i = 0$ implies $\tilde{p} \circ \tilde{i} = 0$.

$\supset$: If $u' \in \text{Ker } \tilde{p}$, then $u' \in \text{Ker } F'$ and $pu' = 0$.

By the exactness of $H_1 \to H_1' \to H_1''$, we have $u \in H_1$ with $iu = u'$. It remains to show that $Fu = 0$, but this is clear, since $jFu = F'iu = F'u' = 0$ and j is injective.

Remark. Before we go to part 2 of the proof, we pause for a moment: It is interesting that for $H' = H \oplus H''$, $i = j$ the inclusion, and $p = q$ the projection, we recapture the addition formula of Exercise 2. In this case, the exact sequence (*) then breaks, as shown there, into two parts

$$0 \to \text{Ker } F \to \text{Ker } F' \to \text{Ker } F'' \to 0 \quad \text{and}$$
$$0 \to \text{Coker } F \to \text{Coker } F' \to \text{Coker } F'' \to 0.$$

In the general case, however, we no longer have

$$\dim \text{Ker } F' = \dim \text{Ker } F + \dim \text{Ker } F''$$

and $\dim \text{Coker } F' = \dim \text{Coker } F + \dim \text{Coker } F''$, but instead we must consider the interaction $(\text{Ker } F'' \to \text{Coker } F)$.

From a topological standpoint, the concept of the index of a Fredholm operator is a special case of the general concept of the *Euler characteristic* $\chi(C)$ of a complex

$$C : \xrightarrow{T_{r+1}} C_r \xrightarrow{T_r} C_{r-1} \xrightarrow{T_{r-1}} C_{r-2} \xrightarrow{T_{r-2}} C_{r-3} \xrightarrow{T_{r-3}}$$

of vector spaces and linear maps (with $T_k \circ T_{k+1} = 0$) with finite *Betti numbers*

$$\beta_k = \dim \text{Ker } T_k / \text{Im } T_{k+1}.$$

Here, $\text{Ker } T_k/\text{Im } T_{k+1}$ is called the k-th *homology space* $H_k(C)$. Assuming that all these numbers are finite, as well as the number of nonzero Betti numbers, we define

$$\chi(C) := \sum_k (-1)^k \beta_k,$$

whence

$$\text{index } F = \chi(C)$$

for

$$C : 0 \to 0 \to H \xrightarrow{F} H \to 0 \to 0,$$

where $C_2 = C_1 = H$ and $C_i = 0$ otherwise; thus, $H_2(C) = \text{Ker } F$ and $H_1(C) = \text{Coker } F$. This is the reason why we can follow in the proof of Theorem 1 the well-known topological arguments (see [Greenberg, 100f] = [Eilenberg-Steenrod, 52f]) yielding the "addition-" or "pasting theorem"

$$\chi(C) - \chi(C') + \chi(C'/C) = 0$$

see also Section III.4.D. In particular, (*) is only a special case of the "long exact homology sequence"

$$\to H_{k+1}(C'/C) \to H_k(C) \to H_k(C') \to H_k(C'/C) \to H_{k-1}(C) \to H_{k-1}(C') \to, \quad (**)$$

[Greenberg, 57f] = [Eilenberg-Steenrod, 125-128].

This more general description would have the advantage that we only need to prove exactness of (**) at three adjacent places with an argument independent of k - rather than at six places as in our "simplified" approach where we restricted ourselves to complexes of length two. But back to our proof:

2. For each exact sequence

$$0 \to A_1 \to A_2 \to \ldots \to A_r \to 0, \quad (1)$$

of finite-dimensional vector spaces, we wish to derive the formula

$$\sum_{k=1}^{r} (-1)^k \dim A_k = 0.$$

Notice first that for r sufficiently large ($r > 3$), the formula for the alternating sum for (1) follows, once we know the formula holds for the exact (prove!) sequences

$$0 \to A_1 \to A_2 \to \text{Im}(A_2 \to A_3) \to 0 \quad (2)$$

and

$$0 \to \text{Im}(A_2 \to A_3) \to A_3 \to \ldots \to A_r \to 0 \quad (3)$$

Since these sequences have length less than r, the formula is proved by induction, if we verify it for $r = 1,2,3$.

$r = 1$: trivial, since then, $A_1 \cong 0$.

$r = 2$: also clear, since then $A_1 \cong A_2$.

$r = 3$: clear, since $0 \to A_1 \to A_2 \to A_3 \to 0$ implies $A_3 \cong A_2/A_1$, whence $\dim A_3 = \dim A_2 - \dim A_1$.

Thus, part 2 is finished and combining it with part 1, the snake lemma is proved. Actually, we have proven much more, namely, whenever two of the three maps F, F', F'' have a finite index, the third has finite index given by the snake formula. □

Exercise 4. Combine Theorem 1 and Exercise 3, to show that

$$\text{index } G \cdot F = \text{index } F + \text{index } G.$$

Hint: Consider the diagram, where

$iu := (u, Fu)$

$jv := (Gv, v)$

$p(u,v) := Fu - v$

$q(w,v) := w - Gv.$ □

$$\begin{array}{ccc}
0 & & 0 \\
\downarrow & & \downarrow \\
H & \xrightarrow{F} & H' \\
\downarrow i & & \downarrow j \\
H \oplus H' & \xrightarrow{G \cdot F \oplus \text{Id}} & H'' \oplus H' \\
\downarrow p & & \downarrow q \\
H' & \xrightarrow{G} & H'' \\
\downarrow & & \downarrow \\
0 & & 0
\end{array}$$

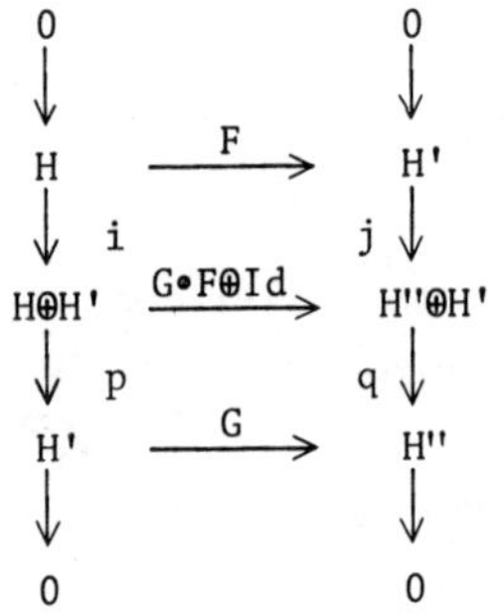

B. Operators of Finite Rank and the Fredholm Integral Equation

Exercise 5. Show that for any operator $K : H \to H$ of finite rank (i.e., $\dim K(H) < \infty$), the sum $\text{Id} + K$ is a Fredholm operator and

$$\text{index}(\text{Id} + K) = 0$$

Here, $\text{Id} : H \to H$ is the identity.

Hint: Set $h := \text{Im } K$ and recall Theorem 1 for the adjacent diagram. To see that the row maps are well defined, we only need

$(\text{Id}+K)(h) \subset h.$

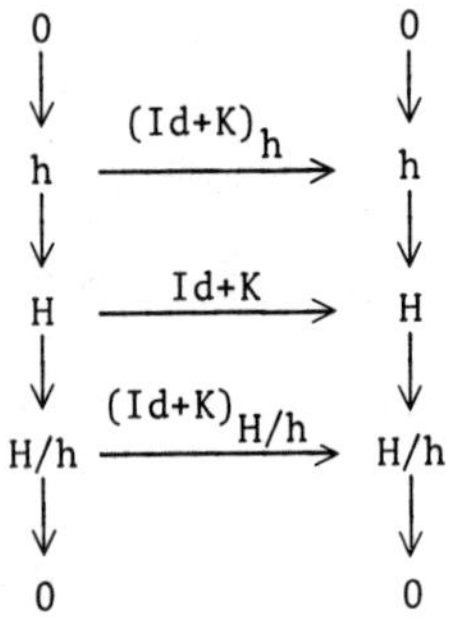

The commutativity of the diagram and the exactness of the columns are clear. Since one can show $\text{Ker}(\text{Id}+K) \subset h$ and $\dim \text{Coker } (\text{Id}+K) \leq \dim h$, the Snake Formula gives us the result once we show

$$\text{index}(\text{Id}+K)_h = 0 \qquad (1)$$

$$\begin{array}{ccc}
0 & & 0 \\
\downarrow & & \downarrow \\
h & \xrightarrow{(\text{Id}+K)_h} & h \\
\downarrow & & \downarrow \\
H & \xrightarrow{\text{Id}+K} & H \\
\downarrow & & \downarrow \\
H/h & \xrightarrow{(\text{Id}+K)_{H/h}} & H/h \\
\downarrow & & \downarrow \\
0 & & 0
\end{array}$$

and

$$\mathrm{index}(\mathrm{Id}+K)_{H/h} = 0 \tag{2}$$

But (1) is clear from Exercise 1 and (2) is trivial because $(\mathrm{Id}+K)_{H/h} = \mathrm{Id}_{H/h}$. Note that one could deduce that $\mathrm{Id}+K$ has finite index by using the observation at the end of the proof of Theorem 1. □

Remark. One may be bothered by the way in which the proposed solution produces the result so directly from the Snake Formula by means of a trick. As a matter of fact, $\mathrm{index}(\mathrm{Id}+K)$ can be computed in a pedestrian fashion by reduction to a system of n linear equations with n unknowns where $n := \dim h$. To do this one verifies that every operator K of finite rank has the form

$$Ku = \sum_{i=1}^{n} \langle u,u_i\rangle v_i$$

with fixed $u_1,\dots,u_n, v_1,\dots,v_n \in H$ (note that every continuous linear functional is of the form $\langle \dots,u_0\rangle$). Whether this "direct" approach, as detailed for example in [Schechter, 88-91], is in fact more transparent than the device used with the Snake Formula, depends a little on the perspective. While in the first approach the key point (namely the use of Exercise 1 for equation (1)) is singled out and separated clearly in the remaining formal argument, we find in the second more constructive approach rather a fusion of the nucleus with its packaging. However, the use of the fairly non-trivial Riesz-Fischer Lemma is unnecessary in the case where K is given in the desired explicit form, as in the following example.

Exercise 6. Consider the *Fredholm integral equation of the second kind*

$$u(x) + \int_a^b G(x,y)u(y)\,dy = h(x)$$

with degenerate (product-) weight function

$$G(x,y) = \sum_{i=1}^{n} f_i(x)\overline{g_i(y)}$$

with fixed $a < b$ real and f_i, g_i square integrable on $[a,b]$. Prove the *Fredholm alternative: Either* there is a unique solution $u \in L^2[a,b]$ for every given right side $h \in L^2[a,b]$, *or* the homogeneous equation $(h = 0)$ has a solution which does not vanish identically. Moreover, the number of linearly independent solutions of the homogeneous equation equals the number of linear conditions one needs to impose on h in order that the inhomogeneous equation be solvable.

Hint: Consider the operator Id + K on the Hilbert space $L^2[a,b]$, where $Ku = \Sigma\langle u,g_i\rangle f_i$, and apply Exercise 5. For the interpretation of the dimension of the cokernel, see Exercise 3.1b below. □

3. ANALYTIC METHODS. COMPACT OPERATORS

A. Analytic Methods

With the last two problems we have reached the limits of our - so far - purely algebraic reasoning where we could reduce everything to the elementary theory of solutions of n linear equations in n unknowns. In fact, the limit process $n \to \infty$ marks the emergence of functional analysis which went beyond the methods of linear algebra while being motivated by its questions and results. This occurred mainly in the study of integral equations.

In 1927 already, Ernst Hellinger and Otto Toeplitz stressed in their article "Integral equations and equations in infinitely many unknowns" in the "Enzyklopädie" that "the essence of the theory of integral equations rests in the analogy with analytic geometry and more generally in the passage from facts of algebra to facts of analysis" [Hellinger-Toeplitz, 1343]. They showed in a concise historical survey how the awareness of these connections progressed in the centuries since Daniel Bernoulli investigated the oscillating string as a limit case of a system of n mass points:

> "Orsus itaque sum has meditationes a corporibus duobus filo flexili in data distantia cohaerentibus; postea tria consideravi moxque quatuor, et tandem numerum eorum distantiasque qualescunque; cumque numerum corporum infinitum facerem, vidi demum naturam oscillantis catenae sive aequalis sive inaequalis crassitiei sed ubique perfecte flexilis."+

The passage to the limit means for the Fredholm integral equation of Exercise 2.6 that more general nondegenerate weight functions G(x,y) are allowed which then can be approximated by degenerate weights (e.g.,

---

+Petropol Comm. 6 (1732/33, ed. 1738), 108-122. In English: "In these considerations I started with two bodies at a fixed distance and connected by an elastic string; next I considered three then four and finally an arbitrary number with arbitrary distances between them; but only when I made the number of bodies infinite did I fully comprehend the nature of an oscillating elastic chain of equal or varying thickness."

polynomials). Such weights play a prominent role in applications, especially when dealing with differential equations with boundary conditions, as we will see below.

For the theory of Fredholm operators developed here we must analogously abandon the notion of an operator of finite rank and generalize it (to compact operator, see below) whereby topological, i.e. continuity considerations, become essential in connection with the limit process. This will bring out the full force of the concept of Fredholm operator. We now turn to this topic.

Exercise 1. For $F : H \to H$ a Fredholm operator, prove:

a) $\operatorname{Im} F$ is closed.

b) There is an explicit criterion for deciding when an element of H lies in $\operatorname{Im} F$. Namely, let $n = \dim \operatorname{Coker} F$; then there are $u_1,\ldots,u_n \in H$ such that for all $w \in H$, we have

$$w \in \operatorname{Im} F \Longleftrightarrow \langle w,u_1\rangle = \ldots = \langle w,u_n\rangle = 0.$$

Hint for a: As a vector subspace of H, naturally $\operatorname{Im} F$ is closed under addition and multiplication by complex numbers. However, here we are interested in topological closure, namely that in passing to limit points we do not leave $\operatorname{Im} F$. This has far-reaching consequences, since only closed subspaces inherit the completeness property of the ambient Hilbert space, and hence have complete orthonormal bases (e.g., [Schechter 1971, 30 and 255]). The trivial fact that finite-dimensional subspaces are closed can be exploited:

Since $\operatorname{Coker} F = H/\operatorname{Im} F$ has finite dimension, we can find $v_1,\ldots,v_n \in H$ whose classes in $H/\operatorname{Im} F$ form a basis. The linear span h of $v_1,\ldots,v_n$ is then an algebraic complement of $\operatorname{Im} F$ in H. Consider the map $F' : H \oplus h \to h$ with $F'(u,v) := Fu + v$. Since F' is linear, surjective, and (by the boundedness of F) continuous, we have that F' is open (according to the "open mapping principle"). It follows that $H \setminus F(H) = F'(H \oplus h \setminus H \oplus \{0\})$ is open.

Hint for b: As a closed subspace of H, $\operatorname{Im} F$ is itself a Hilbert space possessing a countable orthonormal basis $w_1,w_2,\ldots$. Now set $u_i := v_i - Pv_i$, where v_i are as above $(i = 1,\ldots,n)$ and $P: H \to \operatorname{Im} F$ is the projection

$$Pu := \sum_{j=1}^{\infty} \langle u,w_j\rangle w_j .$$

Then $\{u_1,\ldots,u_n\}$ forms a basis for $(\mathrm{Im}\ F)^{\perp}$, the orthogonal complement of Im F in H.

For aesthetic reasons, one can orthonormalize $u_1,\ldots,u_n$ by the Gram-Schmidt process (i.e., without loss of generality, assume they are orthonormal). Then $u_1,\ldots,u_n,w_1,w_2,w_3,\ldots$ is a countable orthonormal basis for H. □

Exercise 2. Once more, prove the chain rule

$$\text{index } G\circ F = \text{index } F + \text{index } G$$

for Fredholm operators $F : H \to H'$ and $G : H' \to H''$.

Hint: In place of the purely algebraic argument in Exercise 2.4, use Exercise 1a to first prove that the images are closed, and then use the technique of orthogonal complements in 1b; e.g., see [Gohberg-Krein 1957, 195]. □

How trivial or nontrivial is it to prove that operators have closed ranges? For operators with finite rank and for surjective operators it is trivial, and for Fredholm operators it was proved in Exercise 1a. Is it perhaps true that all bounded linear operators have closed images? As the following counterexample explicitly shows, the answer is "no". Moreover, we will see below that all compact operators with infinite-dimensional image are counterexamples.

Exercise 3. For a Hilbert space H with orthonormal system $e_1,e_2,e_3,\ldots$, consider the operator

$$Au := \sum_{j=1}^{\infty} \frac{1}{j} \langle u,e_j\rangle e_j .$$

Show that Im A is not closed.

Hint: Clearly A is linear and bounded ($||A|| = ?$), and furthermore, we have the criterion for Im(A)

$$v \in \mathrm{Im}\ A \iff \Sigma j\langle v,e_j\rangle e_j \in H \iff \Sigma j^2|\langle v,e_j\rangle|^2 < \infty.$$

It follows that for

$$v_0 := \Sigma \frac{1}{j\sqrt{j}} e_j$$

and

$$v_n := \Sigma \frac{1}{j\sqrt{j}\ j^{1/n}} e_j; \quad n = 1,2,\ldots ;$$

we get $v_0 \in H \setminus \operatorname{Im} A$ and $v_n \in \operatorname{Im} A$. (The old trick: $\Sigma \frac{1}{j^a}$ converges for $a > 1$ (e.g., for $a = 1 + \frac{1}{n}$), but diverges for $a = 1$.) To finally prove that the sequence actually converges to v_0, observe that

$$||v_0-v_n||^2 = \sum_{j=1}^{\infty} \frac{1}{j^2} \frac{(\sqrt[n]{j} - 1)^2}{j\ j^{2/n}} .$$

It is clear, that for each j

$$\frac{\sqrt[n]{j} - 1}{\sqrt[n]{j}} \rightsquigarrow 0 \quad \text{as} \quad n \rightsquigarrow \infty ,$$

and then in particular $\langle v_0-v_n, e_j\rangle^2 \rightsquigarrow 0$; cf. [Jörgens 1970/1982, §5.1]. However, to show that $||v_0-v_n||^2$ converges to 0 as $n \to \infty$, one must estimate more precisely. To do this, we exploit the fact that $\sum_{j>j_0} \frac{1}{j^3}$ can be made smaller than any $\varepsilon > 0$ for j_0 sufficiently large, while on the other hand, choosing n sufficiently large (so large that $(1+\varepsilon)^n \geq j_0$), we have

$$\frac{\sqrt[n]{j} -1}{\sqrt[n]{j}} < \varepsilon \qquad \text{for } j \leq j_0. \qquad \square$$

B. The Adjoint Operator

We will draw further conclusions from the closedness of the image of a Fredholm operator and to do this we introduce "adjoint operators". The purpose is to eliminate the asymmetry between "kernel" and "cokernel" - or, in other words, between the theory of the homogeneous equation (questions of uniqueness of solutions) and the theory of the inhomogeneous equation (questions of existence of solutions). This is achieved by representing the cokernel of an operator as kernel of a suitable "adjoint" operator.

Projective geometry deals with a comparable problem via duality: one thinks of space on the one hand as consisting of points, on the other as consisting of planes, and depending on the point of view, a straight line is the join of two points or the intersection of two planes. Analytic geometry passes from a matrix (a_{ij}) to its transpose (a_{ji}) - or $(\overline{a}_{ji})$ in the complex case - to technically deal with dual statements. We can do the same successfully for operators (= infinite matrices):

Exercise 4. Show that on the space $\mathcal{B}(H)$ of bounded linear operators of a Hilbert space H, there is a natural isometric (anti-linear) involution

$$*: \mathcal{B}(H) \to \mathcal{B}(H)$$

which assigns to each $T \in \mathcal{B}(H)$ the *adjoint operator* $T^* \in \mathcal{B}(H)$ such that for all $u,v \in H$

$$\langle u,T^*v\rangle = \langle Tu,v\rangle.$$

Hint: It is clear that T^*v is well defined for each $v \in H$, since $u \mapsto \langle Tu,v\rangle$ is a continuous linear functional on H; and so the representation theorem of Ernst Fischer and Friedrich Riesz [Schechter 1971, 28] tells us that the functional is expressed through a unique element of H, which we denote by T^*v. The linearity of T^* is clear by construction. While proving the continuity (= boundedness) of T^*, show more precisely that $||T^*|| = ||T||$; i.e., that $*$ is an isometry.

Just as easily, we have the involution property $T^{**} = T$, the composition rule $(T\circ R)^* = R^*\circ T^*$, and anti-linearity $(aT + bR)^* = \bar{a}T^* + \bar{b}R^*$, where the bars denote complex conjugation. Details can be found for example in [Schechter 1971, 59 and 249 ff.]. Observe that for $T \in \mathcal{B}(H,H')$, where H and H' may differ, the adjoint operator T^* is in $\mathcal{B}(H',H)$. □

Theorem 1. For $F \in \mathcal{F}$, the $u_1,\ldots,u_n$ in Exercise 1b form a basis of $\operatorname{Ker} F^*$, whence

$$\operatorname{Im} F = (\operatorname{Ker} F^*)^{\perp}$$

and $\operatorname{Coker} F \cong \operatorname{Ker} F^*$.

Proof: First we note that $u \in (\operatorname{Im} F)^{\perp}$ exactly when $0 = \langle u,w\rangle = \langle u,Fv\rangle = \langle F^*u,v\rangle$ for all $w \in \operatorname{Im} F$ (i.e., for all $v \in H$); thus, $(\operatorname{Im} F)^{\perp} = \operatorname{Ker} F^*$. By again taking orthogonal complements, we have $(\operatorname{Ker} F^*)^{\perp} = (\operatorname{Im} F)^{\perp\perp} = \operatorname{Im} F$, since $\operatorname{Im} F$ is closed by Exercise 1a. □

Observe that the above argument remains valid for any bounded linear operator with closed range. In this case, we have the criterion that the equation $Fv = w$ is solvable exactly when $w \perp \operatorname{Ker} F^*$.

Theorem 2. A bounded linear operator F is a Fredholm operator, precisely when $\operatorname{Ker} F$ and $\operatorname{Ker} F^*$ are finite-dimensional and $\operatorname{Im} F$ is closed. In this case,

$$\operatorname{index} F = \dim \operatorname{Ker} F - \dim \operatorname{Ker} F^*.$$

Thus, in particular, $\operatorname{index} F = 0$ in case F is self-adjoint (i.e., $F^* = F$).

Proof: Use Theorem 1 and Exercise 1a. □

Remark. Note that $\text{Ker } F^*F = \text{Ker } F$: "$\supset$" is clear; for "$\subset$", take $u \in \text{Ker } F^*F$, and then

$$\langle F^*Fu,v\rangle = \langle Fu,Fv\rangle = 0 \quad \text{for all} \quad v \in H,$$

whence $\langle Fu,Fu\rangle = 0$ and $u \in \text{Ker } F$. If $\text{Im } F$ is closed (e.g., if $F \in \mathcal{F}$), then we also have

$$\text{Im } F^*F = \text{Im } F^*.$$

Here $\subset$ is clear. To prove $\supset$, consider F^*v for $v \in H$, and decompose v into orthogonal components $v = v' + v''$ with $v' \in \text{Im } F$ and $v'' \in \text{Ker } F^*$; then $F^*v = F^*v'$. In this way, we then have represented the kernel and cokernel of any Fredholm operator F as the kernels of the self-adjoint operators F^*F and FF^*, respectively. □

C. Compact Operators

So far we found that $\mathcal{F}$ is closed under composition and passage to adjoints and particularly that all operators of the form $\text{Id} + T$ belong to $\mathcal{F}$ when T is an operator of finite rank. We will now substantially increase this supply of examples, in passing to "compact operators" by taking limits. This, however, does not lead to Fredholm operators of non-zero index.

We begin with an exercise which

- emphasizes a simple topological property of operators of finite rank,
- more generally characterizes the finite dimensional subspaces which are fundamental for the index concept,
- prepares the introduction of "compact operators".

Exercise 5. a) Every operator with finite rank maps the unit ball (or any bounded subset) of H to a relatively compact set.

b) If H is finite-dimensional, then the closed unit ball $B_H := \{u \in H : |u| \leq 1\}$ is compact.

c) If H is infinite-dimensional, then B_H is noncompact.

Hint for a) and b): Recall the theorem of Bernhard Bolzano and Karl Weierstrass that says that every closed bounded subset of $\mathbb{R}^n$ (or $\mathbb{C}^n$) is compact.

For c): Every orthonormal system $e_1, e_2, \ldots$ in H is a sequence in B_H without a convergent subsequence. For instance, how large is $|e_i - e_k|$ for $i \neq k$? □

We denote by K the set of linear operators from H to H which map the open unit ball (or more generally, each bounded subset of H) to a relatively compact subset of H. Such operators are called *compact* (or sometimes "completely continuous") *operators* and, by Exercise 5a, form the largest class of operators that behave (in this respect) like finite rank operators, i.e., like the operators of linear algebra which are defined via matrices.

Despite the risks inherent in pictures, we can perhaps best visualize compact operators as "asymptotically" contracting maps which in the case of operators of finite rank map the ball B_H to a finite dimensional disk, and in general to some sort of elliptical spiral:

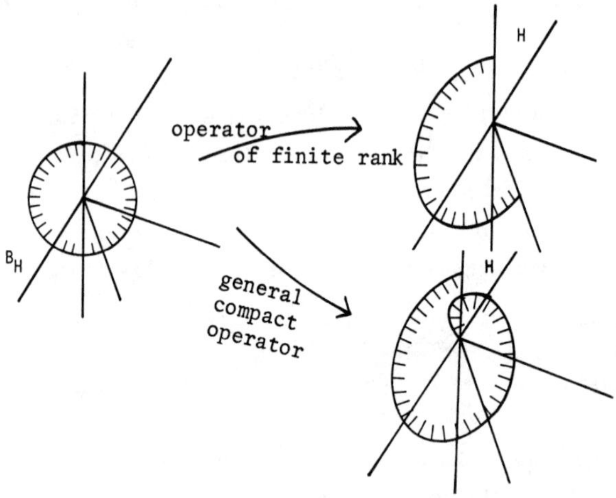

David Hilbert made this visualization precise in his "spectral representation" of a compact operator K. Accordingly (in the "normal" case $KK^* = K^*K$) the value Ku can be expanded into a series in eigenvectors $u_1, u_2, \ldots$ with the corresponding eigenvalues as coefficients, i.e.,

$$Ku = \sum_{j=1}^{\infty} \lambda_j \langle u, u_j \rangle u_j ,$$

whereby the eigenvalues accumulate at 0 and the eigenvectors form an orthonormal system.

In direct analogy to the principal axis transformation of analytic geometry, we thus obtain for the "quadratic form" defined by K the representation

$$\langle Ku,u\rangle = \sum_{j=1}^{\infty} \lambda_j \langle u,u_j\rangle^2 .$$

All proofs and the necessary generalizations to the non-normal case can be found in [Jörgens, 1970/1982, §§6.2-6.4], the historical background in [Hellinger-Toeplitz, §§16, 34 and 40] or more compactly in [Kline, 1064-1066].

We will not use these results any further. However, note that an operator K is bounded (= continuous) if and only if for every point $u \in H$

$$\lim_{|v|\to 0} K(u+v) = K(u),$$

but is compact, if and only if, with respect to a complete orthogonal system, we have

$$\lim_{v} K(u+v) = K(u),$$

provided v approaches 0 componentwise; i.e., provided $\lim_{v} \langle v,e_j\rangle = 0$ for all j, while $|v|^2 = \sum_{j=1}^{\infty} |\langle v,e_j\rangle|^2$ need not go to 0. This describes the context in which Hilbert developed the idea of a compact operator and why he called them "completely continuous".

For our purposes the following result is sufficient.

Theorem 3. a) $\mathcal{K}$ is a "two-sided (non-trivial) ideal" in the Banach algebra $\mathcal{B}$ of bounded linear operators in a separable, infinite-dimensional Hilbert space H. b) $\mathcal{K}$ is closed in $\mathcal{B}$. c) More precisely, $\mathcal{K}$ is the closure of the subset of finite-rank operators. d) $\mathcal{K}$ is invariant under *; i.e., the adjoint of a compact operator is compact.

Proof of a: For bounded T and compact K, the operators $T\circ K$ and $K\circ T$ are compact, by definition. We therefore have

$$\mathcal{B}\circ\mathcal{K} \subset \mathcal{K} \quad \text{and} \quad \mathcal{K}\circ\mathcal{B} \subset \mathcal{K}$$

Moreover, for $\lambda \in \mathbb{C}$, we have λK compact. Now, let K and K' be compact. To prove that K + K' is compact, we use the sequential criterion for compactness (see above). Here this means: A operator is compact, if the image of a bounded sequence of points has a convergent subsequence. Thus, let $u_1, u_2, \ldots$ be a bounded sequence in H. Then, by a two-fold selection of subsequences, we can find $u_{i_1}, u_{i_2}, \ldots,$ such that $Ku_{i_1}, Ku_{i_2}, \ldots$ and $K'u_{i_1}, K'u_{i_2}, \ldots$ both converge in H, whence $(K+K')u_{i_1}, (K+K')u_{i_2}, \ldots$ also converges.

Finally, it is trivial that every compact operator is bounded, since the image of the unit sphere is relatively compact and hence bounded; thus, $\mathcal{K} \subset \mathcal{B}$.

Since $\mathcal{K}$ includes the operators of finite rank (Exercise 5a) but not the identity (Exercise 5c), $\mathcal{K}$ is a nontrivial ideal, and the assertion follows.

Proof of b: Let $T \in \mathcal{B}$ be an operator in the closure of $\mathcal{K}$. In order to show that each open cover (say, without loss of generality, by *all* of the balls of radius $\varepsilon > 0$ [Dugundji 1966, 298]) of the image $T(B_H)$ of the closed unit ball of H has a finite subcover, we use an "$\frac{\varepsilon}{3}$-proof" (as is usual in such situations):

We choose $K \in \mathcal{K}$ with $||T-K|| < \frac{\varepsilon}{3}$ and a finite open covering of $K(B_H)$ by balls of radius $\varepsilon/3$, whose centers we denote by $Ku_1, \ldots, Ku_m$ where $u_1, \ldots, u_m \in B_H$. Then the ε-balls about $Tu_1, \ldots, Tu_m$ form the desired finite covering of $T(B_H)$: For each $u \in B_H$, there is some $i \in \{1, \ldots, m\}$, such that $|Ku-Ku_i| < \varepsilon/3$, and so

$$|Tu-Tu_i| \leq |Tu-Ku| + |Ku-Ku_i| + |Ku_i-Tu_i| < \varepsilon.$$

Proof of c: We now ask which operators are limits in $\mathcal{B}$ of sequences of operators of finite rank. Evidently, these limits typically lie outside of the space of finite rank operators; e.g., see Exercise 6 below. Let $e_0, e_1, e_2, \ldots$ be a complete orthonormal system in H and let

$$Q_n : H \to [e_0, \ldots, e_{n-1}]$$

be the orthogonal projection from H to the linear span of the first n basis elements. This "truncation" yields, for each $T \in \mathcal{B}$, a sequence $Q_1T, Q_2T, \ldots$ of operators of finite rank which converges pointwise to T; i.e.,

$$Q_nTu \rightsquigarrow Tu \quad \text{for each } u \in H.$$

This does *not* mean that, the sequence $Q_1T, Q_2T, \ldots = (Q_nT)_1^\infty$ converges in $\mathcal{B}$ (i.e., in the operator norm) to T. Of course, not every bounded operator can be the limit of a sequence of operators of finite rank (e.g., $||Q_n \mathrm{Id} - \mathrm{Id}|| = 1$ for all n); in fact, we have already proved in b) that limits of compact operators (in particular, finite-rank operators - see Exercise 5a) must be compact. It remains to prove that, for each $K \in \mathcal{K}$, the sequence $(Q_n K)_1^\infty$ converges to K in $\mathcal{B}$. For this, we choose (for each $\varepsilon > 0$) a finite covering of $K(B_H)$ by balls of radius $\varepsilon/3$ with centers at $Ku_1,\ldots,Ku_m$ and some $n \in \mathbb{N}$ so large that

$$|Ku_i - Q_n Ku_i| < \frac{\varepsilon}{3}, \quad i = 1,\ldots,m.$$

This is no problem, since $(Q_n)_1^\infty$ converges pointwise to the identity. For each $u \in B_H$, we then have

$$|Ku-Q_nKu| \leq |Ku-Ku_i| + |Ku_i-Q_nKu_i| + |Q_nKu_i-Q_nKu| \leq \varepsilon/3 + \varepsilon/3 + \varepsilon/3$$

for some $i \in \{1,\ldots,m\}$. For the last term, note that $||Q_n|| = 1$, whence

$$|Q_nKu_i-Q_nKu| = |Q_n(Ku_i-Ku)| \leq |Ku_i-Ku|.$$

Thus, we have proven $||K-Q_nK|| < \varepsilon$ for all sufficiently large n, depending on ε.

Proof of d: Obviously, the space of operators of finite rank is invariant under *, since every such operator T (as in the remark after Exercise 2.5) is of the form

$$T = \sum_{i=1}^{n} \langle .., u_i \rangle v_i,$$

where $u_1,\ldots,v_n \in H$, and then (verify!)

$$T^* = \sum \langle \ldots, v_j \rangle u_j$$

also has finite rank. Now, we can reduce the general case $K \in \mathcal{K}$ to the finite-rank case. Namely, approximate K by a sequence $(T_n)_1^\infty$ of operators of finite rank whose adjoints then approximate K^*, since (by Exercise 4)

$$||T_n^*-K^*|| = ||(T_n-K)^*|| = ||T_n-K||. \quad \square$$

Remark 1. The statements in a, b, and d apply also (admittedly with somewhat different proofs; e.g., see [Schechter 1971, 92-94]) to the more general case of Banach spaces, but not statement c. The search for a counterexample, which led to success in 1973 (P. Enflo, Acta Math.), incidentally supplied many nice examples of the correctness of c in special cases, in particular for almost all well-known Banach spaces; e.g., see [Jörgens 1970/1982, §12.4].

Remark 2. Since the closure of an ideal is again an ideal (as is trivially proved) and since the operators of finite rank obviously form an ideal in $\mathcal{B}$, we note that a) follows from c) - admittedly, somewhat less directly than in the above proof.

Remark 3. For practical needs the sequence $(Q_n K)_1^\infty$ stated in the proof of c) is a poor approximation of K by operators of finite rank. It presupposes the knowledge of K on all of H and works with a completely arbitrary orthonormal system. However K is frequently (see for example Exercises 6 and 7 below) given in a form which suggests a special approximation or which points to a distinguished orthonormal system, namely the eigenvectors of K. The "spectral representation" given in the remark after Exercise 5 is numerically relevant, because it implies that

$$Q_n Ku = KQ_n u = \sum_{j=0}^{n-1} \lambda_j \langle u,u_j\rangle u_j$$

which does indeed permit a "stepwise" approximation. □

Exercise 6. Show that the operator A in Exercise 3 is compact. Namely, give an estimate for $|Au - \Sigma_{j=1}^n \frac{1}{j}\langle u,e_j\rangle e_j|^2$ which is independent of u, for $|u| \le 1$. Recall the Schwartz Inequality $|\langle u,e_i\rangle| \le |u||e_j|$. □

Exercise 7. Show that we obtain a compact operator K on the Hilbert space (see §8) $L^2[a,b]$ for each square-integrable function G on $[a,b] \times [a,b]$ via

$$Ku(x) := \int_a^b G(x,y)u(y)dy, \quad x \in [a,b]$$

where $[a,b]$ is a compact interval in $\mathbb{R}$.

Hint: Approximate the weight function G by step functions, and hence K by operators with finite rank of the kind considered in Exercise 2.6 (integral operators with "degenerate" weight). For details of the argument, see [Schechter 1971, 257-261].

Exercise 8. For the map

$$L^2([a,b] \times [a,b]) \to \mathcal{K}(L^2[a,b])$$

$$G \longmapsto K,$$

defined in Exercise 7, show that:

a) The map is linear and injective.

b) If G is the weight function of K, then

$$||K|| \leq |G| := \left(\int_a^b\int_a^b |G(x,y)|^2 dxdy\right)^{1/2}.$$

c) The adjoint operator K^* then has the weight function $G^*(x,y) = \overline{G(y,x)}$.

d) If K_1 and K_2 are given by weight functions G_1 and G_2, then the operator $K_2 \circ K_1$ belongs to the weight function

$$G(x,y) = \int_a^b G_2(x,z)G_1(z,y)dz.$$

Hint: Linearity is clear. For injectivity, one naturally (Lebesgue integral!) need only show that $\int_\alpha^\beta\int_\gamma^\delta G(x,y)dxdy = \langle 1\!\!1_{[\alpha,\beta]}, K 1\!\!1_{[\gamma,\delta]}\rangle$, where $1\!\!1_{[\alpha,\beta]}$ and $1\!\!1_{[\gamma,\delta]}$ are characteristic functions of subintervals $[\alpha,\beta]$, $[\gamma,\delta] \subset [a,b]$; then, we have $G = 0$ when $K = 0$. Details (and the generalization to the case of unbounded intervals) are in [Jörgens 1970/1982, §11.2]. For the proofs of b), c), and d), one needs to use the theorem of Fubini on iterated integrals; e.g., the details are in [Jörgens 1970/1982, §§11.2-11.3]. □

Clearly, the assertion of Exercise 7 remains true, if we replace $[a,b]$ by a compact submanifold X of Euclidean space, and carry over the concepts of measure and integration to X. Actually, the compactness of the domain of integration plays no role at all, since the only boundedness condition we need holds because the weight function is square-integrable. We can therefore take $\mathbb{R}$ or $\mathbb{R}^n$ in place of $[a,b]$, without having to change any argument to prove the compactness of the integral operator. The same holds for Exercise 8 with the restrictions made in the proof of a).

D. The Classical Integral Operators

The type of operators considered here, i.e., those which are given by weight functions square-integrable on the "product space", are nowadays called Hilbert-Schmidt-operators. By introducing them, the Hungarian mathematician Friedrich Riesz (1907), not only generalized the theory of integral equations with continuous weight function created by Vito Volterra (Turin, 1896), Erik Ivar Fredholm (Stockholm, 1900) and David

Hilbert (Göttingen, 1904 ff.), but drastically simplified it at the same time. Indeed, proofs dealing with L -integration theory on Hilbert space compare very favorably with the cumbersome work with uniform convergence in the Banach space of continuous functions.

Of course, there do exist important integral operators whose weight functions are not square integrable: The best-known example is the Fourier transform (see Ch. 8 below)

$$\hat{f}(x) = \int_{-\infty}^{+\infty} e^{-ixy} f(y)\,dy.$$

It maps $L^2(\mathbb{R})$ bijectively onto itself, but the "L^2-norm" of the weight

$$\int_{-\infty}^{+\infty}\int_{-\infty}^{+\infty} |e^{-ixy}|^2 dxdy = \int_{-\infty}^{+\infty}\int_{-\infty}^{+\infty} 1\ dxdy$$

is "strongly infinite".

Somewhere between these two simplest principal types - the compact Hilbert-Schmidt operators on the one hand and the invertible Fourier transform on the other - lie the "convolution operators", in particular the "Wiener-Hopf operators" and other "singular integral operators". Their weights do not belong to L^2, but frequently they are at least componentwise integrable, for example

$$\int_{-\infty}^{+\infty} |G(x,y)|^2 dy < \infty \quad \text{for} \quad x \in \mathbb{R}.$$

All these operator are highly significant in kinematic as well as stochastic modelling and in solving a multitude of physical, technical and economical problems: as in the method of inverting differential operators into integral operators which goes back to George Green and was developed on a large scale by David Hilbert; and as in the indirect treatment of summation - and aggregation phenomena or more generally in probabilistic situations. We will return to a number of particularly interesting integral operators later in this part and in the following part. A first survey is given in Table 1.

Table 1: Some fundamental integral operators of the form $(Ku)(x) := \int_X G(x,y)\,dy$

No.	X	G(x,y)	u	Key words	Property	Literature or origin
1.	$[a,b]$	$C^0(X \times Y)$	$C^0(X)$	Fredholm integral	$K: C^0(X) \to C^0(X)$ compact	Fredholm (1903)
2.	compact	"	"	"	"	Hilbert (1904-12) [Jörgens, §8.1]
3.	$[a,b]$	G continuous for $y \leq x$; $G = 0$ for $y > x$.	"	Volterra integral	"	Volterra (1896)
4.	"	C^0 on X×X-diag. bounded on X×X.	"	Green's operator	"	[Jörgens, §8.2]
5.	compact	$G = g(x,y)\lvert x-y\rvert^{-\alpha}$ $0 < \alpha < \dim X$	"	pole singularity	"	"
6.	$[a,b]$	$L^2(X \times X)$	$L^2(X)$	Hilbert-Schmidt operator	$K: L^2(X) \to L^2(X)$ compact	Riesz (1907)
7.	$\mathbb{R}^n$	"	"	"	"	[Jörgens, §11.3], here Exercise 7/8
8.	"	$e^{-i(x_1y_1+\ldots+x_ny_n)}$	"	Fourier transform	bijective	[Dym-McKean, §2.10], here Chapter 8
9.	$\mathbb{R}$	$k(x-y)$, $k \in L^1(\mathbb{R})$	"	convolution	see below	[Dym-McKean, §2.1], here Chapter 8
10.	$\mathbb{R}_+$	"		Wiener-Hopf operator	see below	[Dym-McKean, §3.6], here Chapter 9
11.	$\mathbb{R}^n$	$g(x,y)\lvert x-y\rvert^{-\alpha}$ g continuous, $\alpha > n$	"	singular integral operators	see below	Michlin (1948), here Chapter II.3

4. THE FREDHOLM ALTERNATIVE

A. The Riesz Lemma

Lemma. If F is a Fredholm operator on a separable Hilbert space H, then the statement "index $F = 0$" can be expressed in familiar classical terminology as: Either the equation $Fu = v$ has a unique solution $u \in H$ for each $v \in H$, or the homogeneous equation $Fu = 0$ has a nontrivial solution. In the second case, there are at most finitely many linearly independent solutions $w_1,\ldots,w_n$ of $Fw = 0$ and just as many linearly independent solutions $u_1,\ldots,u_n$ of the adjoint homogeneous equation $F^*u = 0$; the inhomogeneous equation $Fu = v$ is solvable exactly when

$$\langle v,u_1\rangle = \ldots = \langle v,u_n\rangle = 0.$$

This statement is called the "Fredholm alternative"; its equivalence with "index $F = 0$" follows from Theorem 3.1.

We saw already that the Fredholm alternative holds for self-adjoint ($F^* = F$), or more generally normal ($F^*F = FF^*$), Fredholm operators. Moreover, in analogy with finite-dimensional linear algebra, it holds for operators of the form $\mathrm{Id} + T$, when T is an operator of finite rank. Even more important for many applications are the following operators for which the Fredholm alternative holds:

Theorem (F. Riesz, 1918): For each compact operator K, $\mathrm{Id} + K$ is a Fredholm operator with vanishing index.

Proof: Using 3.3c, approximate K by a sequence $K_1,K_2,\ldots$ of operators of finite rank and choose n with $||K-K_n|| < 1$. Then $\mathrm{Id}+K-K_n$ is invertible. Indeed, let $Q := K_n-K$ and consider the series $\Sigma_{k=0}^{\infty} Q^k$. Since $||Q|| < 1$ and

$$||\sum_{k=N}^{M} Q^k|| \leq \sum_{k=N}^{M} ||Q^k|| \leq \sum_{k=N}^{M} ||Q||^k,$$

the partial sums form a Cauchy sequence in the Banach algebra $\mathcal{B}$. The series then converges, and one has

$$(\mathrm{Id}-Q)\sum_{k=0}^{\infty} Q^k = \left(\sum_{k=0}^{\infty} Q^k\right)(\mathrm{Id}-Q) = \mathrm{Id}.$$

Thus, we can write $\mathrm{Id}+K$ as a product:

$$\mathrm{Id}+K = (\mathrm{Id}+K-K_n)(\mathrm{Id}+(\mathrm{Id}+K-K_n)^{-1}K_n),$$

where the left factor is invertible and the right is Id + an operator of finite rank, which we know (Exercise 2.5) is a Fredholm operator with index 0. By the composition rule of Exercise 2.4 or Exercise 3.2, the statement is proved. □

Remark. Note that in the above proof of the formula index (Id+K) = 0, it was not needed that K is approximated by a sequence of operators of finite rank. It was sufficient to have a "crudely approximating" operator K_n (with $||K_n - K|| < 1$). Also, for the determination of the index, it was not necessary to compute or know more precisely the inverse operator $(Id+K-K_n)^{-1}$.

This method of proof which reduces the general case to situations permitting explicit or at least iterative solutions goes back to E. Schmidt and works only in Hilbert space. In more general cases where the theorem still holds the proof starts by showing with a compactness argument that Ker(Id+K) (and similarly Ker(Id+K*)) is finite dimensional. Then one needs a careful argument concerning the limit process in order to show that Im(Id+K) is closed, and one gets only that $Id+K \in F$. Finally the homotopy invariant of the index (see Theorem 5.2 below) implies the formula index(Id+K) = 0. □

Exercise 1. Formulate and prove the Fredholm alternative for the linear *Fredholm integral equation of the second kind*

$$u(x) + \int_a^b G(x,y)u(y)dy = h(x),$$

where $G \in L^2([a,b] \times [a,b])$; see Exercise 2.6, Theorem 3.1, Exercise 3.7, and Exercise 3.8c. □

B. The Sturm-Liouville Boundary-Value Problem

We will apply Theorem 1 and Exercise 1 to a classical boundary value problem in the theory of ordinary differential equations which is initially not very transparent. We start with a dynamical system with finitely many degrees of freedom described by a system of ordinary differential equations. Such "ideally simple" models are used in celestial mechanics for the computation of planetary orbits, for investigations of the pendulum and gyroscope, for econometric simulation of economic processes, and for the treatment of many other discrete oscillating systems (N-body problems). Many of these problems are mathematically unsolved, but one knows at least that each set of initial values determines a unique solution curve; i.e., given the differential equation (with half-way reasonable coefficients), the system is completely determined by its state at a single moment in time. The key mathematical tools are the existence- and uniqueness theorems of Emile Picard and Rudolf Lipschitz. See, for example, [UNESCO, 164-170].

We have a different situation with continuous oscillating systems such as the oscillating string or flexible rod, electrical oscillations in wires, acoustical vibrations in tubes, heat conduction, heat propagation and other diffusion processes, particularly the statistical treatment of equilibria and motion. For many such processes, one has partial differential equations (see Part II below) for instance of the form

$$\frac{\partial^2 U}{\partial x^2} = \rho \frac{\partial^2 U}{\partial t^2} + F(x,t). \tag{1}$$

Under suitable assumptions, it can be reduced to an ordinary differential equation by "separating variables". More explicitly, setting

$$U(x,t) = u(x)\psi(t)$$

we obtain

$$u'' + ru = f \tag{2}$$

where r, f are given, and u is to be found. Hereby (2) usually inherits boundary conditions

$$u(0) = u(1) = 0 \tag{3}$$

from (1) which largely is valid only in a closed bounded region. For details see [Courant-Hilbert I, §V.3] or [UNESCO, 193].

We stick with this example which goes back to Bernoulli's "brachistochrone" problem and more generally to the beginnings of the calculus of variations and of geometric optics by Pierre de Fermat [Kline, §24]. For starters let $r = 0$. Evidently the homogeneous differential equation associated with (2) (put $f = 0$) has only the trivial solution $u = 0$ if the boundary conditions (3) are to be satisfied. In this case there is a Green's function (see e.g. [Courant-Hilbert I, §§V.14-15] which, for each (piecewise continuous) f, yields a solution of the differential equation (2) with boundary conditions (3) given by the formula

$$u(x) = \int_0^1 k(x,y)f(y)\,dy. \tag{4}$$

For the present case,

$$k(x,y) = \begin{cases} x(y-1) & \text{for} \quad x \leq y \\ y(x-1) & \phantom{\text{for}} \quad x \geq y \end{cases}$$

[Courant-Hilbert I, §V.15.1].

Now let $f = 0$ and let r be a positive real number. Then the general solution of (2) has the form

$$u = c_1 e^{i\sqrt{r}\,x} + c_2 e^{-i\sqrt{r}\,x},$$

i.e. each of the two conditions of (3) determines a one-dimensional family of solutions where

$$c_1/c_2 = -1 \quad \text{for} \quad u(0) = 0$$

and

$$c_1/c_2 = -e^{-2i\sqrt{r}} \quad \text{for} \quad u(1) = 0.$$

The two families coincide, exactly when $\sqrt{r}$ is a multiple of π. If so, there is in addition to the zero function another solution u_0 of the homogeneous differential equation

$$u'' + ru = 0, \tag{5}$$

which satisfies the boundary conditions (3). Thus we have here - in contrast to the Uniqueness Theorem of Lipschitz - an ordinary differential equation with two conditions imposed (which, however, are not concentrated at an initial point but distributed over two points) whose solution is not unique.

Another peculiarity of this case is that - in contrast to the Existence Theorem of Picard - the inhomogeneous differential equation (2) subject to the conditions (3) need not always have a solution. This happens for example if the driving force f equals the natural oscillation u_0, or more generally [Courant-Hilbert I, §V.14.2] if

$$\int_0^1 f(x)u_0(x)dx \neq 0.$$

This is the case of resonance which means that the system becomes unstable under the influence of an exterior force. The lack of a solution does not mean that nothing happens but from the point of view of the user it is an indication that some critical phenomenon might occur: A short in a wire, the collapse of a bridge, extreme concentration of light beams which is technically utilized in a laser. (Also, mathematically speaking, the lack of a solution means only that no solution of the given or desired type exists, in our example no bounded function which has a piecewise continuous second derivative. Frequently, this is an indication that a reformulation or refinement of the question is necessary.)

We note finally that in the other case when the nontrivial solution sets with $u(0) = 0$ and those with $u(1) = 0$ are distinct, the equations (2) and (3) always have a unique solution, and a Green function can be constructed which carries the essential information of (2) and (3) and yields the solution for each right hand side f in the integral form (4) [Courant-Hilbert I, §V.14.1].

Combining the two cases we obtain a kind of Fredholm alternative:

Either the differential equation (2) together with the boundary conditions (3) possesses a unique solution u for every given f, or else the homogeneous equation (5) has a solution which does not vanish identically. In the second case the equations (2) and (3) have a solution, if and only if the orthogonality condition

$$\int_0^1 f(x)u_0(x)dx = 0$$

holds for each solution u_0 of the homogeneous equation (5), where the right hand side of (2) is f.

The analogy with the Fredholm alternative for integral equations (Exercise 1) is not accidental. When the solution is unique, the Green function makes the connection via formula (4). But even when solutions do not necessarily exist or when they are not unique, then the classical theory manages to work with "generalized Green functions". We will not deal with these questions in detail, but refer the reader to the quoted literature. The fundamental methodological and, in our context, particularly interesting point of view is perhaps best made precise as follows:

Exercise 2. Consider the differential equation

$$u'' + pu' + qu = f \tag{*}$$

on the interval $[0,1]$ with $p,q,f \in C^0[0,1]$ and with the boundary conditions

$$u(0) = a \quad \text{and} \quad u(1) = b. \tag{**}$$

Show that the integral equation

$$v - Kv = g \tag{***}$$

is equivalent to the boundary-value problem (*), (**), if

$$Kv(x) := \int_0^1 G(x,y)v(y)dy, \quad x \in [0,1]$$

$$G(x,y) := \begin{cases} y(q(x)(1-x) - p(x)) & \text{for } y \le x \\ (1-y)(q(x)x + p(x)) & \text{for } y \ge x \end{cases}$$

$$g := ph' + qh - f$$

$$h(x) := a(1-x) + bx \quad (\text{hence, } h' = b-a)$$

Hint: Show that every twice continuously differentiable solution u of (*) and (**) yields a solution $v := u''$ of (***), and conversely, every continuous solution v of (***) gives a twice continuously differentiable solution of (*) and (**) by means of

$$u(x) := h(x) + \int_0^1 k(x,y)v(y)dy,$$

where

$$k(x,y) := \begin{cases} x(1-y) & \text{for } x \le y \\ y(1-x) & \text{for } x \ge y \end{cases} .$$

Details may be found in [Jörgens, 1970/1982, §9.1]. For a first calculation and in order to maintain continuity with the preliminary remarks, it is recommended that one first try $q = 0$, p positive and constant, and set $a = b = 0$. □

Remark 1. With Exercise 1, the "Fredholm alternative" for the boundary-value problem follows from the equivalence proved in Exercise 2. More precisely, Id-K is a Fredholm operator on the Hilbert space $L^2[0,1]$, and index(Id-K) = 0. Thus, we have a Fredholm alternative relative to $L^2[0,1]$. Actually, from the "Closed Graph Theorem" and a regularity theorem (see Chapters II.5/II.8; incidentally, we see here that the Banach space theory is genuinely more difficult than the Hilbert space theory), we have that each square integrable solution of equation (***) is continuous, provided the right side is continuous. Thus, the Fredholm alternative in $C^0[0,1]$ holds: *Either* dim Ker(Id-K) = 0 and so (because index (Id-K) = 0, and hence dim Coker(Id-K) = 0) the equation (***) has a unique solution for each $g \in C^0[0,1]$ (whence, the boundary-value problem (*), (**) also has a unique solution for each $f \in C^0[0,1]$ and fixed boundary values a,b) *or* the homogeneous equation $v-Kv = 0$ has nontrivial solutions.

In the second case, the adjoint integral equation also has a nontrivial solution; i.e., there is a $w \in L^2[0,1]$ with

$$w(x) = (1-x)\int_0^x (q(y)y + p(y))w(y)dy$$
$$+ x\int_x^1 (q(y)(1-y) - p(y))w(y)dy.$$

We can differentiate with respect to the upper and lower bounds, obtaining that w is continuously differentiable and

$$w'(x) = \begin{cases} -\int_0^x (q(y)y + p(y))w(y)dy \\ +(1-x)(q(x)x + p(x))w(x) \\ +\int_x^1 (q(y)(1-y) - p(y))w(y)dy \\ - x(q(x)(1-x) - p(x))w(x) \end{cases}$$

$$= p(x)w(x) - \int_0^x \ldots + \int_x^1 \ldots .$$

We bring $p(x)w(x)$ to the left side, differentiate once more, and obtain

$$(w' - pw)' + qw = 0. \qquad (****)$$

From the integral equation for w, we have $w(0) = w(1) = 0$. Hence, every solution w of the homogeneous adjoint integral equation is a solution of the formal-adjoint homogeneous differential equation (****) with the homogeneous boundary conditions $w(0) = w(1) = 0$. By *formal-adjoint*, we mean that for all $u,w \in C^2[0,1]$ with the homogeneous boundary condition, we have

$$\int_0^1 u((w'-pw)' + qw)dx = \int_0^1 (u'' + pu' + qu)w\,dx,$$

which one can verify through integration by parts. For brevity, we have taken all functions to be real-valued.

In the second case of the Fredholm alternative, the problem (*), (**) is solvable exactly when (***) is solvable; i.e., when (****) has a nontrivial solution w with $\langle g,w\rangle = 0$, which means that in terms of f, we have

$$\int_0^1 f(x)w(x)dx = aw'(0) - bw'(1).$$

Details of this argument and similar treatments of other boundary-value problems for ordinary differential equations of second order (Sturm-Liouville problems) can be found in [Jörgens 1970/1982, §9.2]. □

Remark 2. If the boundary value problem (*) and (**) is equivalent with the integral equation (***), what is the special nature of the presentation (***) in comparison with (*) and (**)? We bring out three points:

1. The integral equation succeeds in combining two equations, the differential equation and the boundary conditions, into one.

2. Let

$$L : C^2[0,1] \to C^0[0,1]$$

denote the differential operator defined by the left side of (*) and let

$$B : C^2[0,1] \to \mathbb{C} \oplus \mathbb{C}$$

denote the boundary operator defined by the left sides of (**). Then Exercise 2 says that the operators

$$L \oplus B : C^2[0,1] \to C^0[0,1] \oplus \mathbb{C} \oplus \mathbb{C}$$

and

$$\text{Id} - K : L^2[0,1] \to L^2[0,1]$$

are equivalent in the sense that $\text{Ker } L \oplus B \cong \text{Ker Id-K}$ and $\text{Coker } L \oplus B \cong \text{Coker Id-K}$. Here Id-K is a bounded operator on a Hilbert space to itself, while the function-analytic structure of $L \oplus B$ is much less clear. The equivalence of the differential and the integral equation is a formal one, while the equivalence of the C^0 -/C^2 - theory and the L^2-theory is fairly elementary, but by no means obvious: While nature poses its problems usually in the spaces C^2 or $C^{2(\text{piecewise})}$, the mathematician decides freely in which spaces he wants to solve these problems. Hilbert spaces are used, not because of their intrinsic beauty, but because integral equations on L^2 can be treated more efficiently and more transparentally than on C^0. The "Regularity Theorem" provides the justification for this procedure and shows at the same time that the "freedom" of the mathematician is not arbitrary.

3. Numerically, the integral operator Id-K is dealt with by approximating the compact operator K by operators of finite rank or by approximating the weight function $G(x,y)$ of K by degenerate weights of the form $\phi(x)\psi(y)$. Jacques-Charles-Francois Sturm and his friend Joseph Liouville first and successfully undertook the systematic investi-

gation of boundary value problems for ordinary differential equations of second order. It is quite characteristic that they also arrived at their algebraic solution methods by an approximation principle: They investigated related difference equations and then passed to the limit.* The difference is that the approximation of the integral equation can in some cases (e.g., if G is continuous and non-negative) be done very naturally by development into a series in eigenfunctions (analogous to principal axis transformation of quadratic forms) so that the integral equation becomes immediately clearer ([Courant-Hilbert I, §III.5.1]). In contrast, the approximation of a differential equation by difference equations is done "blindly" - so to speak. It requires the ingenuity of a Sturm and Liouville - or extensive free computer time at present - in order to regain the necessary information about the boundary value problem from the discrete pieces. (Of course, the "blind" approximation always works, while for many Sturm-Liouville problems no explicit eigenfunctions are known.)

In the final analysis the three viewpoints arise from the duality between local and global terms and operations. This duality pervades large parts of analysis (see the Fourier Inversion Formula in Chapter 8 or the Index Formula itself): While differentiation of a function is a purely local operation, the solution of a differential equation with initial or boundary conditions always requires a certain global operation. This circumstance is illustrated already by the Newtonian formula relating derivative and integral. It may explain also why the "local" theory of (e.g. elliptic) differential equations is so difficult (one has to do global theory anyhow, namely in $\mathbb{R}^n$), and why at times a purposely global approach, say starting with differential operators on closed manifolds, leads more quickly and easily to fundamental local results. We resume this thought in Part II. □

5. THE MAIN THEOREMS

A. The Calkin Algebra

So far, we have introduced compact operators for purely practical reasons: Within pure mathematics, they came from the search for a (closed) class of operators that exhibit properties analogous to those of the operators of finite rank. In applied mathematics, they enter through the theory of integral equations associated with study of oscillations.

---

*Jour. de Math. 1(1936), 106-186 and 373-444.

Actually, the compact operators have yet a deeper significance in the representation of Fredholm operators:

We recall our notation: H is a complex, separable Hilbert space; $\mathcal{B}$ is the Banach algebra of bounded linear operators on H (in modern terminology, $\mathcal{B}$ is even a C^*-algebra; see Exercise 3.4, where one needs to verify the additional axiom $||T^*T|| = ||T||^2$); $\mathcal{K} \subset \mathcal{B}$ is the closed two-sided ideal of compact operators (see Theorem 3.3); and $\mathcal{F} \subset \mathcal{B}$ is the space of Fredholm operators.

We begin with a simple exercise.

<u>Exercise 1</u> (<u>J.W. Calkin, 1941</u>): Show that the quotient space $\mathcal{B}/\mathcal{K}$, consisting of equivalence classes $\pi(T) := \{T-K: K \in \mathcal{K}\}$, where $T \in \mathcal{B}$, forms a Banach algebra.

<u>Hint</u>: Since $\mathcal{K}$ is a linear subspace, clearly $\mathcal{B}/\mathcal{K}$ is a vector space. To prove that $\mathcal{B}/\mathcal{K}$ is an algebra, one must use the fact that $\mathcal{K}$ is a two-sided ideal. Then show that since $\mathcal{K}$ is closed, $\mathcal{B}/\mathcal{K}$ can be made into a Banach space by defining a norm on $\mathcal{B}/\mathcal{K}$ by

$$||\pi(T)|| := \inf\{||T-K||: K \in \mathcal{K}\} = \inf\{||R||: R \in \pi(T)\}.$$

It remains to show that

$$||\pi(Id)|| = 1 \quad \text{and} \quad ||\pi(T)\pi(S)|| \le ||\pi(T)||\ ||\pi(S)||.$$

To prove the left equation, assume that there is a $K \in \mathcal{K}$ with $||Id-K|| < 1$ and show that K is invertible (using the argument in the proof of Theorem 4.1 involving geometric series); this contradicts the compactness of K. To prove the right inequality, apply the trick

$$\inf_{K \in \mathcal{K}} ||TS-K|| \le \inf_{K_1,K_2 \in \mathcal{K}} ||(T-K_1)(S-K_2)||. \quad \square$$

<u>Theorem 1</u> (<u>F.V. Atkinson, 1951</u>): If $(\mathcal{B}/\mathcal{K})^\times$ is the group of units (i.e., elements which are invertible with respect to multiplication) of $\mathcal{B}/\mathcal{K}$ and $\pi: \mathcal{B} \to \mathcal{B}/\mathcal{K}$ is the natural projection, then we have

$$\mathcal{F} = \pi^{-1}((\mathcal{B}/\mathcal{K})^\times).$$

<u>Exercise 2</u>. Show that this theorem of Frederick Valentine Atkinson can also be written as: An operator $T \in \mathcal{B}$ is a Fredholm operator exactly when there are $S \in \mathcal{B}$ and $K_1, K_2 \in \mathcal{K}$ such that

$$ST = Id + K_1 \quad \text{and} \quad TS = Id + K_2.$$

Such an S is called a *parametrix* (or *quasi-inverse*) for T. One also says that T is *essentially invertible*; i.e., invertible modulo $\mathcal{K}$. □

Proof of Theorem: For "$\subset$", let $F \in \mathcal{F}$. We show that $\pi(F)$ is invertible. For this, consider the operator $F^*F + P$, where

$$P : H \to \text{Ker } F \text{ is orthogonal projection.}$$

In the remark for Theorem 3.2, we have already shown that $\text{Ker } F^*F = \text{Ker } F$ and $\text{Im } F^*F = \text{Im } F^*$; thus, $F^*F + P$ is bijective and hence invertible in $\mathcal{B}$. Since P is compact (being of finite rank), it follows that $\pi(F^*F) = \pi(F^*)\pi(F)$ is invertible in $\mathcal{B}/\mathcal{K}$. Similarly, one shows with the help of the orthogonal projection

$$Q : H \to \text{Ker } F^*,$$

that $FF^* + Q$ in $\mathcal{B}$ and $\pi(F)\pi(F^*)$ in $\mathcal{B}/\mathcal{K}$ are invertible. With a left-inverse for $\pi(F^*)\pi(F)$ and a right-inverse for $\pi(F)\pi(F^*)$, it follows that $\pi(F)$ is invertible in $\mathcal{B}/\mathcal{K}$. For "$\supset$", let $T \in \mathcal{B}$ with $\pi(T)$ invertible in $\mathcal{B}/\mathcal{K}$; i.e., there is $S \in \mathcal{B}$ such that TS and ST lie in $\pi(\text{Id})$. Now, $\pi(\text{Id}) = \{\text{Id}+K : K \in \mathcal{K}\}$ consists of Fredholm operators by Theorem 4.1 (indeed, of index zero, but that does not concern us here). In particular, we then have that Ker ST and Coker TS are finite-dimensional. Since

$$\text{Ker } T \subset \text{Ker } ST \quad \text{and} \quad \text{Im } T \supset \text{Im } TS,$$

it follows that $T \in \mathcal{F}$. □

Remark. The trick in the first part of the above proof consists of first considering F^*F and FF^* (whose invertibility modulo $\mathcal{K}$ is trivial) rather than F, and only then drawing conclusions about $\pi(F)$. This has the advantage that one need not explicitly exhibit the "parametrix" (i.e., inverse modulo $\mathcal{K}$) for F. An explicit, if somewhat cumbersome, proof of the theorem of Atkinson can be found in [Schechter 1971, 106-108].

Exercise 3. Show that the set of Fredholm operators is open in the Banach algebra of bounded linear operators on a fixed Hilbert space H.

Hint: Because of the continuity of π (π is even contracting), it suffices to show that $(\mathcal{B}/\mathcal{K})^\times$ is open in $\mathcal{B}/\mathcal{K}$. For this, show in general that the group of units $A^\times$ in any Banach algebra A is open; more precisely, show that about each $a \in A^\times$ there is a ball of radius $1/||a^{-1}||$ contained in $A^\times$. For this, apply again the geometric series argument in the proof of Theorem 4.1 or from Exercise 1 in this section. □

Exercise 4. Conclude from the theorem of Atkinson that the space of Fredholm operators is closed under composition, the adjoint operation, and addition of compact operators. Show that such a conclusion is not circular, since the earlier proofs of the same results (e.g., Exercise 2.3 and Theorem 3.2) were not needed in the proof of Atkinson's theorem. □

Exercise 5. Illustrate Atkinson's theorem with the shift^+ operator (in Exercise 1.1) on $L^2(\mathbb{Z}_+)$. In particular, show that the similarly defined shift^- is a "parametrix" (= an inverse modulo K) for shift^+. Which compact operators do we get for $(\text{shift}^- \circ \text{shift}^+) - \text{Id}$ and for $(\text{shift}^+ \circ \text{shift}^-) - \text{Id}$? □

Exercise 6. Using the theorem of Riesz (Theorem 4.1), show that each parametrix G for a Fredholm operator F is itself a Fredholm operator, and we have

$$\text{index } G = -\text{index } F. \quad \square$$

Exercise 7. From Exercise 4, we know already, that $\mathcal{F}$ is closed under addition of compact operators. Now show that the index is invariant:

$$\text{index } F+K = \text{index } F \quad \text{for all} \quad F \in \mathcal{F} \quad \text{and} \quad K \in \mathcal{K}.$$

Hint: Show that each parametrix for F is also a parametrix for $F+K$, and apply Exercise 6. □

B. Perturbation Theory

The result of Exercise 7, which we obtained as an easy corollary of the Theorem of Riesz and of Theorem 1, is also due to Frederick Valentine Atkinson. It represents a fundamental result of "perturbation theory" which asks how the properties of a complicated system are related to those of an easy, ideal system "close by" whose properties are more easily computed or known. The idea comes from celestial mechanics which tries to determine the deviations of planetary orbits from the "unperturbed" Keplerian paths due to the gravitational forces of other celestial bodies. While the methods used there point in a different direction, it is the perturbation theory of Lord Rayleigh (concerned with continuously extended oscillating systems) which leads frequently and typically to operators perturbed by the addition of a compact operator. This happens for example, when in elasticity the passage is made from constant mass density to variable density. See for example [Courant-Hilbert I, §V.13]. That these are as a rule compact perturbations, is due to the fact that in the underlying partial and ordinary differential equations the terms of highest order remain unchanged and only the coefficients of the derivatives of lower degree are modified. A theorem of Franz Rellich (see below Theorem II.4.3) explains why this produces compact perturbations.

Quantum mechanics poses farther reaching perturbation problems which are in parts mathematically unsolved. An example is the quantitative determination of energy levels of complicated systems of quantum mechanics.

The oscillations and motions of quantum mechanical systems are largely determined by the eigenvalues and eigenfunctions of the corresponding operators. Therefore, the "perturbation theory" usually amounts to applying approximation methods to solving the eigenvalue problem of a complicated linear operator $T + K$ which differs "little" from a simpler T with a solved eigenproblem. Perturbation theory becomes spectral theory which studies the different constituents of the "spectrum" of an operator. A reference is the comprehensive exposition in T. Kato: Perturbation theory for linear operators. Die Grundlehren der mathematischen Wissenschaften, Bd. 132, Springer-Verlag, Heidelberg-New York, 1966.[+]

However, we will not pursue the physical applications any further, since there is abundant motivation for perturbation theory within mathematics. Consider for instance the calculus of variations of which local perturbations are an actual principle, or geometric questions which ask

---

[+]Kato's perturbation theory is incomparatively deeper than our investigation: While we consider a single invariant, the index, Kato's theory is interested in countably many real parameters associated with the power series expansion of the eigenvalues of a perturbed (symmetric) operator $T + \varepsilon K$ where the parameters depend analytically on the perturbation. Just as one can classify symmetric matrices in linear algebra

- according to their rank
- projectively, according to their index of inertia ("Sylvester index") and
- orthogonally according to their diagonal elements (after "principal axis transformation")

we have in the perturbation theory of operators in Hilbert space several levels of stability: index/essential spectrum (see below)/perturbation parameters of the power series expansion.

In the crude mirror of finite-dimensional linear algebra, Kato's theory is closest to the principal axis transformation, while we restrict ourselves to consideration of the rank.

Examples:

a) T := shift$^+$

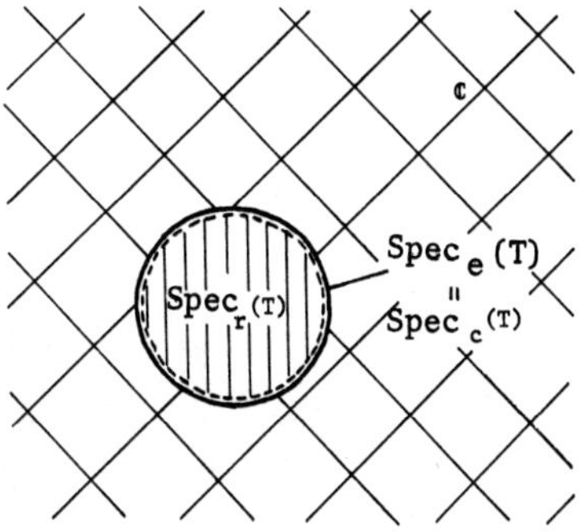

$\mathrm{Spec}_p(T) = \emptyset$

$\mathrm{Spec}_r(T) = \{z : |z| < 1\}$

$\mathrm{Spec}_e(T) = \{z : |z| = 1\}$

[Jörgens 1970/1982, §5.3]

b) T compact and self adjoint

$\mathrm{Spec}(T) \subset \mathbb{R}$

$\mathrm{Spec}_p(T) = \{\lambda_n : n = 1,2,\ldots\}$

$\mathrm{Spec}_r(T) = \{0\}$

$\mathrm{Spec}_e(T) = \{0\}$

[Jörgens 1970/1982, §6.2]

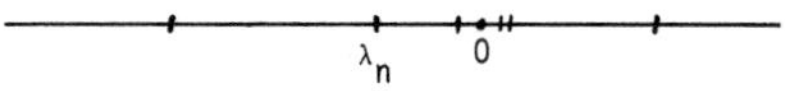

c) T = Fourier transformation on $L^2(\mathbb{R})$

$\mathrm{Spec}\ T = \mathrm{Spec}_e T$

$= \mathrm{Spec}_p T$

$= \{1,i,-1,i\}$

[Dym-McKean 1972, 98]

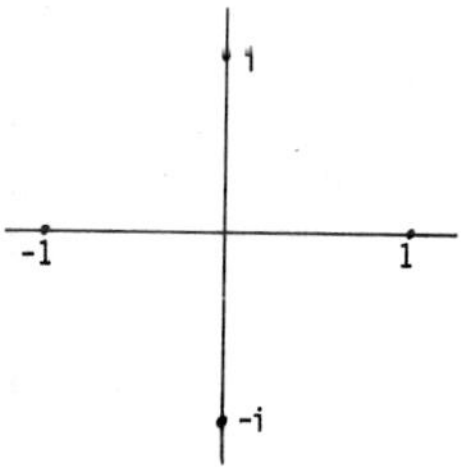

Table 2: The spectra of bounded linear operators $T : H \to H$

	Symbol	Name	Definition	Remarks
1.	Res(T)	resolvent set	$\{z \in \mathbb{C} : T-zId \text{ is invertible}\}$	open in $\mathbb{C}$
2.	Spec(T)	spectrum	$\mathbb{C} \setminus \mathrm{Res}(T)$	closed and bounded
3.	Fred(T)	Fredholm points	$\{z \in \mathbb{C} : T-zId \in \mathcal{F}\}$	contained in Res(T); it is the union of at most countably many disjoint, open, connected components
4.	$\mathrm{Spec}_e(T)$	essential spectrum	$\mathbb{C} \setminus \mathrm{Fred}(T)$	$= \mathrm{Spec}\ \pi(T)$ (where $\pi: \mathcal{B} \to \mathcal{B}/\mathcal{K}$ is the projection), contained in Spec(T)
5.	$\mathrm{Spec}_p(T)$	point spectrum = {eigenvalues}	$\{z \in \mathbb{C} : \mathrm{Ker}(T-zI) \neq 0\}$	contains the limit points of Spec(T). This set consists only of isolated points
6.	$\mathrm{Spec}_c(T)$	continuous spectrum	$\{z \in \mathbb{C} : \mathrm{Ker}(T-zI) = 0$ and $\mathrm{Im}(T-zI)$ properly contained in H, but dense$\}$	contained in $\mathrm{Spec}_e(T)$
7.	$\mathrm{Spec}_r(T)$	residual spectrum	$\{z \in \mathbb{C} : \mathrm{Ker}(T-zI) = 0$ and $\mathrm{codim}(\overline{\mathrm{Im}(T-zI)}) > 0\}$	is equal to $\mathrm{Spec}(T) \setminus (\mathrm{Spec}_p(T) \cup \mathrm{Spec}_c(T))$

how much a curve (asymptotically or in its shape) or a surface etc. changes if relevant parameters in their equations are modified. In particular, we are interested in the degree to which our quantitative invariants dim Ker, dim Coker, and index are independent of "small" perturbations. Here "small" does not exclusively mean that the dimension of the image of the perturbing operator is small - as with operators of finite rank and in a sense with compact operators - but may mean the perturbation is small in operator norm.

To get a feeling for the complications, we put together a list of established results:

1. The group of invertible elements of a Banach algebra is open, by Exercise 3. In particular, for each invertible, bounded, linear operator T, there is an ε $(:= ||T^{-1}||^{-1})$ such that for all $S \in \mathcal{B}$ with $||S|| < \varepsilon$, we have:

   (i) $T+S \in \mathcal{F}$
   (ii) index $T+S$ = index T (= 0)
   (iii) dim Ker $T+S$ = dim Ker T (= 0)
   (iv) dim Coker $T+S$ = dim Coker T (= 0).

2. Further, by Exercise 7, for all $T \in \mathcal{F}$ and $K \in \mathcal{K}$

   (i) $T+K \in \mathcal{F}$
   (ii) index $T+K$ = index T.

3. On the other hand, one can always find a perturbation of the identity by a compact operator K such that

   $$\dim \operatorname{Ker} \mathrm{Id}-K > 10^{80},$$

   making Ker Id-K unimaginably large, since its dimension could not be matched by the atoms in a universe of "only" 10^{11} galaxies. Namely, select an orthonormal basis for H and define K as the orthogonal projection onto the linear span of the first $10^{80} + 1$ basis elements. However, index Id-K = index Id (= 0) by the Riesz Theorem.

4. In each arbitrarily small neighborhood of the zero operator there are Fredholm operators (namely, iterates of the shift operators multiplied by a small constant ε) with any large or small index; e.g.,

   $$\operatorname{index}(0 + \varepsilon(\mathrm{shift}^{+})^{k}) = k.$$

   Hence, in the neighborhood of 0, the index behaves as a holomorphic function in the neighborhood of an essential singularity (Theorem of Felix Casorati and Karl Weierstrass).

5. For the boundary-value problem

$$u'' + ru = 0, \quad u(0) = u(1) = 0$$

treated in Chapter 4, or the equivalent problem

$$v - rKv = 0,$$

where $r > 0$ and

$$Kv = (1-x)\int_0^x yv(y)\,dy + x\int_x^1 (1-y)v(y)\,dy,$$

it was already shown that

$$\dim \operatorname{Ker}(Id-rK) = \begin{cases} 1 & \text{for } r = n^2\pi^2 \text{ and } n \in \mathbb{N} \\ 0 & \text{otherwise} \end{cases}$$

While $\dim \operatorname{Ker}(Id-rK)$ is not perturbable if it is zero (this is also clear because $Id-rK$ is then invertible by the Riesz Theorem, whence Exercise 3 or the above result 1 applies), it is very prone to change when $r = n^2\pi^2$ - however, only in one direction: the dimension can only decrease. In other words, $\dim \operatorname{Ker}(Id-rK)$ is *semi-continuous;* e.g.,

$$\dim \operatorname{Ker}(Id-rK) \leq \dim \operatorname{Ker}(Id-r_0K)$$

for all r sufficiently close to r_0.

The following theorem shows, for arbitrary small (in the operator norm sense) perturbations, what we already proved in Exercise 7 for compact perturbations: Even though the dimensions of the kernel of an operator and of its adjoint are not invariant under perturbations, the two jump by the same amount, so that their difference (the index) remains constant. The perturbation-invariance of the index is its most remarkable property. Together with the composition rule (Exercise 2.4 or Exercise 3.2), it shows that the index is defined "geometrically"; it has properties analogous to homotopy-invariants in algebraic topology such as the Euler characteristic.

C. Homotopy-Invariance of the Index

After these heuristic considerations, we now come to the aforementioned main theorem.

Theorem 2 (J. Dieudonné, 1943): The index: $\mathcal{F} \to \mathbb{Z}$ is locally constant.

Proof: Let $F \in \mathcal{F}$. By Exercise 3, we already know that there is a neighborhood of F in $\mathcal{B}$ which is contained in $\mathcal{F}$. We now amplify the argument used there: By Theorem 1, we first choose a parametrix G for F; i.e., a $G \in \mathcal{B}$ such that

$$FG = \mathrm{Id} + K_1 \quad \text{and} \quad GF = \mathrm{Id} + K_2,$$

where $K_1, K_2 \in \mathcal{K}$. We now show that for all $T \in \mathcal{B}$ with $||T|| < ||G||^{-1}$, we have

$$F + T \in \mathcal{F}.$$

Recall the "geometric series" argument (see the hint for Exercise 3), whereby the operators $\mathrm{Id} + TG$ and $\mathrm{Id} + GT$ are invertible, since $||TG||$ and $||GT||$ are less than 1. Thus $(\mathrm{Id}+GT)^{-1}G$ is a left inverse of $F + T$ modulo $\mathcal{K}$, since

$$(\mathrm{Id}+GT)^{-1}G(F+T) = \mathrm{Id} + (\mathrm{Id}+GT)^{-1}K_2. \tag{*}$$

Similarly, $G(\mathrm{Id}+TG)^{-1}$ is a right inverse of $F + T$ modulo $\mathcal{K}$. Thus, by Theorem 1, we have $F+T \in \mathcal{F}$.

Applying the composition rule and the Theorem of Riesz, we easily obtain from (*) the index formula

$$\mathrm{index}(\mathrm{Id}+GT)^{-1} + \mathrm{index}\, G + \mathrm{index}(F+T) = 0.$$

Hence $\mathrm{index}(F+T) = \mathrm{index}\, F$, since the index of an invertible operator vanishes, and $\mathrm{index}\, G = -\mathrm{index}\, F$ by Exercise 6. □

Exercise 8. Show that the index is constant on the connected components of $\mathcal{F}$. □

Exercise 9. Show that

$$\dim \mathrm{Ker} : \mathcal{F} \to \mathbb{N} \cup \{0\}$$

is semi-continuous, or more precisely,

$$\dim \mathrm{Ker}(F+T) \leq \dim \mathrm{Ker}\, F$$

for $F \in \mathcal{F}$ and $||T||$ sufficiently small.

Hint: Show that $\mathrm{Ker}(F+T) \cap (\mathrm{Ker}\, F)^{\perp} = \{0\}$. Details are found in [Schechter 1971, 116]. □

As an aside, we mentioned that Dieudonné* proved Theorem 2 only implicitly without use of the index concept. The numerous interrelations

*J.D.: Sur les homomorphismes d'espace normés. Bull. Sci. Math. (2) 67 (1943), 72-84.

of this theorem can be seen from the fact that by now a number of quite diverse proofs exist: All proofs have in common the reduction to the geometric series argument or the openness of the group of units of a Banach algebra.

This idea is most apparent in [Douglas, 36f. and 133-138] where it is first shown that $A^\times/A_0^\times$ is discrete where A is an arbitrary Banach algebra (with identity), $A^\times$ its group of units, and $A_0^\times$ the connected component containing the identity. There is an abstract index

$$i : A^\times \to A^\times/A_0^\times$$

defined in a natural way, and whose continuity and hence local invariance is clear from the definition. The main task consists in making the connection between this ideally simple algebraic object and the "real" index.

Less algebraic proofs can be found in [Jörgens 1970/1982, §5.4], where the reduction to the openness of $B^\times$ is achieved in a sequence of explicit extensions and projections which are computed in detail.

The trick of fixing one dimension is carried out particularly elegantly in [Atiyah 1969, 104]. As in [Jörgens] and in contrast to our proof above which is inspired by [Cartan-Schwartz, 12/06] and [Schechter, 115 f.], the Atiyah proof does not use the non-trivial theorems of Riesz and Atkinson and thus may be the most transparent proof in the whole. We will render it next. □

Alternative Proof of Theorem 2: Let $e_0, e_1, \ldots$ be an orthonormal basis for the Hilbert space H. We take H_n to be the closure of the linear span of the e_i with $i \geq n$, and we let P_n be the orthogonal projection of H onto H_n.

Step 1: Clearly P_n is self-adjoint and $P_n \in \mathcal{F}$, since Ker P_n and Coker P_n are finite-dimensional. Hence index $P_n = 0$, and for each $F \in \mathcal{F}$, we have

$$\text{index } P_nF = \text{index } F.$$

Step 2: Since dim Coker $F < \infty$ $(F \in \mathcal{F})$, we can find n_0 such that $e_0, e_1, \ldots, e_{n_0-1}$ and F(H) span H; in particular

$$P_nF(H) = H_n \quad \text{and} \quad \dim \text{Coker } P_nF = n,$$

for all $n \geq n_0$. (Incidentally, we see that dim Coker P_nF and also dim Ker P_nF can be made arbitrarily large with n.)

Step 3: Although the function dim Ker is only semi-continuous on $\mathcal{F}$, we can show that

$$\dim \operatorname{Ker} P_nG = \dim \operatorname{Ker} P_nF \quad \text{and} \quad \dim \operatorname{Coker} P_nG = \dim \operatorname{Coker} P_nF$$

for G sufficiently near to F and n sufficiently large (as in step 2): For $G \in \mathcal{B}$, consider the operator

$$\hat{G} : H \to H_n \oplus \operatorname{Ker} P_nF$$
$$u \mapsto (P_nGu, pu),$$

where $p : H \to \operatorname{Ker} P_nF$ is the projection. Then $\hat{F}$ is bijective, and hence has a bounded inverse by the Open Mapping Principle. Identifying $H_n \oplus \operatorname{Ker} P_nF$ with H, we then have $\hat{F} \in \mathcal{B}^{\times}$. By the familiar argument in the hint to Exercise 3, there is a neighborhood of $\hat{F}$ contained in $\mathcal{B}^{\times}$. Since "^" is continuous, there is also a neighborhood $\mathcal{V}$ of F such that the operator $\hat{G}$ is an isomorphism for all $G \in \mathcal{V}$. From the surjectivity of $\hat{G}$, it follows that $P_nG(H) = H_n$, whence $\dim \operatorname{Coker} P_nG = \dim \operatorname{Coker} P_nF = n$. Moreover, $\operatorname{Ker} P_nG = \hat{G}^{-1}(\operatorname{Ker} P_nF)$, since by definition of $\hat{G}$, a point u is mapped to $\operatorname{Ker} P_nF$ by $\hat{G}$ exactly when the first component (i.e., P_nGu) of $\hat{G}u$ vanishes. Since $\hat{G}$ is an isomorphism, we then also have

$$\dim \operatorname{Ker} P_nG = \dim \operatorname{Ker} P_nF,$$

which establishes the above claim.

Summary: We have shown that for each $F \in \mathcal{F}$, there is a natural number n and $\eta > 0$ such that for all $G \in \mathcal{B}$ with $||F-G|| < \eta$, we have

$$\operatorname{index} F = \operatorname{index} P_nF = \operatorname{index} P_nG = \operatorname{index} G.$$

Here, one could replace "index" by "Ker" or "Coker" in the inner equality, but not in the outer equalities. Moreover, $\operatorname{Im}(P_nF) = \operatorname{Im}(P_nG) = H_n$ and $\operatorname{Ker} P_nG = \hat{G}^{-1}(\operatorname{Ker} P_nF)$. □

Exercise 10: Formulate Theorem 2 for a *continuous family* of Fredholm operators, by which we mean a continuous map

$$G : X \to \mathcal{F},$$

where X is any topological space. More precisely, show that for all $x_0 \in X$, there is a neighborhood U and a natural number n such that for all $x \in U$,

$$\operatorname{Im} P_nG(x) = H_n.$$

Then prove that the function

$$\text{Dim Ker } P_n G : X \to \mathbb{N} \cup \{0\}$$

is constant (say k) on U, and that there are k continuous functions

$$f_i : X \to H; \quad i = 1,\dots,k$$

such that for all $x \in U$, the points $f_1(x),\dots,f_k(x)$ form a basis of Ker $(P_n G(x))$.

<u>Hint</u>: Imitate the preceding alternate proof, replacing G by $G(x)$ and F by $G(x_0)$. Then define $f_1(x_0),\dots,f_k(x_0)$ to be a basis of Ker $P_n G(x_0)$ and set $f_i(x) := \hat{G}(x)^{-1}(f_i(x_0))$. □

The concept of a continuous family of operators comes from classical analysis with its investigation of "families of operators" or "operators depending on a parameter." In the simplest examples, the parameter space X is the unit interval, all $\mathbb{R}$, or a bordered domain in a higher-dimensional Euclidean space (e.g., the domain of permissible control variables). With somewhat more complex problems of analysis (e.g., as in the study of elliptic boundary-value problems), we are quickly forced to consider families with more general parameter spaces: An elliptic differential equation defines a continuous family of Fredholm operators, where the parameter space is the sphere bundle of covectors of the underlying manifold restricted to the boundary; see Part II, Chapter 8.

We will first treat these questions not from the standpoint of applications, but rather out of natural topological-geometric considerations, namely interest in *deformation invariants.** We assign to each compact

---

*Motivated by problems of optics (and also questions in astronomy, surveying, and architecture), the projective geometry of the 17th century originated in the idea of searching for properties of geometric figures which are invariant under transformations (central projection and cross section). In addition to these linear transformations, the concept of a deformation (i.e., the continuous change of a mathematical object) existed, for instance, when Johannes Kepler in 1604 noted that if the plane is compactified, then ellipse, hyperbola, parabola and circle can be transformed into one another by a continuous relocation of the foci (See [Kline, 299]). But it was not until well into the 19th and 20th centuries that deformation invariants were found for a greater variety of mathematical objects. These include the homology and cohomology theories, as presented axiomatically in [Eilenberg-Steenrod] for example, as well as the so-called K-theory, another branch of algebraic topology which was developed by Michael F. Atiyah and Friedrich Hirzebruch and is specifically aimed at the needs of analysis; see Part III below.

parameter space X a group and to each continuous family of Fredholm operators

$$G : X \to F$$

we assign a group element, which is indeed invariant under deformation. This means that another continuous family of Fredholm operators

$$G' : X \to F$$

is assigned to the same group element, if G and G' are homotopic. By this, we mean that G can be continuously deformed to G' (see Chapter III.1); i.e., there is a continuous family g of Fredholm operators parametrized by the product space $X \times I$, where $I = [0,1]$,

$$g : X \times I \to F$$

such that

$$g_{|X\times\{0\}} = G \quad \text{and} \quad g_{|X\times\{1\}} = G'.$$

If X consists of a single point, then a continuous family of Fredholm operators is just a single Fredholm operator, and the homotopy of G and G' clearly means that G and G' lie in the same component of F. In this way, we can answer fundamental questions concerning the nature of the connected components of F (e.g., via approximation theory).

Before we study continuous families of Fredholm operators (i.e., the geometry of F or the group of units $(B/K)^{\times}$ by Theorem 2), we first turn to a simpler problem, the geometric investigation of the group of units $B^{\times}$.

6. FAMILIES OF INVERTIBLE OPERATORS. KUIPER'S THEOREM

This and the following chapter may be skipped and then later read in conjunction with Part III. In particular, here we use some concepts from topology which will be made precise only in Part III below.

A. Homotopies of Operator-Valued Functions

Exercise 1. Let X and Y be topological spaces and f, g, and h continuous maps from X to Y. Show: If f is homotopic to g and g is homotopic to h, then f is homotopic to h.

1. Two continuous maps f and g from X to Y are called *homotopic* ($f \sim g$), if one can continuously deform one into the other; i.e.,

there is a continuous map $F: X \times I \to Y$ (where $I = [0,1]$) such that

$$F \bullet i_0 = f \quad \text{and} \quad F \bullet i_1 = g;$$

here, $i_t : X \to X \times \{t\}$ is the canonical inclusion ($t \in I$). We write F_t for $F \bullet i_t$, and can then roughly regard F as a 1-parameter family (over I) of continuous maps from X to Y. We call F a homotopy of f to g.

2. By the transitivity (Exercise 1) and obvious symmetry and reflexivity of the relation "homotopic", it follows that the *homotopy classes*

$$\overline{f} := \{g : g: X \to Y \text{ continuous and } f \sim g\}$$

and the *homotopy set*

$$[X,Y] := \{\overline{f} : f: X \to Y \text{ continuous}\}$$

are well defined. Note that $[\text{point},Y]$ corresponds to the set of path-wise connected components of Y.

3. Two topological spaces X and Y are *homeomorphic*, if there is a bijective map $f : X \to Y$ which is continuous in both directions.

4. Two topological spaces are *homotopy-equivalent*, if there are continuous maps $f : X \to Y$ and $g : Y \to X$ such that $f \bullet g \sim Id_Y$ and $g \bullet f \sim Id_X$. Clearly, the real line $\mathbb{R}$ and the plane $\mathbb{R}^2$ are homotopy-equivalent and have the same "cardinality" (i.e., there is a bijection between them), but they are not homeomorphic, as we show in Part III.

5. Y is called a *retract* of X, if $Y \subset X$ and there is a continuous map $f : X \to Y$ with $f|Y = Id$. Such an f is called a *retraction*. If, in addition, $i \bullet f \sim Id$, where $i : Y \to X$ is inclusion, then Y is called a *deformation retract* of X and X and Y are homotopy-equivalent. Each $P \in X$ trivially constitutes a retract of X (but the sphere, as the boundary of the solid ball, is not a retract of the ball - see Part III below). If $\{P\}$ is a deformation retract of X, then X is called *contractible*. The shape of X must then be "starlike" in a certain sense. □

Here we are interested in the homotopy type of operator spaces. Let $\mathcal{B}^\times(H)$ be the group of invertible operators on a Hilbert space H, which is allowed to be a finite-dimensional complex vector space, say $\mathbb{C}^N$, and so $\mathcal{B}^\times(H) = GL(N,\mathbb{C})$.

Exercise 2. Investigate the group $\mathcal{B}^\times(H \times H)$ (or $GL(2N,\mathbb{C})$) of invertible operators on the product space $H \times H$, which can be written as 2×2 (block) matrices. For $R,S \in \mathcal{B}^\times(H)$ - or more generally for

$R,S : X \to \mathcal{B}^\times(H)$ continuous with X a given topological space - show that

$$\begin{pmatrix} SR & 0 \\ 0 & Id \end{pmatrix} \sim \begin{pmatrix} R & 0 \\ 0 & S \end{pmatrix}.$$

Hint: Consider the map $F : X \times [0, \frac{\pi}{2}] \to \mathcal{B}^\times(H \times H)$, which is given by

$$F_t := \begin{pmatrix} \cos t & -\sin t \\ \sin t & \cos t \end{pmatrix} \begin{pmatrix} S & 0 \\ 0 & Id \end{pmatrix} \begin{pmatrix} \cos t & \sin t \\ -\sin t & \cos t \end{pmatrix} \begin{pmatrix} R & 0 \\ 0 & Id \end{pmatrix}.$$

Here, we have for brevity written $\cos t$ and $\sin t$ for the operators $\cos t \, Id$ and $\sin t \, Id \in \mathcal{B}(H)$. Show that the image of F really lies in $\mathcal{B}^\times(H \times H)$ - naturally not entirely in $\mathcal{B}^\times(H) \times \mathcal{B}^\times(H)$ - and investigate F_0 and $F_{\pi/2}$. Why can we use an interval of $\mathbb{R}$ different from I in the definition of a homotopy? □

The trigonometric functions which appeared in the preceding problem are typical of homotopy investigations of linear spaces in which rotations and compressions or dilations are the most important deformations. This considerably simplifies the explicit statement of homotopies. Of course, it does not simplify the demonstration of the non-existence of a homotopy since this forces one to consider "all" homotopies, a task which in general is solvable only with the "crude" means of algebraic topology; see III.1 below.

We add another exercise in computation with continuous families of matrices. Recall the fact of linear algebra that the group $GL(N,\mathbb{C})$ of invertible complex $N \times N$-matrices contains the compact subgroup $U(N)$, where $U(N)$ consists of the *unitary matrices* of rank N, i.e., of all those complex $N \times N$ matrices whose ("complex-conjugate") adjoint is just its inverse. The corresponding operators in $\mathbb{C}^N$ to $\mathbb{C}^N$ are exactly those preserving norms (with respect to the Hermitian norm in Hilbert space defined by the inner product). Obviously, $U(N)$ and $GL(N,\mathbb{C})$ are pathwise connected; this follows at once (for instances with the tools of Exercise 2) from a homotopy analysis of the transformations by means of which a complex invertible matrix can be put in diagonal form and finally made into the identity. For the rest, the homotopy type of $U(N)$ is only partially known; see Part III below. In contrast, we can show that $\mathcal{U}(H) := \{T \in \mathcal{B}^\times(H) : T^{-1} = T^*\}$, for infinite dimensional H, is contractible; see Remark 2 following Theorem 2.

Exercise 3. Show that the homotopy used in Exercise 2 does not leave $\mathcal{U}(H \times H)$, if $R,S \in \mathcal{U}(H)$.

Exercise 4. Regard $S^1 := \{z : z \in \mathbb{C} \text{ and } |z| = 1\}$ as a subset of $\mathbb{C} \times \{0\} \subset \mathbb{C}^2$ (via $z \longmapsto (z,0)$) and choose $a \in S^1$ (e.g., $a = (1,0)$). Construct a continuous map

$$g : S^1 \to U(2)$$

with the properties

(i) $(g(z))(z) = a$ for all $z \in S^1$

(ii) $g \sim f$, where $f(z) = \mathrm{Id}_{\mathbb{C}^2}$ for all $z \in S^1$.

Warning: The exercise would be trivial and solvable without using the 2nd dimension (i.e., within $U(1)$ rather than $U(2)$) if S^1 were contractible. In that case the maps

$$f : z \longmapsto 1$$

$$g : z \longmapsto az^{-1}$$

would be homotopic as maps on S^1 to S^1 (or equivalently in $U(1)$).

Hint: Reduce to Exercise 3 setting

$$g : z \longmapsto \begin{pmatrix} az^{-1} & 0 \\ 0 & za^{-1} \end{pmatrix}. \qquad \square$$

Theorem 1. The group $\mathcal{B}^\times$ of invertible bounded linear operators on a Hilbert space H is pathwise connected.

It is not true that the group of units of a Banach algebra is pathwise connected. The group of units $(\mathcal{B}/\mathcal{K})^\times$ in the Calkin algebra is a counterexample; its connected components - the connected components of Fredholm operators - are mapped bijectively to $\mathbb{Z}$ by the index, as the following paragraph shows.

This theorem is usually proved by means of deeper results of spectral theory (see, e.g., [Douglas 1972, 134 ff.]). (One first shows that every unitary operator U has a spectral decomposition $U = \cos A + i \sin A$ where A is a self-adjoint operator. Then, by the Spectral Theorem,

$$t \longmapsto U_t := \cos tA + i \sin tA, \quad t \in I$$

is a continuous path in $\mathcal{U}(H)$ from U to Id. One shows further that each invertible operator R can be factored as R = UB where U is unitary, and $B = \sqrt{R^*R}$ is self-adjoint, positive and invertible. Then one again connects U with Id using U_t and B with Id with the path

$$t \longmapsto B_t := t\ \mathrm{Id} + (1-t)B,$$

which does not go outside $\mathcal{B}^\times$ by the Spectral Theorem, since B is positive and invertible. In this fashion $t \longmapsto U_t B_t$ defines a path from R to Id.)

We will give here - following an idea of Nicolaas Kuiper - a completely elementary proof of the theorem which perhaps is not as elegant as the proof outlined above and which - as all elementary proofs - requires more calculation and perhaps some more geometric imagination. The decisive advantage for us is that the elementary proof generalizes effortlessly to a proof of Kuiper's Theorem (Theorem 2) according to which $[X,\mathcal{B}^\times] = 0$, even if X does not consist of a single point as in Theorem 1 but is an arbitrary compact topological space.

While the content of Theorem 1 remains unchanged in passing from $\mathbb{C}^N$ to the infinite dimensional Hilbert space H, Theorem 2 exhibits a fundamental difference (see the Bott Periodicity Theorem in III.1) between the linear algebra of finite-dimensional vector spaces and the functional analysis of Hilbert space. This aspect we can bring out clearly in the following proof of Theorem 1.

Proof of Theorem 1: Let $R_0 \in \mathcal{B}^\times$. We seek a continuous path in $\mathcal{B}^\times$ connecting R_0 with Id. We proceed in two stages: In the first stage, we connect R_0 with a operator R_2 which is the identity on a cleverly constructed infinite dimensional subspace. In the second stage, we connect R_2 with the identity of H.

Stage 1 of proof, step 1: We begin by recursively constructing a sequence of unit vectors $a_1, a_2, \ldots \in H$ and a sequence of 2-dimensional subspaces $A_1, A_2, \ldots \subset H$, such that

$$A_i \perp A_j \quad \text{for} \quad i \neq j$$

and

$$a_i \in A_i \quad \text{and} \quad R_0 a_i \in A_i \quad \text{for all} \quad i = 1,2,\ldots\ .$$

Start with any unit vector $a_1 \in H$ and a 2-dimensional subspace A_1 which contains a_1 and $R_0 a_1$. Then choose a unit vector

$$a_2 \in A_1^\perp \cap R_0^{-1}(A_1^\perp)$$

and let A_2 be a 2-dimensional subspace containing a_2 and $R_0 a_2$. By construction, we have $A_1 \perp A_2$. Then proceed with

$$a_3 \in A_1^{\perp} \cap A_2^{\perp} \cap R_0^{-1}(A_1^{\perp}) \cap R_0^{-1}(A_2^{\perp}),$$

etc. The construction never breaks down, since the intersection of finitely many subspaces of finite codimension is never trivial.

Stage 1, step 2: Now we deform the operator R_0 to R_1 so that $R_1 a_i$ is a unit vector in the direction of $R_0 a_i$. Thus, define (for $t \in I$)

$$R_t u := \begin{cases} R_0 u & \text{for } u \in (\overset{\infty}{\underset{i=1}{\oplus}} A_i)^{\perp} \\ (1-t+\frac{t}{|R_0 a_i|})R_0 u & \text{for } u \in A_i. \end{cases}$$

Stage 1, Step 3: Deform the operator R_1 to an operator R_2 with the desired property

$$R_2 a_i = a_i \quad \text{for all } i.$$

This is done by rotating the vector $R_1 a_i$ to the position a_i by an element of $U(2)$ in each complex plane A_i, leaving the vectors orthogonal to the rotation plane fixed.

This simple geometric argument (in which we have given up spacial intuition, since a complex plane, of $\mathbb{C}$-dimension 2, has $\mathbb{R}$-dimension 4) can be formalized, and reduced to Exercise 4. Indeed, we can map each A_i by an isometry α_i onto $\mathbb{C}^2$, in such a way that the complex line $\{\lambda R_1 a_i : \lambda \in \mathbb{C}\}$ is mapped to $\mathbb{C} \times \{0\}$. Then let $g_i : S^1 \to U(2)$ be a map with the properties (guaranteed by Exercise 4):

(i) $(g_i(z))(z) = \alpha_i(a_i)$, for all $z \in S^1 \subset \mathbb{C} \times \{0\}$

(ii) There is a continuous map $F_i : S^1 \times I \to U(2)$ with $F_i(\ldots,1) = g_i$ and $F_i(z,0) = \mathrm{Id}_{\mathbb{C}^2}$ for all $z \in S^1$.

Without loss of generality, we have assumed that $\alpha_i(a_i) \in S^1$ (otherwise, we rotate first into an arbitrary point $a \in S^1$ and, from there, to $\alpha_i(a_i)$). For $t \in I$, we now set

$$T_t u := \begin{cases} u & \text{for } u \in (\overset{\infty}{\underset{i=1}{\oplus}} A_i) \\ \alpha_i^{-1} F_i(\alpha_i R_1 a_i, t)\alpha_i u & \text{for } u \in A_i \end{cases}$$

thereby obtaining through

$$t \longmapsto R_{1+t} := T_t \circ R_1$$

a continuous path in $\mathcal{B}^\times$ from R_1 to a R_2 with

$$R_2|H' = \mathrm{Id},$$

where $H' \subset H$ is the infinite-dimensional closed subspace spanned by $a_1, a_2, \ldots$

Stage 2, step 1: Relative to the decomposition $H = H_1 \oplus H'$, where $H_1 := (H')^\perp$ is the orthogonal complement of H' in H, R_2 has the form $\begin{pmatrix} Q & 0 \\ * & \mathrm{Id} \end{pmatrix}$, where $Q \in \mathcal{B}^\times(H_1)$ and the perturbation term $*$ can be deformed to zero by a continuous path in $\mathcal{B}^\times(H)$

$$R_{2+t} = \begin{pmatrix} Q & 0 \\ (1-t)* & \mathrm{Id} \end{pmatrix}, \quad t \in I;$$

with

$$R_3 = \begin{pmatrix} Q & 0 \\ 0 & \mathrm{Id} \end{pmatrix}.$$

Stage 2, step 2: By the classical 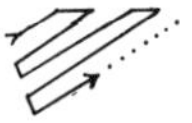 -argument (which one uses in set theory to count $\mathbb{Q}$ or to demonstrate the equipotence of $\mathbb{N}$ and $\mathbb{N} \times \mathbb{N}$), we can decompose H' into an infinite sum of Hilbert spaces $H_2, H_3, \ldots$. Explicitly: Let $a_1, a_2, \ldots$ form an orthonormal basis of H'. Decompose $\mathbb{N}$ into the infinite disjoint subsets

$$\mathbb{N}_j := \{2^{j-2}(2n-1) : n \in \mathbb{N}\}, \quad j = 2,3,4,\ldots$$

and take H_j to be the closed subspace spanned by the a_i with $i \in \mathbb{N}_j$. In this way we have $H = \bigoplus_{j=1}^{\infty} H_j$ and

$$R_3 = \begin{pmatrix} Q & & & & 0 \\ & \mathrm{Id} & & & \\ & & \mathrm{Id} & & \\ & 0 & & \mathrm{Id} & \\ & & & & \ddots \end{pmatrix}.$$

If we now identify H_j and H_1 for $j \neq 1$ (all infinite-dimensional, separable, Hilbert spaces are trivially isomorphic), we can also write:

$$R_3 = \begin{pmatrix} Q & & & & & \\ & QQ^{-1} & & & & 0 \\ & & \mathrm{Id} & & & \\ & & & QQ^{-1} & & \\ & 0 & & & \mathrm{Id} & \\ & & & & & \ddots \end{pmatrix}.$$

With the rotation of Exercise 2, we obtain a continuous path in $\mathcal{B}^\times(H_1) \times \mathcal{B}^\times(H_1 \times H_1) \times \mathcal{B}^\times(H_1 \times H_1) \times \ldots \subset \mathcal{B}^\times(H)$ from R_3 to an operator

$$R_4 = \begin{pmatrix} Q & & & & \\ & Q^{-1} & & & 0 \\ & & Q & & \\ & & & Q^{-1} & \\ 0 & & & & \ddots \end{pmatrix}$$

and with one more rotation (this time in $\mathcal{B}^\times(H_1 \times H_1) \times \mathcal{B}^\times(H_1 \times H_1) \times \ldots$) a continuous path from R_4 to

$$R_5 = \begin{pmatrix} \mathrm{Id} & & 0 \\ & \mathrm{Id} & \\ 0 & & \ddots \end{pmatrix} = \mathrm{Id}_H. \qquad \square$$

B. The Theorem of Kuiper

The idea of the last step of the preceeding proof can be found in Albert Solomonovich Schwarz (The homotopy-topology of Banach spaces (Russian); Dokl. Akad. Nauk SSSR 154 (1964), 61-63. English translation: Sov. Math. Dokl. 5 (1964), 57-59) and in Klaus Jänich (Vektorraumbündel und der Raum der Fredholm-Operatoren; Math. Ann. 161 (1965), 129-142). It bares a "secret" which separates fundamentally - from the topological standpoint - the linear functional analysis in Hilbert space from the linear algebra of finite dimensional vector spaces: It is the possibility to - figuratively speaking - escape any squeeze by moving aside into a new dimension. If we had decomposed H into only finitely many components $H_1 \oplus \ldots \oplus H_m$ we would have gotten stuck in the H_m at the latest either with the homotopy from R_3 to R_4 or with the homotopy from R_4 to R_5 (depending on whether m is even or odd). We meet a similar phenomenon when investigating the geometry of unitary matrices where e.g. the homotopy set $[S^i, U(N)]$ (the well-known "homotopy groups" $\pi_i(U(N))$; S^i is i-sphere of unit vectors in $\mathbb{R}^{i+1}$) is determined for $2N \geq i+1$

by the Periodicity Theorem of Raoul Bott but not known for all smaller values of N. Details are in Chapter III.1.

Finally we wish to remark that generally in topology low-dimensional structures, particularly 3- and 4-dimensional manifolds (recall the key words "knot theory", "Poincaré conjecture", "signature") are among the most difficult areas, while analogous questions for higher-dimensional objects are either solved or at least pose no fundamentally new problems.

This is the background which may help in understanding the following basic theorem.

Theorem 2 (N. Kuiper, 1964): For any compact, topological space X, the homotopy set $[X,\mathcal{B}^\times(H)]$ consists of a single element, where $\mathcal{B}^\times(H)$ is the group of bounded invertible operators in the Hilbert space H.

Remark 1. Just like Theorem 1 (X = point), Theorem 2 holds for non-separable Hilbert spaces (see [Illusie, 284/02 f.] and for real Hilbert spaces (see [Kuiper, 19-30]). However, according to our earlier convention we restrict ourselves always to separable complex Hilbert space.

Remark 2. It is a corollary of Theorem 2 that $\mathcal{B}^\times(H)$ is contractible. This would be completely trivial if Theorem 2 were valid for non-compact X, for example for $X = \mathcal{B}^\times(H)$. But it is not this simple. Still there is a way (by studying the "nerve" of an open cover of $\mathcal{B}^\times(H)$ as in Stage 0 of the following proof) of reducing the question of contractibility to Theorem 2. See [Kuiper, 27 f.] and [Illusie, 284/01 f.].

Remark 3. Contrary to the topological investigation of the general linear group $GL(N,\mathbb{C})$ there is no gain in restricting attention first to the group $\mathcal{U}(H)$ of unitary operators ($TT^* = T^*T = Id$). In the "classical case" (see III.1 below) the advantage results from the compactness of $U(N) := \mathcal{U}(\mathbb{C}^N)$ but for $\dim H = \infty$, $\mathcal{U}(H)$ is not compact.

Proof: We adopt the proof of Theorem 1 with the necessary modifications and several additional observations. In order that we can make a reasonable analogy between a continuous family

$$f : X \to \mathcal{B}^\times(H)$$

of operators and a single operator $R \in \mathcal{B}^\times(H)$, say

$$R : \{point\} \to \mathcal{B}^\times(H),$$

we must guarantee that $\operatorname{Im} f$ is at least contained in a finite dimensional subspace of $\mathcal{B}(H)$. Thus, before we consider the analogy, we establish

Stage 0: Each continuous $f_0 : X \to \mathcal{B}^\times(H)$ with X compact is homotopic to a continuous map $f_1 : X \to \mathcal{B}^\times(H)$ with $f_1(X) \subset V$, where V is a finite-dimensional subspace of $\mathcal{B}(H)$.

To prove this, we use the openness of $\mathcal{B}^\times(H)$ (see Exercise 5.3) and place an open ball contained in $\mathcal{B}^\times(H)$ about each operator $T \in f_0(X)$. This gives us an open cover $\mathcal{U}$ of $f_0(X)$. For safety reasons (see below), we replace each ball $U \in \mathcal{U}$ by a ball U' with the same center, but with 1/3 the radius. Clearly, $\mathcal{U}' := \{U' : U \in \mathcal{U}\}$ is an open cover of $f_0(X)$, since each operator in the image of f_0 is the center of some U'. As the continuous image of a compact set, $f_0(X)$ is also compact and it is covered by a finite subset of $\mathcal{U}'$. Thus, we have $f_0(X) \subset U_*$, where $U_* = \bigcup_{i=1}^N K(T_i,\varepsilon_i)$ is the union of the finite set of open balls

$$K(T_i,\varepsilon_i) := \{T \in \mathcal{B}(H) : ||T-T_i|| < \varepsilon_i\}, \quad i = 1,\dots,N.$$

The balls are contained in $\mathcal{B}^\times(H)$ and are so small that $K(T_i,3\varepsilon_i)$ is also contained in $\mathcal{B}^\times(H)$.

Then we are essentially done with the verification of stage 0: Although U_* is still infinite-dimensional, it is visibly contractible to a simplicial complex with vertices $T_1,\dots,T_N$, meaning a structure that lies entirely in the subspace of dimension $\leq N$ of $\mathcal{B}(H)$ spanned by $T_1,\dots,T_N$; see the illustration on the following page.

Since intuition - particularly for infinite-dimensional spaces - can be deceiving, we will write out this argument precisely: For each $t \in [0,1]$ and $T \in U_*$, we define an operator

$$g_t(T) := (1-t)T + t\sum_{i=1}^N \phi_i(T)T_i.$$

Here, ϕ_i $(i = 1,\dots,N)$ is a "partition of unity" on U_*; i.e., $\phi_i : U_* \to \mathbb{R}$ is continuous and

(i) support $\phi_i \subset \overline{K(T_i,\varepsilon_i)}$; i.e., $\phi_i(T) = 0$ for $||T-T_i|| \geq \varepsilon_i$.

(ii) $0 \leq \phi_i(T) \leq 1$

(iii) $\sum_{i=1}^N \phi_i(T) = 1$ for $T \in U_*$.

For example, set

$$\phi_i(T) := \frac{\psi_i(T)}{\sum_{k=1}^N \psi_k(T)} \quad \text{for } T \in U_*,$$

$f_0(X)$
(Subset of $\mathcal{B}^\times(H) \subset \mathcal{B}(H)$)

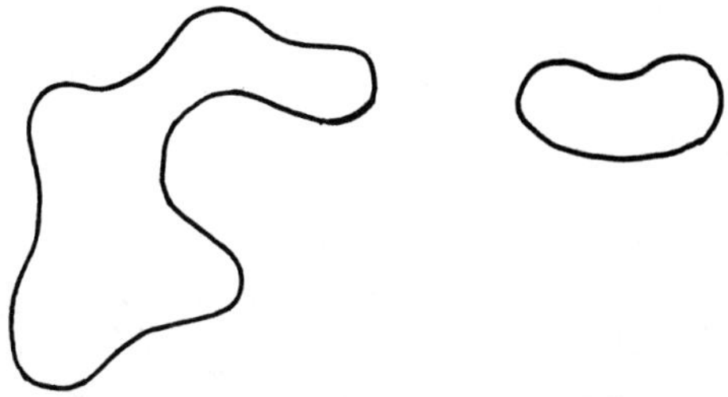

The finite covering U_*

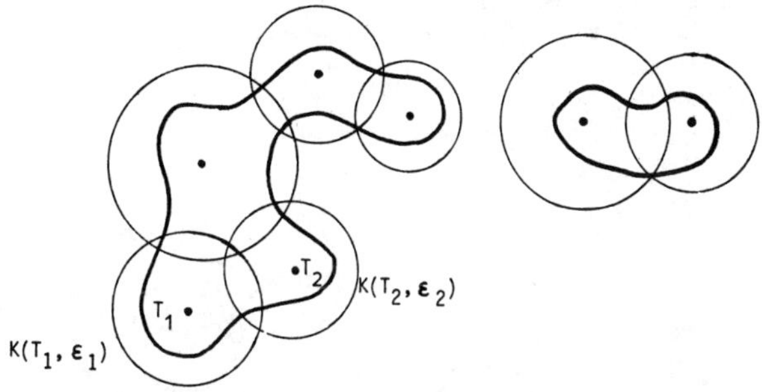

Contraction onto simplicial complex

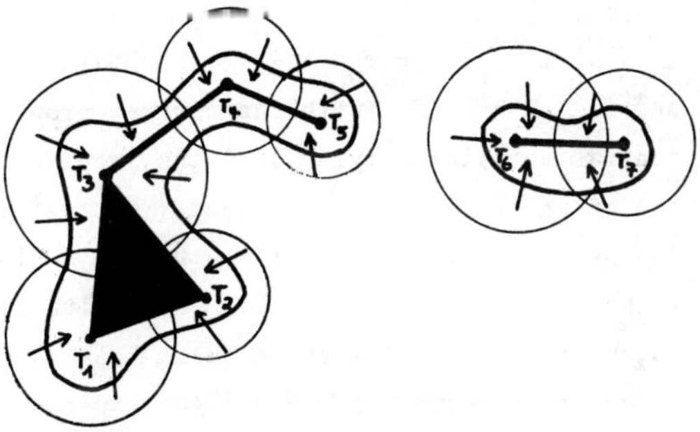

where

$$\psi_i(T) := \begin{cases} \varepsilon_i - ||T-T_i|| & \text{for } T \in U_i \\ 0 & \text{otherwise} \end{cases} .$$

In this way, $g_0 : U_* \to B^\times(H)$ is the inclusion and $g_1 : U_* \to B^\times(H)$ is a retraction of U_* onto a simplicial complex (composed of points, segments, triangles, tetrahedra and corresponding higher-dimensional simplices) with vertices $T_1,\ldots,T_N$.

To be sure that the homotopy of g_0 to g_1 does not leave $B^\times(H)$, we use an "$\frac{\varepsilon}{3}$-argument": Let $T \in U_*$. In considering $g_t(T)$, we are (because of (i)) only interested in those summation indices i for which $T \in K(T_i,\varepsilon_i)$. We compare these balls and let $K(T_m,\varepsilon_m)$ be the one with the largest radius. By the triangle inequality, each of these balls is contained in $K(T_m,3\varepsilon_m)$. Thus, the convex hull of T and *these* T_i (and hence $g_t(T)$, $0 \leq t \leq 1$) is contained in the ball $K(T_m,3\varepsilon_m)$, which by construction lies in $B^\times(H)$.

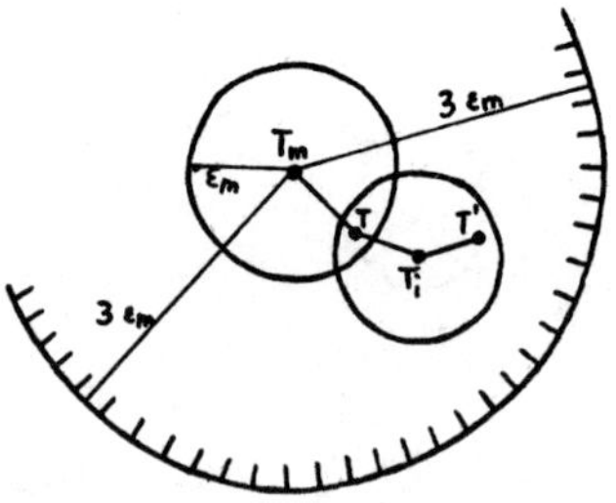

With $f_t := g_t f_0$, $t \in I$, we obtain a homotopy in $B^\times(H)$ of f_0 to an $f_1 : X \to B^\times(H)$ with the desired properties. This completes the essential work in carrying over the proof of Theorem 1 to the more general situation of Theorem 2. Indeed, the rest of the proof now is quite analogous; we no longer consider R_0 as a single element of $B^\times(H)$, but rather as the intersection of the finite-dimensional subspace V (in which $f_1(X)$ lies) with $B^\times(H)$. Then we need only show that $\{Id\}$ is a deformation retract of $V \cap B^\times(H)$, just as we showed in the proof of Theorem 1 that R_0 and Id are connected by a continuous path in $B^\times(H)$, and we are done. The proofs are identical in principle. In the particulars, we need the following modifications:

Stage 1, step 1: We construct as above a sequence of unit vectors $a_1, a_2, \ldots \in H$ and a sequence of pairwise orthogonal $n+1$-dimensional subspaces $A_1, A_2, \ldots \subset H$ such that $a_i \in A_i$ and $Ra_i \in A_i$ for all $R \in \mathcal{V}$ and $i = 1,2,\ldots$; here, $n := \dim \mathcal{V}$ and $\mathcal{V}$ is the vector space constructed in stage 0, except for containing in addition the operator Id.

Stage 1, steps 2 and 3: Now we show that the canonical inclusion

$$\gamma_0 : \mathcal{V} \cap \mathcal{B}^\times(H) \hookrightarrow \mathcal{B}^\times(H)$$

is homotopic to a map

$$\gamma_2 : \mathcal{V} \cap \mathcal{B}^\times(H) \to \mathcal{B}^\times(H)$$

with

$$\gamma_2(R)a_i = a_i \quad \text{for} \quad i = 1,2,\ldots \quad \text{and} \quad R \in \mathcal{V} \cap \mathcal{B}^\times(H).$$

As above, with step 2 we first go from γ_0 to γ_1 with

$$\gamma_1(R)u := \begin{cases} Ru & \text{for} \quad u \in (\oplus A_i)^\perp \\ \\ \dfrac{1}{|Ra_i|} Ru & \text{for} \quad u \in A_i. \end{cases}$$

For step 3, one must again take up the rotation argument of Exercise 4 and generalize it somewhat by induction: We regard $S^{2n-1} = \{z = (z_1,\ldots,z_n) \in \mathbb{C}^n : z_1\bar{z}_1 + \ldots + z_n\bar{z}_n = 1\}$ as a subset of $\mathbb{C}^n \times \{0\} \subset \mathbb{C}^{n+1}$ and construct, for each $a \in S^{2n-1}$, a continuous map

$$g : S^{2n-1} \to U(n+1) \tag{0}$$

with the properties

(i) $g(z)(z) = a$ for all $z \in S^{2n-1}$

(ii) $g \sim h$, where $h(z) = \mathrm{Id}_{\mathbb{C}^{n+1}}$ for all $z \in S^{2n-1}$

(see also [Illusie, 284/04]).

We reduce our problem to this situation by mapping each A_i by an isometry α_i onto $\mathbb{C}^{n+1}$ so that $\{Ta_i : T \in \mathcal{V}\}$ is mapped into $\mathbb{C}^n \times \{0\}$. Then, if $F_i : S^{2n-1} \times I \to U(n+1)$ is the corresponding homotopy for $a := \alpha_i(a_i)$, then the following is a homotopy from γ_1 to γ_2 with the desired properties:

$$\gamma_{1+t}(T)u := \begin{cases} \gamma_1(T)u & \text{for} \quad u \in (\overset{\infty}{\underset{i=1}{\oplus}} A_i)^{\perp} \\ \alpha_i^{-1} \; F_i(\alpha_i\gamma_1(T)a_i,t)\alpha_i \; \gamma_1(T)u & \text{for} \quad u \in A_i. \end{cases}$$

In particular, we have an infinite-dimensional H' (with orthonormal basis $a_1, a_2, \ldots$) such that $\gamma_2|H' = \mathrm{Id}$.

Stage 2: In the proof of Theorem 1, we have already implicitly shown that $B^{\times}((H')^{\perp}) \times \mathrm{Id}_{H'}$ is a deformation retract of the group of all invertible operators on H which are the identity on H' (step 1), and that $B^{\times}((H')^{\perp}) \times \mathrm{Id}_{H'}$ can be contracted to $\{\mathrm{Id}_H\}$ (step 2). The proof of Theorem 2 is now complete. □

7. FAMILIES OF FREDHOLM OPERATORS. INDEX BUNDLES

Like the preceeding one, this section may be skipped and read later in connection with the study of the topology of the general linear group (Bott's Periodicity Theorem, Chapter III.1) and the topological interpretation of elliptic boundary value problems (Chs. II. 7/8 and Section III.4.H).

A. The Topology of F

Our point of departure is the principal theorem on the "homotopy invariance" of the index saying that the index is defined in a neighborhood of a Fredholm operator and is constant there, and consequently on each connected component (Theorem 5.2). Regarding the topology of F, the space of Fredholm operators on a given Hilbert H, the following questions arise:

1. What is the number of connected components of F?
2. What can be said about the "structure" of the individual connected components? How many "holes" are there of each type?

Conceptually imagine a serving of Swiss cheese: Not only do we note the numbers of slices - Question 1 - but also the kind of holes - Question 2 - those that cross the cheese like a channel and those which are enclosed like air bubbles.

We already have a partial answer for question 1: there are at least $\mathbb{Z}$ connected components, since the right-handed and left-handed shift, together with its iterates (the natural powers), show that every integer can be the index of a Fredholm operator. We denote as in Ch. 6 the path connected components of F by [point,F]. Then the map

$$\text{index} : [\text{point}, F] \to \mathbb{Z}$$

is well-defined by Theorem 5.2 (homotopy invariance of the index) and bijective. In fact, the map is injective also as we will prove in this chapter. This answers Question 1 completely. In particular, it is utterly impossible that F or the group of units $(B/K)^\times$ of the quotient algebra is contractible, in contrast to $B^\times$ (Theorem 6.1).

In addition, the individual connected components of F differ as to homotopy type - Question 2 - radically from $B^\times$, which has an extremely simple structure according to the Theorem of Kuiper (Theorem 6.2). In fact, we will show that the "holes" of F, its "fissures", can be arbitrarily complex in some sense, that $\overline{F}$ is a kind of model for "all possible" topological structures (distinguishable via the functor K; see below).

We can sketch this aspect before starting with formal definitions and theorems, as follows: In algebraic topology, for instance in the "K-theory" treated in Part III, one has methods which assign to certain topological spaces (compact or triangulable spaces, differentiable manifolds, etc.) certain algebraic objects (groups, rings, algebras, etc.) and to continuous (differentiable etc.) maps certain homomorphisms between the associated algebraic objects. In this fashion certain topological-geometric phenomena which are very difficult to distinguish in the concrete visualization - can be reduced to algebraic terms ("made discrete", F. Waldhausen) for which there is a well-developed formalism and which are easier to comprehend. For example, it is immediate that there is no epimorphism of the group $\mathbb{Z}$ onto the group $\mathbb{Z} \oplus \mathbb{Z}$, while the non-existence of a retraction of the n-dimensional ball B^n onto its boundary S^{n-1} is not immediately clear (see Theorem III.1.8).

The construction of the "index bundle" which will be treated in this chapter is the basis of a further step from algebraic topology to functional analysis or to "elliptic topology" (Atiyah). We will be able to explain this term only in the following chapters which deal with the connection between elliptic differential equations and Fredholm operators.

In this way, deep geometric questions in the proof of Bott's Periodicity Theorem (generalization of the concept "winding number") find an algebraic formalism in K-theory. Thereby the "Bott isomorphism $K(X \times \mathbb{R}^2) \to K(X)$" is described by means of families of Fredholm operators, and in this form, can be understood more easily and elementarily by a symmetric use of classical results of functional analysis. See Chapter III.1. This is an example where the function-analytic interpretation makes the understanding of geometric or algebraic situations easier or at least possible.

Conversely, the topological and algebraic problems and methods serve analysis: for example, when the index or the index bundle yield algebraic-topological invariants for Fredholm operators or families of Fredholm operators which turn up concretely in problems of analysis.

B. The Construction of Index Bundles

We now come to the construction of "index bundles". Let $T : X \to \mathcal{F}$ be a continuous family of Fredholm operators in a Hilbert space H, where X is a compact topological space. If X is connected, Theorem 5.2 gives

$$\text{index } T_x = \text{index } T_{x'}, \quad \text{for all} \quad x,x' \in X.$$

In this way, we can assign to each T an integer, which is independent of possible "small continuous perturbations" of T. For X connected, we therefore have a map

$$\text{index} : [X,\mathcal{F}] \to \mathbb{Z}$$

which is well defined on the homotopy set $[X,\mathcal{F}]$; see Ch. 6 above. Actually, we can extract much more information out of T.

Exercise 1. Show that for each continuous family $T : X \to \mathcal{F}$ with constant kernel dimension, i.e.,

$$\dim \operatorname{Ker} T_x = \dim \operatorname{Ker} T_{x'}$$

for all $x,x' \in X$, one can assign a vector bundle $\operatorname{Ker} T$ over X, in a natural way.

A *vector bundle* E over X is a "continuous locally-trivial family" of complex vector spaces E_x of finite dimension, parametrized by the "base space" X. For the details of the definition, we refer to the Appendix.

Hint: Set $\operatorname{Ker} T := \bigcup_{x \in X} \{x\} \times \operatorname{Ker} T_x$, and give this the induced topology that it inherits as a subset of $X \times H$. Then show, as in the alternative proof of Theorem 5.2, the property of local triviality; i.e., locally there is a basis of $\operatorname{Ker} T_x$ which depends continuously on x. See also Exercise 2c of the Appendix, where $X = S^1$. □

Exercise 2. Under the same assumptions as in Exercise 1, show that there are naturally defined vector bundles Ker T* and the "quotient bundle" Coker T. Moreover, show that these bundles are "isomorphic" (see Appendix). □

We denote the set of all isomorphism classes of vector bundles over X by Vect (X). By Exercises 1 and 2, we have a map ι from the set of continuous families of Fredholm operators with constant kernel dimension to Vect (X) × Vect (X):

$$T \xrightarrow{\iota} ([\text{Ker } T], [\text{Coker } T]).$$

If X consists of a single point, then a vector bundle over X is simply a single vector space, and Vect(X) can be identified with $\mathbb{Z}_+$, since vector spaces are isomorphic precisely when their dimensions are equal; symbolically,

$$[\cdots] \cong \dim (\cdots\cdot).$$

In this case, we then have the maps

$$\begin{array}{ccccccc} T & \overset{\iota}{\longmapsto} & ([\text{Ker } T], [\text{Coker } T]) & \overset{\delta}{\longmapsto} & [\text{Ker } T] - [\text{Coker } T] \\ \cap & & \cap & \delta & \cap \\ F & \longrightarrow & \mathbb{Z}_+ \times \mathbb{Z}_+ & \longrightarrow & \mathbb{Z} \end{array}$$

where δ is the difference mapping and $\delta \circ \iota$ = index.

In the more general case where X is not a single point, we can formally write such a difference

$$[\text{Ker } T] - [\text{Coker } T],$$

at least when the family T has constant kernel dimension. Admittedly, this is meaningless for the moment: While one can naturally introduce an addition ⊕ in Vect (X) by forming the direct sum pointwise (see Exercise 3 of the Appendix), this only makes Vect (X) a semigroup. However, one can go from the abelian semigroup Vect (X) to an abelian group (just as from $\mathbb{Z}_+$ to $\mathbb{Z}$), which we denote by K(X) and define as follows. An equivalence relation on the product space Vect(X) × Vect(X) is defined by means of

$$(E,F) \sim (E \oplus G, F \oplus G) \quad \text{for} \quad G \in \text{Vect}(X),$$

and then

$$K(X) := (\text{Vect}(X) \times \text{Vect}(X))/\sim.$$

The equivalence class of the pair (E,F) is then written as E-F; these we call "difference bundles".

The details of this construction, and its basic importance for algebraic topology, is explained in Section III.1.C. Here we only need to establish that the difference map δ extends from $\mathbb{Z}_+ \times \mathbb{Z}_+$ to $\mathrm{Vect}(X) \times \mathrm{Vect}(X)$ in a canonical way so that its values form an abelian group $K(X)$. (To be careful, X is always compact in this chapter, but many of the steps carry over easily to more general cases.)

Warning: In spite of the close analogy between the construction of $K(X)$ and $\mathbb{Z}$, notice that (for $X \neq$ point) the natural map

$$\mathrm{Vect}(X) \to K(X)$$

$$E \mapsto E - 0$$

need not be injective. In Ch. III.1 we will get to know vector bundles E and F over the two-dimensional sphere S^2 which are not isomorphic, but when the same vector bundle G is added to each of them, the results are isomorphic. (Visually, one may note that the tangent bundle of S^2 and the real two-dimensional trivial bundle over S^2 are not isomorphic, but the addition of a trivial one-dimensional bundle to each yields isomorphic bundles; see the Appendix, Exercise 7a).

With the help of the maps ι and δ just introduced, we can now deduce from Exercise 1 and Exercise 2:

Exercise 3: To each continuous family $T : X \to F$ of Fredholm operators with constant kernel dimension, there can be assigned a difference bundle ("index bundle")

$$\text{index } T := [\mathrm{Ker}\, T] - [\mathrm{Coker}\, T] \in K(X).$$

Moreover, for a point space X (T is then a single Fredholm operator and $K(X) \cong \mathbb{Z}$), the concepts of "index" and "index bundle" coincide.

Hint: Set $\text{index} := \delta \circ \iota$. □

Now we will show that the unrealistic condition that the kernel dimension be constant can be dropped, and the index bundle is invariant under continuous deformation just as the index (Theorem 5.2).

Theorem 1. For each continuous family $T : X \to F$ of Fredholm operators in a Hilbert space H (X compact) there is an index bundle

$$\text{index } T \in K(X)$$

assigned in a canonical way.

Proof: As in the alternative proof of Theorem 5.2, we first choose an orthonormal basis $e_0, e_1, e_2, \dots$ for H and consider the Fredholm operator $P_n T_x$ (which has the same index as T_x), where P_n is again the projection of H onto the closed subspace H_n spanned by $e_n, e_{n+1}, \dots$. Since X is compact, we can choose n such that

$$\operatorname{Im}(P_n T_x) = H_n \quad \text{for all} \quad x \in X,$$

and then

$$\dim \operatorname{Ker} P_n T_x = \dim \operatorname{Ker} P_n T_{x'} \quad \text{for all} \quad x, x' \in X.$$

Indeed, first we show that for each $y \in X$ there is a natural number n_y and a neighborhood U_y so that the assertion holds for all $x, x' \in U_y$. Then we pass to a finite subcover $\{U_y : y \in Y\}$, where Y is a finite subset of X, and set $n := \max\{n_y : y \in Y\}$. Relative to the family $P_n T$, we can therefore (as in Exercise 3) set

$$\begin{aligned}\operatorname{index} T := \operatorname{index} P_n T &= [\operatorname{Ker} P_n T] - [\operatorname{Coker} P_n T] \\ &= [\operatorname{Ker} P_n T] - [X \times H_n^{\perp}].\end{aligned}$$

Thus, we have assigned an index bundle in $K(X)$ to T. Indeed, we have expressed the index bundle in "normal form", in the sense that the bundle subtracted is trivial.

We must show that the definition is independent of the sufficiently large natural number n: Without loss of generality, replace n by $n+1$. Then, by construction,

$$[\operatorname{Coker} P_{n+1} T] = [X \times (H_{n+1})^{\perp}] = [\operatorname{Coker} P_n T] \oplus [X \times \mathbb{C} e_n].$$

To calculate $[\operatorname{Ker} P_{n+1} T]$, we note that for all $x \in X$

$$P_n T_x \mid (\operatorname{Ker} P_n T_x)^{\perp} : (\operatorname{Ker} P_n T_x)^{\perp} \to H_n$$

is bijective. Hence, by the closure of H_n and the "open mapping principle", there is a bounded inverse operator

$$\tilde{T}_x : H_n \to (\operatorname{Ker} P_n T_x)^{\perp} \subset H.$$

We set

$$v_x := \tilde{T}_x(e_n) \in (\operatorname{Ker} P_n T_x)^{\perp}$$

so that

$$P_n T_x v_x = e_n \quad \text{for all} \quad x \in X.$$

Then for all $x \in X$, we have

$$\operatorname{Ker} P_{n+1} T_x = \operatorname{Ker} P_n T_x \oplus \mathbb{C} v_x$$

and

$$[\operatorname{Ker} P_{n+1} T] = [\operatorname{Ker} P_n T] \oplus \{(x, z v_x) : x \in X,\ z \in \mathbb{C}\}.$$

As with T, the family $\tilde{T}$ is also continuous (prove!). Thus, $\tilde{T}$ yields an isomorphism of bundles

$$X \times \mathbb{C} e_n \to \{(x, z v_x) : x \in X,\ z \in \mathbb{C}\}$$

(i.e., $x \mapsto v_x$ is a continuous, nowhere zero vector field over X in $X \times H$). Thus, the index bundles index $(P_{n+1}T)$ and index $(P_n T)$ (in $K(X)$) are equal.

Finally, we must show the independence of the choice of orthonormal basis. For this problem we use the "independence of n" just shown: Let $f_0, f_1, \ldots$ be another complete orthonormal basis for H. We first choose $m \in \mathbb{N}$ such that the linear span of $f_0, \ldots, f_{m-1}$ contains the elements $e_0, \ldots, e_{n-1}$ (and hence $(T_x(H))^\perp$ for all $x \in X$, by the previous choice of n).

If $\tilde{P}_m$ is the orthogonal projection of H onto the closed subspace spanned by $f_m, f_{m+1}, \ldots$, then we have (by the same extension argument as above)

$$\operatorname{index} \tilde{P}_m T = \operatorname{index} P_n T,$$

because $\tilde{H}_m$ is contained in H_n. Then we are done, since again by the preceding argument index $\tilde{P}_{m'} T =$ index $\tilde{P}_m T$, if m' is so large that $(T_x(H))^\perp \subset (\tilde{H}_{m'})^\perp$ for all x. □

Remark 1. For simplicity, we have assumed that the continuous family

$$T : X \to F$$

of Fredholm operators is such that T_x is defined on the same Hilbert space for all $x \in X$.

In applications, we actually often have a different situation. For example, in analysis the arising Hilbert spaces are function spaces with values in a vector space V_x which can vary with the parameter $x \in X$. Then T_x is a Fredholm operator on the Hilbert space $H \otimes V_x$; i.e., the Hilbert space on which the continuous family of Fredholm operators operates is not constant.

We will encounter such examples mainly in Chapter II.8, but also in Part III, when we interpret homomorphisms of K-theory, in particular products of algebraic topology, with tools of analysis.

What can be done in such cases? In general, using the Theorem of Kuiper, one can show that every Hilbert space bundle is trivial (i.e., isomorphic to a product bundle $X \times H$, where H is a single Hilbert space). This follows from ([Steenrod, 54 f.]) and the contractibility of the "structure group" $\mathcal{B}^\times$, and more simply from $[A,\mathcal{B}^\times] = 1$ for *all* compact A (Theorem 6.2) following [Steenrod, 148 f.] (e.g., in the case that X is a compact, triangulable manifold).

For the construction of index bundles, it suffices to have only a local product structure which is already part of the definition of a Hilbert bundle and always holds (by the "naturality" of definitions) in our applications.

Remark 2. In the proof of Theorem 1, we have strongly used the fact that H is a Hilbert space. We treated the general case with variable $\operatorname{Ker} T_x$ by reducing it, through composition with an orthogonal projection, to the simple special case of constant kernel dimension. Instead, we can study the family T on the quotient spaces $H/\operatorname{Ker} T_x$ which produces the constant kernel dimension 0, or we can make the operators T_x surjective by extending their domain of definition (to $H \oplus \operatorname{Ker}(T_x^*)$ say) which produces constant cokernel dimension 0. Details for these two alternatives, which are already indicated in the usual proofs of the homotopy invariance of the index (e.g., cf. [Jörgens 1970/1982, §5.4] and our comments after Theorem 5.2), can be found for the first alternative in [Atiyah 1967a, 155-158] and for the second alternative - for instance if H is a Fréchet space and T_x is an elliptic operator (see below) - in [Atiyah-Singer 1971a, 122-127]. □

Warning: In connection with "Wiener-Hopf operators", we will later meet families of Fredholm operators

$$T : X \to \mathcal{F}$$

for which it is quite possible that

$$\operatorname{index} T_x = 0 \quad \text{for all} \quad x \in X,$$

while the global index bundle $\operatorname{index} T \in K(X)$ does not vanish. The index bundle is simply a much sharper invariant than just an integer. Thus one must be very careful if one wishes to infer properties of the index bundle from those of the index. The following exercise is comforting.

Exercise 4. Show that the index bundle of a continuous family of self-adjoint Fredholm operators,

$$T : X \to F \quad \text{with} \quad T_x = T_x^* \quad \text{for all} \quad x \in X,$$

is zero; i.e.,

$$\text{index } T = 0.$$

Hint: Consider the homotopy $tT + (1-t)\text{Id}$. Show that a self-adjoint operator A has a real spectrum; i.e.,

$$A - z\text{Id} \in B^{\times} \quad \text{for} \quad z \in \mathbb{C} - \mathbb{R}:$$

Step 1: $u \in \text{Ker}(A-z\text{Id})$ implies $zu = Au = A^*u = \bar{z}u$, and so $u = 0$ since $z \neq \bar{z}$.

Step 2: If v is in the orthogonal complement of $\text{Im}(A-z\text{Id})$, then $\langle Au-zu,v\rangle = 0$ and so

$$\langle u,Av\rangle = \langle Au,v\rangle = \langle zu,v\rangle = \langle u,\bar{z}v\rangle$$

for all $u \in H$. Thus, $Av = \bar{z}v$, and $v = 0$ by step 1. Hence, $\text{Im}(A-zI)$ is dense in H.

Step 3. Let $v \in H$ and let $v_1,v_2,\ldots \in \text{Im}(A-z\text{Id})$ be a sequence converging to v. Then show that the sequence of unique preimages u_1, $u_2,\ldots$ is a Cauchy sequence, and set $u := \lim u_i$. Then note that $Au - zu = v$. □

We will now investigate the construction in the preceding Theorem 1 more gnerally, and show that it has all the properties that one might reasonably expect. We begin with the following exercise.

Exercise 5. Show that our definition of index bundle is functorial: Let $f : Y \to X$ be a continuous map (X and Y compact) and let

$$T : X \to F$$

be a continuous family of Fredholm operators. Then (see Exercise 1d of the Appendix), we have

$$\text{index } Tf = f^*(\text{index } T).$$

Hint: If $e_0,e_1,\ldots$ is an orthonormal basis for H and the projection P_n is chosen so that $\dim \text{Ker } P_nT_x$ is independent of x, then $\dim \text{Ker } P_nT_{f(y)}$ is also independent of $y \in Y$. Thus, one does not need to choose another projection P_n to exhibit the index bundle for the family $Tf : Y \to F$. □

Exercise 6. Show that for each continuous Fredholm family

$$T : X \to F$$

we have

$$\text{index } T^* = -\text{index } T,$$

where $T^* : X \to F$ is the adjoint family

$$(T^*)_x := (T_x)^* \quad \text{for} \quad x \in X.$$

Warning: One cannot deduce Exercise 4 from Exercise 6, since it is possible that $K(X)$ has a finite cyclic ("torsion-") factor; e.g., possibly $a+a = 0$, but $a \neq 0$. □

Theorem 2. The construction of the index bundle in Theorem 1 depends only on the homotopy class of the given family of Fredholm operators, and yields a homomorphism of semigroups

$$\text{index} : [X,F] \to K(X)$$

for X compact.

Proof: 1. We show first the *homotopy-invariance* of the index bundle. Let

$$T : X \times I \to F$$

be a homotopy between the families of Fredholm operators $T_0 := Ti_0$ and $T_1 := Ti_1$ parametrized by X, where

$$i_t : X \to X \times \{t\} \hookrightarrow X \times I, \quad t \in I$$

are the mutually homotopic natural inclusions. The functoriality of index bundles (Exercise 5) then yields

$$\text{index } T_0 = i_0^* (\text{index } T) = i_1^*(\text{index } T) = \text{index } T_1,$$

where the middle equality follows from the homotopy invariance of $K(X)$; namely, $f^* = g^*$ for $f \sim g : Y \to X$ (Theorem 1, Appendix).

2. For the *homomorphism property* of the index we point out first that for two Fredholm families

$$S,T : X \to F,$$

we have a well defined product family

$$TS : X \to F$$

given by composition in F ($(TS)_x := T_x S_x$), and that $[X,F]$ is then a semigroup. On the other hand, K(X) has an addition defined via the direct sum of vector bundles which makes it a group (Section III.1.C).

The index is a semigroup homomorphism now, since:

$$\begin{aligned}\text{index } TS &= \text{index } TS + \text{index Id}\\ &= \text{index } (TS \oplus \text{Id})\\ &= \text{index } (S \oplus T)\\ &= \text{index } S + \text{index } T.\end{aligned}$$

In the second and fourth equalities, we have used the fact that (by the construction in Theorem 1) the index bundle for a family of Fredholm operators on a product space $H \times H$, which can be written in the form $S \oplus T := \begin{pmatrix} S & 0 \\ 0 & T \end{pmatrix}$, coincides with the direct sum of the index bundles of the two diagonal elements. Moreover, for the third equality, we have applied the usual trick of homotopy theory, where $\begin{pmatrix} TS & 0 \\ 0 & \text{Id} \end{pmatrix}$ can be deformed into $\begin{pmatrix} S & 0 \\ 0 & T \end{pmatrix}$ not only in $\mathcal{B}^\times(H \times H)$, but also in $\mathcal{F}(H \times H)$. This is clear by the definition of this homotopy in Exercise 6.2. □

Remark. Note that in the proof of homotopy invariance, we did not need to again investigate the topology of $\mathcal{F}$, but rather everything followed from the entirely different aspect of "homotopy invariance" of vector bundles.

Why did it take more work to prove the homotopy-invariance of the index of a single operator in the proof of Theorem 5.2, while in the proof of Theorem 2 we did not use this result, even though we can deduce it for X := point? The solution of this paradox lies in the fact that the actual generalization of the homotopy invariance of the index is already contained in the *construction* of index bundles in Theorem 7.1!

Exercise 7. The proof of Theorem 7.2 yields (for X = point) a further proof of the composition rule

$$\text{index } TS = \text{index } S + \text{index } T$$

for Fredholm operators. We have already seen three proofs (Exercise 2.4, Exercise 3.2, and Exercise 5.4). Do these three proofs carry over without difficulty to the case of families of Fredholm operators? Does one need a homotopy argument each time? What is the real relationship between the four proofs?

Exercise 8. Show that the set

$$\mathcal{F}_0 := \{T \in \mathcal{F} : \text{index } T = 0\}$$

is pathwise-connected.

Hint: For $T \in F_0$, choose an isomorphism

ϕ: Ker $T \to (\text{Im } T)^{\perp}$

(vector spaces of the same finite dimension!) and set

$$\Phi := \begin{cases} \phi & \text{on } \operatorname{Ker} T \\ 0 & \text{on } (\operatorname{Ker} T)^{\perp}. \end{cases}$$

By construction, we have $T+\Phi \in B^{\times}$ and $T+t\Phi \in F$ for $t \in I$ (Why? What kind of operator is Φ?). Now apply Theorem 6.1. □

C. The Theorem of Atiyah-Jänich

The sets $F_i := \{T \in F : \text{index } T = i\}$ are bijectively (modulo compact operators) mapped onto each other in a continuous fashion by shift operators (see Exercise 1.1). Thus, Exercise 8 with the homotopy-invariance of the index already proved in Theorem 5.2 gives us a bijection from the pathwise-connected components of F to $\mathbb{Z}$.

Naturally, this result still does not say much about the topology of F, and we do not want to carry it out in detail. Much more informative is the following theorem which gives the result,

index : Pathwise-comps. of $F \to \mathbb{Z}$ bijective, as the special case X = point.

Theorem 3 (M. F. Atiyah, K. Jänich 1964):

index : $[X,F] \to K(X)$ is an isomorphism.

Proof: We show that the natural sequence of semigroups

$$[X,B^{\times}] \to [X,F] \xrightarrow{\text{index}} K(X) \to 0$$

is exact. From this result of M. Atiyah and K. Jänich along with the theorem of N. Kuiper (Theorem 6.2), the present theorem follows. (For the concept of "exactness", see the material following Exercise 2.3).

Step 1: The index bundle of a continuous family $T : X \to B^{\times}$ is trivially zero.

Step 2: Now let $T : X \to F$ be a Fredholm family with vanishing index bundle. In Exercise 8, we have stated what one must do in the case where X = {point}; i.e., T is a single Fredholm operator. Namely, one chooses an isomorphism

ϕ: Ker $T \to (\text{Im } T)^{\perp}$

and then

$$\Phi := \begin{cases} \phi & \text{on} \quad \text{Ker } T \\ 0 & \text{on} \quad (\text{Ker } T)^{\perp} \end{cases}$$

is an operator of finite rank with $T + t\Phi$ the desired homotopy in $\mathcal{F}$ to $T+\Phi \in \mathcal{B}^{\times}$.

To generalize this to $X \neq \{\text{point}\}$, we now must deal with two difficulties:

1. $\bigcup_{x\in X} \text{Ker } T_x$ is not always a vector bundle.
2. In general, there is no canonical choice of

   $$\phi_x : \text{Ker } T_x \to (\text{Im } T)^{\perp}$$

   which depends continuously on x.

The first problem, we can quickly solve: We choose again an orthogonal projection

$$P_n : H \to H_n,$$

such that $\dim \text{Ker } P_n T_x$ is constant and

$$\text{index } T = [\text{Ker } P_n T_x] = [X \times (H_n)^{\perp}] \in K(X).$$

Then "index T = 0" means

$$[\text{Ker } P_n T] = [X \times (H_n)^{\perp}] \quad \text{in} \quad K(X),$$

which in turn means (see Section III.1.C) that, for N sufficiently large, there is an isomorphism of vector bundles

$$\phi : \quad (\text{Ker } P_n T) \oplus (X \times \mathbb{C}^N) \cong (X \times (H_n)^{\perp}) \oplus (X \times \mathbb{C}^N).$$

This means that by "augmenting" with the trivial bundle $X \times \mathbb{C}^N$, we can also cure the second difficulty in principle. Actually we can avoid this K-theoretical argument. Indeed, we do not need to "augment", if we take n to be large enough: As shown in the proof of Theorem 7.1, we have

$$\text{Ker } P_{n+1} T \cong \text{Ker } P_n T \oplus (X \times \mathbb{C}).$$

Thus, for $m := n+N$, the map ϕ can be regarded as an isomorphism

$$\phi: \operatorname{Ker} P_m T \cong X \times (H_m)^{\perp}.$$

In this way, the construction of Φ in Exercise 8 carries over pointwise, whence we obtain a homotopy of Fredholm families between

$$P_m T : X \to F(H)$$

and

$$P_m T + \Phi : X \to B^{\times}(H).$$

Since the index of the orthogonal projection $P_m : H \to H$ vanishes (and hence coincides with the index of Id), we can connect the "constants" P_m with Id in $F(H)$ (Exercise 8), and hence also connect $P_m T$ with T in a further homotopy of maps $X \to F(H)$.

Each Fredholm family with vanishing index bundle is therefore homotopic to a continuous family of invertible operators, and, in conjunction with step 1, exactness in the middle of the short sequence is now proved.

Step 3: We must now show the surjectivity of

$$\text{index} : [X,F] \to K(X).$$

By the construction in Exercise 6 of the Appendix (see also Section III.1.C), every element of $K(X)$ can be written in the form $[E] - [X \times \mathbb{C}^k]$, where E is a vector bundle over X and $k \in \mathbb{Z}_+$.

Hence we will be finished with the proof if, for each vector bundle E, we can find a continuous family S of (surjective) Fredholm operators on a suitable Hilbert space such that

$$\text{index } S = [E].$$

Then, the homomorphism of the index bundle construction (Theorem 7.2) yields

$$\text{index}(\text{shift}^{+})^k S = [E] - [X \times \mathbb{C}^k],$$

where shift^{+} (as in Exercise 1.1) is the displacement to the right relative to an orthonormal basis of the Hilbert space. Note that, since $\text{index shift}^{+} = -1$, the constant family $x \mapsto \text{shift}^{+}$ gives the index bundle $- [X \times \mathbb{C}]$.

Thus, let E be a vector bundle over X. If X consists of a single point, then we just complete an orthonormal basis of E (regarded as a subspace of an infinite-dimensional separable Hilbert space) to an orthonormal basis of the whole Hilbert space and set $S := (\text{shift}^{-})^{\dim E}$. Now, we consider the general case. By Exercise 6 of the Appendix, there is a vector bundle F over X and a finite-dimensional (complex) vector space

V such that $E \oplus F \cong X \times V$. Let $\pi_x : V \to E$ be the projection.

Let H be an arbitrary Hilbert space. We will see that it is simplest to consider the desired operators S_x, $x \in X$, to be defined on the space $\mathrm{Hom}(V,H)$ of linear transformations from V to H. We choose for the vector space V (which we can identify with $\mathbb{C}^N$, $N = \dim V$, via a basis) a Hermitian scalar product $\langle ..,.. \rangle : V \times V \to \mathbb{C}$. For every pair $(f,u) \in V \times H$, we have an element of $\mathrm{Hom}(V,H)$ defined by

$$v \longmapsto \langle f,v\rangle u, \quad v \in V,$$

which we will denote by $f \otimes u$; recall the isomorphism $\mathrm{Hom}(V,H) \cong V' \otimes H$ of linear algebra, where V' ($\cong V$) is the vector space of linear maps from V to $\mathbb{C}$. Then, we have

$$\mathrm{Hom}(V,H) = \left\{\sum_{i=1}^{m} f_i \otimes u_i : m \in \mathbb{N},\ f_i \in V,\ u_i \in H\right\}$$

where

$$zf \otimes u = f \otimes zu, \quad z \in \mathbb{C}$$

and

$$(f+f') \otimes (u+u') = f\otimes u + f\otimes u' + f'\otimes u + f'\otimes u'.$$

We define a scalar product

$$\left\langle \sum_{i=1}^{m} f_i\otimes u_i, \sum_{j=1}^{m} f_j'\otimes u_j' \right\rangle := \sum_{i=1}^{m}\sum_{j=1}^{m} \langle f_i,f_j'\rangle\langle u_i,u_j'\rangle,$$

which makes $\mathrm{Hom}(V,H)$ a Hilbert space (since V is finite-dimensional, nothing can go wrong; addition and scalar multiplication by complex numbers are defined naturally). One easily checks that $\{f_j \otimes e_i\}_{j=1,\dots,N,\ i=0,1,2,\dots}$ is an orthonormal basis for $\mathrm{Hom}(V,H)$, if $f_1,\dots,f_N$ is an orthonormal basis for V and $e_0,e_1,e_2,$ is one for H.

Each bounded linear operator T on H and endomorphism τ of V clearly define a bounded linear operator $\tau \otimes T$ on $\mathrm{Hom}(V,H)$, by means of

$$(\tau\otimes T)(f\otimes u) := \tau(f) \otimes T(u).$$

For a fixed chosen basis $e_0,e_1,\dots$ of H, we set

$$S_x := (\pi_x \otimes \mathrm{shift}^-) + (\mathrm{Id}_V-\pi_x) \otimes \mathrm{Id}_H .$$

Then, for $i \geq 1$,

$$S_x(f \otimes e_i) = \pi_x(f) \otimes e_{i-1} + f \otimes e_i - \pi_x(f) \otimes e_i,$$

and particularly for $f \in E_x$,

$$S_x(f \otimes e_1) = f \otimes 0 + f \otimes e_1 - f \otimes e_1 = 0.$$

Hence, we have $\operatorname{Im} S_x = \operatorname{Hom}(V,H)$ and $\operatorname{Ker} S_x = \{f \otimes e_1 : f \in E_x\}$. Thus, $\operatorname{Ker} S_x$ is isomorphic to E_x in a natural way, and index $S = [\operatorname{Ker} S] - 0 = [E]$. □

The preceding theorems are significant on various levels: In the following chapters, in dealing topologically with boundary value problems as well as in proving analytically the Periodicity Theorem of the topology of linear groups, we will use repeatedly Theorem 7.1 and 7.2, i.e., the construction of the index bundle and its elementary properties, but we will not use explicitly Theorem 7.3, our actual "main theorem". Nevertheless, the last theorem has fundamental significance for our topic, as it provides the reasons for the theoretical relevance of the notion "index bundle" and explains why this concept proved suitable to express deep relations in analysis as well as topology.

D. Homotopy and Unitary Equivalence

The power and limitations of the homotopy theoretic technique in analysis, specifically in the theory of Fredholm operators, are demonstrated in recent results of Lawrence Brown, Lewis Coburn, Ronald Douglas, Peter Fillmore, William Helton, Roger Howe and others. We will review them briefly; for details see the collection [Fillmore 1973].

1. Let $\mathcal{B}$ be the Banach algebra of linear bounded operators on the Hilbert space H with the closed ideal $\mathcal{K}$ of compact operators and the canonical projection $\pi: \mathcal{B} \to \mathcal{B}/\mathcal{K}$. For $S,T \in \mathcal{B}$, we define in $\mathcal{B}/\mathcal{K}$ the "unitary equivalence" $\pi(S) \approx \pi(T)$ by either of the following equivalent ([Fillmore 1973, 77]) conditions:

(i) There is a unitary operator $U \in \mathcal{B}$ (i.e., $U^* = U^{-1}$), such that

$$S - UTU^* \in \mathcal{K}.$$

(ii) There is a unitary element v in $\mathcal{B}/\mathcal{K}$ such that

$$\pi(S) = v\pi(T)v^*.$$

Now let $S \in \mathcal{F}$, i.e. (Theorem 5.1), $\pi(S)$ is an invertible element of $\mathcal{B}/\mathcal{K}$. What is the relationship in $\mathcal{F}$ between the topological relation

> $S \sim T$, i.e., S and T can be connected by a continuous path in F, or equivalently: index S = index T (S and T are "homotopic")

and the numerical-analytic relation

> $S \approx T \bmod K$, i.e., $\pi(S) \approx \pi(T)$ (S and T are "modulo K unitarily equivalent")?

Since $B^\times$ and even more U are connected (Theorem 6.1), the homotopy follows trivially from the unitary equivalence modulo K. The converse is not immediately clear. Rather, by looking, for example, at the homotopic operators Id and -Id, it is apparent that Fredholm operators from the same path-connected component of F may well differ by "more" than a compact operator.

However, Brown, Douglas and Fillmore showed in 1970: In the group of unitary elements of B/K the relation $\sim$ and $\approx$ coincide; i.e., "essentially unitary" operators in Hilbert space can be joined by a continuous path in F - they have the same index - if and only if they are "modulo K unitarily equivalent." In still another way:

The unitarily equivalent unitary elements of B/K form an infinite cyclic group whose elements are given by

$$\pi(\mathrm{Id}) \quad \text{or} \quad \pi((\mathrm{shift}^+)^n) \quad \text{or} \quad \pi((\mathrm{shift}^-)^n),\ n \in \mathbb{N},$$

[Fillmore 1973, 71].

For all "essentially-unitary" operators $T \in B$ (i.e., $\pi(T)$ is unitary in B/K), the "essential spectrum" $\mathrm{Spec}_e(T) = \mathrm{Spec}\ \pi(T)$ is contained in $S^1 = \{z : z \in \mathbb{C} \text{ and } |z| = 1\}$, since $||\pi(T)|| = 1$ (see Exercise 5.3). Since $\mathrm{index}(T - z\,\mathrm{Id}) = 0$ whenever $|z| > 1$, the homotopy invariance of the index implies that the inclusion $\mathrm{Spek}_e(T) \subset S^1$ is proper, only if index T = 0. Thus each class of unitarily equivalent unitary elements in B/K is characterized by the index, where for fixed T the index of $T - z\mathrm{Id}$ is a map on $\mathbb{C} \setminus S^1$ into $\mathbb{Z}$ with value 0 everywhere outside S^1 and constant value n = index T inside S^1. In geometric language: The index is a "complete" unitary invariant for the unitary elements in B/K.

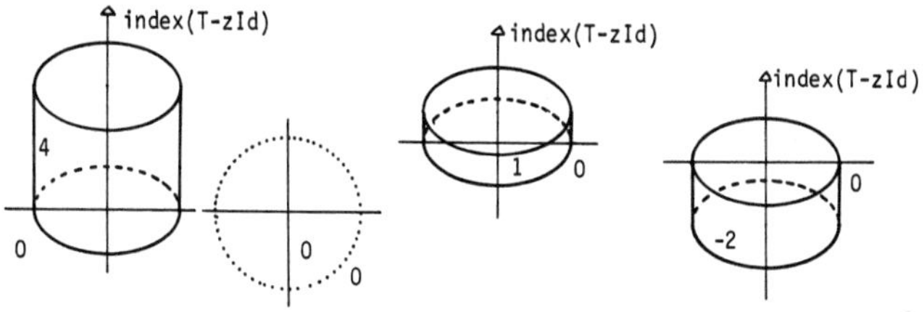

2. For normal operators $(T \in \mathcal{N} \Longleftrightarrow T^*T - TT^* = 0)$, however, the essential spectrum is the "complete" unitary invariant* while the index is identically 0 in the complement of the essential spectrum (see the remark following Theorem 3.2).

The class of operators which is the most natural next object of study after the normal and Fredholm operators are - working modulo $\mathcal{K}$ at any rate - the "essentially normal" operators, i.e., the operators $T \in \mathcal{B}$ for which $\pi(T)$ is normal, i.e., $TT^* - T^*T \in \mathcal{K}$. This includes in particular the compact perturbations of normal operators (nothing new since their index vanishes) and the essentially-unitary operators. We have the following theorem of Brown-Douglas - Fillmore (1973): Two essentially normal operators S and T are unitarily equivalent modulo $\mathcal{K}$, if and only if they have the same essential spectrum X and if on every connected component of $\mathbb{C}\setminus X$ we have $\operatorname{index}(S - z\mathrm{Id}) = \operatorname{index}(T - z\mathrm{Id})$ [Fillmore 1973, 73-122].

The proof of this theorem with its dependence on delicate questions of topological algebra and K-theory is by no means trivial. Let us consider the situation once more: The essential spectrum X of an essentially normal operator T is a compact subset of $\mathbb{C}$. The expression $\operatorname{index}(T - z\mathrm{Id})$ defines a $\mathbb{Z}$-valued function on $\mathbb{C}\setminus X$. The connected components of $\mathbb{C}\setminus X$ with non-vanishing index are - so to speak - the "obstructions" to the normality of T.

---

*I.D. Berg, Trans. Amer. Math. Soc. 160 (1971), 365-371, shows that every normal operator can be diagonalized by a compact perturbation, and that $S,T \in \mathcal{N}$ are unitarily equivalent modulo $\mathcal{K}$ (i.e., there is a unitary operator U with $S - UTU^* \in \mathcal{K}$), if and only if $\operatorname{Spec}_e(T) = \operatorname{Spec}_e(S)$. For self-adjoint S,T this result is due to John Neumann, whose point of departure was a lemma by Hermann Weyl (1909) saying that the accumulation points of the spectrum of a self-adjoint operator remain unchanged under perturbation by a compact operator.

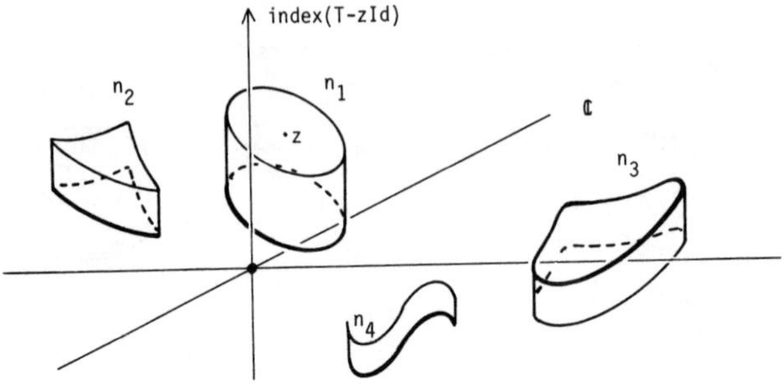

3. While pathwise connectedness in F yields nothing but a decomposition into the $\mathbb{Z}$ connected components - a classification by index T - zId at the point z = 0 - we have in 2. a finer classification by "index towers" on the connected components of the complement of the essential spectrum.

We state some concrete consequences which form a transition to the chapter after the next.

(i) An essentially normal operator T with index T - zId = 0 for all z outside the essential spectrum of T belongs to $N + K$, i.e., can be written as a sum of a normal and a compact operator [Fillmore, 118].

(ii) The family $N + K$ is topologically closed (in the operator norm); thus elements of the complement of $N + K$ in B - even if their index vanishes - cannot be approximated by a sequence in $N + K$ [Fillmore, 119]. Unfortunately, there is not yet an elementary proof for this remarkable result.

(iii) It is another very interesting fact, that essentially normal operators whose essential spectrum is described by the image of a simple closed curve are modulo K unitarily equivalent to Wiener-Hopf operators with the same "characteristic curve". Details are in the chapter after next and in [Fillmore, 73].

8. FOURIER SERIES AND INTEGRALS (FUNDAMENTAL PRINCIPLES)

The following fundamentals and elementary facts are standard mathematical knowledge today, and can be found in a great number of text books in analysis. As a general reference, we mention [Dym-McKean 1972].

A. Fourier Series

$C^0(S^1)$ Banach space of continuous (complex-valued) functions on the circle $S^1 = \{z \in \mathbb{C} : |z| = 1\}$ with the norm $||f||_\infty :=$ $\sup \{|f(z)| : |z| = 1\}$ for $f \in C^0(S^1)$.

$L^1(S^1)$ Banach space of (complex-valued) integrable functions on S^1 with the norm $||f||_1 := \int_{S^1} |f|$. Often one represents $L^1(S^1)$ via the space $L^1[0,1]$ of integrable functions on the interval $[0,1]$ and then has $||f||_1 = \int_0^1 |f(x)|dx$.

$L^2(S^1)$ Hilbert space of square-integrable functions on S^1 with the inner product $\langle f,g\rangle := \int_0^1 f(x)\overline{g(x)}dx$ and the associated norm $||f|| := (\int_0^1 |f(x)|^2 dx)^{1/2}$. Here we have identified the space $L^2(S^1)$ with $L^2[0,1]$. □

<u>Warning</u>: Functions in $L^1(S^1)$ or $L^2(S^1)$ are identified if they agree outside a set of measure zero. In particular, f is identified with the zero function, if f is zero "almost everywhere"; i.e., f is nonzero on at most a set of measure zero. In this way, we have $||f|| = 0$ precisely when $f = 0$.

Thus, strictly speaking, the elements of $L^1(S^1)$ or $L^2(S^1)$ are not functions, but rather equivalence classes of functions. While this is true, in practice it is much simpler and generally harmless to disregard this fine distinction, and we will do this in what follows.

<u>Exercise 1</u>. Show that $L^2(S^1)$ with $\langle\cdot\cdot,\cdot\cdot\rangle$ is indeed a Hilbert space. We need to show:

a) $\langle\cdot\cdot,\cdot\cdot\rangle : L^2(S^1) \times L^2(S^1) \to \mathbb{C}$ is well defined. <u>Hint</u>: The pointwise estimate

$$z|f(z)\overline{g(z)}| = 2|f(z)|\ |g(z)| \leq |f(z)|^2 + |g(z)|^2$$

shows that $f\bar{g} \in L^1(S^1)$ for $f,g \in L^2(S^1)$.

b) Symmetry (i.e., $\langle f,g\rangle = \overline{\langle g,f\rangle}$), bilinearity (linearity in the first slot, and conjugate linearity in the second), and positivity (i.e.,

$\langle f,f\rangle \geq 0$ and $\langle f,f\rangle = 0$ exactly when $f = 0$) hold. (This is trivial.)

c) $L^2(S^1)$ is a complex vector space. Hint: For closure under addition, prove Hermann Minkowski's inequality

$$(\int |f+g|^2)^{1/2} \leq (\int |f|^2)^{1/2} + (\int |g|^2)^{1/2}$$

(= Triangle Inequality).

d) $L^2(S^1)$ is complete. In order to prove that a Cauchy sequence $\{f_n\}_{n=1,2,\ldots}$ with $f_n \in L^2(S^1)$ and $||f_n - f_m||_2 \rightsquigarrow 0$ possesses a limit $f \in L^2(S^1)$ with $||f_n - f|| \rightsquigarrow 0$, one must apply the fundamental convergence theorems which distinguish the Lebesgue integral from the Riemann integral. The rather technical proof can be found in [Dym-McKean, 16-20].

e) $L^2(S^1)$ is separable. Hint: Show that the family of piecewise constant functions, having rational real and imaginary parts and with jumps at finitely many rational points, is dense in $L^2(S^1)$.

Supplement: With the help of the smoothing functions of the kind

$$g(x) := \begin{cases} 0 & \text{for } x \leq 0 \\ e^{(1/x)(1/x-1)} & \text{for } 0 < x < 1 \\ 0 & \text{for } 1 \leq x, \end{cases}$$

it follows that even $C^\infty(S^1)$ is dense in $L^2(S^1)$. □

Convolution: In $L^1(S^1)$, there is a commutative and associative product ("convolution") given by

$$(f*g)(x) := \int_0^1 f(x-y)g(y)dy,$$

which makes $L^1(S^1)$ an algebra (without identity). In particular, we have

$$||f*g||_1 \leq ||f||_1 \cdot ||g||_1.$$

Moreover, one can show that, relative to $*$, $L^2(S^1)$ is an ideal in $L^1(S^1)$; hence, $f*g$ is in $L^2(S^1)$, whenever one of the factors lies in $L^2(S^1)$. See [Dym-McKean, 41].

Orthonormal systems: The family $\{z^n : n \in \mathbb{Z}\}$, where $z^n : S^1 \to \mathbb{C}$ is the function that assigns to each $z \in S^1$ the value z^n, is an orthonormal system in $L^2(S^1)$. Regarding $L^2(S^1)$ as $L^2[0,1]$, the corresponding functions have the form $x \mapsto e^{2\pi inx}$.

Fourier series: This orthonormal system is complete; i.e., its linear span is dense in $L^2(S^1)$. Because of this, each function $f \in L^2(S^1)$ can be expanded in a Fourier series

$$f = \sum_{-\infty}^{\infty}{}_n \hat{f}(n) z^n$$

with the Fourier coefficients

$$\hat{f}(n) := \langle f, z^n \rangle = \int_0^1 f(x) e^{-2\pi inx} dx.$$

Note that f equals its infinite Fourier series, in the sense that the partial sums $\sum_{|n| \leq k} \hat{f}(n) z^n$ converge to f in the $L^2(S^1)$-norm as $k \to \infty$. One can construct an isomorphism from $L^2(S^1)$ to the space $L^2(\mathbb{Z})$ of absolute square-summable sequences of complex numbers. The isomorphism is an isometry

$$||f||_2^2 = \sum_{-\infty}^{\infty}{}_n |\hat{f}(n)|^2 \quad \text{(Plancherel's Identity)}.$$

Details are in [Dym-McKean]. The Fourier coefficients are also defined for $f \in L^1(S^1)$, and from the theorem of Guido Fubini on interated integrals, it then follows [Dym-McKean, 42] that

$$(f*g)\hat{}(n) = \hat{f}(n)\hat{g}(n).$$

Incidentally, the algebra A of those sequences which appear as the Fourier coefficients of integrable functions has been barely investigated: "The best information available to date indicates that A has no decent description at all." [Dym-McKean, 43].

On the other hand, when the product fg (in the sense of pointwise multiplication) is integrable (e.g., when $f, g \in L^2(S^1)$, by Exercise 1a), then the Fourier coefficients satisfy

$$\begin{aligned} (fg)\hat{}(n) &= (\hat{f} * \hat{g})(n) \\ &:= \sum_{-\infty}^{\infty}{}_k \hat{f}(n-k)\hat{g}(k). \end{aligned}$$

For the proof, we do not need Fubini's Theorem as above, but rather we insert the Fourier series of g in the formula for $(fg)\hat{}(n)$, and then use the usual limit theorems for the Lebesgue integral, to interchange the integral and sum.

B. The Fourier Integral

One can proceed from the standard representation of functions on a circle as functions of period 1 on the real line, to the more general case of period T, and then let T go to ∞. This leads in a natural way to the concept of Fourier integrals [Dym-McKean, 86 f.]:

$$\hat{f}(\xi) := \int_{-\infty}^{\infty} f(x)e^{-i\xi x}dx.$$

We consider the Fourier transformation on the following spaces:

$C_{\downarrow}^{\infty}(\mathbb{R})$ This is the space of "rapidly decreasing" C^{∞} functions on $\mathbb{R}$ (with complex values).

"Rapidly decreasing" means that these functions and all their derivatives tend to 0 at infinity, even when they are multiplied by arbitrary polynomials.

$L^1(\mathbb{R})$ This is the Banach space of integrable functions with

$$||f||_1 := \int_{-\infty}^{\infty} |f(x)|dx < \infty.$$

It is an algebra (without identity) under convolution

$$(f*g)(x) := \int_{-\infty}^{\infty} f(x-y)g(y)dy,$$

and we have [Dym-McKean, 87 f.]:

$$||f*g||_1 \leq ||f||_1 \cdot ||g||_1.$$

$L^2(\mathbb{R})$ This is the Hilbert space (proved as in Exercise 1) of square integrable functions with

$$\langle f,g\rangle := \int_{-\infty}^{\infty} f(x)\overline{g(x)}dx$$

$$||f||_2 := \sqrt{\langle f,f\rangle}.$$

By the argument of Exercise 1a, it follows that $C_{\downarrow}^{\infty}(\mathbb{R})$ is dense in $L^1(\mathbb{R})$, as well as in $L^2(\mathbb{R})$. Naturally one cannot expect, as with functions on the (compact) circle S^1, that $C^{\infty}(\mathbb{R})$ or $C^0(\mathbb{R})$ will be contained in the Lebesgue spaces.

A further complication arises since we no longer have $L^2 \subset L^1$. Counterexamples for the inclusion in both directions can easily be constructed using the zero function and the function $|x|^{-3/4}$. □

<u>Exercise 2</u>. The following statements are easy to prove:

a) For each $f \in L^1(\mathbb{R})$ and each $\xi \in \mathbb{R}$, $\hat{f}(\xi)$ is well defined. (Trivial.)

b) If f is differentiable in addition, and the derivative $f' \in L^1(\mathbb{R})$, then we have the formulas

$$(f')^\wedge = i\xi\hat{f} \qquad (*)$$

and

$$(\hat{f})' = (-i\xi f)^\wedge \qquad (**)$$

where ξ is the identity function on $\mathbb{R}$. The proof of each is via integration by parts.

Using the following notation for the various operations above

$$Df := \frac{1}{i} f'$$

$Mf := xf$, the function assigning to x the product $xf(x)$.

$$Ff := \hat{f},$$

we have

$$FDf = MFf \qquad (*)$$

and

$$DFf = -FMf \qquad (**)$$

More generally, by induction, we have (for $f \in C_\downarrow^\infty(\mathbb{R})$ and $p,q \in \mathbb{N}$)

$$M^pD^qF = (-1)^qFD^pM^q \qquad (***)$$

This formula is of fundamental importance for the treatment of differential operators (with constant coefficients) which are converted into simple multipliers. See II.3 on pseudo-differential operators. From the topological viewpoint these formulas are most remarkable, because they express a deep duality between local and global properties: Thus (*) relates the smoothness of f with the rate of decay (asymptotic behavior) of $\hat{f}$, and (**) relates the smoothness of $\hat{f}$ with the decay of f. In fact, $\hat{f}$ is differentiable (a local property) when f decreases so fast that the Fourier integral converges even for $-i\xi f$. This local global duality is also a feature of the index formula for elliptic operators, and we will deal with it further in that context.

c) The *Fourier Inversion Formula*

$$f(x) = \frac{1}{2\pi}\int_{-\infty}^{\infty} \hat{f}(\xi)e^{ix\xi}d\xi$$

holds for $f \in C_\downarrow^\infty(\mathbb{R})$.

In direct analogy with the role of Fourier coefficients in Fourier series, $\hat{f}(\xi)$ is the density of the frequency ξ in the harmonic decomposition of f.

Hint: Prove the formula first for functions with compact support (i.e., vanishing outside a compact subset of $\mathbb{R}$). In this case there are no difficulties with the limit process which reduces to functions of period T and then lets T go to infinity. Use smoothing functions as in Exercise 1e in order to approximate rapidly decreasing C^∞ functions by functions of compact support. The $L^1(\mathbb{R})$ estimates needed next are somewhat tricky, but can be looked up in [Dym-McKean, 89 f.]. A shorter direct proof can be found in [Hörmander 1963, 18 f.].

d) As a corollary to the proof of c), one obtains the *Plancherel Formula*

$$||\hat{f}||_2 = (2\pi)^{-1/2}||f||_2,$$

and that

$$F : C_\downarrow^\infty(\mathbb{R}) \to C_\downarrow^\infty(\mathbb{R})$$

is linear and bijective, where F again denotes the Fourier transformation. By the Fourier Inversion Formula, we obtain the inverse transformation

$$F^{-1}f = \frac{1}{2\pi} F(- \,..).$$

e) Extend F to $L^2(\mathbb{R})$! Hint: Approximate $f \in L^2(\mathbb{R})$ in the $L^2(\mathbb{R})$-norm by a sequence $\{f_n\}$ with $f_n \in C_\downarrow^\infty(\mathbb{R})$. Using the additivity of F and the Plancherel Identity, show that $\{\hat{f}_n\}$ is a Cauchy sequence in $L^2(\mathbb{R})$, whence $\hat{f} := \lim_{n\to\infty} \hat{f}_n$ defines an element of the Hilbert space $L^2(\mathbb{R})$. Finally check that $\hat{f}$ indeed depends only on f and not on the choice of the sequence. In this way, one obtains an isomorphism from $L^2(\mathbb{R})$ to $L^2(\mathbb{R})$, which we denote by F again.

f) The spaces $C_\downarrow^\infty(\mathbb{R})$ and $L^2(\mathbb{R})$ share the property that they are mapped into themselves by F. This is not true for $L^1(\mathbb{R})$. Still, one can easily show [Dym-McKean, 102] that then

(i) $\hat{f} \in C^0(\mathbb{R})$

(ii) $\lim_{|\xi|\to\infty} \hat{f}(\xi) = 0$

(iii) $(f*g)\hat{} = \hat{f}\,\hat{g}$. □

C. Higher Dimensional Fourier Integrals

By the theorem of Guido Fubini, the closed linear span of the n-fold products

$$f_1(x_1)\dots f_n(x_n)$$

of functions in $L^2(\mathbb{R})$ is $L^2(\mathbb{R}^n)$ (Prove!). Thus, the preceding results carry over directly to the case of several variables:

Exercise 3. As above, define the spaces $C_{\downarrow}^{\infty}(\mathbb{R}^n)$, $L^1(\mathbb{R}^n)$, and $L^2(\mathbb{R}^n)$ and investigate the Fourier transform

$$\hat{f}(\xi) := \int_{\mathbb{R}^n} f(x)e^{-i<x,\xi>}dx,$$

where $x = (x_1,\dots,x_n)$, $\xi = (\xi_1,\dots,\xi_n)$ and $<x,\xi> = x_1\xi_1 + \dots + x_n\xi_n$. Show:

a) $f(x) = (2\pi)^{-n}\int_{\mathbb{R}^n} \hat{f}(\xi)e^{i<x,\xi>}d\xi$

(Fourier Inversion Formula).

b) $M^pD^qF = (-1)^{|q|}FD^pM^q$, p,q multi-indices, $Ff := \hat{f}$ and $|q| := q_1 + \dots + q_n$.

c) $f \in L^1(\mathbb{R}^n) \Rightarrow \hat{f} \in C^0(\mathbb{R}^n)$.

d) $\int f(x)\overline{g(x)}dx = (2\pi)^{-n}\int \hat{f}(x)\overline{\hat{g}(x)}dx$ (Parseval)

e) $\widehat{fg} = (2\pi)^{-n}(\hat{f}*\hat{g})$ and $\hat{f}\hat{g} = (f*g)^\wedge$.

Details are in [Dym-McKean 1972, 132 ff.] or [Hörmander, 1963, 17-19].

9. WIENER-HOPF OPERATORS

A. The Reservoir of Examples of Fredholm Operators

We already proved some deep theorems on Fredholm operators, but our supply of examples is still very small, even trivial, as we only studied the following types of Fredholm operators:

1. The identity operator Id.

2. The shift operator shift^+ (with respect to an orthonormal basis); see Example 1.1.

3) The Riesz operators Id + K, where K is an operator with finite rank or, more generally, a Hilbert-Schmidt integral operator of the form

$$(Ku)(x) := \int_{X\times X} G(x,y)u(y)\,dy$$

with square integrable weight function G; see Exercise 3.7.

All the other Fredholm operators which appeared so far were elementary function-analytic modifications of the above three basic types: For instance, the left-handed shift is the adjoint of the right-handed shift, i.e.,

$$\text{shift}^- = (\text{shift}^+)^*.$$

Further, the (unitary) Fourier transformation (Exercise 8.2e)

$$F : L^2(\mathbb{R}) \to L^2(\mathbb{R})$$

can be written as a direct sum

$$F = i\,\mathrm{Id} \oplus -\mathrm{Id} \oplus -i\mathrm{Id} \oplus \mathrm{Id}$$

by decomposing $L^2(\mathbb{R})$ into a direct sum of four closed subspaces H_1, H_2, H_3, H_4 which are the eigenspaces of F to the eigenvalues i^n. (A proof which explicitly exhibits the eigenfunctions, the "Hermite functions", can be found in [Dym-McKean, 97-101].) Viewed in this fashion, from the standpoint of our abstract operator theory on Hilbert space, the Fourier transformation is nothing but a "trivial" modification of the identity.

We will now enlarge our supply of examples by a class of operators which is connected to all three of the basic types: The Wiener-Hopf operators of the form $\mathrm{Id} + K$, defined on the Hilbert space $L^2[0,\infty]$ by

$$Ku(x) := \int_0^\infty k(x-y)u(y)\,dy, \quad x \geq 0,$$

where $k \in L^1(\mathbb{R})$. We will give some background information before developing the mathematical theory of these operators and the analogous "discrete" operators

$$S : L^2(\mathbb{Z}_+) \to L^2(\mathbb{Z}_+) : u \longrightarrow Su$$

$$(Su)_n := \sum_{k\geq 0} f_{n-k}u_k,$$

whereby the f_m are, for example, the Fourier coefficients of a continuous function $f \in C^0(S^1)$ on the circle S^1.

B. Origin and Fundamental Significance of Wiener-Hopf Operators

Norbert Wiener wrote (1954) in his autobiography:

"However, the best of the work which he (Eberhard Hopf) and I undertook together concerned a differential equation occurring in the study of the radiation equilibrium of the stars. Inside a star there is a region where electrons and atomic nuclei coexist with light quanta, the material of which radiation is made. Outside the star we have radiation alone, or at least radiation accompanied by a much more diluted form of matter. The various types of particles which form light and matter exist in a sort of balance with one another, which changes abruptly when we pass beyond the surface of the star. It is easy to set up the equations for this equilibrium, but it is not easy to find a general method for the solution of these equations.

The equations for radiation equilibrium in the stars belong to a type now known by Eberhard Hopf's name and mine. They are closely related to other equations which arise when two different physical regimes are joined across a sharp edge or a boundary, as for example in the atomic bomb, which is essentially the model of a star in which the surface of the bomb marks the change between an inner regime and an outer regime; and, accordingly, various important problems concerning the bomb receive their natural expression in Hopf-Wiener equations. The question of the bursting size of the bomb turns out to be one of these.

From my point of view, the most striking use of Hopf-Wiener equations is to be found where the boundary between the two regimes is in time and not in space. One regime represents the state of the world up to a given time and the other regime the state after that time. This is the precisely appropriate tool for certain aspects of the theory of prediction, in which a knowledge of the past is used to determine the future. There are however many more general problems of instrumentation which can be solved by the same technique operating in time. Among these is the wave-filter problem, which consists in taking a message which has been corrupted by a simultaneous noise and reconstructing the pure message to the best of our ability.

Both prediction problems and filtering problems were of importance in the last war and remain of importance in the new technology which has followed it. Prediction problems came up in the control of anti-aircraft fire, for an anti-aircraft gunner must shoot ahead of his plane as does a duck shooter. Filter problems were of repeated use in radar design, and both filter and prediction problems are important in the modern statistical techniques of meteorology." (N.W.: I am a Mathematician, Victor Follancz Ltd., London, 1956.)

We cannot treat here the three main areas of application of Wiener-Hopf operators mentioned by Norbert Wiener, viz.

(i) analysis of boundary-value problems

(ii) filter problems in information theory

(iii) time series analysis in statistics.

We have to concentrate on the aspect (i) (see II.8 and III.1, where we intend to clarify the connection with topological-geometric questions). But it is useful for this purpose to have an idea of the other applications, since it simplifies the transfer of the methods in (ii) and (iii) to our area (i).

C. The "Characteristic Curve" of a Wiener-Hopf Operator

From information sciences we are interested in the stance taken by electrical engineers: The computation in $\mathbb{C}$ and the Fourier analysis of electric oscillations with the classification of filters - or more generally control circuits - by the geometric shape of the "characteristic curve". Imagine a "filter" K acting on an input signal u resulting in an output

$$Ku(x) = \int_{-\infty}^{+\infty} k(x-y)u(y)dy$$

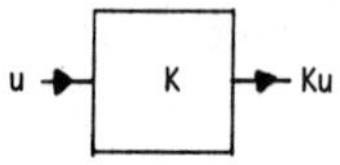

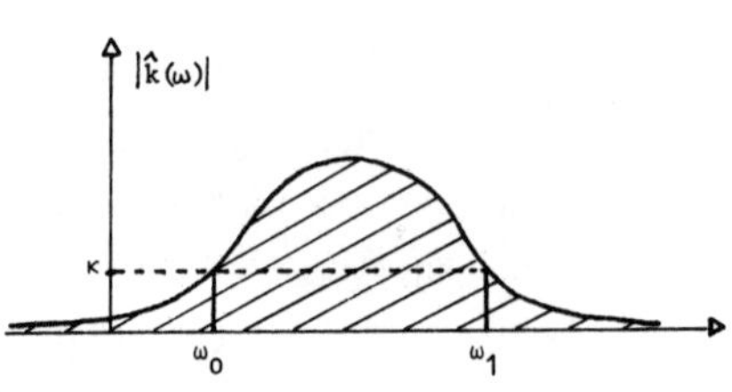

Such "linear", "time independent" and - if $k(x) = 0$ for $x < 0$ - "purely past-dependent" filters are good models for many devices of physics and technology. The information scientist "measures" such channels of information by processing a pure sine wave $u(x) = e^{-i\omega x}$ through the filter

$$Ku(x) = \int_{-\infty}^{+\infty} k(x-y)e^{-i\omega y}dy = \hat{k}(\omega)e^{-i\omega x} = \hat{k}(\omega)u(x)$$

and sketching the "characteristic values" $\hat{k}(\omega)$ as a function of the "phase" ω or the "frequency" $1/\omega$. The "amplitude ratio" $|\hat{k}(\omega)|$ is only one measure for the "linear distortion" indicating its "reinforcement" or "weakening". From it the "transmission region" $[\omega_0, \omega_1]$ may be found via the condition $|\hat{k}(\omega)| \geq \kappa$. However, a real "harmonic analysis" is achieved only if the "phase shift", i.e., the argument of the complex number $\hat{k}(\omega)$, is taken into account. The "nonlinear" distortion is given essentially by the shape of the curve $\{\hat{k}(\omega) : \omega \in \mathbb{R}\}$. This "characteristic curve", "filter characteristic" or "periodogram" coincides

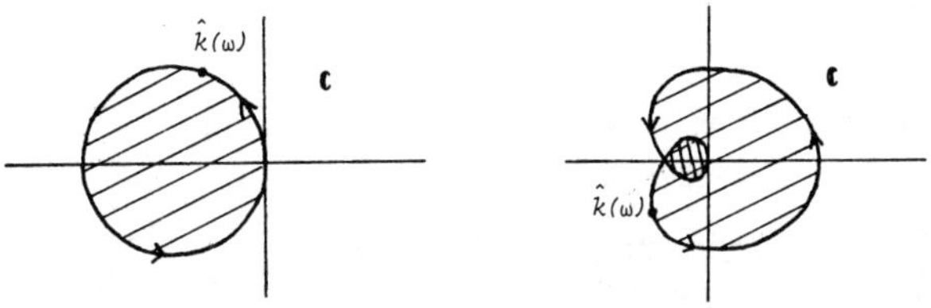

under certain conditions with the essential spectrum of the operator K; see Theorem 9.2 below. (On the left we have the "characteristic curve" of a filter without "feedback", while on the right "feedback" is possible in a certain region, although the "linear distortion" of both filters may be the same.)

For details, in particular for the relationship with the general theory of electric circuits we refer to [Dym-McKean, 170-176] and the literature quoted there.

D. Wiener-Hopf Operators and Harmonic Analysis

We have just pointed out how functional analysis and complex analysis, with its varied geometric-topological aspects, enters markedly into operator theory through information theory. The real reason is - roughly - that the Fourier transform of a square-integrable function k which vanishes identically on the left half-line is holomorphic on the upper half-plane $\mathbb{C}_+$ [Dym-McKean, 161 f.]. Formulated differently, the reason is that the situation of singularities in $\mathbb{C}$ of certain functions associated with dynamical systems carries information about the asymptotic behavior of the oscillating system; see II.6 below and the literature listed there.

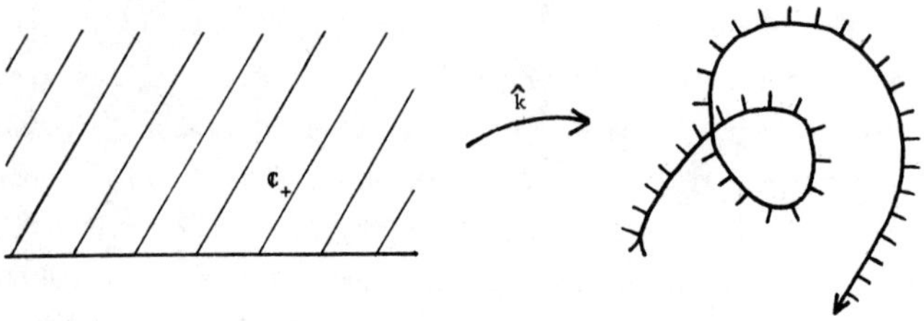

The methods of complex analysis thus introduced are based on the idea (founded in the notion of a holomorphic function expandable in a power series) of quantities which vary smoothly and continuously and which are ultimately completely determined through the knowledge of the function value and those of the derivatives at a single point. In contrast, the statistical theory of time series analysis rests on the theory of real functions and thus enters into functional analysis an experience of dealing mathematically (in the framework of harmonic analysis) with curves which are pieced together from unrelated parts.

With the terminology of C we have - roughly - that every operator on the past of $u(x)$ which is linear and invariant under translation of the time origin can be represented as a filter

$$Ku(x) = \int_0^{\infty} k(y)u(x-y)dy$$

or as the limit of a sequence of such operators. If K is defined in this fashion as a linear statistical prediction operator, for example, then the method of "least squares" yields an "optimality criterion"

$$\int_{-\infty}^{+\infty} |u(x+a) - (Ku)(x)|^2 dx \to \text{Min},$$

where a is a given prediction period and the function k which defines K is sought. However, in a statistical theory no statements are made about single occurences but only about large numbers of such. Correspondingly, the prediction or extrapolation based on a single "time series" u, the determination of k from a single u, does not make any sense. The optimality criterion itself must be interpreted statistically, and the goodness of the operator must be measured not by a single sample but by its average effect. Hence the "stochastic processes" which appear are classified by their "autocorrelation"

$$\phi(a) := \lim_{T\to\infty} \frac{1}{2T}\int_{-T}^{T} u(x-a)\overline{u(x)}\,dx, \quad a \in \mathbb{R}.$$

When passing from u to the function ϕ, a certain part of the "information content" of the time series u is isolated, while for the rest the specific features of u are ignored. For a class of time series with known autocorrelation ϕ the optimality criterion can be written as a Wiener-Hopf equation

$$\phi(x+a) - \int_0^{\infty} \phi(x-y)k(y)\,dy = 0, \quad x \geq 0,$$

whereby ϕ and a are given and k is sought.

These methods have become standard fare in the statistical time series analysis through the pioneering works [Kolmogorov 1943] and [Wiener 1949], and can be found in any of the textbooks on statistics and probability theory, frequently under the title "Spectral theory of stochastic processes". A survey with an abundance of examples from economics and technology is provided by [Ivachnenko-Lapa 1967], which includes "non-stationary processes" also.

The details of these methods are not always interesting from our point of view (computation of the index of Fredholm operators). Conversely, the computation of the index is as a rule uninteresting for correlation theory, since the Wiener-Hopf operators which show up usually have vanishing index, see [Ivachnenko-Lapa 1967, 75]; but also note [Tsypkin 1970, 147 f.] who warns about the "illusion of an easy computability of the optimal kernel functions" and points out the "large computational effort necessary for the determination of the correlation functions..." in spite of their uniqueness and explicit solvability in principle. He suggests "adaption algorithms" as an alternative. These are associated with other types of Fredholm operators, and the uniqueness of the solution is lost. In our context, we want to retain the probabilistic method which - roughly - consists in forming averages by means of Lebesgue integration, and in compressing and selecting information. The relevant information is then that which - as autocorrelation and the prediction operator itself - yields statements on the kind of connections and transitions between one curve segment (time series) and the next ("transition probabilities").

This is exactly the same strategy that is practical in algebraic topology which investigates - again roughly - how geometric structures are composed of simpler "pieces", see Part III. On this background, the explanation takes shape of why the Wiener-Hopf operators, which originated in boundary value problems of analysis and gained significance in probability theory, most recently turned out to be relevant for the representation of operations in K-theory (see III.1.E). It is simply because they are - as all Fredholm operators - a function analytic tool in the treatment of "seams", "transitions", and "relations".

E. The Discrete Index Formula

In Chapter 8, we already became familiar with the Hilbert space $L^2(S^1)$ of measurable, square-integrable functions on the circle, and we saw that the functions

$$z \longmapsto z^n, \quad n \in \mathbb{Z}$$

form an orthonormal basis for $L^2(S^1)$.

Exercise 1. Let H_n be the subspace of $L^2(S^1)$ spanned by the functions z^k with $k \geq n$. Show that the functions $z^0, z^1, \ldots, z^{n-1}$ form a basis of the orthogonal complement $(H_n)^\perp$ of H_n in H_0.

Exercise 2. Let P be the orthogonal projection $L^2(S^1) \to H_0$, and let f be a continuous complex-valued function on the circle: $f \in C^0(S^1)$.

a) Show that

$$T_f := PM_f|H_0$$

defines a bounded linear operator on the Hilbert space H_0, where M_f denotes multiplication by f.

b) One verifies that for $u \in H_0$ and $n \in \mathbb{Z}_+$

$$(\widehat{T_f u})(n) = \sum_{k=0}^{\infty} \hat{f}(n-k)\hat{u}(k),$$

where $\hat{f}(m) := \langle f, z^m \rangle$ is the m-th Fourier coefficient of f (see Chapter 8). T_f is the (discrete) *Wiener-Hopf-Operator* assigned to f. □

Exercise 3. Show that

$$f \longmapsto T_f$$

defines a continuous linear map

$$T : C^0(S^1) \to \mathcal{B}(H_0)$$

where the Banach algebra $C^0(S^1)$ has norm $||f|| := \sup\{|f(z)| : z \in S^1\}$.

Hint: $||T_f|| \leq ||f||$. Incidentally, is T a Banach algebra homomorphism; i.e., does it respect the ring structure? See Step 2 in the proof of Theorem 1 below. □

Theorem 1 ("Discrete" Gohberg-Krein Index Formula, 1956): If $f \in C^0(S^1)$ and $f(z) \neq 0$ for all $z \in S^1$, then

a) $T_f : H_0 \to H_0$ is a Fredholm operator,
b) $\text{index } T_f = -W(f,0)$.

For the definition of winding number $W(f,0)$, see Section III.1.A.

Proof: We begin with a). Step 1: Let $\mathcal{B} := \mathcal{B}(H_0)$ be the Banach algebra of bounded linear operators on the Hilbert space H_0, and let $K \subset \mathcal{B}$ be the closed ideal of compact operators on H_0 with

$$\pi : \mathcal{B} \to \mathcal{B}/K$$

the canonical projection onto the quotient algebra; see Chapter 3 also. From Exercise 3, it follows that

$$\pi \circ T : C^0(S^1) \to \mathcal{B}/K$$

is linear and continuous.

Step 2: Let $\check{C}$ be the subalgebra of $C^0(S^1)$ consisting of the continuous functions representable by *finite* Fourier series. Let $f,g \in \check{C}$, say

$$f(z) = \sum_{-n}^{n} {}_k \hat{f}(k) z^k, \quad g(z) = \sum_{-m}^{m} {}_k \hat{g}(k) z^k,$$

for $n,m \in \mathbb{N}$. In Chapter 8, we already calculated the Fourier coefficients of fg:

$$(\widehat{fg})(j) = \sum_{-\infty}^{\infty} {}_k \hat{f}(j-k)\hat{g}(k), \quad j \in \mathbb{Z},$$

where the sum is actually taken over only finitely many k. Thus, we have (see also Exercise 2b)

$$T_f T_g(z^k) = T_{fg}(z^k) \quad \text{for } k \geq m+n.$$

The operators $T_f T_g$ and T_{fg} coincide on the subspace H_{m+n} of H_0.

Since the codimension of H_{m+n} in H_0 is finite (= m+n), this means that $T_f T_g - T_{fg}$ is an operator of finite rank, and hence is compact.

While T is *not* a homomorphism of Banach algebras (give a counter-example with f := ... and g := ...), by passing to the quotient algebra $\mathcal{B}/\mathcal{K}$, we have

$$\pi T(fg) = \pi T(f) \pi T(g).$$

Thus, $\pi \circ T$ is a homomorphism, when restricted to the subalgebra $\check{C}$.

Step 3: By the Approximation Theorem of Karl Weierstrass (see Chapter 8 or, for a direct proof, [Dym-McKean, 49]), each continuous function on a compact interval can be uniformly approximated (i.e., in the sup-norm) by polynomials, and even more so by rational functions. Thus, $\check{C}$ is dense in $C^0(S^1)$. Since $\pi \circ T$ is continuous, the multiplicative property carries over; i.e., $\pi \circ T : C^0(S^1) \to \mathcal{B}/\mathcal{K}$ is a homomorphism of Banach algebras.

Step 4: Since $\pi T(1) = 1$ (where the 1 on the left is the constant function $z \to 1$ and the 1 on the right is the class $\{Id+K; K \in \mathcal{K}\}$), it follows that $\pi \circ T$ takes invertible functions into invertible elements of $\mathcal{B}/\mathcal{K}$. Hence, if $f(z) \neq 0$ for all $z \in S^1$, then $\pi(T_f)$ is invertible in $\mathcal{B}/\mathcal{K}$, and so T_f is a Fredholm operator by the Theorem of Atkinson (Theorem 5.1).

b) We begin with the simplest case, the function $f(z) = z^m$. Relative to the canonical orthonormal basis of H_0 consisting of the functions z^n, $n \in \mathbb{Z}$, the Wiener-Hopf operator T_{z^m} assigned to f (Exercise 2b) has the form of the one-sided shift operator $(\text{shift}^+)^m$ for $m \geq 0$ and $(\text{shift}^-)^{|m|}$ for $m < 0$. By Exercise 1.1, we then have $\text{index}\, T_{z^m} = -m$. From the continuity of T (Exercise 3) and the continuity = homotopy invariance = local constancy of the index (Theorem 5.2), it follows with a) that $\text{index}\, T_g = -m$ for any $g \in C^0(S^1)$ with values in $\mathbb{C}^\times = \mathbb{C} - \{0\}$ which can be connected to the function z^m by a continuous path of functions in $C^0(S^1)$ with values in $\mathbb{C}^\times$. Now, the winding number of the curve $S^1 \to \mathbb{C}^\times$ (defined by z^m) about the point 0 is m. Since curves in $C^\times$ are homotopic through curves in $C^\times$ exactly when they have the same winding number (see Section III.1.A.), we have $\text{index}\, T_g = -W(g,0)$, and the index formula is proved. □

Exercise 4. In the construction of the index bundle (Theorem 7.1), we have seen that for each prescribed orthonormal basis $e_0, e_1, e_2, \ldots$ and

Fredholm operator $S \in F(H_0)$, there is an $n \in \mathbb{N}$ such that

$$P_n S : H_0 \to H_n$$

is surjective, where P_n is the orthogonal projection of H_0 onto the closed subspace of H_n spanned by the basis elements $e_n, e_{n+1}, e_{n+2}, \ldots$. Now show that (in the case of Wiener-Hopf operators) to each $f \in C^0(S^1)$ with $f(S^1) \subset \mathbb{C} \setminus \{0\}$, one can explicitly give an n for which $P_n T_f : H_0 \to H_n$ will be surjective.

Hint: One naturally exploits the fact that here we deal not with an arbitrary Hilbert space, but rather with function spaces, where there is additional "structure": Approximate the function $z \to 1/f(z)$ by a finite Fourier series

$$g(z) = \sum_{-n}^{n}{}_k \, \hat{g}(k) z^k,$$

with n chosen large enough so that $\sup\{|f(z)g(z)-1| : z \in S^1\} < 1$. □

Theorem 1 and Exercise 4 demand a detailed topological discussion - in relation to Chapter 1 and in view of Part III. However, the families of Wiener-Hopf operators which we will encounter in the following are, in fact, not of such elementary type; so we need some generalizations.

F. The Case of Systems

The first generalization is apparent, if we interpret the Wiener-Hopf operator T_f as a prediction operator for a time series $\ldots u_{-4}, u_{-3}, u_{-2}, u_{-1}, u_0$ of (say geophysical) measurements:

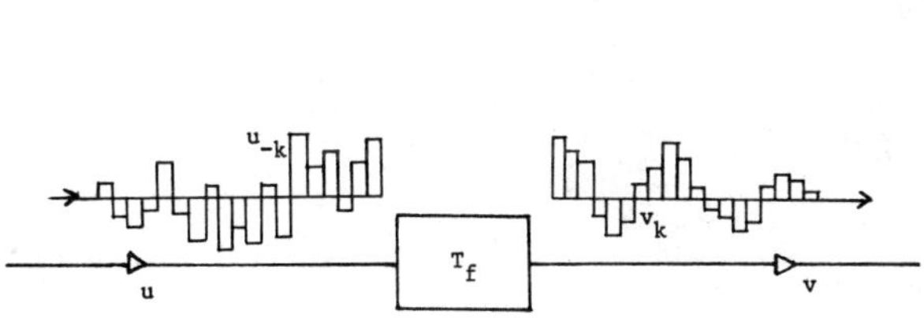

where, e.g., $v_n = \Sigma_{k=0}^{\infty} f_{n-k} u_{-k}$ with $u_{-k} = \hat{u}(k)$ is the given time series, $f_{n-k} = \hat{f}(n-k)$ is the "weighting" and $v_n = (T_f u)\hat{}(n)$ the predicted time series.

From the standpoint of the statistician, it is now perfectly obvious (even if one is interested in the weather in Frankfurt exclusively) that the inclusion of additional series of meteorological measurements (from Iceland or the Azores, say) can result in more information than the most sophisticated evaluation of a single series of data (of Frankfurt, for example) could provide. While for a single time series, the "weights" f_{n-k} are numbers, they must be matrices in the statistical analysis of multiple time series. Hereby, the condition $f(z) \neq 0$ which implies the Fredholm property must be replaced by $\det(f(z)) \neq 0$ where $f(z) \in GL(N,\mathbb{C})$, if we deal with an N-fold time series.

Exercise 5. Let H be a Hilbert space of complex-valued functions (e.g., $H = L^2(S^1)$ or other examples in Chapter 8). Show that the well-known notion of tensor product from multilinear algebra for finite-dimensional vector spaces also yields a sensible definition $H \otimes \mathbb{C}^N$. Convince yourself that $H \otimes \mathbb{C}^N$ is again a Hilbert space and (for the concrete examples) is related to the scalar-valued function space H, in such a way that one can regard $H \otimes \mathbb{C}^N$ as being the "corresponding" function space with values in $\mathbb{C}^N$.

Hint: Compare the analogous considerations in the proof of Theorem 7.3 with regard to the Hilbert space $\mathrm{Hom}(\mathbb{C}^N,H)$ isomorphic to $H \otimes \mathbb{C}^N$. How does one obtain a basis for $H \otimes \mathbb{C}^N$ from bases of H and $\mathbb{C}^N$? Details of the algebraic construction are in [UNESCO 1970, 116 f.], and the peculiarities of infinite-dimensional spaces (which are indeed no problem when one factor of the tensor product is finite dimensional) are found in [Douglas 1972, 31 and 79 f.]. □

Exercise 6. For a continuous map $f : S^1 \to GL(N,C)$, define the Wiener-Hopf operator

$$T_f := PM_f|H_0 \otimes \mathbb{C}^N : H_0 \otimes \mathbb{C}^N \to H_0 \otimes \mathbb{C}^N,$$

where $P : H \otimes \mathbb{C}^N \to H_0 \otimes \mathbb{C}^N$ is the projection, and M_f is multiplication by the matrix function f. Show:

a) T_f is a Fredholm operator,

b) index T_f depends only on the homotopy class of f in the homotopy set $[S^1,GL(N,\mathbb{C})]$.

Hint: Repeat the arguments from Exercise 2a, 3, and Theorem 1. Because of b, we can identify index $T_f \in \mathbb{Z}$ with the element $[f]$ in the fundamental group $\pi_1(GL(N,\mathbb{C})) \cong \mathbb{Z}$ that f represents. Each continuous map of S^1 into $GL(N,\mathbb{C})$ is homotopic to a continuous map of S^1 into the space of invertible diagonal matrices of rank N. Therefore, set

$$[f] := W(\det f, 0),$$

where $\det f(z)$ is the determinant of the matrix $f(z)$. See also under Section III.1.B. □

Exercise 7. In the next generalization, let X be a compact parameter space. Assign to each continuous map $f : S^1 \times X \to GL(N,\mathbb{C})$ a Fredholm family $T_f : X \to F$ and also an index bundle $\text{index}\, T_f \in K(X)$. Show that $\text{index}\, T_f$ only depends on the homotopy class of f.

Hint: Note that $f(z,x)$ is an invertible matrix that depends continuously on the variables z and x. Apply Exercise 6, noting that we obtain $T_{f(\cdot\cdot,x)} \in F$, for each $x \in X$. Here F is the space of Fredholm operators on the Hilbert space $H_0 \otimes \mathbb{C}^N$. Show that $T_{f(\cdot\cdot,x)}$ depends continuously on x, and then apply the construction from Theorem 7.1. □

Exercise 8. For a further generalization let E be a complex vector bundle over X of fiber dimension N. Figuratively speaking, one allows the vector space $\mathbb{C}^N$ to change from point to point. Given a function $f(z,x) \in \text{Iso}(E_x, E_x)$ which depends continuously on z and x and therefore defines a family of automorphisms of the vector bundle E, construct a family of Fredholm operators (in the variable Hilbert space $H \otimes E_x$), and finally an index bundle $\text{index}\, T_f \in K(X)$ that again only depends on the homotopy class of f.

Hint: See Theorem 7.1, Remark 1, where we may take the base X to be "sufficiently nice" (e.g., triangulable). Question: Do we really need the Theorem of Kuiper here in Exercise 8 - as in the above mentioned remark or can we proceed directly because of the particular structure of the problem? See [Atiyah 1969, 115]. □

G. The Continuous Analogue

In connection with elliptic boundary-value problems (Chapter II.8) and topological investigations of the general linear group $GL(N,\mathbb{C})$ (Chapter III.1, the Periodicity Theorem of Raoul Bott), we will return to the preceding construction. For the moment, we will only consider the continuous analogue of Theorem 1:

Exercise 9. Let $L^1(\mathbb{R})$ be the space of measurable, absolutely integrable functions. Show that each $\phi \in L^1(\mathbb{R})$ defines, via

$$(K_\phi u)(x) := \int_0^\infty \phi(x-y)u(y)dy, \quad x \in \mathbb{R}_+,$$

a bounded linear operator $K_\phi : L^2(\mathbb{R}_+) \to L^2(\mathbb{R}_+)$.

Hint: Regard $L^2(\mathbb{R}_+)$ as a subspace of $L^2(\mathbb{R})$, and then apply the results of Chapter 8 on the "convolution". For detailed estimates, see [Titchmarsh 1937, 90 f.]. □

Theorem 2. Let $\phi \in L^1(\mathbb{R})$ with $\hat{\phi}(t) + 1 \neq 0$ for all $t \in \mathbb{R}$ and let K_ϕ be as in Exercise 9. Then

$$\mathrm{Id} + K_\phi : L^2(\mathbb{R}_+) \to L^2(\mathbb{R}_+)$$

is a Fredholm operator, and we have

$$\text{index } (\mathrm{Id} + K_\phi) = W(\hat{\phi}+1,0),$$

where $W(\hat{\phi}+1,0)$ is the winding number of the oriented curve $\{\hat{\phi}(t)+1: t \in \mathbb{R}\}$ about the origin (see Section III.1.A).

Remark 1. In this index formula, one always must be aware of the dependence of the orientation in the definition of winding number (for us, $W(z,0) = 1$) and the orientation in the Fourier transformation (for us, $\hat{\phi}(x) = \int_{-\infty}^{\infty} e^{-ixy}\phi(y)dy$). If one removes the minus sign in the exponent (e.g., as does Mark Krein), then one obtains a minus sign in the index formula.

Remark 2. More exactly, for any $\phi \in L^1(\mathbb{R})$:

(i) $\mathrm{Spec}_e(K_\phi) = \{\hat{\phi}(t) : t \in \mathbb{R}\}$

(ii) $\mathrm{index}(z\mathrm{Id}-K_\phi) = W(\hat{\phi},z)$ for $z \in \mathrm{Spec}_e(K_\phi)$

(iii) $z\mathrm{Id}-K_\phi$ is $\begin{cases} \text{injective for index } z\mathrm{Id}-K_\phi \geq 0 \\ \text{surjective for index } z\mathrm{Id}-K_\phi \leq 0 \end{cases}$

Proofs for these results discovered by Mark Krein are found in [Jörgens 1970/1982, 13.4], for example.

Remark 3. If we regard $\mathrm{Id} + K_\phi$ as a map of $L^1(\mathbb{R})$, then under the assumptions of Theorem 2, we have that $\mathrm{Id} + K_\phi$ is an isomorphism [Wiener 1933,...]. Here, we then have no index problem.

Proof: One can prove Theorem 2 in very different ways, on which - in place of presenting a real proof - we will comment.

Approach 1: Reduce to Theorem 1 with the Cayley transformation

$\kappa z := \frac{z-i}{z+i}$, which maps the upper half-plane conformally onto the open unit disk.

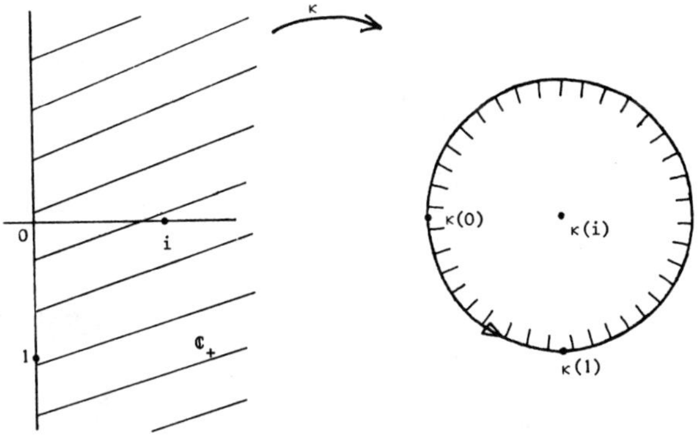

For $v \in L^2(S^1)$,

$$(Uv)(x) := \sqrt{2}\,\frac{v(\kappa x)}{x+i}, \qquad x \in \mathbb{R}$$

defines an isometry from $L^2(S^1)$ to $L^2(\mathbb{R})$, which carries the Hilbert space

$$H_0(S^1) := \{v : v \in L^2(S^1) \text{ and } \hat{v}(n) = 0 \text{ for } n < 0\}$$

to the Hilbert space

$$H_0(\mathbb{R}) := \{u : u \in L^2(\mathbb{R}) \text{ and } \hat{u}|(-\infty,0) = 0\}.$$

Equivalently, $H_0(\mathbb{R})$ consists of the square-integrable functions on $\mathbb{R}$ which can be analytically continued to the lower half-plane $\mathbb{C}_-$; e.g., see [Devinatz 1967, 82-84].

Instead of working with the projection $P : L^2(S^1) \to H_0(S^1)$ (see Exercise 2), we utilize the corresponding projection $Q : L^2(\mathbb{R}) \to H_0(\mathbb{R})$, where $Q = UPU^{-1}$. To each continuous complex-valued function $f \in C^0(\mathbb{R})$ of the form $f = \text{const} + \hat{\phi}$, where $\phi \in L^1(\mathbb{R})$, we assign a (continuous)

*Wiener-Hopf Operator** $W_f := QM_f|H_0(\mathbb{R}) \to H_0(\mathbb{R})$, where M_f means multiplication by f.

We now have defined three different operators:

- the discrete Wiener-Hopf operator T_g, $g \in C^0(S^1)$,
- the convolution operator K_ϕ, $\phi \in L^1(\mathbb{R})$,
- the continuous Wiener-Hopf operator W_f for $f = c + \hat{\phi}$.

From the properties of U, it then follows [Devinatz, 91] that

(i) $T_g = U^{-1}W_fU$, if $f = g\circ\kappa$.

(ii) $\hat{W}_f(u) = \hat{f} * \hat{u}$, for all $u \in H_0(\mathbb{R})$, i.e.,

$$\hat{h}(x) = c\hat{u}(x) + \int_0^\infty \phi(x-y)\hat{u}(y)dy, \quad x \in \mathbb{R}_+,$$

if $W_f(u) := h$ and $f = c + \hat{\phi}$ with $c \in \mathbb{C}$ constant and $\phi \in L^1(\mathbb{R})$.

Denoting the Fourier transformation by $F : L^2(\mathbb{R}) \to L^2(\mathbb{R})$, we can also write

(ii') $FW_f = (c\mathrm{Id} + K_\phi)F$; $f = c + F\phi$.

(i) expresses the unitary equivalence of discrete and continuous Wiener-Hopf operators, which is trivial by the definition of W_f here. (ii) requires some caution with the Fourier transformation: In Chapter 8, we dealt only with the "harmonic analysis" of periodic processes or of processes which in some sense abate with increasing or decreasing time. However, in the kinematic and statistical analysis of most natural, physical, technical, economic etc. processes, the classical machinery is, in fact, not sufficient, since these processes oscillate about some "mean", without being strictly periodic. The formally analogous Fourier analysis requires "functions" on $\mathbb{R}$ which are identically 0 away from a point but are so "strongly infinite" at this one point that the integral over all of $\mathbb{R}$ does not vanish. Physicists, such as Paul Dirac, used this idea in their computations long before Norbert Wiener rigorously provided the necessary generalizations of harmonic analysis [Wiener 1933], and which Laurent Schwartz later placed on an even broader foundation with his theory of distributions. In the sense of distributions the Fourier transformation of the constant function $\mathbb{1}$ is just the Dirac distribution δ at the point 0. See [Hörmander 1963, 21f.] or [Schwartz II 1959, 111].

*In contrast to this, one often refers to the discrete Wiener-Hopf Operators as Toeplitz operators.

Theorem 2 follows immediately from Theorem 1 with (i) and (ii').

Approach 2: When Allen Devinatz proved the unitary equivalence between discrete and continuous Wiener-Hopf operators, he showed more than is actually necessary for the proof of Theorem 2. Alternatively, one can reduce Theorem 2 to Theorem 1 in "pedestrian fashion" via approximating $f - \mathbb{1}$ by functions which are identically zero outside a bounded interval. If g is such a function, then $g + \mathbb{1}$ can be considered a continuous periodic function, i.e., an element of $C^0(S^1)$. The approximation is done in such a way that $g(z) + 1 \neq 0$ for all $z \in S^1$ and then Theorem 1 applies to $g + \mathbb{1}$.

We now proceed as in the passage from Fourier series to Fourier integrals (see Chapter 8) whereby the convergence questions must be considered very carefully. For an indication of the computations involved, see e.g. [Gelfand-Raikov-Shilov, 129-132].

Approach 3: Avoidance of Theorem 1 and basically new proof, e.g. [Krein 1958, Theorem 9.2], [Jörgens 1970/1982, §13.4], also for "systems", i.e., matrix valued f [Gohberg-Krein 1958, Theorem 4.1]. These proofs employ the famous "factorization method" introduced by Eberhard Hopf and Norbert Wiener in their original paper (Über eine Klasse singulärer Integralgleichungen. Sitzber. Preuss. Akad. Wiss., Sitzung der phys.-math. Klasse, Berlin 1931, 696-706). Its idea and essential content are presented very comprehensibly in [Wiener 1949, 153-157] (Norman Levinson's heuristic addendum), [UNESCO 1970, 47-48] (Friedrich Sommer's survey paper on complex analysis) or [Dym-McKean 1972, 176-184]. The last source contains an altogether good introduction to the function theoretic properties of "Hardy functions", the elements of our spaces $H_0(S^1)$ and $H_0(\mathbb{R})$.

Approach 4: Projection methods of a more general sort which comprise the discrete as well as the continuous case. A detailed exposition is found in [Gohberg-Feldmann 1974]. Here - as in Approach 3 - the stress is on finding explicit solutions so that statements about the index enter more frequently in the "opposite direction", since "the applicability of one or another projection method to the Wiener-Hopf integral equation is determined by the index" (p. 91). Complete proofs of our Theorem 2, in the manner of these projection methods, are in [Prössdorf 1974/1978, §§2.1.5, 2.4.1, 3.2]. Hereby Theorem 2 is not only proved for square integrable functions, but at once for broad varieties of more general function spaces - as do the authors of Approach 3. □

Finally we return another time to the discrete Wiener-Hopf operators whose totality $\{T_f : f \in C^0(S^1)\}$ we will denote by T, after adjoining the compact operators to H_0.

Exercise 10. Show that the following short sequence of Banach spaces is exact:

$$0 \to K(H_0) \to T \to C^0(S^1) \to 0,$$

where $K(H_0)$ denotes the compact operators on the Hilbert space $H_0(S^1)$. How are the arrows defined?

Hint: One best begins with the fact, established in the proof of Theorem 1, that the commutator ideal $\{T_\phi T_\psi - T_{\phi\psi} : \phi,\psi \in C^0(S^1)\}$ is contained in $K(H_0)$. Show then that the quotient algebra $T/K(H_0)$ is mapped isomorphically and isometrically via T onto $C^0(S^1)$, where the maps of the short sequence are defined using algebraic generalities. The proof is not entirely simple. One may consult [Douglas 1972, 184]. Note the similarity to the exact symbol sequence in the theory of partial differential equations (see Part II below). Also compare it to the "tensorial sequence" in the case of systems [Douglas 1972, 202 f.]. □

With these classes of Wiener-Hopf operators, we have greatly enlarged our reservoir of examples. One can even show that - modulo "unitary equivalence modulo compact operators" (see our comment after Theorem 7.3) - every "essentially normal" operator R on a separable Hilbert space H can be written in the form of a Wiener-Hopf operator. More precisely:

1. If the essential spectrum of R has the form of a simple, closed curve (say the image of the circle S^1 under an orientation - preserving homeomorphism or "parameterization" $\eta: S^1 \to ℂ$) and $\text{index}(R-z\text{Id}) = n$ for z interior to the curve, then R is unitarily equivalent to a compact perturbation of the multiplication operator M_η or the discrete Wiener-Hopf operator $T_{\eta\circ\kappa^{-n}}$, where $\kappa(z) := z$ [Fillmore 1973, 73].

2. Even when the essential spectrum cannot be parametrized so nicely, classes of generalized Wiener-Hopf operators (namely on "generalized Hardy spaces", where the domain of holomorphy need not be the upper half-plane or the open disk but may be any bounded region of ℂ) exist from which a "model" for R can be patched together; see [Fillmore 1973, 122] and the original papers quoted there.

Part II. Analysis on Manifolds

> "But, we ask, will not the growth of mathematical knowledge eventually make it impossible for a single researcher to embrace all parts of this knowledge? In answer let me point out how thoroughly, by the very nature of the mathematical sciences, any true progress brings with it the discovery of more incisive tools and simpler methods which at the same time facilitate the understanding of earlier theories and eliminate older more awkward developments. By acquiring these sharper tools and simpler methods the individual researcher succeeds more easily in orienting himself in the different branches of mathematics. In no other science is this possible to the same degree."
>
> (D. Hilbert, 1900)

1. PARTIAL DIFFERENTIAL EQUATIONS

A. Linear Partial Differential Equations

The theory of partial differential equations serves the characterization of motions and equilibria "with infinitesimal interactions" and constitutes "the mathematics of all quantities varying in space and time" (Norbert Wiener) which makes up a good part of mathematical physics and of applied mathematics altogether.

We distinguish *ordinary* and *partial* differential equations. In ordinary differential equations, the unknown is a function or a system of functions which depend on a single independent variable. In most applications, this variable is time. In partial differential equations, the one or more unknown functions depend on several variables. In applications, these variables are usually the coordinates of a point in space, but one of them may be time. A differential equation expresses relations between measurable quantities and their changes in space and/or time (rates of change).

In geometric language solving an ordinary differential equation means finding a curve and solving a partial differential equation means finding a family of curves or a surface or a manifold of higher dimension, whereby

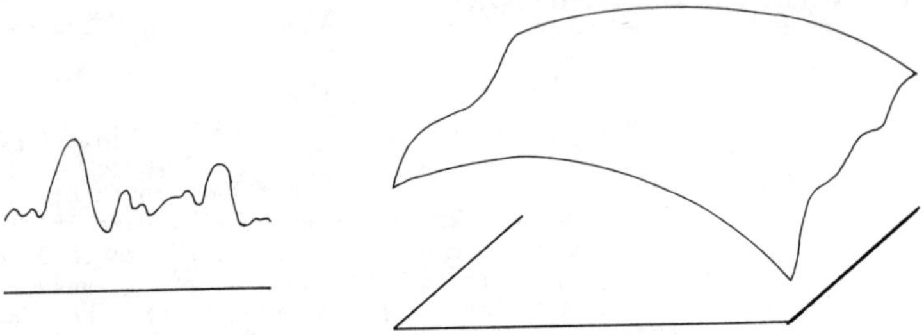

the curvatures of the curves or surfaces must satisfy the conditions expressed by the differential equation.

We restrict ourselves here to the treatment of *linear* differential equations of the form

$$Pu = f$$

where u and f are infinitely differentiable complex-valued functions on $\mathbb{R}^n$ and

$$Pu(x) := \sum_\alpha a_\alpha(x)(D^\alpha u)(x), \quad x \in \mathbb{R}^n.$$

Here, $\alpha = (\alpha_1,\dots,\alpha_n) \in \underbrace{\mathbb{Z}_+ \times \cdots\cdots \times \mathbb{Z}_+}_{n\text{-times}}$ is a multi-index to specify the partial derivative; e.g.

$$D^{(1,0,\dots,0)} := \frac{1}{i}\frac{\partial}{\partial x_1}$$

$$D^{(2,0,\dots,0)} := \left(\frac{1}{i}\right)^2 \frac{\partial^2}{\partial x_1^2}$$

$$D^\alpha := \left(\frac{1}{i}\right)^{|\alpha|} \frac{\partial^{|\alpha|}}{\partial x_1^{\alpha_1} \cdots \partial x_n^{\alpha_n}}$$

where $|\alpha| = \alpha_1 + \dots + \alpha_n$.

One carries the factor of $\frac{1}{i}$, because then integration by parts can be done symmetrically. For example, when $n = 1$

$$\int_a^b (Df)\overline{g} = -\int_a^b f\,\overline{Dg} + \text{boundary terms, for } D = \frac{d}{dx}$$

and

$$\int_a^b (Df)\overline{g} = \int_a^b f\,\overline{Dg} + \text{boundary terms, for } D = \frac{1}{i}\frac{d}{dx}.$$

In this way we achieve that the differential operators D^α are formally self-adjoint (see Exercise 2.7 below) and yield better expressions under Fourier transformation (see Exercise I.8.2b).

The coefficients a_α are always taken to be infinitely differentiable; moreover, $a_\alpha = 0$ for all but finitely many α. P is then called a *differential operator of order* $\max\{|\alpha| : a_\alpha \neq 0\}$.

We give the space $C^\infty(\mathbb{R}^n)$ of infinitely differentiable (complex-valued) functions on $\mathbb{R}^n$ the topology defined by the following family of semi-norms (for $k \in \mathbb{Z}_+$, $K \subset \mathbb{R}^n$ compact):

$$||f||_{k,K} := \sum_{|\alpha| \leq k} \sup\{|D^\alpha f(x)| : x \in K\}.$$

Accordingly, a sequence $f_1, f_2, \ldots$ of C^∞ functions converges to the constant function 0, if and only if the functions f_m and all their derivatives converge to 0 uniformly on each compact subset of $\mathbb{R}^n$.

Exercise 1. Show that a linear (scalar) differential operator P is a continuous linear, and local map $P : C^\infty(\mathbb{R}^n) \to C^\infty(\mathbb{R}^n)$; here "local" means that

$$\text{Support } Pf \subset \text{Support } f \quad \text{for all} \quad f \in C^\infty(\mathbb{R}^n),$$

where

$$\text{Support } f := \text{the closure of the set of points } x \in \mathbb{R}^n \text{ with } f(x) \neq 0.$$

Remark. Conversely, one can show that every continuous, linear, local map $P : C^\infty(\mathbb{R}^n) \to C^\infty(\mathbb{R}^n)$ is a differential operator (if one allows the order to be infinite and only finite on compact subsets). In fact, for each $x \in \mathbb{R}^n$, the map $f \mapsto (Pf)(x)$ is a continuous linear form on $C^\infty(\mathbb{R}^n)$ with one-point support $\{x\}$, whence [Schwartz I, Ch. III, Theorem XXXV] it is a finite linear combination of derivatives (in the distributional sense) of the Dirac delta at x. As x varies in $\mathbb{R}^n$, one can piece together these distributions to obtain the desired differential operator with C^∞ coefficients. The details are in [Cartan-Schwartz, 1-03 f.]. In 1960, Peetre showed that one can drop the continuity assumption. One can find the completely elementary proof, avoiding distribution theory, in [Narasimhan 1973, 172-175]. □

Exercise 2. Show that the linear differential operators with coefficients in $C^\infty(\mathbb{R}^n)$ form a non-commutative algebra. Verify that the commutator $PQ - QP$ is a differential operator of order at most $m+m'-1$, if P has order m and Q has order m'. □

Exercise 3. Show that the space of linear differential operators with constant coefficients forms a commutative subalgebra which is isomorphic to the polynomial algebra $\mathbb{C}[\xi_1,\dots,\xi_n]$ in the variables $\xi_1,\dots,\xi_n$. □

Exercise 4. Study the connection between the following partial differential equations appearing most frequently in mathematical physics texts:

a) The *wave equation* is the differential equation for the spreading of vibrations in a homogeneous medium

$$\frac{\partial^2 u}{\partial t^2} - a^2\left(\frac{\partial^2 u}{\partial x_1^2} + \frac{\partial^2 u}{\partial x_2^2} + \frac{\partial^2 u}{\partial x_3^2}\right) = f(x_1,x_2,x_3,t),$$

where the right side vanishes if no force intervenes, and u denotes the displacement (e.g., of a vibrating membrane).

b) The *heat equation* (which governs many other "diffusion processes") in a homogeneous isotropic body is

$$\frac{\partial u}{\partial t} - a^2\left(\frac{\partial^2 u}{\partial x_1^2} + \frac{\partial^2 u}{\partial x_2^2} + \frac{\partial^2 u}{\partial x_3^2}\right) = f(x_1,x_2,x_3,t).$$

Here the right side vanishes when no sources or sinks are present; u is the temperature.

c) The *potential (or Poisson) equation* for the potential of an electric field (for example) is

$$\frac{\partial^2 u}{\partial x_1^2} + \frac{\partial^2 u}{\partial x_2^2} + \frac{\partial^2 u}{\partial x_3^2} = -4\pi f(x_1,x_2,x_3)$$

where f is the given charge density and u is the potential whose gradient is the electric field. □

One can easily classify the (scalar) second order linear differential equations in several independent variables. For the corresponding differential operator

$$P = \sum_{|\alpha|\le 2} a_\alpha D^\alpha$$

and a point $x \in \mathbb{R}^n$, consider the *characteristic form*

$$\sum_{|\alpha|=2} a_\alpha(x)\xi_1^{\alpha_1}\dots\xi_n^{\alpha_n}$$

which is a quadratic form in $\xi_1,\dots,\xi_n$, since $|\alpha| = \sum_{j=1}^{n} \alpha_j = 2$. In analogy with the classification of conic sections in affine geometry, P is called *elliptic* at the point x, if the form is definite

$$\sum_{|\alpha|=2} a_\alpha(x)\xi_1^{\alpha_1} \cdots \xi_n^{\alpha_n} \neq 0 \quad \text{for} \quad (\xi_1,\ldots,\xi_n) \in \mathbb{R}^n \setminus \{0\}.$$

In this case, by a change of variable $(\xi_i) \to (\eta_i)$ (not necessarily orthogonal), one can express the form as

$$\pm(\eta_1^2 + \cdots + \eta_n^2).$$

We call P *hyperbolic* at x, if the characteristic form can be expressed as

$$\eta_1^2 + \cdots + \eta_{n-1}^2 - \eta_n^2.$$

by a change of variables; and P is *parabolic* at x, if we can express the form as

$$\eta_1^2 + \cdots + \eta_{n-1}^2.$$

The wave equation is then hyperbolic (n = 4), the heat equation is parabolic (n = 4), and the potential equation is elliptic (n = 3). □

B. Elliptic Differential Equations

Roughly speaking the elliptic differential equations of second order differ from the other classical types, in that there is no distinguished coordinate (e.g., "time"). More precisely, if P is not elliptic at x_0, then in general (i.e., except in certain degenerate cases which can cause difficulties when using Hamilton-Jacobi methods) there is a function $f \in C^\infty(\mathbb{R}^n)$ with $f(x_0) = 0$ and

$$\sum_{|\alpha|=k} a_\alpha(x_0)\left(\frac{\partial f}{\partial x_1}(x_0)\right)^{\alpha_1} \cdot\ldots\cdot \left(\frac{\partial f}{\partial x_n}(x_0)\right)^{\alpha_n} = 0, \qquad (*)$$

where the gradient $(\frac{\partial f}{\partial x_1},\ldots,\frac{\partial f}{\partial x_n})(x_0) \in \mathbb{R}^n \setminus \{0\}$; i.e., the directional derivatives of f at x_0 do not all vanish. By the Implicit Function Theorem, the set $S := \{x : f(x) = 0\}$ is an (n-1)-dimensional submanifold of $\mathbb{R}^n$ in a neighborhood of x_0. The manifold S is called *characteristic* for the differential operator P at the point x_0.

Solutions which are otherwise smooth can have jumps of their second derivatives only along these "characteristic surfaces". (In physics the characteristic surfaces are possible "wave fronts".) Furthermore, one

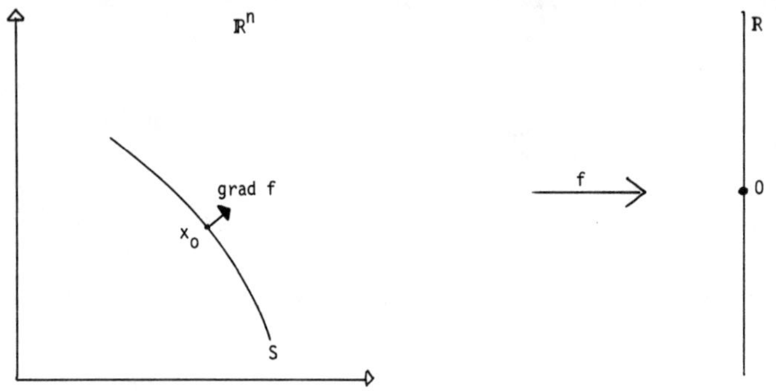

obtains from them certain curves along which ("separation of variables") the partial differential equation reduces to a simpler differential equation of first order, the so-called "transport equation". For these reasons the study of characteristic surfaces is a central task in the theory of non-elliptic differential equations. However, we will deal with elliptic differential operators only which have no (real) characteristic manifolds. The above may well be the reason why, in the theory of elliptic differential equations, it is not initial value problems but boundary value problems and problems on compact curved manifolds involving global questions which are at the center of interest. Slightly exaggerated: Since elliptic operators look locally the same everywhere (there are no characteristic manifolds, no "distinguished directions" etc.), and since the local solvability presents no problems (according to [Hörmander 1963, Theorem 7.2.1] there are no local singularities which could cause trouble globally) and because there are "sufficiently many" local solutions (e.g. the "large" spaces of harmonic functions for the Laplace operator and of holomorphic functions for the Cauchy-Riemann operator), interesting global problems can be formulated immediately and at times solved. We will come back to this philosophy later.

We refer to [Atiyah-Bott-Patodi 1973], for the connection between elliptic equations and parabolic initial value problems, which we will not discuss. The solutions of the parabolic heat equation, with arbitrary initial values, solve the potential equation asymptotically. This fact is made the starting point for a new proof of the Atiyah-Singer Index Formula. □

Until now we have considered only single differential equations (i.e., scalar differential operators). The treatment of simultaneous differential equations (where the interacting unknowns cannot be decoupled) requires the concept of *vectorial differential operators*. These are operators of the form $P = \sum_\alpha a_\alpha(x)D^\alpha$, where (for each $x \in \mathbb{R}^n$) $a_\alpha(x)$ is a linear map from a complex vector space V to a complex vector space W. Relative to bases of V and W, one can regard the $a_\alpha(x)$ as matrices with complex entries. The differential equation $Pu = f$, where u and f are C^∞ vector-valued functions on $\mathbb{R}^n$ ($u(x) \in V \cong \mathbb{C}^N$ and $f(x) \in W \cong \mathbb{C}^M$), can be regarded as a system of M differential equations in N unknown functions. Then, we have

$$P : \overbrace{C^\infty(\mathbb{R}^n) \times \ldots\ldots\ldots \times C^\infty(\mathbb{R}^n)}^{N \text{ times}} \to \overbrace{C^\infty(\mathbb{R}^n) \times \ldots\ldots\ldots \times C^\infty(\mathbb{R}^n)}^{M\text{-times}}. \qquad \square$$

Exercise 5. Show that Exercises 1-3 carry over to vectorial differential operators. A differential operator P of order k is said to be *elliptic*, if (for all $x \in \mathbb{R}^n$ and $(\xi_1,\ldots,\xi_n) \in \mathbb{R}^n \setminus \{0\}$) the "characteristic polynomial" or the "principal part"

$$\sum_{|\alpha|=k} a_\alpha(x)\, \xi_1^{\alpha_1} \cdot \ldots \cdot \xi_n^{\alpha_n}$$

is an isomorphism from V to W; in particular, $\dim V = \dim W$. $\square$

In the following paragraphs we will investigate the concept of ellipticity more fully; in particular we will work out the geometric meaning of the "principal part". Here, we give only a few hints for why one is interested in elliptic differential operators and what kinds of related questions come to the forefront. $\square$

C. Where Do Elliptic Differential Operators Arise?

Linear elliptic differential operators emerge in many different contexts.

(i) Modeling of equilibrium states of oscillating systems. A typical example from mathematical physics is the Laplace equation of potential theory; see Exercise 4c above. More complicated problems, require more complicated operators: Operators with variable coefficients which only pointwise resemble Laplace operators (e.g., when the material is not isotropic); operators of higher order; and operators on those function spaces, where the individual functions are not the concrete distributions of a continuous quantity (e.g., temperature), but for instance, the probability amplitudes ("wave functions") describing a discrete quantum mechanical system consisting of single electrons, atoms, or molecules.

(ii) Investigation of classical operators on more complicated geometric surfaces. In analogy with the Laplace operator $\Delta = (\partial^2/\partial x_1^2) + \dots + (\partial^2/\partial x_n^2)$, one can construct an operator on any Riemannian manifold; see Ch. 2. Various properties of such operators depend on the form of the manifold and serve to classify such surfaces and manifolds to some degree; see Chapter III.4 below.

(iii) Probabilistic characterization of diffusion processes. In contrast to discrete decay processes with transition probabilities, e.g., on lattices, we deal here with infinitesimal descriptions of flows and other processes, whereby the transition probabilities are given in the form of vector fields. Depending on the model, the random growth, the mean exit time (for problems with boundary), the expectation of some other given quantity, etc. appear as solutions of "characteristic operators" which are associated with the "Markov process" via some infinitesimal consideration. Conceptually, imagine a particle which performs a "symmetrical random motion" on the lattice points of $\mathbb{Z}^n$ by moving in equal time intervals one unit to one of the $2n$ neighboring lattice points with transition probability $1/2n$ always, i.e., the transition probability is "equidistributed" and "history independent". If f is a ("payoff") function defined on the lattice points then the expectation of the payoff f after one time unit is given by the "mean"

$$(Pf)(x) := \frac{1}{2n} \sum_{k=-n}^{n} f(x + e_k),$$

where the random motion placed the particle one unit ago at the point $x \in \mathbb{Z}^n$, and the $e_{-n},\dots,e_n$ are the lattice points adjacent to the origin, i.e., $e_1,\dots,e_n$ are the canonical basis vectors of $\mathbb{R}^n$ and $e_{-k} := -e_k$. The linear operator $P - \mathrm{Id}$ is then a discrete analogue to the operator $\frac{1}{2}\Delta$: One can show that the "statistical" operator $P - \mathrm{Id}$ yields the Laplace operator, when the distances between lattice points approach zero. The reason is the identity

$$(\Delta f)(x) = \lim_{h\to 0} \frac{\Sigma f(x+he_k) - 2nf(x)}{h^2}$$

which holds for sufficiently smooth functions. In this fashion, the Laplace operator is linked with the Wiener process which models the random motion of very small particles suspended in some fluid. The Wiener process is characterized probabilistically by the fact that the random change $x(t+s) - x(t)$ of a trajectory x possesses a normal distribution with a particularly simple density function. Other probability distributions yield different characteristic operators, but again elliptic ones if the underlying random process is a "diffusion process". A very elementary and clear exposition can be found in [Dynkin-Yushkevich 1960]. Further details are in [Karoui-Reinhard 1973].

(iv) Branching of solutions of non-linear differential equations. It should be noted that physical, biological, or social systems rarely contain intrinsic justifications for the linearity assumption of mathematical models: The supposition that the effect on a system under study is exactly proportional to the effect contradicts the presence of friction and, more generally, the laws of thermodynamics. Linear models are therefore used exclusively for pragmatic reasons, "either in order to facilitate computation or on account of the present imperfection of engineering techniques of realization" (of models). [Wiener 1949, 12]. There are a multitude of situations which unquestionably warrant the use of linear models, for example, in the theory of elasticity of materials, whose deformations are "fairly" proportional to the forces acting on them, or for many questions of stability theory and of control theory, for which the underlying machinery has been made "fairly" linear by man. On the other hand, some situations require non-linear modeling, since the essential phenomenon of "branching of solutions" cannot be described in any other way. (Some examples from mechanics are the bending of a straight rod under a constant force, the buckling of a flexible plate, the oscillations of a satellite in its orbital plane, and the surface waves of a heavy fluid.)

These facts in no way render the study of linear models superfluous. Rather it is true that very many non-linear systems can be approximated by so-called "implicit operators" which are linear, and in many cases also elliptic differential operators. In these cases the "index" of the implicit linear elliptic differential operator plays an important role for the derivation of the "branching equation". The following example illustrates why the theory of the branching of solutions of a non-linear equation, with an "analytic variety" as solution manifold, is a natural analogue of the Fredholm theory with affine spaces as solution manifolds. Consider the non-linear operator $(x,\lambda) \longmapsto Tx - \lambda x$ on $H \times \mathbb{R}$ where H is a Hilbert space and T a (linear) compact operator. The solution set $\{(x,\lambda) : Tx - \lambda x = 0\}$ is made out of the $\mathbb{R}$-axis and the kernels of the Fredholm operator $T - \lambda \mathrm{Id}$, $\lambda \neq 0$. Here the jumps of the kernel dimension of $T - \lambda \mathrm{Id}$ (i.e., the eigenvalues of T) are of special interest. See [Vaynberg-Trenogyn 1973, Chapters VII/VIII, particularly §27] and [Ize 1976] for this fast developing theory.

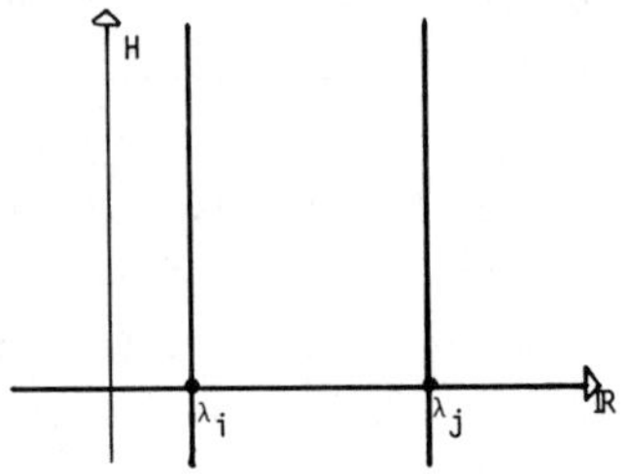

(v) Problems of optimization theory. Frequently elliptic differential equations are solved by solving the "associated" variational problem, i.e., a problem of optimization. Conversely, many complicated problems of optimization, particularly those occurring in "control theory", can be reduced to elliptic differential equations and in this way made clearer and more accessible for particular questions. A comprehensive exposition of this aspect can be found in [Morrey 1966].

(vi) Non-elliptic boundary value problems. Another area of applications is the treatment of systems of non-elliptic differential equations which sometimes can be represented as a family of elliptic differential operators in space coordinates parametrized by time. This is true for instance for the important type of parabolic differential equations which describes a multitude of spacial growth and differentiation processes. Here the connection between parabolic initial value problems and families of elliptic operators is well researched (see above Section B). □

D. Boundary-Value Conditions

Notice that in (i), (iii) and (iv) boundary-value conditions play an essential role while in (ii) interesting and deep results can be found by considering operators on "closed manifolds" (see Ch. 2 below), thereby avoiding the analytic difficulties of boundary-value problems. We will see below how closely connected boundary-value problems are with problems on closed manifolds. In fact, in the geometric expressions of K-theory, every boundary-value problem on a region X with boundary has a "corres-

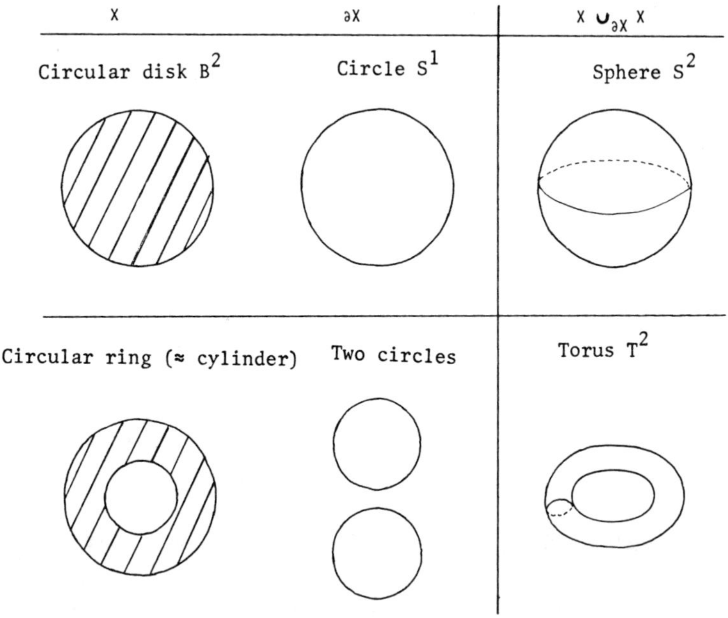

ponding" problem on the boundary ∂X of X and a problem on the "double" $X \cup_{\partial X} X$ of X (see Chapters II.7/8 and Section III.4.H below). Conversely, elliptic operators over a closed manifold reflect in this fashion how complicated manifolds are built from "macromolecules", the classical regions with boundary of Euclidean space $\mathbb{R}^n$.

Warning: Conceptually, the term "boundary-value problems" first brings to mind the boundary-value problems of the theory of elasticity, where an oscillating membrane is held fast along its border. This is mathematically the Dirichlet problem $u|\partial X = 0$, or more generally $u|\partial X = g$, where g is a function on ∂X. But in many applications, we deal with much more general types of boundary-value conditions. Good examples for all that can occur on the boundary of a region are furnished by the theory of diffusion processes described in (iii). We list just a few of the simplest phenomena following [Dynkin-Yushkevich 1969, 137-139]:

(i) Backward jump of the particle upon reaching the boundary to a fixed point x' inside X, possibly according to a certain probability distribution π generally depending on the boundary point y.

(ii) Absorption: The particle stays for good at the boundary point first reached.

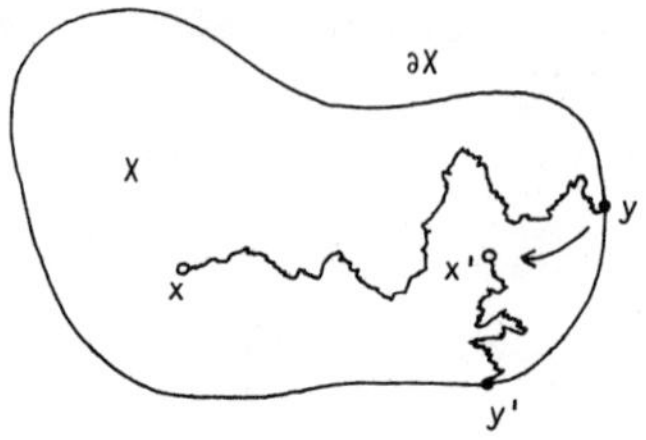

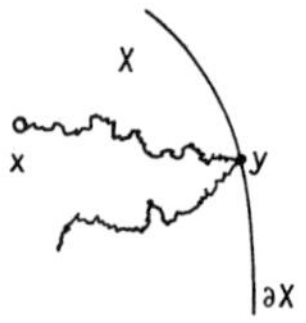

(iii) Extinction: The particle is annihilated upon first reaching the boundary.

(iv) Reflection: Symmetric reflection of the trajectory in the boundary.

These different boundary-value problems only serve as a supply of conceptual examples, and we will not pursue them further. But we wish to stress that it is lastly the investigation and classification of the various boundary-value problems (just like the investigation and classification of various manifolds) that yield the most interesting results. A simple but meaningful example is the Theorem of Ilya Nestorovich Vekua (Theorem 1). □

E. Main Problems of Analysis and the Index Problem

Let X be a region in $\mathbb{R}^n$ (or a C^∞- manifold; see below) and P a differential operator on X with C^∞-coefficients. Consider the equation

$$Pu = f,$$

where u and f are functions (not necessarily C^∞) on X. Somewhat vaguely, we can -- following [Hörmander 1971c] -- formulate the following questions:

(i) Under which conditions on P and X can one obtain local or global existence results?

(ii) Given X and P, how are the singularities of u and of f related?

Lars Hörmander shows in detail [loc. cit.] that these questions "are in fact so closely related that they can be considered different forms of the same problem." We are interested in the index of elliptic problems that is in questions of type (i). We will show in Part II that, for $Pu = f$ to have a solution at all, every elliptic problem with "suitable" boundary-value conditions must possess an index, i.e., a finite number of linearly independent solutions of the homogeneous equation $(f = 0)$ and a finite number of linear conditions for f. In Part III, we will introduce methods for computing the index from the coefficients of P and from numerical invariants of the structure of X, and conversely, for representing topological invariants of manifolds as indices of elliptic operators. □

F. Numerical Aspects

"Much of the modern work in partial differential equations looks highly esoteric, and only a few years ago such work would have been considered of no interest for applications, where one wants a solution expressed in a workable form, say by a sufficiently simple formula. The advent of the modern computing machines has changed this. If a problem involving a differential equation is sufficiently understood theoretically, then, in principle at least, a numerical solution can be obtained on a machine. If the mathematics of the problem is not understood, then the biggest machine and an unlimited number of machine-hours may fail to yield a solution." (COSRIMS Report of the National Science Foundation, 1969).

Sometimes the choice of numerical methods can cleverly be based on previous knowledge of the index. While dealing with Wiener-Hopf operators in Chapter I.9, we pointed out such results of I. Z. Gohberg and I. A. Feldman (after Theorem 9.2). Similarly, in the numerical treatment of non-linear problems, different methods have been recommended, depending on the index of the associated linear problem ([Vaynberg-Trenogyn 1973], [Ize 1976]).*

*Volker Strassen and other mathematicians uncovered the importance of the Theorem of Riemann-Roch (see Section III.4.G) and of other quantitative (index-) formulas of algebraic geometry for basic questions of computational complexity (e.g., for the calculation of the computational steps needed for inverting a matrix). Thus, there is an indirect relevance of index calculations on computer oriented numerical mathematics in this setting as well.

G. Elementary Examples

After these general remarks we will work out in detail some elementary examples.

Exercise 6. Investigate the (trivially elliptic) ordinary differential operator on the unit interval $I = [0,1]$ defined by

$$P : C^\infty(I) \times C^\infty(I) \to C^\infty(I) \times C^\infty(I)$$
$$(f \quad , \quad g) \to (f' \quad , \quad -g')$$

with the boundary conditions

$$\text{(i)} \quad R_1 : C^\infty(I) \times C^\infty(I) \to C^\infty(\partial I) \cong \mathbb{C} \times \mathbb{C}$$
$$(f \quad , \quad g) \to (f-g)|\partial I$$
$$\text{(ii)} \quad R_2 : (f \quad , \quad g) \to f|\partial I$$
$$\text{(iii)} \quad R_3 : (f \quad , \quad g) \to (f+g')|\partial I$$
$$\text{(iv)} \quad R_4 : (f \quad , \quad g) \to \ldots \quad \text{(your choice)}.$$

Determine the index of the operators

$$(P,R_i) : C^\infty(I) \times C^\infty(I) \to C^\infty(I) \times C^\infty(I) \times C^\infty(\partial I)$$

for $i = 1,\ldots,4$.

Hint: Clearly, $\dim \operatorname{Ker}(P,R_i) \leq 1$; to determine the cokernel, one writes $P(f,g) = (F,G)$ and $R_i(f,g) = h$ with $F,G \in C^\infty(I)$ and $h \in \mathbb{C} \times \mathbb{C}$, obtaining

$$f(t) = \int_0^t F(\tau)d\tau + c_1, \qquad g(t) = -\int_0^t G(\tau)d\tau + c_2$$

and two more equations for the boundary condition. The dimension of $\operatorname{Coker}(P,R_i)$ is then the number of linearly independent conditions on F, G, and h which must be imposed in order to eliminate the constants of integration.

For a more comprehensive treatment of the existence and uniqueness of boundary-value problems for ordinary differential equations (including systems), we refer to [UNESCO 1970, 195-197] and [Hartman 1964, 322-403]. Does the index always vanish?

We now consider the *Laplace operator* $\Delta := \frac{\partial^2}{\partial x^2} + \frac{\partial^2}{\partial y^2}$, as a linear elliptic differential operator from $C^\infty(X)$ to $C^\infty(X)$, where X is the unit disk $\{z : |z| \leq 1\}$ in the plane of complex numbers $z = x + iy$.

Exercise 7. For the boundary-value problem (named after Peter Gustav Dirichlet) with

$$R : C^{\infty}(X) \to C^{\infty}(\partial X)$$
$$u \to u|\partial X,$$

show that

a) $\operatorname{Ker}(\Delta,R) = \{0\}$ and

b) $(\operatorname{Image}(\Delta,R))^{\perp} = \{0\}$ (orthogonal complement in $L^2(X) \times L^2(\partial X)$).

In particular, it follows* that $\operatorname{index}(\Delta,R) = 0$.

Hint for a): $\operatorname{Ker}(\Delta,R)$ consists of functions of the form $u + iv$, where u and v are real-valued. Since the coefficients of the operators Δ and R are real, we may assume $v = 0$ without loss of generality. Thus, consider a real solution u with $\Delta u = 0$ in X and $u = 0$ on ∂X. Then (relative to the standard measure on X): $0 = -\int_X u\Delta u \overset{(\#)}{=} \int_X \langle \operatorname{grad} u, \operatorname{grad} u\rangle$, whence $\operatorname{grad} u = (\frac{\partial u}{\partial x}, \frac{\partial u}{\partial x}) = 0$, since u is real. Thus, u is constant and hence zero (since $u = 0$ on ∂X).

The trick lies in the equality (#), an integration by parts which is perhaps most simply derived from the integral theorem of George Gabriel Stokes in the calculus of differential forms (see Appendix, Exercise 8, and [UNESCO 1970, 133 f.]). The Stokes formula reads $\int_X d\omega = \int_{\partial X} \omega$, where ω is a 1-form. We set $\omega := u \wedge *du$ (see Appendix, Ex. 8) and obtain

$$d\omega = du \wedge *du + u \wedge d*du \tag{*}$$

and furthermore

$$\int_X d\omega = \int_{\partial X} u \wedge *du = 0$$

since $u|\partial u = 0$. From $\Delta u = 0$, conclude that

---

*Here, consider that the intersection of the orthogonal complement of $\operatorname{Im}(\Delta,R)$ relative to the usual inner product in $L^2(X) \times L^2(\partial X)$ with the space $C^{\infty}(X) \times C^{\infty}(\partial X)$ is isomorphic to $\operatorname{Coker}(\Delta,R)$. This is true, since the image of (the natural Sobolev extension of) (Δ,R) is closed in the L^2-norm, and its orthogonal complement (relative to the L^2 inner product) is contained in $C^{\infty}(X) \times C^{\infty}(\partial X)$. For a proof of these two important properties, which apply to elliptic boundary-value problems in general, see Chapter 8.

$$0 = -\int_X u \wedge d * du \overset{(*)}{=} \int_X du \wedge * du,$$

whence $du = 0$, and so u is constant.

Hint for b): Choose $L \in C^\infty(X)$ and $\ell \in C^\infty(\partial X)$ with (L,ℓ) orthogonal to $\mathrm{Im}(\Delta,R)$, whence (relative to the usual measure on X and ∂X):

$$\int_X (\Delta u)L + \int_{\partial X} u\ell = 0 \quad \text{for all } u \in C^\infty(X).$$

Then, show that

$$\int_X u\Delta L \overset{(\#\#)}{=} 0.$$

for all $u \in C^\infty(X)$ whose support (:= the closure of $\{z : z \in X$ and $u(z) \neq 0\}$) is contained in the interior of X. (Use a 2-fold integration by parts; in the exterior calculus, $\int_X u(d * dL) = \int_X (d * du)L$.) Thus, $\Delta L = 0$. Now show that a further integration by parts yields

$$\int_{\partial X} (x \frac{\partial L}{\partial x} + y \frac{\partial L}{\partial y} + \ell)u - L(x \frac{\partial u}{\partial x} + y \frac{\partial u}{\partial y}) = 0$$

for all $u \in C^\infty(X)$, and then $L|\partial X = 0$ and $\ell = -(x \frac{\partial L}{\partial x} + y \frac{\partial L}{\partial y})$; finally apply a). Details are in [Hörmander 1963, 264]. □

We now consider a C^∞ "*vector field*" $\nu: \partial X \to \mathbb{C}$ on the boundary $\partial X = \{z : |z| = 1\}$. For $u \in C^\infty(X)$, $z \in \partial X$, and $\nu(z) = \alpha + i\beta$, one defines the "*directional derivative*" of the function u relative to the vector field ν at the point z to be the number

$$\frac{\partial u}{\partial \nu}(z) := \alpha \frac{\partial u}{\partial x}(z) + \beta \frac{\partial u}{\partial y}(z).$$

(From the standpoint of differential geometry it is better, either to denote the vector field by $\frac{\partial}{\partial \nu}$ -- or to write the directional derivative simply as $(\nu \cdot u)(z)$; since $\frac{\partial}{\partial x}$ and $\frac{\partial}{\partial y}$ can be regarded as vector fields also; see Chapter 2 below.) The pair $(\Delta, \frac{\partial}{\partial \nu})$ defines a linear operator

$$(\Delta, \frac{\partial}{\partial \nu}) : C^\infty(X) \to C^\infty(X) \oplus C^\infty(\partial X)$$

$$u \mapsto (\Delta u, \frac{\partial u}{\partial \nu}).$$

Theorem 1 (I. N. Vekua 1948). For $p \in \mathbb{Z}$ and $\nu(z) := z^p$, we have that $(\Delta, \frac{\partial}{\partial \nu})$ is an operator with finite-dimensional kernel and cokernel, and

$$\mathrm{index}\,(\Delta, \frac{\partial}{\partial \nu}) = 2(1-p).$$

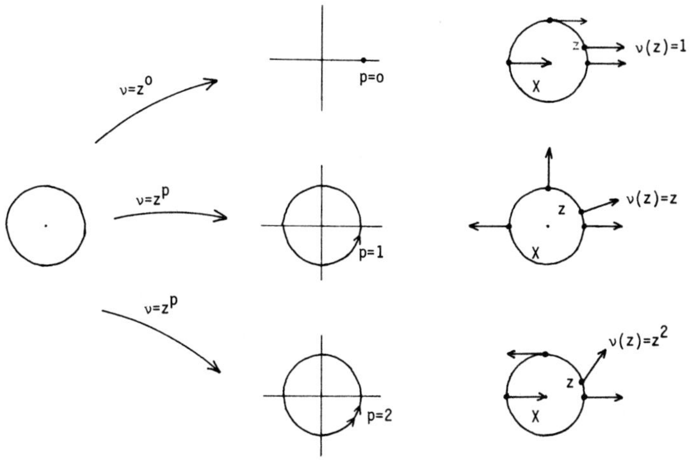

Remark 1. The theorem of Vekua remains true, if we replace z^p by any nonvanishing "vector field"

$$\nu\colon\ \partial X \to \mathbb{C} \smallsetminus \{0\}$$

with "winding number" p:

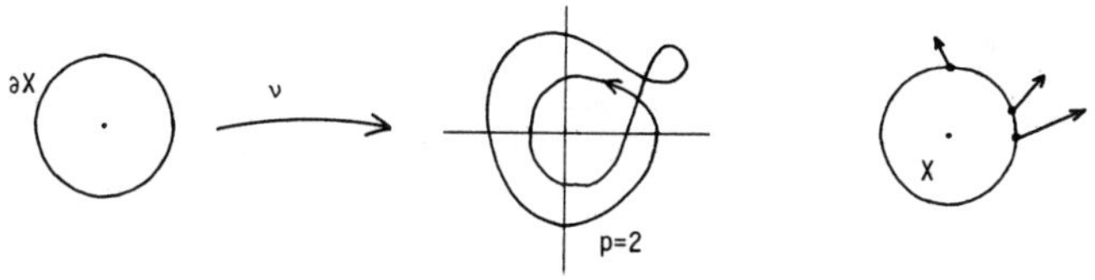

Moreover, in place of the disk, we can take X to be any simply connected (i.e., without "holes") domain in $\mathbb{C}$ with a "smooth" boundary ∂X; see Chapter 2 below. The reason is the homotopy invariance of the index (see Theorem I.5.2) which holds for elliptic differential operators in general (see Chapter 8 below).

Remark 2. One encounters the number $2(1-p)$ also in the theory of Riemann surfaces of genus p; e.g., in the theorem of Bernhard Riemann and Gustav Roch (see Ch. III.4). This is no accident, but rather it is connected with the relation between elliptic boundary-value problems and elliptic operators on closed manifolds, as mentioned above in C.

Remark 3. Motivated by the method of replacing a differential equation by difference equations, David Hilbert and Richard Courant expected "linear problems of mathematical physics which are correctly posed to behave like a system of N linear algebraic equations in N unknowns... If for a correctly posed problem in linear differential equations the corresponding homogeneous problem possesses only the "trivial" solution zero then a uniquely determined solution of the general inhomogeneous system exists. However, if the homogeneous problem has a non-trivial solution, the solvability of the non-homogeneous system requires the fulfillment of certain additional conditions." This is the "heuristic principle" which [Courant-Hilbert II, 179/231] saw in the Fredholm Alternative (see Chapter I.4). Ilya Nestorovich Vekua disproved it with his example where the principle fails for $p \neq 1$. We remark that in addition to these "oblique-angle" boundary-value problems, "coupled" oscillation equations (see for example Exercise 9 below), as well as suitable restrictions of boundary-value problems even with vanishing index, furnish further elementary examples for index $\neq 0$.

Proof (after [Hörmander 1963, 266 f.]): Since the coefficients of the differential operators $(\Delta, \frac{\partial}{\partial \nu})$ are real, we may restrict ourselves to real functions. Thus, $u \in C^\infty(X)$ denotes a single real-valued function, rather than a complex-valued function (i.e., a pair $u_1 + iu_2$ of real-valued functions u_1 and u_2).

Ker$(\Delta, \frac{\partial}{\partial \nu})$: It is well-known that Ker Δ consists of the real (or imaginary part of holomorphic functions on X (e.g., see [Ahlfors 1953, 175 ff.]). Such functions are called harmonic. Hence, $u \in \text{Ker}(\Delta)$, exactly when $u = \text{Re}(f)$ where $f = u + iv$ is holomorphic; i.e., the Cauchy-Riemann equation $\frac{\partial f}{\partial \bar{z}} = 0$ holds, or explicitly

$$\frac{\partial u}{\partial x} = \frac{\partial v}{\partial y} \quad \text{and} \quad \frac{\partial u}{\partial y} = -\frac{\partial v}{\partial x}. \tag{*}$$

Every holomorphic (= complex differentiable) function f is twice complex differentiable and its derivative is given by

$$f' = \frac{\partial f}{\partial z} = \frac{1}{2}\left(\frac{\partial u}{\partial x} + \frac{\partial v}{\partial y}\right) + \frac{i}{2}\left(\frac{\partial v}{\partial x} - \frac{\partial u}{\partial y}\right) = \frac{\partial u}{\partial x} - i\,\frac{\partial u}{\partial y} \quad \text{(by (*))}.$$

In this way we have a holomorphic function $\phi := f'$ for each $u \in \text{Ker}(\Delta)$. The boundary condition $\frac{\partial u}{\partial \nu} = 0$ $(\nu = z^p)$ then means that the real part $\text{Re}(\phi(z)z^p)$ vanishes for $|z| = 1$. For $p \geq 0$, $\phi(z)z^p$ is holomorphic as well as ϕ, hence

$$\left.\begin{array}{l} u \in \operatorname{Ker}(\Delta, \frac{\partial}{\partial\nu}) \\ \nu = z^p \\ p \geq 0 \end{array}\right\} \Rightarrow \left\{\begin{array}{l} \operatorname{Re}(\phi(z)z^p) \in \operatorname{Ker}(\Delta, R) \\ \text{where} \quad \phi := \frac{\partial u}{\partial x} - i\frac{\partial u}{\partial y} \\ \text{and} \quad R(\cdot\cdot) := (\cdot\cdot)|\partial X \end{array}\right.$$

Thus, we succeed in associating with the "oblique-angle" boundary-value problem for u a Dirichlet boundary-value problem for $\operatorname{Re}(\phi(z)z^p)$, which has only the trivial solution by Exercise 7a. Since $\phi(z)z^p$ is holomorphic with $\operatorname{Re}(\phi(z)z^p) = 0$, the partial derivatives of the imaginary part vanish, and so there is a constant $C \in \mathbb{R}$ such that $\phi(z)z^p = iC$ for all $z \in X$. If $p > 0$, then we have $C = 0$ (set $z = 0$). Hence $\phi = 0$, and (by the definition of ϕ) the function u is constant (i.e., $\dim \operatorname{Ker}(\Delta, \frac{\partial}{\partial\nu}) = 1$). If $p = 0$, then $\phi(z) = iC$; and so $u(x,y) = -Cy+\tilde{C}$, whence $\dim \operatorname{Ker}(\Delta, \frac{\partial}{\partial\nu}) = 2$ in this case.

We now come to the case $p < 0$, which curiously is not immediately reducible to the case $q > 0$ where $q := -p$. One can try to look for a solution by simply turning $\frac{\partial}{\partial\nu}$ around to $-\frac{\partial}{\partial\nu}$, as illustrated:

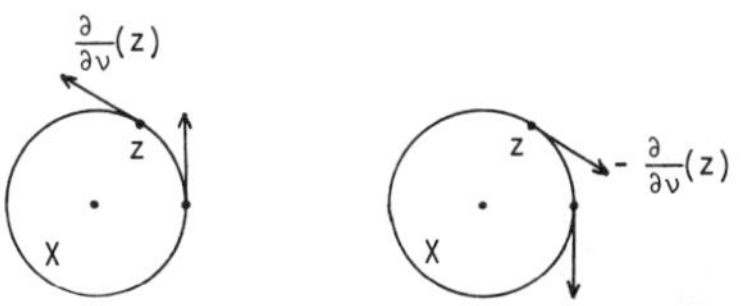

However, this leads to a dead end, since the winding numbers of ν and $-\nu$ about 0 are the same. In order to reduce the boundary-value problem with $p < 0$ to the elementary Dirichlet problem, we must now go through a more careful argument. ($\phi(z)z^p$ can have a pole at $z = 0$, whence $\operatorname{Re}(\phi(z)z^p)$ is not necessarily harmonic...). We write the holomorphic function ϕ as a finite Taylor series

$$\phi(z) = \sum_{j=0}^{q} a_j z^j + g(z)z^{q+1},$$

where $q := -p$ and g is holomorphic. We define a holomorphic function ψ with $\psi(0) = 0$:

$$\psi(z) := g(z)z + \sum_{j=0}^{q-1} \bar{a}_j \ z^{q-j}.$$

Then one can write

$$\phi(z)z^p = a_q + \sum_{j=0}^{q-1} (a_j z^{j-q} - \bar{a}_j z^{q-j}) + \psi(z).$$

The boundary condition $\frac{\partial u}{\partial \nu} = 0$ implies $\mathrm{Re}(\phi(z)z^p) = 0$ for $z = 1$; this means, by the above equation, that $\mathrm{Re}(\psi(z) + a_q) = 0$ for $|z| = 1$. Since ψ is holomorphic, we have again arrived at a Dirichlet boundary-value problem; this time for the function $\mathrm{Re}(\psi(z) + a_q)$. From Exercise 7a, it follows again that $\psi(z) + a_q$ is an imaginary constant, whence $\psi(z) = \psi(0) = 0$ and a_q is pure imaginary. We have

$$\phi(z) = \phi(z)z^p z^q = a_q z^q + \sum_{0}^{q-1} (a_j z^j - \bar{a}_j z^{2q-j}),$$

for arbitrary $a_0, a_1, \ldots, a_{q-1} \in \mathbb{C}$ and $a_q \in i\,\mathbb{R}$. As a vector space over $\mathbb{R}$, the set

$$\{\frac{\partial u}{\partial x} - i\,\frac{\partial u}{\partial y} : u \in \mathrm{Ker}(\Delta, \frac{\partial}{\partial \nu})\}$$

has dimension $2q+1$; here we have restricted ourselves to real u, according to our convention above. Since u is uniquely determined by ϕ up to an additive constant, it follows that for $\nu = z^p$ and $p < 0$,

$$\dim \mathrm{Ker}(\Delta, \frac{\partial}{\partial \nu}) = 2q+2 = 2-2p.$$

$\underline{\mathrm{Coker}(\Delta, \frac{\partial}{\partial \nu})}$: As Exercise 7b shows, the equation $\Delta u = F$ has a solution for each $F \in C^\infty(X)$. Thus we can identify $\mathrm{Coker}(\Delta, \frac{\partial}{\partial \nu})$ (i.e., the quotient space $C^\infty(X) \times C^\infty(\partial X) / \mathrm{Im}(\Delta, \frac{\partial}{\partial \nu})$) with the "simpler" quotient space $C^\infty(\partial X)/\frac{\partial}{\partial \nu}(\mathrm{Ker}\ \Delta)$, as follows: We assign to each representative pair $(F,h) \in C^\infty(X) \times C^\infty(\partial X)$ the class of $h - \frac{\partial u}{\partial \nu} \in C^\infty(\partial X)$, where u is chosen so that $\Delta u = F$. This map is clearly well defined on the quotient space of pairs; the inverse map is given by $h \to (0,h)$. Hence, we have found a representation for $\mathrm{Coker}(\Delta, \frac{\partial}{\partial \nu})$ in terms of the "boundary functions" $\{\frac{\partial u}{\partial \nu} : u \in \mathrm{Ker}\ \Delta\}$, rather than the cumbersome pairs in $\mathrm{Im}(\Delta, \frac{\partial}{\partial \nu})$. (This trick can always be applied for the boundary-value problems (P,R), when the operator P is surjective.)

We therefore investigate the existence of solutions of the equation $\Delta u = 0$ with the "inhomogeneous" boundary condition $\frac{\partial u}{\partial \nu} = h$, where h is a given C^∞ function on ∂X. According to the trick introduced in the first part of our proof, it is equivalent to ask for the existence of a holomorphic function ϕ with the boundary condition $\mathrm{Re}(z^p\phi(z)) = h$, $|z| = 1$, i.e., for a solution of a Dirichlet problem for $\mathrm{Re}(z^p\phi(z))$. By Exercise 7b, there is a unique harmonic function which restricts to h on the boundary ∂X; hence, we have a (unique up to an additive imaginary constant) holomorphic function θ with $\mathrm{Re}\,\theta(z) = h$ for $|z| = 1$.

In the case $p \leq 0$, the boundary problem for ϕ is always solvable; namely, set $\phi(z) := z^{-p}\theta(z)$. Hence, we have (for $\nu(z) = z^p$ and $p \leq 0$)

$$\dim \mathrm{Coker}(\Delta, \tfrac{\partial}{\partial\nu}) = 0.$$

For $p > 0$, we can construct a solution of the boundary-value problem in ϕ from θ, if and only if there is a constant $C \in \mathbb{R}$ such that $(\theta(z) - iC)/z^p$ is holomorphic (i.e., the holomorphic function $\theta(z) - iC$ has a zero of order at least p at $z = 0$. Using the Cauchy Integral Formula, these conditions on the derivatives of θ at the point $z = 0$ correspond to conditions on line integrals around ∂X. In this way, we have $2p-1$ linear (real) equations that h must satisfy in order that the boundary-value problem have a solution.

We summarize our results in the following table ($\nu(z) = z^p$);

p	$\dim \mathrm{Ker}(\Delta, \frac{\partial}{\partial\nu})$	$\dim \mathrm{Coker}(\Delta, \frac{\partial}{\partial\nu})$	$\mathrm{index}(\Delta, \frac{\partial}{\partial\nu})$
> 0	1	$2p-1$	$2-2p$
$0 \leq$	$2-2p$	0	

<u>Warning 1</u>: We already noted in the proof the pecularity that the case $p < 0$ cannot simply be played back to the case $p > 0$. This is reflected here in the asymmetry of the dimensions of kernel and cokernel and the index. It simply reflects the fact that there are "more" rational functions with prescribed poles than there are polynomials with "corresponding" zeros. See also Section III.4.G, the Riemann-Roch Theorem.

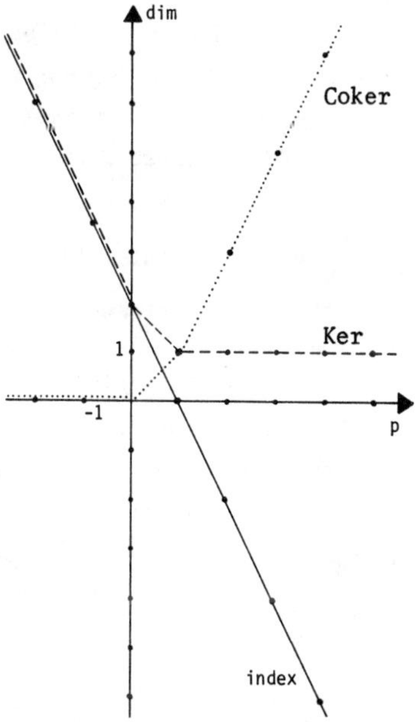

Warning 2: In contrast to the Dirichlet Problem, which we could solve via integration by parts (i.e., Stokes' Integral Theorem), the above proof is function-theoretic in nature and cannot be used in higher dimensions. This is no loss in our special case, since the index of the "oblique-angle" boundary-value problem must vanish anyhow in higher dimensions for topological reasons; see [Hörmander 1963, 265 f.] or Chapter 8 below. The actual mathematical challenge of the function-theoretic proof arises less from the restriction $\dim X = 2$ than from the certain arbitrariness, the tricks and devices of the definitions of the auxiliary functions ϕ, ψ, θ, by means of which the oblique-angle problem is reduced to the Dirichlet problem. Is there not a canonical, straightforward general method for finding the index of a boundary-value problem? We will return to this question below (II.8 and III.4).

Warning 3: The theory of ordinary differential equations easily conveys the impression that partial differential equations also possess a "general solution" in the form of a functional relation between the unknown function ("quantity") u, the independent variables x and some arbitrary constants or functions, and that every "particular solution" is obtained by substituting certain constants or functions f, h, etc. for the arbitrary constants and functions. (Corresponding to the higher degree of freedom in partial differential equations, we deal not only with constants of integration but with arbitrary functions.) The preceding calculations, regarding the boundary value problem of the Laplace operator, clearly indicate how limited this notion is which was conceived in the 18th century on the basis of geometric intuition and physical considerations. The classical recipe of first searching for general solutions and only at the end determining the arbitrary constants and functions fails. The concrete form of boundary conditions -- for example -- must enter the analysis to begin with.

Exercise 8. Without using Theorem 1, show that $\mathrm{Index}(\Delta, \frac{\partial}{\partial\nu}) = 0$ for $\nu = z$. (This boundary-value problem, where $\frac{\partial}{\partial\nu}$ is the field normal to the boundary is named after Carl Neumann. From the topological viewpoint it is equivalent (modulo constant functions) to the Dirichlet boundary-value problem defined by a tangent vector field.) □

NEUMANN:

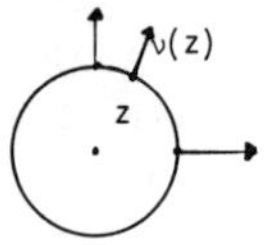

DIRICHLET:

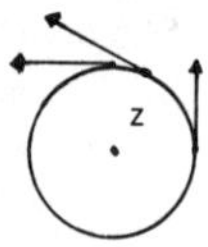

Exercise 9. Let $X := \{z = x+iy; |z| \leq 1\}$ be the unit disk and define an operator

$$T : C^\infty(X) \oplus C^\infty(X) \to C^\infty(X) \oplus C^\infty(X) \oplus C^\infty(\partial X)$$
$$(u\ ,\ v) \mapsto \left(\frac{\partial u}{\partial \bar{z}}, \frac{\partial v}{\partial z}, (u-v)|\partial X\right)$$

where $\frac{\partial}{\partial z} = \frac{1}{2}(\frac{\partial}{\partial x} - i\frac{\partial}{\partial y})$ is complex differentiation and $\frac{\partial}{\partial \bar{z}} = \frac{1}{2}(\frac{\partial}{\partial x} + i\frac{\partial}{\partial y})$ is the Cauchy-Riemann differential operator "formally adjoint" to $\frac{\partial}{\partial z}$. (See before Exercise 2.7.) Prove that $\mathrm{index}(T) = 1$.

Hint: Show first that $\dim(\mathrm{Ker}\ T) = 1$: Since $\frac{\partial u}{\partial \bar{z}} = 0$ means that u is holomorphic, it follows from $\frac{\partial v}{\partial z} = 0$ and $v = u$ on ∂X, that $v = u$ on all X and v is holomorphic. From $\frac{\partial v}{\partial z} = 0$, we then know v is constant. Then show $\mathrm{Coker}(T) = \{0\}$, or more precisely $(\mathrm{Im}\ T)^\perp = \{0\}$; see the footnote to Exercise 7. For this, choose arbitrary $f,g \in C^\infty(X)$ and $h \in C^\infty(\partial X)$ and prove that f, g and h must identically vanish, if (for all $u,v \in C^\infty(X)$)

$$\int_X \left(\frac{\partial u}{\partial \bar{z}} f + \frac{\partial v}{\partial z} g\right) + \int_{\partial X} (u-v)h = 0$$

Remark. In engineering one calls a system of separate differential equations

$$Pu = f$$
$$Qv = g,$$

which are related by a "transfer condition" $R(u,v) = h$, a "coupling problem"; when the domains of u and v are different, but have a common boundary (or boundary part) on which the transfer condition is defined, then we have a "transmission problem"; e.g., see [Booss 1972, 7 ff.]. Thus, we may think of T as an operator for a problem on the spherical surface $X \cup_{\partial X} X$ (see Exercise 2.12 below) with different behavior on the upper and lower hemispheres, but with a fixed coupling along the equator. □

2. DIFFERENTIAL OPERATORS OVER MANIFOLDS

A. Motivation

The modern concept of a closed "manifold" allows us to generalize and simultaneously drastically simplify the index problem by eliminating boundary conditions. For example, the homogeneous Laplace equation $\Delta u = 0$ on the disk has infinitely many linearly independent solutions, viz. the harmonic functions, while the corresponding Laplace equation on the sphere has a one-dimensional solution space consisting of the constant functions. In this respect the notion of differentiable manifold does not make the mathematics more complicated, but is a genuine first approximation to the "difficult"* boundary value problems in Euclidean space $\mathbb{R}^n$.

But also from the point of view of immediate applications, the geometric concept of a manifold played an important role. In fact, space-time problems defined initially and canonically in Euclidean space frequently do not have unrestricted independent variables, but these variables are restricted by side conditions to certain submanifolds of Euclidean space. Examples are the "constraints" in mechanics; the path equations of electrodynamics into which enter essentially the shape and surface of the "conductor"; or the symmetry conditions of elementary particle physics which replace the high dimensional Euclidean state spaces by low dimensional state spaces in the form of manifolds.

B. Differentiable Manifolds - Foundations

We begin with a compilation of the basic notions and elementary relations of the concept of a "differentiable manifold". As a general reference, we refer to [Singer-Thorpe 1967] and [Bröcker-Jänich 1973/1982].

Exercise 1. Recall the following two classical theorems of differential calculus, which form the foundation of the concept of a differentiable manifold.

*The development of mathematics shows again and again how, in the growth of knowledge, the conceptual and non-conceptual form a unit, alternating and fading into one another. A most striking example is furnished by the famous four-color problem, which characteristically still presents many puzzles in the plane, even after its computer aided solution, while the corresponding questions for closed manifolds have long been disposed of. "Most of the early attempts at solving this problem were based on direct attack, and they not only failed, but did not even contribute any useful mathematics." Only "a new and highly indirect approach to the coloring problem based on a generalization of Kirchhoff's laws of circuit theory in a completely unforeseen direction", proved to be "successful in understanding a variety of combinatorial problems." (Gian-Carlo Rota, The Mathematical Sciences: A Report, 1969. Reprinted with the permission of the National Academy of Sciences.)

a. (Inverse Function Theorem). If $f = (f_1,\ldots,f_n)$ is a C^∞ map from $\mathbb{R}^n$ to $\mathbb{R}^n$ whose Jacobian matrix $((\partial f_i/\partial x_j)(p))_{i,j=1,\ldots,n}$ has rank n at $p \in \mathbb{R}^n$ (i.e., its determinant is nonzero), then there is a neighborhood U of p in $\mathbb{R}^n$ which is diffeomorphic by f to a neighborhood V of $f(p)$ in $\mathbb{R}^n$.

b. (Implicit Function Theorem). Let $f = (f_1,\ldots,f_n)$ be a C^∞ map from $\mathbb{R}^m$ to $\mathbb{R}^n$ $(m \geq n)$, whose Jacobian matrix $\left(\frac{\partial f_i}{\partial x_j}(p)\right)_{\substack{i=1,\ldots,n \\ j=1,\ldots,m}}$ at the point $p = (p_1,\ldots,p_m) \in \mathbb{R}^m$ has maximal rank n; say, for some permutation of the x_j coordinates, we have

$$\det\left(\frac{\partial f_i}{\partial x_j}(p)\right)_{i,j=1,\ldots,n} \neq 0.$$

Then there is a differentiable map ("implicit function") $g = (g_1,\ldots,g_n)$ defined in a neighborhood V of $(p_{n+1},\ldots,p_m) \in \mathbb{R}^{m-n}$ with values in a neighborhood W of $(p_1,\ldots,p_n) \in \mathbb{R}^n$, such that for all $x \in W \times V \subset \mathbb{R}^m$ we have:

$$f(x) = f(p) \iff x_i = g_i(x_{n+1},\ldots,x_m)$$

for all $i \in \{1,\ldots,n\}$.

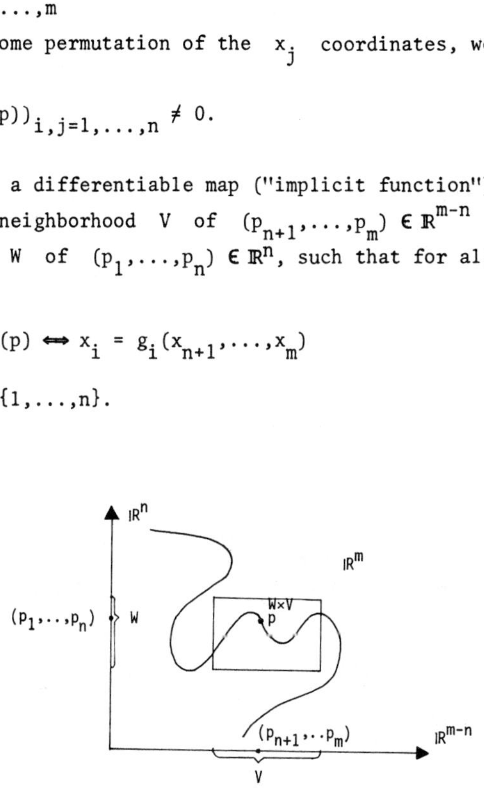

□

A *topological manifold* without boundary is a locally Euclidean, Hausdorff topological space X. By locally Euclidean, we mean that each point has a neighborhood U which is homeomorphic to $\mathbb{R}^n$ (or an open subset V or $\mathbb{R}^n$), for some choice of $n \in \mathbb{N}$. If $u\colon U \to V$ is such a bijective, bicontinuous function, then u is called a *chart* for X.

One might concretely think of geography, where a curved and uneven piece of the earth's surface is mapped onto a flat piece of paper. A chart is also known as a local *coordinate system*, when one wishes to stress the computational point of view. A set A of charts, whose domains of definition form an open covering of X, is called an *atlas*.

The number n, the "local dimension", is constant on each connected component of X. In our applications, n will not vary from component to component; so we may speak of the dimension of the manifold X.

Now let X be an n-dimensional manifold and A an atlas for X. Of geometric and analytic interest is the study of the *coordinate changes* $u \circ v^{-1}$, for two charts $u, v \in A$ whose domains have non-void intersection. The change $u \circ v^{-1}$ is a continuous function from one open subset of $\mathbb{R}^n$ to another. This is trivial, by definition. However, in $\mathbb{R}^n$ one has a much richer structure, which permits to impose further restrictions on the coordinate changes: The atlas A is called a C^∞-*atlas*, if all the coordinate changes are C^∞ maps.

Exercise 2. Show: Each atlas A for X induces on an open subset W of X an atlas $A|W := \{u|W : u \in A\}$. If A is a C^∞-atlas, then so is $A|W$. □

Exercise 3. Show that for each C^∞-atlas A on a n-dimensional manifold X, there is a commutative subalgebra (consisting of "C^∞ functions on X") of the algebra $C^0(X)$ of continuous complex-valued functions on X, namely

$$C^\infty(X) := \{\phi : \phi \in C^0(X) \text{ and } \phi \circ u^{-1} \text{ is a } C^\infty \text{ map from an open subset of } \mathbb{R}^n \text{ to } \mathbb{C}, \text{ for all } u \in A\}.$$

Moreover, show that the usual properties hold; e.g.,:

a) For $\phi \in C^\infty(X)$, we have $\phi|W \in C^\infty(W)$, where $W \subset X$ is open and $C^\infty(W)$ corresponds to the C^∞-atlas $A|W$.

b) Suppose ϕ is a fixed complex-valued function that is "locally" C^∞ (i.e., $\phi|W \in C^\infty(W)$ for all $W \in \mathcal{W}$, where $\mathcal{W}$ is an open covering of X); then $\phi \in C^\infty(X)$.

c) For $\phi \in C^\infty(X)$ and $\psi \in C^\infty(\mathbb{C})$, we have $\psi \circ \phi \in C^\infty(X)$.

d) The constant functions on X are in $C^\infty(X)$. □

A C^∞-*manifold* is a topological manifold X with a "C^∞-structure" $C^\infty(X)$, defined by a C^∞-atlas A.

A C^∞-*map* from a C^∞-manifold X to a C^∞-manifold Y is a function $f : X \to Y$ with $\phi \circ f \in C^\infty(X)$ for all $\phi \in C^\infty(Y)$. One can express this condition in terms of local coordinates as follows: For each u in the atlas for X and v in the atlas for Y, we have $v \circ f \circ u^{-1}$ C^∞, as a map from an open subset of $\mathbb{R}^n$ to $\mathbb{R}^m$, where $n = \dim X$ and $m = \dim Y$.

The C^∞-manifolds X and Y are called *diffeomorphic* when there is a *diffeomorphism* from X to Y; i.e., a bijective C^∞-map $f : X \to Y$ whose inverse is also a C^∞-map. If $X = Y$, one also calls such a map an *automorphism*.

For technical reasons one frequently requires that a C^∞-manifold be *paracompact*; i.e., every covering of X by open subsets $\{U_\iota\}_{\iota \in J}$ possesses a locally finite refinement $\{V_\kappa\}_{\kappa \in J'}$ ("refinement" means each V_κ is an open subset of some U_ι.). Unlike the condition of compactness, we do not require that J' is finite, but rather that the set $\{\kappa : x \in V_\kappa\}$ is finite for each $x \in X$. It is well-known that every metric space is paracompact [Dugundji 1966, 186]. □

Theorem 1. For every open covering $\{U_\iota\}_{\iota \in J}$ of a paracompact C^∞-manifold X, there is a "C^∞ partition of unity"; i.e., a family $\{\phi_\iota\}_{\iota \in J}$ with

(i) $\phi_\iota \in C^\infty(X)$

(ii) $\phi_\iota \geq 0$

(iii) support ϕ_ι (:= closure of $\{x : \phi_\iota(x) \neq 0\}) \subset U_\iota$ for all $\iota \in J$, such that the family $\{\text{support } \phi_\iota\}_{\iota \in J}$ is locally finite, and

(iv) $\sum_{\iota \in J} \phi_\iota(x) = 1$ for all $x \in X$.

Proof: Without loss of generality, we may assume that the covering $\{U_\iota\}_{\iota \in J}$ is locally finite and each U_ι is contained in the domain of a coordinate chart which maps U_ι to a relatively compact subset of $\mathbb{R}^n$ $(n = \dim X)$.

Step 1. We select an open neighborhood W_x about each point $x \in X$, such that the closure $\overline{W}_x$ is contained in some U_ι. We may assume that there is a subset $Y \subset X$, such that $\{W_y\}_{y \in Y}$ is a locally finite covering of X. By defining

$$V_\iota := \bigcup_{y \in Y_\iota} W_y, \quad \text{where} \quad Y_\iota := \{y : W_y \subset U_\iota\},$$

we have a covering $\{V_\iota\}_{\iota \in J}$ with $\overline{V}_\iota \subset U_\iota$ for all $\iota \in J$.

Step 2. For each $\iota \in J$, we now construct a C^∞ function ψ_ι which is positive on $\overline{V}_\iota$ and identically zero on $X - U_\iota$. Then we set $\psi := \Sigma_\iota \psi_\iota$ and $\phi_\iota = \psi_\iota / \psi$. With the help of the coordinate functions, we can carry out the construction in $\mathbb{R}^n$. In the special case when V_ι is mapped to a ball of radius $c < 1$, we may simply take ψ_ι to be the C^∞ function

$$\eta(x) := s(x_1^2 + \ldots + x_n^2), \quad x \in \mathbb{R}^n,$$

where s is the well-known C^∞ function on $\mathbb{R}$ defined by

$$s(r) := \begin{cases} \exp(-1/c-r) & \text{for } r < c \\ 0 & \text{for } r \geq c. \end{cases}$$

More generally, let the image of $\overline{V}_\iota$ be the compact set $K \subset \mathbb{R}^n$, let the image of U_ι be the open set U, and let δ be the distance from K to $\mathbb{R}^n - U$. Cover K with finitely many balls as in the special case, where the size of c plays no role. More precisely, for each $a \in K$, set

$$\mathcal{O}_a := \{x \in \mathbb{R}^n : \eta(\tfrac{x-a}{\delta}) > 0\},$$

so that in particular $a \in \mathcal{O}_a \subset U$. By the compactness of K, we have a finite subset $A \subset K$ such that $K \subset \bigcup_{a \in A} \mathcal{O}_a$. Then one defines ψ_ι to be the function $\Sigma_{a \in A}\, \eta(\frac{\ldots - a}{\delta})$. □

<u>Remark 1</u>. The preceding proof is typical for many set-theoretic and non-constructive arguments in analysis. Actually, one almost always has canonically given charts relative to which an explicit partition of unity can be provided as above.

<u>Remark 2</u>. The C^∞ partition of unity are an important tool to globally piece together locally given data. Although in applications we deal with analytic manifolds as a rule, we will stay in the "category" of C^∞ manifolds, since we do not want to do without this useful C^∞ tool.

C. Geometry of C^∞ Mappings

In elementary differential calculus, many geometrical questions (e.g., the location of extreme values, inflection points, etc.) can be answered by investigating the pointwise-defined linear approximation (i.e., derivative) of the respective function. By means of linear algebra, one can also study C^∞ mappings. The essential concepts for this are:

1. The *directional derivative*. Let x be a point of a C^∞ manifold X, $\phi \in C^\infty(X)$, and $c : \mathbb{R} \to X$ a C^∞ map (a C^∞ curve) with $c(0) = x$. Then the derivative of the function ϕ in the direction of the curve c is defined to be $(\phi \circ c)'(0)$, the derivative of $\phi \circ c$ at 0 in

the sense of elementary calculus. Two such curves are *equivalent,* when the directional derivatives of each function relative to the two curves are the same. In analogy with the usual notation in physics, we denote such an equivalence class by $\dot{c}(0) := \{\tilde{c} : \tilde{c} : \mathbb{R} \to X$ a C^∞ curve with $\tilde{c}(0) = x$ and $(\phi\circ\tilde{c})'(0) = (\phi\circ c)'(0)$ for all $\phi \in C^\infty(X)\}$. Note that $\dot{c}(0)$ only depends on how c is defined near 0.

2. The *tangent space*. The set of directional derivatives, and hence the set of equivalence classes of curves, forms a real vector space, $(TX)_x := \{\dot{c}(0) : c : \mathbb{R} \to X$, a C^∞ curve with $c(0) = x\}$ called the tangent space of X at x. Clearly, the multiplication of the directional derivative $\dot{c}(0)$ by a real number λ is given by a λ-fold increase in the speed; i.e.,

$$\lambda\dot{c}(0) = \dot{\tilde{c}}(0), \text{ where } \tilde{c}(t) := c(\lambda t) \text{ for } t \in \mathbb{R}.$$

Also, for two curves c_1 and c_2, one can find a curve c_3 with

$$c_1(0) = c_2(0) = c_3(0) = x$$

and

$$\dot{c}_1(0) + \dot{c}_2(0) = \dot{c}_3(0)$$

(This is easily seen in $\mathbb{R}^n$, where c_3 can be chosen to be a line.) One can verify that the axioms for a vector space hold, and check that the dimensions of the vector space $(TX)_x$ and the manifold X coincide. For this, one chooses a C^∞ chart $u : U \to \mathbb{R}^n$, from the open neighborhood U of x to an open subset of $\mathbb{R}^n$ ($n = \dim X$). Then, for each positively directed coordinate line through $u(x)$, there is a corresponding C^∞ curve in X; the corresponding directional derivatives are denoted by $\left.\frac{\partial}{\partial u_1}\right|_x, \ldots, \left.\frac{\partial}{\partial u_n}\right|_x$ and form a basis for $(TX)_x$.

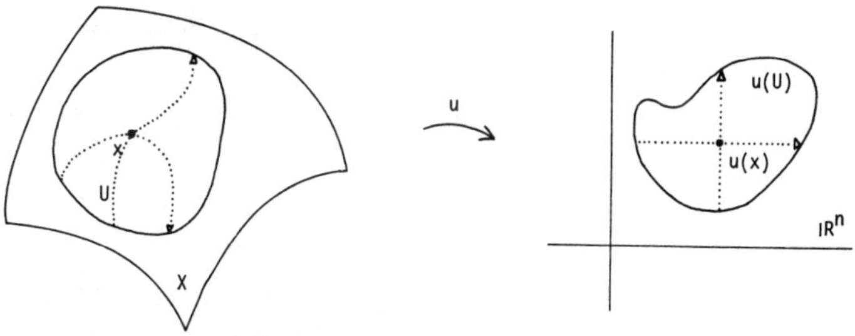

3. The *tangent bundle*. The disjoint union $TX := \cup_{x\in X}(TX)_x$ of tangent spaces has the structure of a real C^∞ vector bundle over X; see Appendix, Exercise 7. A map $s : X \to TX$ is a C^∞ section, exactly when $s(x) \in (TX)_x$ for all $x \in X$ and (for each $\phi \in C^\infty(X)$) the function

$$x \longmapsto \text{derivative of } \phi \text{ at } x \text{ in the direction of } s(x)$$

is in $C^\infty(X)$.

4. The *differential*. A C^∞ map $f : X \to Y$ determines a linear map (the differential of f at x):

$$f_{*|x} : (TX)_x \longrightarrow (TY)_{f(x)}$$
$$\dot{c}(0) \longmapsto \dot{\widehat{f\circ c}}(0),$$

which is given, in terms of coordinates, by the Jacobian matrix $(\partial f_i/\partial u_j)$. Note that $f_* : TX \to TY$ is a bundle map. Moreover, f is called

an *immersion*, if f_* is injective,
an *embedding*, if f and f_* are injective
a *submersion*, if $f_{*|x}$ is surjective at each $x \in X$.

5. *Submanifolds*. A subset $Y \subset X$ is called a submanifold, if Y is a C^∞ manifold with the induced topology and the inclusion $i : Y \to X$ is a (C^∞) embedding. (One can also consider the image sets of immersions as "submanifolds with self-intersections (multiple points)", a concept that we will not pursue further.) For all $y \in Y$, $(TY)_y$ is a linear subspace of $(TX)_y$ in a natural way.

Theorem 2. Let X and Y be C^∞ manifolds of dimensions m and n $(m \geq n)$.

a) ("Definition of manifolds through equations") Let $q \in Y$ and $f : X \to Y$ a C^∞ map with $f_{*|x}$ surjective for all $x \in f^{-1}\{q\}$. Then $f^{-1}\{q\}$ has the structure of a $(m-n)$-dimensional submanifold, in a natural way.

b) (Representation of submanifolds of $\mathbb{R}^n$) If Y is a submanifold of $\mathbb{R}^N$ and $y \in Y$, then (for a certain numbering of the Euclidean coordinates $x_1,\ldots,x_N$), the projection to the space $x_{n+1} = \ldots = x_N = 0$ is a local coordinate system for Y in some neighborhood U of y. Conversely, there is a neighborhood V of y in $\mathbb{R}^N$, such that (within V) Y is the set of points satisfying the following system of equations for certain C^∞ functions g_i:

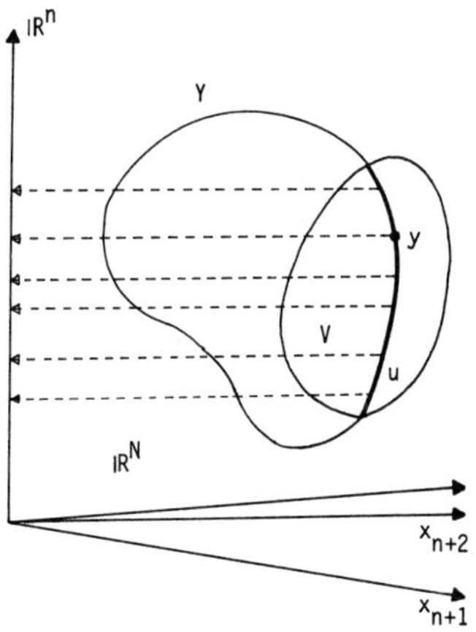

$$g_{n+1}(x_1,\ldots,x_n) - x_{n+1} = 0$$
$$\vdots$$
$$g_N(x_1,\ldots,x_n) - x_N = 0$$

c) (Embedding Theorem) Every compact C^∞ manifold X can be embedded in $\mathbb{R}^N$, for N sufficiently large.

Remark 1. The proof of a) follows without difficulty from the Implicit Function Theorem (Exercise 1b). The complete and elementary proofs for b) and c) can be found in [Wallace 1968, 35-43]. We will not repeat the proofs here, since Theorem 2 is stated only to illustrate some closely related representations of C^∞ manifolds.

Remark 2. One might consider the special cases of part a) of Theorem 2 where the sphere is represented as a level set of the distance function, or where the matrix manifold $SL(n, \mathbb{R})$ is represented as a level set of the determinant function. One can visualize b) with $Y = S^2$ and $N = 3$. For c), we remark that $N = 2mk + m$ suffices, where $m = \dim X$ and k is the number of charts in some atlas (which can be made finite, since X is compact) for X. □

6. The *cotangent bundle*. Let X be a C^∞ manifold with $x \in X$. In place of the tangent space $(TX)_x$, one can consider its dual space $(T^*X)_x$ of linear maps from $(TX)_x$ to $\mathbb{R}$. As an element of $(TX)_x$ can be represented by a curve $c : \mathbb{R} \to X$ with $c(0) = x$, so can an element of $(T^*X)_x$ be represented by a real valued function $\phi \in C^\infty(X)$; two functions ϕ and ψ are considered equivalent, exactly when their

directional derivatives coincide for all curves c (with $c(0) = x$) : $(\phi\circ c)'(0) = (\psi\circ c)'(0)$. One writes $d\phi|_x$ for this equivalence class and calls it the "differential" of ϕ at x. In the terminology of 4), we have $d\phi|_x = \phi_*|_x$. If u is a chart for X in a neighborhood of x, then the differentials $du_1|_x,\dots,du_n|_x$ form a basis for $(T^*X)_x$, where u_i is the i-th coordinate function. One can also give the disjoint union $T^*X = \cup_{x\in X}(T^*X)_x$ a bundle structure, and indeed, T^*X is exactly the "dual bundle" of TX; see Appendix, Exercise 3. For short, T^*X is called the "dual bundle", "covariant bundle", or most commonly, the "cotangent bundle". This is because under a coordinate change $v = \kappa\circ u$, the differentials change "covariantly",

$$dv_i = \sum_{j=1}^{n} \frac{\partial\kappa_i}{\partial x_j}\, du_j,$$

while the tangent vectors transform "contravariantly" by means of the Jacobian of κ^{-1}. From the standpoint of category theory, the cotangent bundle is contravariant, however, since the differential $f_* : TX \to TY$ of $f : X \to Y$ is covariant, while the "*pull back*" $f^* : T^*Y \to T^*X$ (defined when f is a diffeomorphism by $f^*\alpha := \alpha\circ f_{*x}$ for $\alpha \in (T^*Y)_{f(x)}$) is contravariant.

If X is a submanifold of Y with the embedding $f : X \to Y$, then we can easily define $f^* : T^*Y|f(X) \to T^*X$, and then $\operatorname{Ker} f^*$ is a subbundle (of $T^*Y|f(X)$) known as the normal bundle of the embedding.

D. Integration on Manifolds

First, suppose that X is an n-dimensional submanifold of $\mathbb{R}^{n+1}$ (i.e., a "hypersurface"), and moreover assume that X is the boundary of a bounded open subset of $\mathbb{R}^{n+1}$. From the notion of integration on $\mathbb{R}^{n+1}$ where one has a canonical volume element, we have a "surface element" on X, whence integration over X is well-defined. In principle, one can use the same recipe for a compact C^∞ Riemannian manifold X:

1. A C^∞ manifold X is *Riemannian*, if for all $x \in X$ the tangent space $(TX)_x$ is equipped with a fixed Euclidean metric $\langle\cdot\cdot,\cdot\cdot\rangle_x$ (positive, symmetric, nondegenerate bilinear form), such that for two C^∞ sections s_1 and s_2 of the tangent bundle TX, the function $\langle s_1,s_2\rangle$ is in $C^\infty(X)$. With the help of a C^∞ partition of unity (see above Theorem 1), one can furnish every paracompact manifold with a "Riemannian metric". Details are found in [Singer-Thorpe 1967, 134ff/151]. The Euclidean space $\mathbb{R}^N$ induces a Riemannian structure on each

submanifold in a natural way.

2. A C^∞ manifold X is *oriented* when an atlas for X has been chosen such that the Jacobian matrix for each coordinate change is positive (i.e., $\det(\partial\kappa_i/\partial x_j) > 0$, for charts u and v, with $u = \kappa \circ v$ on the intersection of their domains). More simply, without recourse to differential calculus, we can also express this as follows. The bases of a finite-dimensional vector space are divided into two "orientation" classes; two bases belong to the same class, if their transformation matrix has positive determinant. By means of an orientation of a C^∞ manifold, one may select a class of bases from each tangent space in such a way that in a neighborhood of each point the choice is given by a continuous or differentiable choice of basis.

The familiar Möbius band is an example of a non-orientable manifold. A submanifold Y of an orientable manifold X (even of codimension only 1) is therefore not always orientable. On the other hand, X is automatically orientable if it is the boundary of an open submanifold, since then at each point $y \in Y$ one can define a basis of $(TY)_y$ to be positively oriented when a positive basis of $(TX)_y$ is obtained by adjoining an "outward pointing" vector. Since we will only be interested in "bounding" manifolds with classically defined surface elements in most of our applications, we state without proof that with the help of a suitable partition of unity and local charts in which the metric can be expressed in terms of curvilinear coordinates, the concept of integration on $\mathbb{R}^n$ carries over to the case of oriented Riemannian manifolds; orientability is required so that the signs of n-forms will not vary under change of chart - see Appendix, Exercise 8.

Remark 1. For some computations on manifolds it is impractical and confusing to constantly revert back to local coordinates. In such cases "intrinsic" coordinate-invariant concepts of integration are available, and they require in part weaker hypotheses. Always the existence of a C^∞ volume element is essential, or, more generally in modern terminology, the existence of a distinguished n-form where $n = \dim X$. See, e.g., [Singer-Thorpe 1967, 134/150].

Remark 2. We will use here only the above integration aspects of the Riemannian metric, and so do not advance below the surface of Riemannian geometry which unfolds in the great classic theorems on parallel displacement, curvature and rigidity with their varied computations. In fact, there is a close relationship between the integral of the "Gaussian curvature" and the topological form, the "genus" of a surface, and between certain curvature tensors and the Euler characteristic, see below Chapters III.3/4.

E. Differential Operators on Manifolds

Let X be a compact, oriented, C^∞ Riemannian manifold of dimension n. Let E and F be complex C^∞-vector bundles of fiber-dimension N, i.e., families of N-dimensional complex vector spaces E_x resp. F_x (with the parameter x ranging over X) whose disjoint union carries a C^∞ structure in a natural way; see Appendix, Exercise 7b. We denote the complex vector space of C^∞ sections of E by $C^\infty(E)$. For example, if E is the "trivial" product bundle $X \times \mathbb{C}^N$, where we write $\mathbb{C}^N_X$ when we wish to emphasize the bundle point of view, then $C^\infty(E)$ denotes the systems of N complex-valued C^∞ functions on X. (Strictly speaking we should write $C^\infty(\mathbb{C}_X)$ instead of $C^\infty(X)$, but when the context is clear, this is not necessary.)

Since manifolds and vector bundles can be locally described in terms of coordinate functions, the following definition makes sense:

A *differential operator* P (of order k) from E to F is a linear mapping $P : C^\infty(E) \to C^\infty(F)$, which can be represented via local coordinates as a vectorial differential operator (see before Exercise 1.5) in which no derivatives of order $\geq k+1$ appear. We write $P \in \mathrm{Diff}_k(E,F)$. P is called *elliptic*, if all of the locally-defined vectorial differential operators are elliptic (Exercise 1.5); i.e., for each local representation of P over the chart domain U, the "characteristic polynomial of the principal part" (associated to the term of highest order)

$$p_k(x,\xi) := \sum_{|\alpha|=k} a_\alpha(x)\xi^\alpha$$

is an invertible linear map for all x in U and all $\xi \in \mathbb{R}^n - \{0\}$.

For the further treatment of elliptic differential operators, this definition is somewhat opaque, since the "characteristic polynomial" is defined in a piecewise unrelated fashion. Actually, the principal part of a differential operator on a manifold admits a geometric interpretation: Let $T'X$ be the bundle T^*X without the zero section (i.e., the bundle of nonzero covectors), and let $T'(X) \xrightarrow{\pi} X$ be the basepoint map. Then the locally given "characteristic polynomials" collectively define a bundle homomorphism $\sigma(P) : \pi^*(E) \to \pi^*(F)$ via $p_k(x,\xi) : E_x \to F_x$, where $\pi^*(E)$ resp. $\pi^*(F)$ are the bundles "pulled back" to $T'X$:

Exercise 4. Show that $\sigma(P)$ is a well-defined vector bundle homomorphism, which does not depend on the choice of charts and local trivializations.

Hint: Write the mapping in a coordinate-free manner via the formula (see [Wells 1973, 118-119])

$$\sigma(P)(x,v)(e) = \frac{i^k}{k!} P(\phi^k g)|_x,$$

where $x \in X$, $v \in (T^*X)_x$, $v \neq 0$, $e \in E_x$, ϕ is a real-valued C^∞-function on X with $d\phi|_x = v$ and $\phi(x) = 0$, and $g \in C^\infty(E)$ with $g(x) = e$. (Such choices are always possible.) □

In this way, we have defined a linear map $\sigma : \mathrm{Diff}_k(E,F) \to \mathrm{Smbl}_k(E,F)$, where $\mathrm{Smbl}_k(E,F) := \{\sigma \in \mathrm{Hom}(\pi^*E,\pi^*F) : \sigma(x,\lambda v) = \lambda^k\sigma(x,v)$ for all $\lambda > 0$ and $(x,v) \in T'X\}$.

$\sigma(P)$ is called the *symbol* (or also the k-symbol) of the differential operator P. P is elliptic exactly when $\sigma(P)$ is a vector bundle isomorphism. □

What is the meaning of the geometric description of the symbol of an elliptic operator? This is the basic question which will concern us from now on. For a first vague answer, imagine an electron microscope aimed at a point x of our manifold. Under enlargement the neighborhood of x and the vector bundle above it become linear, of course, while the hole in the cotangent bundle becomes large like a sphere. With even greater magnification, we see, instead of a highly complicated differential operator on infinite dimensional spaces of cross-sections (the object of study of functional analysis), families of maps on spheres S^{n-1} into the general linear group $GL(N,\mathbb{C})$ (the object of study of linear algebra and of topology). The Atiyah-Singer Index Formula is - crudely - a manifestation of this change of the planes of investigation.

The term "symbol" recalls the "symbolic method" of Oliver Heaviside, which owes its power to the transition between the plane of operator theory and the plane of polynomial algebra. For this reason it is characterized in [Courant-Hilbert II, 187/518] pointedly as the "separation of the algebraic part from the mathematical-conceptual part."

Initially the relationship between the study of operators and the study of their symbols was penetrated more deeply not by mathematicians but by physicists with the ideas of Erwin Schrödinger and Paul Adrien Maurice Dirac on the "quantization" of classical mechanical systems according to which classical mechanics deals with symbols and quantum mechanics with operators (for quantum mechanical systems with spin, these are differential operators for nontrivial vector bundles). Thus first one considers a problem of classical physics (mechanics, electrodynamics), establishes the classical Hamiltonian function, changes to position- and momentum coordinates, and obtains e.g., for the "harmonic oscillator" the function

$$h(x,p) = \frac{1}{2m} p^2 + \frac{k}{2} x^2.$$

(Interpretation: Consider on the real axis the motion of a particle of mass m with the kinetic energy $\frac{m}{2}\dot{x}^2 = \frac{1}{2m}p^2$, where $x(t)$ is the location at time t and $p = m\dot{x}$ is the momentum. Then one supposes that the particle moves in a force field whose potential energy is $\frac{k}{2}x^2$.) According to the "quantization rule", we now choose a suitable Hilbert space, as a rule an L^2- or Sobolev space (see Ch. 4 below), replace the position coordinate x by multiplication by x and the momentum coordinate p by the differential operator $\frac{\hbar}{i}\frac{d}{dx}$ and we obtain in the Schrödinger representation the operator H with

$$Hf = -\frac{\hbar^2}{2m}\frac{d^2f}{dx^2} + \frac{k}{2}x^2 f,$$

with the 2-symbol

$$\sigma_2(H)(x,\xi) = \frac{\xi^2}{2m}$$

and the "total" symbol (which is not invariant and therefore not defined here)

$$\sigma_{2,1,0}(H)(x,\xi) = \frac{\xi^2}{2m} + \frac{k}{2}x^2.$$

So much about this simple example; see e.g. [Hermann II, 257-293], who strongly advocates this point of view which goes far beyond the scope of our topic. □

Exercise 5. Show that the following sequence of vector spaces

$$0 \to \mathrm{Diff}_{k-1}(E,F) \xrightarrow{j} \mathrm{Diff}_k(E,F) \xrightarrow{\sigma} \mathrm{Smbl}_k(E,F)$$

is exact, where j is the natural inclusion.

Hint: This is clear from the representation in coordinates; see Exercise 4. Replacing $\mathrm{Smbl}_k(E,F)$ by the subspace of "polynomial" symbols (see [Palais 1965, 63]), one gets a surjective symbol map, and the exact sequence may be extended by zero on the right. □

Exercise 6. Show that for $P \in \mathrm{Diff}_k(E,F)$ and $Q \in \mathrm{Diff}_j(F,G)$, the operator QP is in $\mathrm{Diff}_{k+j}(E,G)$ with $\sigma(QP) = \sigma(Q) \circ \sigma(P)$.

Hint: Carry out the proof using the chain rule first for the local vector-valued differential operators (see Chapter 1 above), and then generalize. □

In the following, we assume that the vector bundle E is equipped with a Hermitian metric; i.e., each fiber E_x has a non-degenerate, conjugate-symmetric bilinear form $(\cdot\cdot,\cdot\cdot)_{E_x}$ which is C^∞ in the sense that $(e_1,e_2)_E \in C^\infty(X)$ for any two sections $e_1,e_2 \in C^\infty(E)$. Since X is assumed to be compact, orientable, and Riemannian, we can form the integral $\int_X(e_1,e_2)_E$, obtaining a Hermitian bilinear form on the vector space $C^\infty(E)$. Assume that the vector bundle F is also given a Hermitian metric.

In analogy with Hilbert space theory (see Chapter I.3), we say that the operators $P \in \mathrm{Diff}_k(E,F)$ and $P^* \in \mathrm{Diff}_k(F,E)$ are *formally adjoint*, if $\int_X(Pe,f)_F = \int_X(e,P^*f)_E$ for all sections $e \in C^\infty(E)$ and $f \in C^\infty(F)$.

Exercise 7. Show:

a) There is at most one formally adjoint differential operator P^* for a given P.

b) $(P+Q)^* = P^* + Q^*$, $(Q\circ P)^* = P^* \circ Q^*$, $P^{**} = P$. □

Exercise 8. Show that for each $P \in \mathrm{Diff}_k(E,F)$ there is an adjoint differential operator $P^* \in \mathrm{Diff}_k(F,E)$ such that $\sigma(P^*) = \sigma(P)^*$, where $\sigma(P)^* : \pi^*(F) \to \pi^*(E)$ is the homomorphism pointwise adjoint to $\sigma(P)$.

Hint: 1. Begin with the special case $k = 0$, where $P \in \mathrm{Diff}_0(E,F)$ is given by a vector bundle homomorphism $h : E \to F$ (i.e., a family of linear maps $h_x : E_x \to F_x$ parametrized smoothly by $x \in X$) : $P(e)(x) = h_x(e(x))$ and $\sigma(P)(x,v) = h_x$, where $e \in C^\infty(E)$ and $v \in (T'X)_x$. Let $h_x^* : F_x \to E_x$ be the linear map adjoint to h_x relative to the Hermitian metrics on E_x and F_x; then, for $f \in C^\infty(F)$

$$(P^*f)(x) = h_x^*(f(x)) \quad \text{and} \quad \sigma(P^*)(x,v) = h_x^*.$$

Thus, the statement is proven for this trivial case.

2. For each $\chi \in C^\infty(TX)$ define an operator $P \in \mathrm{Diff}_1(\mathbb{C}_X,\mathbb{C}_X)$ by

$$P\phi := \frac{1}{i}(\chi\cdot\phi), \quad \phi \in C^\infty(X),$$

where $\chi\cdot\phi$ is the function assigning to each point x the derivative of

ϕ in the direction $\chi|_x$. Then we have

$$\sigma(P)(x,v) = v(\chi|_x), \quad \text{where} \quad v \in (T'X)_x.$$

Furthermore, by the Stokes Theorem in the classical Green form (e.g., see [Guillemin-Pollack 1974, 152 and 182-187] or the Cartan calculus in our Appendix, Exercise 8 which we have already used in Exercise 1.7 and can also of course, apply here), we have for all $\phi,\psi \in C^\infty(X)$

$$\int_X (\chi\cdot\phi)\overline{\psi} + \phi\,\overline{\operatorname{div}(\psi\chi)} = 0,$$

where $\operatorname{div}(\psi\chi) \in C^\infty(X)$ is the divergence of the vector field $\psi\chi$. Thus, $P^*\psi = \frac{1}{i}\operatorname{div}(\psi\chi)$. One further checks that $\sigma(P^*)(x,v) = \sigma(P)(x,v) = v(\chi|_x)$, and since $v(\chi|_x)$ is real and hence self-adjoint as a linear map from $\mathbb{C}$ to $\mathbb{C}$, we have $\sigma(P^*) = \sigma(P)^*$.

3. One may now show that every global differential operator can be constructed from the two preceding types via sums and compositions (locally, this is entirely trivial), and thus Exercise 8 reduces to Exercise 7. □

Remark. In contrast to Exercise 7, the solution of the preceding exercise is not so trivial, even though we only applied Stokes' theorem in the weak form. Alternatively, one can first assign to each vector-valued differential operator

$$P(\cdot\cdot) = \sum_{|\alpha|\le k} a_\alpha(x)D^\alpha(\cdot\cdot)$$

$$P : C^\infty(\mathbb{C}^N_U) \to C^\infty(\mathbb{C}^N_U)$$

over an open set $U \subset \mathbb{R}^n$ the operator

$$P^*(\cdot\cdot) := \sum_{|\alpha|\le k} D^\alpha(a^*_\alpha(x)\cdot\cdot),$$

where $a^*_\alpha(x)$ is the adjoint (conjugate transpose) of the $N \times N$ matrix $a_\alpha(x)$. Using integration-by-parts it then follows at once that $\int_U (Pu,v) = \int_U (u,P^*v)$ for all $u,v \in C^\infty_0(\mathbb{C}^N_U)$, where $(\cdot\cdot,\cdot\cdot)$ is the canonical Hermitian scalar product on $\mathbb{C}^N$. The major work consists of globabilizing this result; see [Palais 1965, 70-73], [Narasimhan 1973, 181-183], or [Wells 1973, 121f]. □

Exercise 9. P elliptic $\Rightarrow P^*$ elliptic. □

F. Manifolds with Boundary

Instead of modeling a manifold locally on open subsets of $\mathbb{R}^n$, we can also work with charts that map open subsets of the topological space X homeomorphically onto open subsets of the half-space $\mathbb{R}^n_+ := \{(x_1,\ldots,x_n) \in \mathbb{R}^n : x_1 \geq 0\}$ such that the "coordinate changes" are again C^∞. In this way, we introduce the concept of a C^∞ *manifold with boundary,* in the same way as we have done above for (unbounded) manifolds. We call $x \in X$ an *interior point,* if it has a neighborhood which is mapped by a chart onto an open subset of $\mathbb{R}^n$ (i.e., contained in the interior of $\mathbb{R}^n_+$). On the other hand, if there is a chart mapping x to a point on the boundary of $\mathbb{R}^n_+$, then x is called a *boundary point*; we write ∂X for the set of all boundary points. As examples, the closed solid sphere or torus are three-dimensional C^∞ manifolds with boundaries being the 2-sphere or 2-torus, respectively.

Exercise 10. Let X be a C^∞ manifold with boundary.

a) Carry the concepts $C^\infty(X)$, TX, T*X, orientation, Riemannian metric, etc., over to this case.

b) Construct a C^∞ atlas for ∂X from a C^∞ atlas for X, showing that ∂X is a C^∞ manifold of dimension n-1, when dim X = n. Show that $\partial\partial X = \emptyset$ and that ∂X inherits Riemannian structure and orientation from those on X. □

For short, we write $Y = \partial X$ here. Since each C^∞ path in Y is a C^∞ path in X, we have a canonical embedding of TY into $TX|Y$. Over Y, we have the following diagram of tangent and cotangent bundles

$$\begin{array}{ccc} TY & \cong & T^*Y \\ \cap & & \cap \\ (TX)|Y & \cong & (T^*X)|Y . \end{array}$$

Only the left inclusion is canonical; the other isomorphisms and the right inclusion depend on the choice of a Riemannian metric.

Exercise 11. With the help of a Riemannian metric $(\cdot\cdot,\cdot\cdot)$, define a "normal field" $\nu \in C^\infty(TX|Y)$ such that $(\nu(y),w) = 0$ and $(\nu(y),\nu(y)) = 1$ for all $y \in Y$ and $w \in TY_y$. Show that there are two such normal fields, and characterize the "inner" one through the condition $\nu(y)\cdot\phi \geq 0$ for all real-valued $\phi \in C^\infty(X)$ which are positive except at y. Characterize the "dual normal field" $\nu^* \in C^\infty((T^*X)|Y)$. □

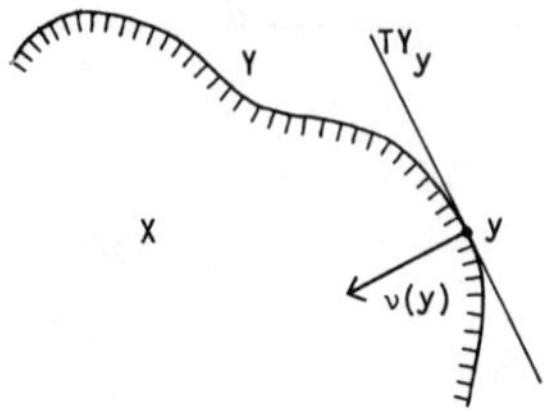

Exercise 12. Let X_1 and X_2 be C^∞ manifolds with boundaries, and let $f : Y_1 \to Y_2$ be a diffeomorphism of their boundaries. Show that one can construct a C^∞ manifold $X_1 \cup_f X_2$ in a canonical way by identifying the boundaries of X_1 and X_2 via f.

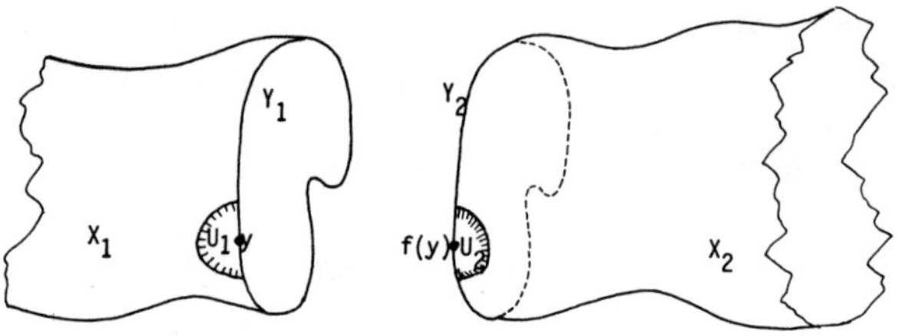

Hint: Form the disjoint union of $X_1 \setminus Y_1$, $X_2 \setminus Y_2$, and $\{(y,f(y)) : y \in Y_1\}$. It is clear what a neighborhood of $x \in X_i - Y_i$ $(i = 1,2)$ should be. For $x = (y,f(y))$, choose a neighborhood U_1 of y in X_1 and a neighborhood U_2 of f(y) in X_2 with $f(U_1 \cap Y_1) = U_2 \cap Y_2$. Then U_1 and U_2 form a neighborhood of $(y,f(y))$. By this recipe, define an atlas for $X_1 \cup_f X_2$ from C^∞ atlases for X_1 and X_2, such that $X_1 \cup_f X_2$ becomes a topological manifold. It is not entirely easy to prove that the coordinate changes are C^∞. Without loss of generality, assume $X_1 = X_2 =: X$ and $f = \mathrm{id}.$; in our applications, we always will have this situation. In order to avoid the difficulties with the "corners" originating from the way in which the charts are joined, choose a Riemannian metric and extend the above-mentioned normal vector field (Exercise 11) to a neighborhood of $\partial X =: Y$. The integral curves of the vector field provide a diffeomorphism ("collar") of $Y \times [0,1]$ with a neighborhood of Y in X. The differentiable "doubling" of $Y \times [0,1]$ along $Y \times \{0\}$ is trivial; see also [Bröcker-Jänich 1973/1982, §§13.5-13.11]. □

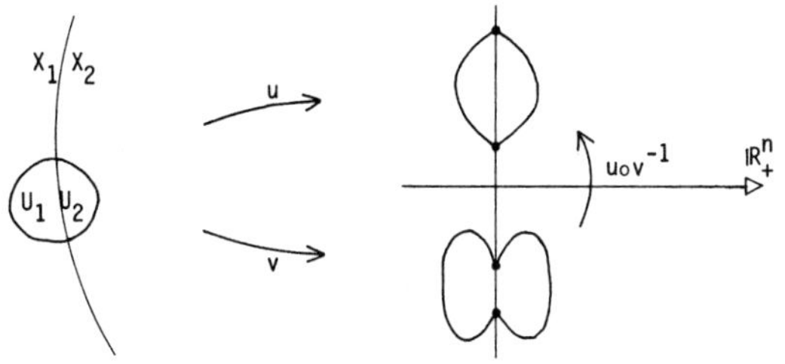

Remark. The definition of differential operators on manifolds with boundary does not require any modification in relation to the case discussed above. The following difference is essential, however: While every differential operator over a "closed" (i.e., compact, without boundary) manifold has a formal adjoint by Exercise 8, one always has an extra term for manifolds with boundary

$$\int_X (Pe,f) - (e,P^*f) = \int_{\partial X} ***,$$

involving a differential operator over the boundary; e.g., see our discussion above on the Sturm-Liouville boundary-value problem (Ch. I.4), or more generally [Palais 1965, 73-75]. We will come back to this problem in Chapter 8. Thus, one of the advantages of the computations on unbounded compact manifolds is the existence of formal adjoints. In passing to the boundary-value problems of interest in applications, Exercise 12 comes into play. More generally, in Chapter 9 and Section III.4.H, we will assign to each boundary-value problem over X "associated" operators over the two closed manifolds ∂X and $X \cup_{\partial X} X$. Incidentally, these are manifolds which can be rather complicated topologically, even in the classical case when X is a bounded domain in $\mathbb{R}^n$. (See the "Heegaard diagrams", whereby each three-dimensional oriented manifold can be represented as such a doubling with boundary diffeomorphism, not necessarily the identity [Seifert-Threlfall 1980, 219].)

3. PSEUDO-DIFFERENTIAL OPERATORS

A. Motivation

We will now turn to a class of operators which - roughly speaking (details below) - are locally presentable in the form

$$(Pu)(x) := \int e^{i<x,\xi>} p(x,\xi)\hat{u}(\xi)d\xi,$$

where

$$\hat{u}(x) := \int e^{-i<x,\xi>} u(\xi)d\xi$$

is the Fourier transform of u (see the crash course above in I.8), p is the "amplitude" of the operator and $<x,\xi> = x_1\xi_1 + \ldots + x_n\xi_n$ its "phase function".

There are a number of reasons why these "pseudo-differential operators" have commanded increasing attention during the last 15 years after the appearance of the pioneering studies by Solomon Grigoryevich Mikhlin on "Singular Integral Equations" (1948). We mention the following not completely disjoint aspects.

1. This class is large enough to contain in addition to the differential operators (Exercise 1) the Green operators (see also I.4 above) and other "singular" integral operators which play a role in solving partial differential equations. In particular this class of pseudo-differential operators contains, with each elliptic operator, its "parametrix", i.e., a quasi-inverse modulo an operator of "low order". In Theorem 5.2 below we will incorporate this operator calculus into Hilbert space theory, and in this fashion, we will be able to derive easily the classical results on elliptic operators (regularity theorems, finiteness of the index), using the elementary theory of Fredholm operators developed in I.1-5. Thereby "some of the techniques used in the case of differential operators appear here as general properties of the class of integro-differential operators considered" (Seeley).

2. The class is small enough, close enough to the differential operators to allow convenient computations. This standpoint is important, particularly because the progress in functional analysis of the past 30 - 50 years permitted the definition of more and more general and involved operator classes and "phantom spaces" (Thom), while the exploration of their properties was too difficult and lagged behind. In contrast, turning

to pseudo-differential operators for which an exact calculus was developed signalled "a trend in the theory of general partial differential equations towards essentially constructive methods" (Hörmander).

3. A special aspect is the attempt to deal with differential operators with variable coefficients, by means of pseudo differential operators "in first approximation", the way differential operators with constant coefficients are treated by means of the Fourier transform: For example, for $f \in C^\infty(\mathbb{R}^n)$ with compact support and $n > 2$, consider the inhomogeneous equation $\Delta u = f$, where $\Delta = \frac{\partial^2}{\partial x_1^2} + \ldots + \frac{\partial^2}{\partial x_n^2}$ is the Laplace operator. With the Fourier transform (see I.8 and the multiplication rule stated there, we obtain $-(x_1^2 + \ldots + x_n^2)\hat{u}(x) = \widehat{\Delta u}(x) = \hat{f}(x)$, i.e., $\hat{u}(x) = -\frac{1}{|x|^2}\hat{f}(x)$, as an L^2-function, and further with the Fourier inversion formula,

$$(Qf)(x) = u(x) = -(2\pi)^{-n}\int_{\mathbb{R}^n} e^{i\langle x,\xi\rangle}|\xi|^{-2}\hat{f}(\xi)d\xi,$$

where Q is the inverse operator (= "fundamental solution") of Δ. In general, suppose P is a differential operator with constant coefficients which can be written as a polynomial $P = p(D)$ where $D = (-i\frac{\partial}{\partial x_1}, \ldots, -i\frac{\partial}{\partial x_n})$, and consider the inhomogeneous equation $p(D)u = f$, $f \in C^\infty(\mathbb{R}^n)$ of compact support. We obtain in the same way at least formally a solution $u = Qf$, where

$$(Qf)(x) := (2\pi)^{-n}\int e^{i\langle x,\xi\rangle}q(\xi)\hat{f}(\xi)d\xi$$

and $q(\xi) := (p(\xi))^{-1}$ is the amplitude. In the process, a number of difficulties arise (Qf is in general not C^∞, and possibly only a distribution; the integral must be interpreted, since the zeros of p can cause divergences) but these can be resolved almost completely. See, for example, [Hörmander 1963, Chapters III and IV].

Now, if - as in Chapter 1 - $U \subset \mathbb{R}^n$ is open and

$$P = p(x,D) = \sum_{|\alpha|\le k} a_\alpha(x)D^\alpha; \quad a_\alpha \in C^\infty(U)$$

is a differential operator with variable coefficients, then all these methods fail initially. But we can, "as a good physicist would" (Atiyah), formally invert the operator P by "freezing" the coefficients at a point $x_0 \in U$ and considering P as a perturbation of $p(x_0,D)$ which is a differential operator with constant coefficients. In this way we obtain as an approximate inverse of P a pseudo differential operator with the amplitude $q(\xi) = p(x_0,\xi)^{-1}$. In order to get a better approximate inverse, it is natural to slowly "thaw" the coefficients, i.e., to let the point x_0 vary in U. This yields the operator

$$(Qf)(x) := (2\pi)^{-n} \int e^{i<x,\xi>} q(x,\xi)\hat{f}(\xi)d\xi$$

with the amplitude $q(x,\xi) = p(x,\xi)^{-1}$ $(x \in U,\ f \in C^{\infty}(X))$ with compact support. This basic "perturbation argument", which was applied in the study of elliptic differential equations by the Italian mathematician Eugenio Elia Levi already in the year 1907, thus finds its theoretical frame with the class of pseudo-differential operators.

We remark (see Theorem 5.2 below) that in the elliptic case an "equally good approximation" is obtained by choosing as amplitude the inverse of the principal part (symbol), i.e., the function $p_k(x,\xi)^{-1}$ which is homogeneous of degree $-k$ in ξ. There is a particularly simple calculus of such operators, since the asymptotic expansions of $p(x,\xi)$ and $q(x,\xi)$ and the underlying iteration - usually necessary - is avoidable here.

4. The theory of pseudo-differential operators allows a certain "relaxing of customary precision", a precision which is "senseless" or at least "exaggerated" in a number of practical problems. Thus, in order to investigate regularity and solvability of the differential equation $Pu = f$, no "fundamental solution", i.e., an actual inverse operator, is needed but - in the framework of Fredholm theory - only a "parametrix", i.e., a "quasi-inverse" modulo certain elementary operators (see above I.5). This has considerable computational advantages: In the case of a differential operator $P = p(D)$ with constant coefficients (as in 3), we may take the amplitude to be

$$q(\xi) := \chi(\xi)p(\xi)^{-1}$$

where χ is a fixed C^{∞}-function which is identically zero in a disk about the origin and identically 1 for large ξ. In this fashion we avoid the delicate convergence problems for the integral

$$(Qf)(x) := (2\pi)^{-n} \int e^{i<x,\xi>} q(\xi)\hat{f}(\xi)d\xi$$

which are required for the amplitude $p(\xi)^{-1}$ because of singularity at the zeros of p. While $PQ = \mathrm{Id}$ and $QP = \mathrm{Id}$ are not valid, we still have

$$PQf = f + Rf \quad (f \in C^{\infty}(U)),$$

where

$$(Rf)(x) := \int r(x-\xi)f(\xi)d\xi$$

and $\hat{r} = \chi - 1$; so $r \in C^{\infty}$, and R is a "smoothing operator". Since very much is known about such simple "correction" or "residue" operators, such a "parametrix" Q serves as well as a true "fundamental solution" for which R vanishes. At any rate, "fundamental solutions" do not usually exist when passing to variable coefficients in the "perturbation method" (sketched in 3) and when replacing operators and their amplitudes by "symbols" (the terms of highest order of the amplitudes, see 5 below). However, the simple computations modulo "smoothing operators" and other "operators of low order" (see 5 below) can be used very efficiently and arise naturally in the theory of pseudo-differential operators.

5. In II.2 (p. 137) we pointed out, in connection with the "symbol", the basic significance of the "change of planes" in passing from operators to the functions which characterize them in approximation. The pseudo-differential operators form a class - and this is tied to the perturbation argument - whose operators can at least in approximation be described by their "amplitudes" and "symbols". The latter are functions satisfying simple rules of computation resulting in a particularly simple approximation theory for the corresponding operators; see for example the composition rules in Theorem 5 below. Surely, mathematicians such as Vito Volterra, Erik Ivar Fredholm, David Hilbert and Friedrich Riesz had this goal in mind when they developed the theory of integral equations as a means of dealing with differential operators. However, starting with the classical representation

$$(Qf)(x) = \int K(x,z)f(z)dz,$$

it turns out that the formulation of the "correct" conditions for the "weights" K are by far not as simple and natural as are those for the "amplitudes" and "symbols" in the representation via Fourier transform. The same is true for the transformation and composition rules (see Theorems 1 and 2 below).

6. From the topological standpoint, the following issues are particularly important: In the "passage between the planes", we prefer to operate with symbols (they are more accessible by topological means)

rather than operators. In Part III below, the decisive advantage of the larger class of pseudo-differential operators will be this: the associated extension of the "symbol space" beyond the polynomial maps to arbitrary C^∞-functions permits lifting homotopies of the symbol of a differential operator to homotopies of the operator itself in the space of pseudo-differential operators. This is generally impossible in the space of differential operators. Using heavier topological machinery, this difficulty can be dealt with without the use of pseudo-differential operators. However, the difficulties which occur are not to be underestimated: For example, not much is known about the "simplest" question of when there is an elliptic system (in $\mathbb{R}^n$) of N differential equations of order k with constant coefficients. The answer for $k = 1$ is: Exactly, when the $(N-1)$-sphere S^{N-1} admits $(n-1)$ linearly independent vector fields; hence, for $N = n$, (according to a famous theorem of John Frank Adams) exactly for the values 2, 4, and 8. More about this is in [Atiyah 1970a].

7. Finally, we point out that the class of pseudo-differential operators originally was developed only in connection with elliptic differential equations, and only there (and with the closely related "hypoelliptic" differential equations) the beautiful properties listed above unfold fully. However, the Swedish mathematician Lars Hörmander and other authors succeeded, in a series of papers, in generalizing the concept of a pseudo-differential operator in such a way that the theory of "Fourier integral operators" so created leads to new results also in heat transfer and wave operators, for example. For this aspect, which we cannot pursue further, see [Hörmander 1971b] and [Taylor 1974, Chapters II, IV, VI and VII]. □

B. "Canonical" Pseudo-Differential Operators

We begin with the definition of the prototypes of our pseudo-differential operators - in local form over the open subset $U \in \mathbb{R}^n$:

$$(Pu)(x) = \int_{\mathbb{R}^n} e^{i\langle x,\xi\rangle} p(x,\xi)\hat{u}(\xi)d\xi,$$

where $x \in U$, $u \in C_0^\infty(U)$, (i.e., $u \in C^\infty(U)$ and the support of u is compact). P is called a *canonical pseudo-differential operator* of order $k \in \mathbb{Z}$, if the amplitude $p \in C^\infty(U \times \mathbb{R}^n)$ satisfies the following "asymptotic conditions" of growth as ξ approaches ∞: For each compact subset $K \subset U$ and multiindices $\alpha = (\alpha_1,\ldots,\alpha_n)$, $\beta = (\beta_1,\ldots,\beta_n) \in (\mathbb{Z}_+)^n$, there is a $C \in \mathbb{R}$ such that for all $x \in K$ and $\xi \in \mathbb{R}^n$, we have

$$|D_x^\beta D_\xi^\alpha p(x,\xi)| \leq C(1 + |\xi|)^{k-|\alpha|} \qquad (*)$$

Here, we recall $D_x^{(0,\ldots,1,\ldots 0)} = -i \frac{\partial}{\partial x_j}$, when "1" stands in the j-th place of the multiindex; $|\alpha| = \alpha_1 + \ldots + \alpha_n$. □

The estimate (*) plays a key role in the derivation of many useful properties of pseudo-differential operators, as in the following

Theorem 1. Each canonical pseudo-differential operator is a linear map from $C_0^\infty(U)$ to $C^\infty(U)$.

Proof: For all $\xi \in \mathbb{R}^n$, the integrand

$$x \longmapsto e^{i\langle x,\xi\rangle} p(x,\xi)\hat{u}(\xi)$$

is obviously a C^∞ function. To show that the function

$$x \longmapsto (Pu)(x) := \int e^{i\langle x,\xi\rangle} p(x,\xi)\hat{u}(\xi)d\xi$$

is also C^∞, we must show that the integral converges "sufficiently well" so that integration and differentiation may be switched. More precisely: By the theorem of Henri Lebesgue, a function which is the limit of a sequence of measurable functions, uniformly bounded by an integrable function, is itself integrable and the "limit" and "integral" may be interchanged. To apply this to our situation, we must show that, for each $x \in U$ and each multiindex β, the function

$$\xi \longmapsto |D_x^\beta p(x,\xi)\hat{u}(\xi)|, \quad \xi \in \mathbb{R}^n$$

can be estimated by an integrable function. Since the support of u is compact, we have that (see Chapter I.8)

$$\xi^\alpha \hat{u}(\xi) = \int e^{-i\langle x,\xi\rangle} D^\alpha u(x)dx,$$

which goes to 0 as $\xi \to \infty$; hence the function $\xi \mapsto |\xi^\alpha \hat{u}(\xi)|$ is bounded for each multiindex α. Thus, $\hat{u}$ decreases faster than any power of $|\xi|$ as ξ goes to ∞; i.e., for each N there is a constant C_1 such that for all $\xi \in \mathbb{R}^n$,

$$|\hat{u}(\xi)| \leq C_1(1 + |\xi|)^{-N}.$$

By assumption, the amplitude is estimated by $|D_x^\beta p(x,\xi)| \leq C_2(1 + |\xi|)^k$. Thus, we have

$$|D_x^\beta p(x,\xi)\hat{u}(\xi)| < C(1 + |\xi|)^{k-N},$$

the right side being integrable for N sufficiently large. □

Exercise 1. Show that the following operators define canonical pseudo-differential operators modulo smoothing operators (i.e., pseudo-differential operators, whose amplitudes have compact support in the second variable; see also Remark 1 following Exercise 2 below).

a) $P = \sum_{|\alpha|\le k} a_\alpha D^\alpha$, where $a_\alpha \in C^\infty(U)$.

b) $(Pu)(x) = \int_{\mathbb{R}^n} K(x,y)u(y)dy$, where $K \in C^\infty(U \times U)$ and support $K(x,\cdot\cdot)$ is compact for all $x \in U$. For example, the convolution $u \to u*\phi$ with $\phi \in C_0^\infty(U)$ is a canonical pseudo-differential operator ($K(x,y) = \phi(x-y)$).

c) The "Riesz operator" $P = \Sigma a_\alpha R^\alpha$, where $a_\alpha \in C_0^\infty(U)$ ($a_\alpha = 0$ for all but finitely many multiindices α) and $(R^\alpha u)(x) := \int e^{i\langle x,\xi\rangle}(\xi/|\xi|)^\alpha \hat{u}(\xi)d\xi$. (See the remark for Exercise 5 below.)

Hint for a): One applies the differential operator P to

$$u(x) = (2\pi)^{-n}\int_{\mathbb{R}^n} e^{i\langle x,\xi\rangle}\hat{u}(\xi)d\xi,$$

and then obtains a pseudo-differential operator with amplitude

$$p(x,\xi) = \sum_{|\alpha|\le k} a_\alpha(x)\xi^\alpha; \quad x \in U, \quad \xi \in \mathbb{R}^n.$$

For brevity, the factors $(2\pi)^{-n}$ are omitted!

For b: Here also, we begin with the Fourier Inversion Formula. One obtains the amplitude

$$p(x,\xi) = \int e^{i\langle y-x,\xi\rangle}K(x,y)dy,$$

which (after factoring out $e^{-i\langle x,\xi\rangle}$) for each fixed x, is the inverse Fourier transform of a function with compact support, and hence is in $C_\downarrow^\infty(\mathbb{R}^n)$. (Argue as in the proof of Theorem 1 where partial integration interchanges multiplication and differentiation.) Thus, the required conditions on the amplitude hold for each $k \in \mathbb{Z}$.

For c: $p(x,\xi) = \chi(\xi)\ \Sigma a_\alpha(x)(\xi/|\xi|)^\alpha$ where $\chi(\xi) \in C^\infty(\mathbb{R}^n)$ is a "cut-off function", i.e., $\chi(\xi) = 0$ for $|\xi| < \rho$, $\rho > 0$, and $\chi(\xi) = 1$ for $\xi > \rho' > \rho$; the order of P is therefore $k = 0$, and different choices of χ lead modulo smoothing functions to the same pseudo-differ-tial operator. One also calls Riesz operators "singular integral operators", since they can be alternatively represented by singular convolutions. For example, if $\alpha = (1,0,\dots,0)$, then one can (up to a constant factor, which we will ignore) write

$$(a_\alpha R^\alpha u)(x) = \lim_{\varepsilon\to 0} \int_{|x-y|>\varepsilon} K(x,x-y)u(y)dy,$$

where

$$K(x,z) := a_\alpha(x) z_1 |z|^{-n-1},$$

whence the weight function K has a singularity at the point $z = 0$. For the connection between Riesz operators, Hilbert transformations, and Wiener-Hopf operators in the case $n = 1$, see Chapter I.9, [Taylor 1974, 36], and [Prössdorf 1972], where an algebra of "pseudo-multiplication operators" in the half-space $\mathbb{R}^n_+$ is investigated; this algebra is formed with the help of pseudo-differential operators and contains the Wiener-Hopf operators. □

Warning: While differential operators are "local" operators (see Exercise 1.1), a pseudo-differential operator can increase supports. For example, if P is defined as convolution with $\phi \in C_0^\infty(U)$ as in Exercise 1b, we can have support Pu = (support u) + support ϕ. The "translation operator" with amplitude

$$p(x,\xi) = (2\pi)^{-n} e^{i\langle x_0,\xi\rangle}, \quad x_0 \text{ fixed},$$

which sends $u(x)$ to $u(x+x_0)$, is *not* a canonical pseudo-differential operator. Actually, the asymptotic amplitude estimate guarantees "pseudo-locality", a kind of locality modulo operators of "lower order", whereby (rather than the support) the "singular support" (the closure of the set where a function is not C^∞) is not increased. Details are found in [Nirenberg 1970, 151ff] or [Palais 1965, 260]. □

Exercise 2. Show that one can write every canonical pseudo-differential operator P as an integral operator

$$\left.\begin{aligned} (Pu)(x) &= \int_U K(x,x-y)u(y)dy \\ &\text{or} \\ (Pu)(x) &= \int_U K_\lambda(x,x-y)(1-\Delta)^\lambda u(y)dy \end{aligned}\right\} \quad u \in C_0^\infty(U),$$

where the weight function $K(x,z)$ or $K_\lambda(x,z)$ is C^∞ for $z \neq 0$.

Hint: Suppose the operator P has order $k \in \mathbb{Z}$ and amplitude $p \in C^\infty(U \times \mathbb{R}^n)$. The case $k < -n$ is easily analyzed: Without loss of generality, suppose $z_1 \neq 0$ and show as in the proof of Theorem 1 (repeated partial integration) that $K(x,z) := (2\pi)^{-n} \int e^{i<\xi,z>} p(x,\xi)d\xi$ is actually C^∞ for $z \neq 0$. $K(x,z)$ is continuous even at $z = 0$, and differentiable there for k sufficiently negative. In the case $k \geq -n$, formally write the integral for $K(x,z)$ as in the first case. The integral need not converge, since one no longer has the estimate

$$|p(x,\xi)| \leq (1 + |\xi|)^{-n-1}.$$

Hence, insert the factor (= 1)

$$(1 + |\xi|^2)^\lambda (1 + |\xi|^2)^{-\lambda}$$

in the integrand, and note that, using

$$e^{i<\xi,z>}(1 + |\xi|^2)^\lambda = (1 - \Delta_z)^\lambda e^{i<\xi,z>}$$

with $\lambda > k+n$, the case $k \geq -n$ reduces to the first case. Here $\Delta_z = -\Sigma_{j=1}^n D_j^2$, $D_j = -i \frac{\partial}{\partial x_j}$ is the usual Laplace operator. Details are found in [Nirenberg 1970, 152]. □

Remark 1. By definition (see before Theorem 1), a canonical pseudo-differential operator, whose amplitude has compact support in the second variable, is of arbitrarily small order ("$k = -\infty$"), and so it can be represented as an integral operator with C^∞ weight function (i.e., a "smoothing operator").

Remark 2. Contrary to the classical notation for integral operators (where the singularities of the weight function lie on the diagonal of $U \times U$), we write $K(x,x-y)$ instead of $K(x,y)$ under the integral, and achieve by this trick that the weight function $K(x,z)$ is singular only at $z = 0$. □

C. Pseudo-Differential Operators on Manifolds

In order to adapt the theory of pseudo-differential operators to our problem of treating elliptic differential equations, first on closed manifolds and then on bordered domains, we must solve two problems.

1. Instead of the weight function K or the amplitude p, we require the notion of *symbol*, a sort of homogeneous main part of the amplitude: For a differential operator $P = \sum_{|\alpha| \leq k} a_\alpha D^\alpha$, the amplitude was the

polynomial function (Exercise 1a)

$$p(x,\xi) = \sum_{|\alpha|\leq k} a_\alpha(x)\xi^\alpha, \quad x \in U, \quad \xi \in \mathbb{R}^n.$$

From this, the symbol $\sigma(P)(x,\xi)$ of P was taken to be the homogeneous polynomial in ξ of order k obtained by taking the sum only over the terms of highest order ($|\alpha| = k$); see Chapter 1. This process of "separation" does not carry over to the amplitude of an arbitrary canonical pseudo-differential operator. Thus, we make the following additional assumptions about the amplitude p of a pseudo-differential operator of order $k \in \mathbb{Z}$: The limit

$$\sigma_k(p)(x,\xi) := \lim_{\lambda\to\infty} \frac{p(x,\lambda\xi)}{\lambda^k} \tag{**}$$

exists for all $x \in U$ and $\xi \in \mathbb{R}^n \setminus \{0\}$. Moreover, for some cut-off function $\chi \in C^\infty(\mathbb{R}^n)$ with $\chi(\xi) = 0$ for $|\xi|$ small and $\chi(\xi) = 1$ for $|\xi| \geq 1$, we assume $p(x,\xi) - \chi(\xi)\sigma_k(p)(x,\xi)$ is the amplitude of a canonical pseudo-differential operator of order $k-1$ (***). Finally, we assume that $p(x,\xi)$ has compact support in the x-variable (****). For us, these last conditions serve only a technical purpose, since we then obtain convergence of integrals and estimates more easily (e.g., see the above hint to Exercise 2b). Actually, one can forgo these conditions and, as in Theorem 2 below, go over to a "Fourier integral operator" with a three-slot amplitude. For applications, we must drop these further assumptions, and we do so for the additional reason that we define our "global pseudo-differential operators" so that they possess amplitudes with compact support only in their "localized form" (see below).

In contrast to the "canonical pseudo-differential operators", whose amplitudes only satisfy the estimate (*) before Theorem 1 above, we now say that P is a "pseudo-differential operator" (with compact support), if the amplitude p of P meets all four conditions (*), (**), (***), and (****).

2. Our second task consists of defining pseudo-differential operators on a paracompact C^∞ manifold X. Thus, consider a linear map

$$P : C_0^\infty(X) \to C^\infty(X),$$

where $C_0^\infty(X)$ again denotes the space of C^∞ functions with compact support. (We will consider operators on sections of vector bundles below in Exercise 4). For each local coordinate system $\kappa : U \to \mathbb{R}^n$ with U

open in X, P yields a "local operator" $P_\kappa u := P(\overline{u\circ\kappa})\circ\kappa^{-1}$, for $u \in C_0^\infty(\kappa(U))$, where

$$\overline{u\circ\kappa} := \begin{cases} u\circ\kappa & \text{on } U \\ 0 & \text{on } X - U. \end{cases}$$

Now, P is called a *pseudo-differential operator of order* k, $P \in \mathcal{P}\mathrm{Diff}_k(X)$, if P_κ is a pseudo-differential operator (with compact support) for all C^∞ charts κ with relatively compact image.

The definition seems to be analogous to the introduction of differential operators on manifolds. Actually, the situation here is somewhat different and more complex, since pseudo-differential operators do not need to be local (see the Warning after Exercise 1), while differential operators may actually be characterized by their locality (i.e., support $Pu \subset$ support u); see Exercise 1.1 and its remark. In particular, we have the following problems:

(i) How "invariant" is the definition of $\mathcal{P}\mathrm{Diff}_k(X)$? Must one actually show that the induced "local operators" are pseudo-differential operators for *all* charts, or is it enough to check this for an atlas? The difficulty lies in the fact that the formation of the "local operators" is not "transitive"; i.e., in general, one obtains two different operators, if one first restricts a chart κ on $U \subset X$ to a open subset $U' \subset U$ obtaining $P_{(\kappa|U')}$ and then considers the restriction of P_κ to $C_0^\infty(\kappa(U'))$. However, it turns out that the difference is a "smoothing operator" of the simple form treated in Exercise 1b.

(ii) With a differential operator P, the amplitude $p(x,\xi)$ and symbol $\sigma(P)(x,\xi)$ can be obtained "intrinsically" from the action of the operator, without explicitly representing it in terms of local coordinates first. Namely, we have (see Exercise 2.4 above) in a local chart

$$\sigma(P)(x,\xi) = \frac{i^k}{k!} P((\phi - \phi(x))^k)(x)$$

and trivially

$$p(x,\xi) = e^{-i\langle x,\xi\rangle} P(\psi e^{i\langle x,\xi\rangle})(x),$$

where ϕ is a real-valued C^∞ function on X with partial derivatives $\left.\frac{\partial\phi}{\partial u_i}\right|_x = \xi_i$, $i = 1,\ldots,n$, and ψ is a C^∞ function with compact support and which is equal to 1 in a neighborhood of x.

For pseudo-differential operators (in general, k is not positive) the first formula does not make sense, and there is no known simple "invariant" formula for the symbol of a pseudo-differential operator; the second formula holds only in an approximate sense (e.g., see [Nirenberg 1970, 152 ff]); the amplitude of a pseudo-differential operator is not unique, but is only "asymptotically" determined by the operator.

Thus, the task of defining a *global* symbol (for pseudo-differential operators defined on the whole manifold X) lies before us.

(iii) For this, we investigate the behavior of the "local operators" and their symbols under a coordinate change, and determine the transformation rule in order to obtain a global symbol. These calculations are somewhat lengthy, since under a coordinate change, the phase $\langle x,\xi\rangle$ and the amplitude $p(x,\xi)$ cannot be directly expressed in the form $\langle y,\eta\rangle$ and $q(y,n)$ in the new coordinates. One can drastically simplify these computations (Theorem 3), as well as the derivation of the composition rules, the formula for the symbol of the adjoint operator (Theorem 5), and the multiplicative properties under "tensor product" (see [Palais 1965, 206-209] and [Hörmander 1971b, 96]) by passing over to an apparently larger operator class.

Theorem 2 (M. Kuranishi, 1969): Let $U \subset \mathbb{R}^n$ be open and $k \in \mathbb{Z}$. We consider operators of the form

$$(Qu)(x) = \iint e^{i\phi(x,y,\xi)} q(x,y,\xi)u(y)\,dy\,d\xi;$$

$$x \in U, \quad u \in C_0^\infty(U),$$

where the "amplitude" $q \in C^\infty(U \times U \times \mathbb{R}^n)$ meets the following conditions (analogous to the conditions on the amplitude of a pseudo-differential operator):

$$|D_\xi^\alpha D_x^\beta D_y^\gamma q(x,y,\xi)| \leq C_{\alpha,\beta,\gamma}(1 + |\xi|)^{k-|\alpha|}. \qquad (*)$$

$$\sigma_k(q)(x,y,\xi) := \lim_{\lambda\to\infty} \frac{q(x,y,\lambda\xi)}{\lambda^k} \quad \text{exists for } \xi \neq 0. \qquad (**)$$

$q(x,x,\xi) - \chi(\xi)\sigma_k(q)(x,x,\xi)$ (with cut-off function χ) is the amplitude of a canonical pseudo-differential operator of order $k-1$ on U. (***)

$q(x,y,\xi)$ has compact support in the x and y variables. (****)

The "phase function" ϕ is defined on $U \times U \times \mathbb{R}^n$, real-valued, linear in the variable ξ, C^∞ for $\xi \neq 0$, and for each fixed x (resp. y) without critical points (y,ξ) (resp. (x,ξ)). Furthermore, assume

$$\frac{\partial\phi}{\partial\xi_1}(x,y,\xi) = \ldots = \frac{\partial\phi}{\partial\xi_n}(x,y,\xi) = 0 \Longleftrightarrow x = y.$$

Conclusion: Q can be written as a pseudo-differential operator (with compact support).

Remark 1. These operators are special types of "Fourier integral operators". The term is due to the Swedish mathematician Lars Hörmander who, in a series of papers, developed for them a precise theory which can be applied to the general theory of partial differential equations. In doing so, he could resort to ideas of the Dutch mathematician and physicist Christian Huygens (1629-1695) and of the contemporary Soviet mathematicians Vladimir Igorevich Arnold, Yuriy Vladimirovich Egorov, and Venyaminovich Clavdiy Maslov, who dealt with fundamentals of geometric optics and the formalization of its more or less intuitive methods ("aggregation principle", "quantization", etc.). Obviously, every pseudo-differential operator with amplitude $p(x,\xi)$ can be written as a "Fourier integral operator" with $q(x,y,\xi) := p(x,\xi)$ and $\phi(x,y,\xi) := \langle x-y,\xi\rangle$.

Proof: Step 0: First we show the convergence of the integral defining $Qu(x)$. As the integral stands, it is only absolutely convergent when the order k of q is very negative. However, the following integral is absolutely convergent for sufficiently large r:

$$\iint e^{i\phi(x,y,\xi)}(L^t)^r(q(x,y,\xi)u(y))\,dy\,d\xi, \quad x \in U,$$

where L^t is the adjoint of the operator

$$L := -i(|\nabla_y\phi|^2 + |\xi|^2||\nabla_\xi\phi||^2)^{-1}((\nabla_y\phi)\nabla_y + (\nabla_\xi\phi)\nabla_\xi)$$

which is a well defined differential operator by (****) with the oscillation-preserving property

$$Le^{i\phi(x,\cdot,\cdot)} = e^{i\phi(x,\cdot,\cdot)}.$$

In this way, the original integral can be replaced by an absolutely-convergent integral via repeated integration by parts, and $Qu(x)$ is well defined. Without difficulty it follows that Q is a linear map from $C_0^\infty(U)$ to $C^\infty(U)$.

Step 1: We now show that on a neighborhood Ω of the diagonal in $U \times U$, one can find a C^∞ map $\Psi\colon \Omega \to GL(n, \mathbb{R})$ such that for all $(x,y) \in \Omega$ and $\xi \in \mathbb{R}^n$ we have

$$\phi(x,y,\psi(x,y)\xi) = \langle x-y,\xi\rangle.$$

By assumption, we can write ϕ in the form

$$\phi(x,y,\xi) = \sum_{j=1}^{n} \phi_j(x,y)\xi_j, \quad \text{where} \quad \phi_j(x,y) := \frac{\partial\phi}{\partial\xi_j}(x,y,\xi),$$

$\xi \neq 0$ arbitrary.

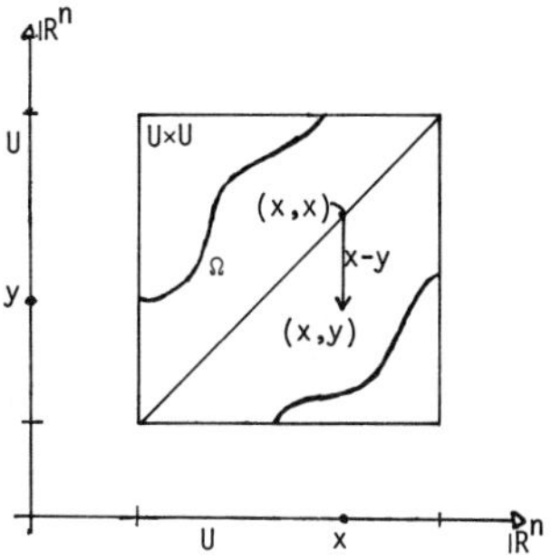

We now show that the functional matrix

$$F(x,y) := \begin{pmatrix} \frac{\partial\phi_1}{\partial x_1}(x,y) & \cdots & \frac{\partial\phi_1}{\partial x_\nu}(x,y) & \cdots & \frac{\partial\phi_1}{\partial x_n}(x,y) \\ \vdots & & & & \vdots \\ \frac{\partial\phi_n}{\partial x_1}(x,y) & \cdots & \frac{\partial\phi_n}{\partial x_\nu}(x,y) & \cdots & \frac{\partial\phi_n}{\partial x_n}(x,y) \end{pmatrix}$$

is invertible for $x = y$. However, this is clear, since (for all $y \in U$) $\phi \mid U \times \{y\} \times (\mathbb{R}^n \setminus \{0\})$ has no critical points, whence

$$\xi \neq 0 \Rightarrow \sum_{\nu=1}^{n} \left|\frac{\partial\phi}{\partial x_\nu}(x,y,\xi)\right| + \sum_{j=1}^{n} \left|\frac{\partial\phi}{\partial\xi_j}(x,y,\xi)\right| \neq 0;$$

i.e., for each $\xi \neq 0$, there is a $\nu \in \{1,\ldots,n\}$ such that

$$\frac{\partial\phi}{\partial x_\nu}(x,y,\xi) = \sum_{j=1}^{n} \xi_j \frac{\partial\phi_j}{\partial x_\nu}(x,y) \neq 0, \quad \text{when} \quad x = y.$$

Thus, the matrix $F(x,x)$ must be invertible, since the rows are linearly independent.

By assumption $\phi_j(x,x) = 0$. Thus we have the short Taylor expansion

$$\phi_j(x,y) = \sum_{\mu=1}^{n} \phi_{\mu j}(x,y)(x_\mu - y_\mu),$$

where the functions $\phi_{\mu j}$ are C^∞ near the diagonal of $U \times U$. The matrix $\Phi(x,y) = (\phi_{\mu j}(x,y))$ is invertible in a neighborhood Ω of the diagonal, because $\Phi(x,x) = F(x,x)^t$. Since

$$\phi(x,y,\xi) = \sum_{j=1}^{n} \phi_j(x,y)\xi_j = \sum_{\mu=1}^{n} (x_\mu - y_\mu) \sum_{j=1}^{n} \phi_{\mu j}(x,y)\xi_j = \langle x-y, \Phi(x,y)\xi \rangle,$$

we have the desired property

$$\phi(x,y,\Psi(x,y)\xi) = \langle x-y,\xi \rangle,$$

where $\Psi(x,y) := \Phi(x,y)^{-1}$ for $(x,y) \in \Omega$.

For later, we note that

$$\Phi(x,x) = \phi''_{x\xi}(x,y,\xi)\big|_{y=x} := \left(\frac{\partial^2 \phi}{\partial x_i \partial \xi_j}(x,y,\xi)\big|_{y=x}\right),$$

whence in particular,

$$\det \Psi(x,x) = 1/\det \phi''_{x\xi}(x,y,\xi)\big|_{y=x}.$$

Step 2: Now we eliminate the phase function ϕ. For this, we first assume that for all ξ,

$$\text{support } q(\cdot\cdot,\cdot\cdot,\xi) \subset \Omega.$$

Then, for all $x \in U$, the integration domain in the formula for $(Qu)(x)$ (see above) is small enough so that the change of variable transformation $\xi = \Psi(x,y)\theta$ can be applied to obtain

$$(Qu)(x) = \int_U \int_{\mathbb{R}^n} e^{i\langle x-y,\theta \rangle} q(x,y,\Psi(x,y)\theta)\,|\det \Psi(x,y)|\,u(y)\,dy\,d\theta.$$

The new amplitude

$$(x,y,\theta) \to a(x,y,\theta)\,|\det \Psi(x,y)|$$

with $a(x,y,\theta) := q(x,y,\Psi(x,y)\theta)$ then automatically satisfies the conditions (**), (***), and (****). To check (*), we must calculate: By the chain rule, we obtain the formula (for $z := (x,y) \in \mathbb{R}^{2n}$)

$$(\partial_z a(z,\theta), \partial_\xi a(z,\theta)) = (\partial_z q(z,\Psi(z)\theta), \partial_\xi q(z,\Psi(z)\theta)) \left(\begin{array}{c|c} \mathrm{Id} & 0 \\ \hline \Psi'(z)\theta & \Psi(z) \end{array}\right),$$

where ∂_z denotes the partial derivatives with respect to the first 2n variables and ∂_ξ denotes those with respect to the last n variables. Hence, we have

$$\left|\frac{\partial a}{\partial \xi_i}(z,\theta)\right| = \left|\sum_{j=1}^{n} \frac{\partial q}{\partial \xi_j}(z,\Psi(z)\theta)\psi^{ij}(z)\right|$$

$$\leq nC_K(1 + |\Psi(z)\theta|)^{k-1} \max_{i,j,z} |\psi^{ij}(z)|,$$

where C_K is a real number from assumption (*) for q when z varies within a compact domain $K \subset U \times U \subset \mathbb{R}^{2n}$. Since we can find positive constants C_1 and C_2 with

$$C_1|\theta| \leq |\Psi(z)\theta| \leq C_2|\theta|$$

for all $z \in K$ and $\theta \in \mathbb{R}^n$, we finally have the estimate

$$\left|\frac{\partial a}{\partial \xi_i}(z,\theta)\right| \leq \tilde{C}_K(1 + |\theta|)^{k-1},$$

where $\tilde{C}_K \in \mathbb{R}$. Similarly one can obtain estimates for the higher derivatives, wherein the factor $|\det \Psi(x,y)|$ of the amplitude can be trivially neglected.

Step 3: Now, we consider the general case, where support $q(\cdot\cdot,\cdot\cdot,\xi)$ is not necessarily contained in Ω. (In our applications of the theorem of Kuranishi - see Theorem 3 below - we are only concerned with a local argument; i.e., we can manage with the case treated in step 2. We will therefore be brief in showing that in general we may assume the first case without loss of generality.) We choose a non-negative C^∞ function χ on $U \times U$ having support in Ω and being equal to 1 in a neighborhood of the diagonal. Then Q can be written as the sum of two "Fourier integral operators", where one has the amplitude χq of the form in step 2, and the other has the form

$$(Ru)(x) = \iint e^{i\theta(x,y,\xi)} r(x,y,\xi)u(y)dyd\xi,$$

where $r = (1-\chi)q$ is a C^∞ function vanishing in a neighborhood of the diagonal in $U \times U$.

Just as in Exercise 2, it follows that R can be written in the form

$$(Ru)(x) = \int_U K(x,y)u(y)dy$$

or

$$(Ru)(x) = \int_U K(x,y)(1-\Delta)^\lambda u(y)\,dy,$$

where the weight function K is C^∞ off the diagonal of $U \times U$ according to Exercise 2, and vanishes in a neighborhood of the diagonal by construction; K is then C^∞ everywhere. By Exercise 1b, R can then be written as a canonical pseudo-differential operator; the corresponding conditions (**), (***), and (****) are met without difficulty.

Step 4: Without loss of generality, we may now assume that the operator Q is given in the form

$$(Qu)(x) = \int_{\mathbb{R}^n}\int_U e^{i<x-y,\xi>} q(x,y,\xi)u(y)\,dy\,d\xi.$$

Then, we have

$$(Qu)(x) = \int e^{i<x,\xi>}\{\int e^{-i<y,\xi>} q(x,y,\xi)u(y)\,dy\}\,d\xi,$$

where the braces enclose the Fourier transform of the product function

$$y \longmapsto q(x,y,\xi)u(y)$$

or - equivalently, by Chapter I.8 - the convolution of the Fourier transforms in the variable ξ:

$$\{\dots\} = (2\pi)^{-n}\int_{\mathbb{R}^n} \hat{q}(x,\xi-\eta,\xi)\hat{u}(\eta)\,d\eta,$$

where $\hat{q}(x,\dots,\xi)$ is the Fourier transform of $y \mapsto q(x,y,\xi)$.

Inserting the factor $e^{i<x,\eta>}e^{-i<x,\eta>}$ and reversing the order of integration, we obtain

$$\begin{aligned}(Qu)(x) &= (2\pi)^{-n}\int_{\mathbb{R}^n} e^{i<x,\eta>}\left(\int_{\mathbb{R}^n} e^{i<x,\xi-\eta>}\hat{q}(x,\xi-\eta,\xi)\,d\xi\right)\hat{u}(\eta)\,d\eta \\ &= \int_{\mathbb{R}^n} e^{i<x,\eta>} p(x,\eta)\hat{u}(\eta)\,d\eta,\end{aligned}$$

where (by a change of variables $\zeta = \xi-\eta$)

$$p(x,\eta) := (2\pi)^{-n}\int_{\mathbb{R}^n} e^{i<x,\zeta>}\hat{q}(x,\zeta,\zeta+\eta)\,d\zeta.$$

We now show that $p(x,\eta)$ is actually the amplitude of a pseudo-differential operator (the support is trivially compact by construction, whence (****) already holds):

(*): Let α and β be multi-indices and let x range over a compact subset of $\mathbb{R}^n$. For the estimation of

$$|D_x^\beta D_\eta^\alpha p(x,\eta)|,$$

we first note that by the Fourier multiplication rule (Chapter I.8), we have

$$|D_x^\beta M_\theta^\gamma D_\eta^\alpha \hat{q}(x,\theta,\eta)| = \left|\int e^{-<y,\theta>} D_x^\beta D_y^\gamma D_\eta^\alpha q(x,y,\eta)\,dy\right| \leq c(1 + |\eta|)^{k-|\alpha|},$$

where γ is a further multi-index and M_θ^γ is multiplication by $\theta^\gamma = \theta_1^{\gamma_1} \ldots \theta_n^{\gamma_n}$; the inequality follows from the assumptions (*) and (****) for q.

Hence, for each positive ν, we have

$$|D_x^\beta D_\eta^\alpha \hat{q}(x,\theta,\eta)| \leq \tilde{C}(1 + |\eta|)^{k-|\alpha|}(1 + |\theta|)^{-\nu}.$$

Thus, by definition of p and by means of differentiation under the integral, we get

$$|D_x^\beta D_\eta^\alpha p(x,\eta)| \leq \tilde{\tilde{C}}(1 + |\eta|)^{k-|\alpha|}.$$

(**): By the Mean-Value Theorem, we obtain for suitable ζ_0 between 0 and ζ

$$p(x,\eta) = (2\pi)^{-n}\int e^{i<\zeta,x>}(\hat{q}(x,\zeta,\eta) + \sum_{|\alpha|=1} D_\eta^\alpha \hat{q}(x,\zeta,\eta+\zeta_0)\zeta^\alpha)\,d\zeta$$

$$= q(x,x,\eta) + \text{a correction term } E(x,\eta).$$

We have already seen in the proof of (*) that

$$|D_\eta^\alpha \hat{q}(x,\zeta,\eta+\zeta_0)| \leq C_\nu(1 + |\eta+\zeta_0|)^{k-1}(1 + |\zeta|)^{-\nu},$$

for arbitrarily large ν. Since $|\zeta_0| \leq |\zeta|$, we have

$$|D_\eta^\alpha \hat{q}(x,\zeta,\eta+\zeta_0)| \leq \tilde{C}(1 + |\eta|)^{k-1}(1 + |\zeta|)^{-\nu+k-1}.$$

By integration as in the proof of (*), we find $|E(x,\eta)| \leq (1 + |\eta|)^{k-1}$. It follows that all of the limits exist in

$$\lim_{\lambda\to\infty} \frac{p(x,\lambda\eta)}{\lambda^k} = \lim_{\lambda\to\infty} \frac{q(x,x,\lambda\eta)}{\lambda^k} + \lim_{\lambda\to\infty} \frac{E(x,\lambda\eta)}{\lambda^k}$$

and

$$\sigma_k(p)(x,\eta) = \sigma_k(q)(x,x,\eta),$$

since the limit on the right is 0.

(***): It follows easily from the estimate for $E(x,\eta)$ and the corresponding assumption for $q(x,x,\eta)$, that $p(x,\eta) - \chi(\eta)\sigma(p)(x,\eta)$ is the amplitude of a canonical pseudo-differential operator of order $k-1$. □

Remark 2. We note that, from the constructive proof of Masatake Kuranishi, a simple formula for the symbol follows from steps 2 and 4:

$$\sigma_k(p)(x,\eta) = \frac{\sigma_k(q)(x,x,\Psi(x,x)\eta)}{|\det \phi''_{x,\xi}(x,y,\xi)|_{y=x}|},$$

where $\Psi(x,x)$ is the inverse of the functional matrix (also denoted by $F(x,x)$) in step 1, namely

$$\phi''_{x,\xi}(x,y,\xi)|_{y=x} = \left(\frac{\partial^2\phi}{\partial x_i \partial \xi_j}(x,y,\xi)|_{y=x}\right).$$

The term Ru in step 3 with amplitude r does not affect the symbol formula, since $r(x,x,\eta) = 0$ for all x and η, whence $\sigma_k(r)(x,x,\eta) = 0$.
□

Now we investigate the behavior of pseudo-differential operators under a coordinate change:

Theorem 3. Let $\kappa : U \to V$ be a diffeomorphism between relatively compact, open subsets of $\mathbb{R}^n$. If P is a pseudo-differential operator (with compact support) of order $k \in \mathbb{Z}$ on V, then the "transported" operator

$$P_\kappa u := P(u \circ \kappa^{-1}) \circ \kappa, \quad u \in C_0^\infty(U),$$

is a pseudo-differential operator (with compact support) of order k over U. If p and q are amplitudes for P and P_κ resp., then their symbols are related by

$$\sigma_k(q)(x,\xi) = \sigma_k(p)(\kappa(x),(\kappa'(x)^t)^{-1}\xi), \quad x \in U, \quad \xi \in \mathbb{R}^n \setminus \{0\},$$

where

$$(\kappa'(x))^t = \begin{pmatrix} \frac{\partial\kappa_1}{\partial x_1}(x) & \cdots & \frac{\partial\kappa_n}{\partial x_1}(x) \\ \vdots & & \vdots \\ \frac{\partial\kappa_1}{\partial x_n}(x) & \cdots & \frac{\partial\kappa_n}{\partial x_n}(x) \end{pmatrix}$$

is the transpose of the functional matrix of κ at x.

Proof: Suppose the operator P is of the form

$$(Pv)(y) = \int e^{i<y,\eta>} p(y,\eta)\hat{v}(\eta)d\eta$$

$$= \iint e^{i<y-\theta,\eta>} p(y,\eta)v(\theta)d\theta d\eta, \quad v \in C_0^\infty(V), \quad y \in V,$$

whence for $v = u\circ\kappa^{-1}$ and $y = \kappa(x)$, $u \in C_0^\infty(U)$ and $x \in U$:

$$(P_\kappa u)(x) = \iint e^{i<\kappa(x)-\theta,\eta>} p(\kappa(x),\eta)u(\kappa^{-1}(\theta))d\theta d\eta$$

$$= \iint e^{i<\kappa(x)-\kappa(\xi),\eta>} p(\kappa(x),\eta)|\det \kappa'(\xi)|u(\xi)d\xi d\eta$$

by means of the change of variable $\kappa(\xi) = \theta$. The "transported" operator is therefore a "Fourier integral operator" with the phase function $\phi(x,\xi,\eta) := <\kappa(x)-\kappa(\xi),\eta>$ and amplitude $q(x,\xi,\eta) := p(\kappa(x),\eta)|\det \kappa'(\xi)|$. Since ϕ and q meet the hypotheses of the Theorem of Kuranishi (Theorem 2), we have

$$(P_\kappa u)(x) = \iint e^{i<x-\xi,\eta>} p(\kappa(x),\Psi(x,\xi)\eta)D(x,\xi)u(\xi)d\xi d\eta,$$

where

$$D(x,\xi) := |\det \kappa'(\xi)| \; |\det \Psi(x,\xi)|$$

and Ψ is the matrix-valued function constructed in step 1 of the proof of Theorem 2; in particular,

$$(\Psi(x,x))^{-1} = \phi''_{x,\eta}(x,\xi,\eta)|_{\xi=x} = \left(\frac{\partial\kappa_j}{\partial x_\nu}\right)_{\substack{j=1,\dots,n\\ \nu=1,\dots,n}} = (\kappa'(\xi))^t$$

and $D(x,x) = 1$.

By the theorem of Kuranishi, P_κ is a pseudo-differential operator for which we derived an explicit formula for the amplitude in the above proof. For the symbol, we have

$$\sigma_k(q)(x,\eta) = \sigma_k(\tilde{q})(x,x,\eta),$$

where

$$\tilde{q}(x,\xi,\eta) = p(\kappa(x),\Psi(x,\xi)\eta)D(x,\xi).$$

Thus by our previous observation:

$$\sigma_k(q)(x,\eta) = \sigma_k(p)(\kappa(x),(\kappa'(x)^t)^{-1}\eta). \qquad \square$$

Exercise 3. Let X be a paracompact C^∞ manifold and $k \in \mathbb{Z}$.

a) Show that the space $\mathcal{P}\mathrm{Diff}_k(X)$, defined by "localization" at the beginning of this section, coincides with the space of pseudo-differential operators of order k on X, when X is a bounded open subset of $\mathbb{R}^n$.

b) Define a canonical vector space structure on $\mathcal{P}\mathrm{Diff}_k(X)$.

c) Show $\mathcal{P}\mathrm{Diff}_k(X) \subset \mathcal{P}\mathrm{Diff}_{k+1}(X)$.

d) Let $\kappa : U \to \mathbb{R}^n$, U open in X, be a local coordinate system for X, $f \in C_0^\infty(U)$, and $P \in \mathcal{P}\mathrm{Diff}_k(X)$. Show that the formula

$$(x,\xi) \longmapsto q_f(x,\xi) := e^{-i<x,\xi>}(P(f(\cdot\cdot)e^{i<\kappa(\cdot\cdot),\xi>}))(\kappa^{-1}(x));$$
$$x \in \kappa(U), \quad \xi \in \mathbb{R}^n,$$

defines the amplitude of a pseudo-differential operator of order k on $\kappa(U)$, and that

$$\sigma_k(q_f)(x,\xi) = (2\pi)^{-n}\sigma_k(p)(x,\xi),$$

when p is the amplitude for the "localized" operator P_κ, and f is identially 1 in a neighborhood of $\kappa^{-1}(x)$.

e) Set $\mathrm{Smbl}_k(X) := \mathrm{Smbl}_k(\mathbb{C}_X,\mathbb{C}_X)$; see Chapter 2 above. Thus, $s \in \mathrm{Smbl}_k(X) :\Leftrightarrow s\colon T'X \to \mathbb{C}$ with $s(x,\lambda v) = \lambda^k s(x,v)$ for all $x \in X$ and $v \in (T^*X)_x$, $v \neq 0$. Show that the linear map

$$\sigma_k : \mathcal{P}\mathrm{Diff}(X) \to \mathrm{Smbl}_k(X)$$

is well defined and coincides with the earlier definition (see Exercise 2.4 above) on $\mathrm{Diff}_k(X) \subset \mathcal{P}\mathrm{Diff}_k(X)$.

Hint: For a: Theorem 3. For b: Proceed by using the vector space structure of $C^\infty(X)$. For c: Use amplitude estimates. For d: One can characterize $\mathcal{P}\mathrm{Diff}_k$ within the space of linear operators from $C_0^\infty(X)$ to $C^\infty(X)$ as those such that, for all local coordinate systems κ and "cut-off functions" f, the associated q_f is the amplitude of a pseudo-differential operator of order k on an open subset of $\mathbb{R}^n$. For details of the computation, see [Hörmander 1971b, 112] and [Nirenberg 1970]. For e: It remains only to show that the locally well defined symbol in d) transforms "correctly" under a coordinate change, so that it forms a global homomorphism from $T'X \times \mathbb{C}$ to $T'X \times \mathbb{C}$ which is homogeneous of degree k in the cotangent vectors. For this, check that the transformation rule in Theorem 3 can be written in the form

$$\sigma_k(q)(x,\kappa^*\eta) = \sigma_k(p)(\kappa(x),\eta),$$

where η lies in $T^*(\mathbb{R}^n)_{\kappa(x)}$, which is the space of covectors at the point $\kappa(x)$ canonically identified with $\mathbb{R}^n$, and $\kappa^*(\eta)$ is the "pull back" covector via κ; see Chapter 2 above; further details are found in [Atiyah-Bott 1967, 404-407].

Warning: In order that σ_k coincide with the previous definition, we must pay regard to the factor $(2\pi)^{-n}$ according to Exercise 1a! Hence, we redefine

$$\sigma_k(P)(x,\eta) := \lim_{\lambda\to\infty} \frac{p(x,\lambda\eta)}{\lambda^k}, \quad \text{if}$$

$$(Pu)(x) = (2\pi)^{-n} \int e^{i<x,\xi>} p(x,\xi)\hat{u}(\xi)d\xi. \quad \square$$

Exercise 4. Define the space $\mathcal{P}Diff_k(E,F)$ when E and F are complex vector bundles over the C^∞ manifold X, and show the existence of a canonical linear map $\sigma_k : \mathcal{P}Diff_k(E,F) \to Smbl_k(E,F)$.

Hint: Represent an operator $P : C_0^\infty(E) \to C^\infty(F)$ locally (choose a chart $\kappa : U \to \mathbb{R}^n$, $U \subset X$ open, and $\kappa(U)$ relatively compact, and trivializations $E|U \cong U \times \mathbb{C}^N$ and $F|U \cong U \times \mathbb{C}^M$) as a $N \times M$ matrix of pseudo-differential operators (with compact support) of order k.

Remark. From a computational point of view the definition of $\mathcal{P}Diff_k(E,F)$ through "localizations" is unsatisfactory. The choice of coordinates is awkward with its avalanche of subscripts which are frequently unavoidable even in fairly simple situations. The method is particularly unsatisfactory, when the operator (as in Exercise 5 below) can be written in closed, global form explicitly and much more clearly. [Hörmander 1971b, 113f] contains the following idea for writing $\mathcal{P}Diff_k(E,F)$ by means of the Kuranishi Theorem directly as a space of "Fourier Integral Operators" with phase function $\phi : G \to \mathbb{R}$ and amplitude $q : G \to Hom(E,F)$: Let G be a real vector bundle of fiber dimension n over a neighborhood of the diagonal in $X \times X$, e.g., $G = \pi^*(T^*X)$ where π is the projection $(x,y) \mapsto y$, and $q(x,y,\xi) \in Hom(E_y,F_x)$. It turns out that one can formulate the necessary conditions on ϕ and q directly, globally and with little difficulty: For example, ϕ is linear in the fibers and the restriction of ϕ to a fiber has a critical point exactly when the fiber lies above a point of the diagonal of $X \times X$. Then (loc. cit.) $\mathcal{P}Diff_k(E,F)$ consists of all operators that can be written as the sum of an operator with C^∞ kernel and one of the form

$$(Pe)(x) := (2\pi)^{-n} \int_{T^*X} e^{i\phi(x,y,\eta)} q(x,y,\eta) e(y)\, dy d\eta$$

where $e \in C_0^\infty(X)$, $dyd\eta$ is the invariant volume element on the covariant covector bundle T^*X, and q an amplitude "of order k" which vanishes for (x,y) outside a small neighborhood of the diagonal of $X \times X$. By step 4 of the proof of Theorem 2, it follows that

$$\sigma_k(P)(x,\eta) = \lim_{\lambda\to\infty} \frac{q(x,x,\lambda\eta)}{\lambda^k}, \quad x \in X, \quad \eta \in (T^*X)_x \setminus \{0\}. \quad \square$$

Exercise 5. Show that the following "singular integral operators" are pseudo-differential operators of order 0 over $\mathbb{R}$ or $S^1 = \mathbb{R}/2\pi\mathbb{Z}$ *, and determine their symbols:

a) The Hilbert transform $Q : C_0^\infty(\mathbb{R}) \to C^\infty(\mathbb{R})$, defined by

$$(Qu)(x) := -\frac{1}{\pi i} (P) \int_{-\infty}^{\infty} \frac{u(y)}{x-y}\, dy, \quad u \in C_0^\infty(\mathbb{R}).$$

b) The projection operator $P : C^\infty(S^1) \to C^\infty(S^1)$, defined by

$$Pe^{im\theta} := \begin{cases} e^{im\theta} & \text{for } m \geq 0 \\ 0 & \phantom{\text{for }} m < 0. \end{cases}$$

c) The "Toeplitz operator" $gP + (\mathrm{Id}-P)$, for $g \in C^\infty(S^1)$.

Hint: For a: Show that $(Qu)(x) = (2\pi)^{-1}\int_{-\infty}^{\infty} e^{ix\xi}\operatorname{sign}(\xi)\hat{u}(\xi)\,d\xi.$

We have

$$\int_{|x-y|>\varepsilon} \frac{u(y)}{x-y}\, dy = \int_{|y|>\varepsilon} \frac{u(x-y)}{y}\, dy = (u * g_\varepsilon)(x)$$

$$= (2\pi)^{-1} \int_{|y|>\varepsilon} e^{ix\xi}\hat{u}(\xi)\hat{g}_\varepsilon(\xi)\, d\xi,$$

where

---

*More precisely: Write them as a sum of a pseudo-differential operator of the kind treated so far and a "smoothing" operator. Many "classical" pseudo-differential operators Q are defined as here via an amplitude q which is homogeneous in the second variable, but has a singularity at the origin. Through multiplication by a C^∞ function χ which is identically 1 in a neighborhood of ∞, we obtain a "singularity-free" amplitude $\tilde{q}(x,\xi) := \chi(\xi)q(x,\xi)$, which defines a pseudo-differential operator $\tilde{Q}$ in our (Hörmander's) sense. Then $\tilde{Q} - Q$ has an amplitude with compact support and consequently (with reasoning as in Remark 1 of Exercise 2) can be represented as an integral operator with a C^∞ weight function.

$$g_\varepsilon(x) := \begin{cases} \frac{1}{x} & \text{for } |x| > \varepsilon \\ 0 & \phantom{\text{for }} |x| < \varepsilon, \end{cases}$$

and for the last equality the well-known convolution formula is used. Thus, it is natural to try to evaluate the improper integral

$$(P)\int_{-\infty}^{\infty} \frac{e^{-i\xi t}}{t}\, dt := \lim_{\varepsilon\to 0} \hat{g}_\varepsilon(\xi).$$

Now distinguish cases according to the sign of ξ! One obtains

$$\hat{g}_\varepsilon(\xi) = -2i\ \mathrm{sign}(\xi) \int_{|\xi|\varepsilon}^{\infty} \frac{\sin t}{t}\, dt.$$

Using function theoretic means prove via integration of the function $f(z) := \frac{e^{iz}}{z}$, $z = t+is$, along the sketched curve, that

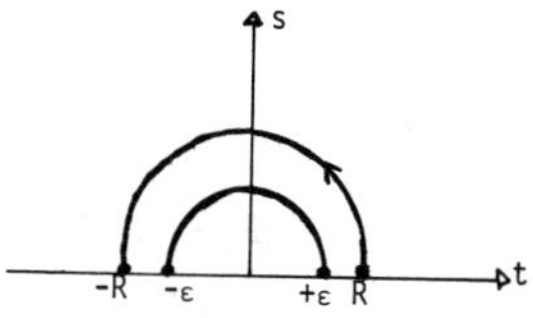

$$(P)\int_0^{\infty} \frac{\sin t}{t}\, dt = \frac{\pi}{2}.$$

Compare also [Dym-McKean 1972, 93 and 150].

<u>For b</u>: Reduce to a) by means of the formula $Pu = \frac{1}{2}Hu + \frac{1}{2}u$, where

$$(Hu)(e^{i\theta}) := \frac{1}{\pi i}(P)\int_{S^1} \frac{u(z)}{z-e^{i\theta}}\, dz, \quad u \in C^\infty(S^1),$$

is the Cauchy-Hilbert transformation (on the circle) which is carried over to the Hilbert transformation Q (on the line $\mathbb{R}$) by means of the Cayley transformation; see Chapter I.8 above. Details for this are in [Taylor 1974, 4-5 and 36].

A more direct way may be found in [Atiyah-Singer 1969a, 525]. For this, as in Exercise 4d form the expression

$$q_f(x,\xi) := e^{-ix\xi} P(f(x)e^{ix\xi}) = \sum_{n=0}^{\infty} \hat{f}(n-\xi)e^{ix(n-\xi)}$$

$$= f(x) - \sum_{n=-1}^{-\infty} \hat{f}(n-\xi)e^{ix(n-\xi)},$$

where $f \in C_0^\infty(\mathbb{R})$ with compact support in an interval of length $< 2\pi$, so that f may be regarded as a function on the circle with support in a "canonical" coordinate domain.

Trick: For $\xi < 0$, estimate $\Sigma_{n=0}^{\infty}$ and its derivatives, showing that for $\xi \to -\infty$ they go to 0 faster than any power of $|\xi|$. Show that for $\xi > 0$, the sum $\Sigma_{n=-1}^{-\infty}$ has the corresponding property. By the hint for Exercise 3d, one is done and obtains

$$\sigma_0(P)(x,\xi) = \begin{cases} 1 & \text{for } \xi > 0 \\ 0 & \text{for } \xi < 0. \end{cases}$$

For c: Reduce to b). Note that in the notation of Chapter I.0

$$gP + (\mathrm{Id}-P) = \begin{cases} T_g & \text{on } C^\infty(S^1) \cap H_0 \\ \mathrm{Id} & \text{on } C^\infty(S^1) \cap H_0^{\perp}\, L_2(S^1), \end{cases}$$

where T_g is the Wiener-Hopf operator induced by g with index $T_g = -W(g,0)$, if g is nowhere zero on S^1. In particular, by Exercise I.2.2,

$$\mathrm{index}(gP + \mathrm{Id}-P) = -W(g,0). \quad \square$$

D. Approximation Theory for Pseudo-Differential Operators

Here we will show that all C^∞ symbols are obtained as symbols of pseudo-differential operators. Moreover, we will show how one can calculate with symbols instead of operators, using some simple rules.

Theorem 4. If E and F are complex vector bundles over a C^∞ manifold X, then the map

$$\sigma_k : \mathcal{P}\mathrm{Diff}_k(E,F) \to \mathrm{Smbl}_k(E,F)$$

is surjective.

Proof: Let $s \in \mathrm{Smbl}_k(E,F)$. In reference to the remark after Exercise 4 it suffices to give a "phase function" $\phi : G \to \mathbb{R}$ and an "amplitude" $a : G \to \mathrm{Hom}(E,F)$ with the properties required by definition (estimates, etc.; see above) such that

$$a(x,y,\eta) = s(x,\eta), \quad x \in X, \quad \eta \in (T^*X)_x \setminus \{0\},$$

where $G = \pi^*(T^*X)$ is a real vector bundle of fiber dimension n and $\pi : X \times X \to X$ is the projection given by $(x,y) \longmapsto y$. For ϕ, we choose a real-valued function with $\phi(x,x,\eta) = 0$ and $d\phi|_{(x,x,\eta)} = \eta \oplus -\eta \oplus 0 : (T^*G)_{(x,x,\eta)} \to \mathbb{R}$, where $(T^*G)_{(x,y,\eta)} \cong (T^*X)_x \oplus (T^*X)_y \oplus (T^*(T^*X))_\eta$ for

$\eta \in (T^*X)_x$. Such a map ϕ, which is also linear on the fiber and only possesses critical points over the diagonal of $X \times X$, is locally easy to construct (relative to a chart κ about the point x, one simply sets $\phi(x,y,\eta) = \langle\kappa x-\kappa y, (\kappa^{-1})^*\eta\rangle$); a global construction may be carried out using a partition of unity (see Theorem 2.1) in a neighborhood of the diagonal.

For "a", we choose an arbitrary extension of $a(x,x,\eta) := s(x,\eta)$ to a neighborhood of the diagonal; then we can smoothly extend "a" by multiplying by a C^∞ function (with support in the neighborhood) which is identically 1 in a smaller neighborhood of the diagonal. An extension can be found, since the diagonal is closed in $X \times X$, and "a" can be regarded as a section of the lift of the bundle $\mathrm{Hom}(E,F)$ by means of the projection $(x,y,\eta) \to (x,y)$. Compare with step 1 of the proof of Theorem 1 in the Appendix, in connection with the Whitney Approximation Theorem; e.g., [Narasimhan 1973, 34f] or [Bröcker-Jänich 1973/1982, §14.8]. With this, the proof (which strongly depends on the Theorem of Kuranishi, Theorem 2) is done. A direct proof can be found in [Wells 1980, 132ff]. □

Theorem 5. The direct sum $\mathcal{P}\mathrm{Diff}(E,F) := \Sigma_k\ \mathcal{P}\mathrm{Diff}_k(E,F)$ forms a graded algebra via composition, and is closed under the operation of taking formal adjoints. For the symbols, we have the following calculation rules:

a) If E, F, and G are complex vector bundles over the C^∞ manifold X, and $P \in \mathcal{P}\mathrm{Diff}_k(E,F)$ and $Q \in \mathcal{P}\mathrm{Diff}_j(E,F)$, then $Q\circ P \in \mathcal{P}\mathrm{Diff}_{j+k}(E,F)$ and

$$\sigma_{j+k}(Q\circ P)(x,\eta) = \sigma_j(Q)(x,\eta) \circ \sigma_k(P)(x,\eta),$$
$$x \in X, \quad \eta \in (T^*X)_x \setminus \{0\}.$$

b) Let $P \in \mathcal{P}\mathrm{Diff}_k(E,F)$, where the bundles E and F are equipped with Hermitian metrics and the manifold X is Riemannian and oriented. Then, there is a unique operator $P^* \in \mathcal{P}\mathrm{Diff}_k(E,F)$ with

$$\int_X \langle Pu,v\rangle_F = \int_X \langle u,P^*v\rangle_E, \quad u \in C_0^\infty(E), \quad v \in C_0^\infty(F);$$

and we have

$$\sigma_k(P^*)(x,\eta) = (\sigma_k(P)(x,\eta))^*, \quad x \in X, \quad \eta \in (T^*X)_x \setminus \{0\}.$$

Proof: We begin with b: The uniqueness of P^* is clear. For the proof of existence, we need only to show that for each $v \in C_0^\infty(F)$ and each open "coordinate domain" $U \subset X$ (with $E|U$ and $F|U$ trivial) there is a $P_U^* v \in C^\infty(E|U)$ such that

$$\int \langle Pu,v\rangle_F = \int \langle u,P_U^* v\rangle_E \quad \text{for all} \quad u \in C_0^\infty(E|U).$$

Indeed, when such a $P_U^* v$ exists, then it is uniquely determined, and so for a second "coordinate domain" U'

$$P_U^* v|U \cap U' = P_{U'}^* v|U \cap U'.$$

Then we have a global C^∞ section $P^*v \in C^\infty(E)$ with

$$\int \langle Pu,v\rangle_F = \int \langle u,P^*v\rangle_E \quad \text{for all} \quad u \in C_0^\infty(E)$$

which is constructed by covering X with finitely many coordinate domains U_j and writing $u = \Sigma u_j$ with $u_j \in C_0^\infty(E|U_j)$ by means of a C^∞ partition of unity.

Hence, let v and U be given. Without loss of generality, we assume that there is a coordinate domain V which contains $\overline{U}$ as well as support(v). (Otherwise one covers support(v) with finitely many coordinate domains and pieces v together from a C^∞ parition of unity). In local coordinates, relative to a C^∞ N-framing for $E|V$ and M-framing for $F|V$, one can write P in the form

$$(Pu)(x) = (2\pi)^{-n} \int_{\mathbb{R}^n} e^{i\langle x,\xi\rangle} p(x,\xi)\hat{u}(\xi)d\xi, \quad u \in C_0^\infty(E|V)$$

where p is an $M \times N$ matrix of amplitude functions with compact support. Hence,

$$\begin{aligned}\langle Pu,v\rangle &= (2\pi)^{-n} \iint e^{i\langle x,\xi\rangle} \langle p(x,\xi)\hat{u}(\xi),v(x)\rangle_{\mathbb{C}^M}\, d\xi dx \\ &= (2\pi)^{-n} \iiint e^{i\langle x-y,\xi\rangle} \langle p(x,\xi)u(y),v(x)\rangle_{\mathbb{C}^M}\, dy d\xi dx \\ &= \int \langle u(y),(2\pi)^{-n} \iint e^{i\langle x-y,\xi\rangle} p^*(x,\xi)v(x)dxd\xi\rangle_{\mathbb{C}^N} dy.\end{aligned}$$

Hence, we obtain

$$\langle Pu,v\rangle = \langle u,P^*v\rangle$$

for

$$\begin{aligned}(P^*v)(y) &:= (2\pi)^{-n} \iint e^{i\langle x-y,\xi\rangle} p^*(x,\xi)v(x)d\xi dx \\ &= (2\pi)^{-n} \iint e^{i\phi(y,x,\xi)} q(y,x,\xi)v(x)dxd\xi\end{aligned}$$

with

$$q(y,x,\xi) := p^*(x,\xi) \quad \text{and} \quad \phi(y,x,\xi) := \langle x-y,\xi\rangle.$$

Here, $p^*(x,\xi)$ denotes the adjoint of $p(x,\xi)$ (i.e., complex-conjugate transpose), an $N \times M$ matrix of complex numbers for each $x \in V$, $\xi \in \mathbb{R}^n$. By the Theorem of Kuranishi, there now exists an amplitude $\tilde{p}$ so that

$$(P^*v)(y) = (2\pi)^{-n} \int e^{i\langle y,\xi\rangle}\tilde{p}(y,\xi)\hat{v}(\xi)d\xi.$$

Hence, we have found a $P^* \in \mathcal{P}\mathrm{Diff}_k(\mathbb{C}_V^N,\mathbb{C}_V^M)$ with the desired property. Also, we have

$$\sigma_k(P^*)(x,\xi) = \lim_{\lambda\to\infty} \lambda^{-k}q(x,x,\lambda\xi)$$

$$= \lim_{\lambda\to\infty} \lambda^{-k}\, p^*(x,\lambda\xi) = (\sigma_k(P)(x,\xi))^*.$$

Furthermore, we remark that with this we obtain (for all $u \in C_0^\infty(\mathbb{C}_V^N)$ and $v \in C_0^\infty(\mathbb{C}_V^M)$)

$$\langle Pu,v\rangle = \int \langle u(y),(2\pi)^{-n} \int e^{i\langle y,\xi\rangle}\tilde{p}(y,\xi)\hat{v}(\xi)d\xi\rangle_{\mathbb{C}^N}dy$$

$$= (2\pi)^{-n} \int \langle \{\int e^{-i\langle y,\xi\rangle}(\tilde{p}(y,\xi)^*u(y)dy\},\hat{v}(\xi)\rangle_{\mathbb{C}^M}\, d\xi.$$

By the Parseval Formula, $\widehat{Pu}(\xi)$ is exactly the expression in the curly braces. Hence, we have the *additional formula*

$$\widehat{Pu}(\xi) = \int e^{-i\langle y,\xi\rangle}(\tilde{p}(y,\xi))^*u(y)dy,$$

where $(\tilde{p}(y,\xi))^*$ is the adjoint matrix of the amplitude $\tilde{p}(y,\xi)$ of the operator P^*.

For a: This time we do not rely on the global representation of P and Q, but rather on their definition by localizations. Without loss of generality, let X be an open, relatively compact subset of $\mathbb{R}^n$. Let p and q be the amplitudes of orders k resp. j belonging to P and Q, respectively. By the preceding "additional formula", we obtain for $u \in C_0^\infty(\mathbb{C}_X^N)$

$$(QPu)(x) = (2\pi)^{-n} \iint e^{i\langle x-y,\xi\rangle}q(x,\xi)(\tilde{p}(y,\xi))^*u(y)dyd\xi,$$

which is thus a pseudo-differential operator (of order $k+j$) by the Theorem of Kuranishi, and

$$\begin{aligned}\sigma_{k+j}(Q\circ P)(x,\eta) &= \lim_{\lambda\to\infty} \frac{q(x,\lambda\eta)(\tilde{p}(x,\lambda\eta))^*}{\lambda^{k+j}} \\ &= \left(\lim_{\lambda\to\infty} \frac{q(x,\lambda\eta)}{\lambda^j}\right)\left(\lim_{\lambda\to\infty} \frac{(\tilde{p}(x,\lambda\eta))^*}{\lambda^k}\right) \\ &= \sigma_j(q)(x,\eta) \circ \sigma_k(p)(x,\eta),\end{aligned}$$

since $\sigma_k(\tilde{p}) = (\sigma_k(p))^*$ by b). □

Remark. The derivation of the formally adjoint operator seems trivial here - in comparison to the lengthy calculation for differential operators; see Exercise 2.8 above. Actually, we have three entirely different problems: By definition, it is trivial that every Fourier-integral operator P possesses a formal adjoint P^*. To prove that P^* is a pseudo-differential operator if P is, we need more (namely, the Theorem of Kuranishi or somewhat long-winded direct computations). It is possible, by the way, to prove the "sharper" result $P \in \mathrm{Diff}_k \Rightarrow P^* \in \mathrm{Diff}_k$ in this fashion, by analyzing carefully the various transformations in the Theorem of Kuranishi with this goal in mind.

4. SOBOLEV SPACES (Crash Course)

A. Motivation

How can we fit the analytic concept of "differential operators" and "pseudo-differential operators" into the function analytic framework of "Hilbert space theory" (see Part I)? There is first of all the L^2-concept of a Lebesgue measurable square integrable function which can be transferred naturally to sections in a Hermitian vector bundle E on a Riemannian manifold X: A section $u : X \to E$ (not necessarily continuous) represents an element of $L^2(E)$ if $\int_X \langle u,u\rangle < \infty$. Here $\langle \ldots,\ldots \rangle$ is a Hermitian metric for the vector bundle E, i.e., $\langle u,u\rangle$ is a complex valued function on X which is integrated with respect to the volume element defined by the Riemannian structure of X (see above Ch. 2, Section G). In this way, $L^2(E)$ becomes a Hilbert space with the usual identifications.

The traditional way of fitting a differential operator $P \in \mathrm{Diff}_k(E,F)$ into the well-understood and powerful Hilbert space theory consists in considering P as a map on $L^2(E)$ to $L^2(F)$ - restricted to functions with sufficient differentiability. We proceeded this way above, when introducing the notion of "formal adjoint operators" (Chapter 2, Section E).

But even restricted to these subspaces, the differential operators are not continuous in the norm topology of L^2. A simple example is the operator d/dt which maps the sequence $\frac{1}{n}\sin nt$ (converging to 0) to the sequence $\cos nt$ which does not converge in L^2. This circumstance leads to the extensive field of classical mathematical research on "unbounded" linear operators.

Thanks to Sergey Lvovich Sobolev, we now have a more potent tool for the definition of Hilbert spaces with various differentiability properties, viz. the Sobolev spaces W^s. They have gained great importance in the theory of partial differential equations, especially for existence questions where precise statements on the "regularity" of solutions are desired which frequently are impossible in the language of C^k-Banach spaces. Since Sobolev spaces are an established part of modern analysis, we may keep it short and refer for the rest to the abundant textbook literature: [Bers-John-Schechter 1964, Chapter III/IV], [Hörmander 1963, 33-63], [Lions-Magenes 1968, 1-118], [Narasimhan 1973, 184-200], [Palais 1965, 125-174], and [Yosida 1965/1974, 55 and 173 ff.].

B. Definition

In the following, we put together some of the various customary equivalent definitions of Sobolev spaces. Here, we restrict ourselves to the "case of functions" ($s \geq 0$). In the framework of distribution theory, the spaces W^s can be treated clearly and uniformally also for $s < 0$; e.g., see [Hörmander 1963], [Lions-Magenes 1968] and [Adams 1975].

Exercise 1. Show that, for $u \in C_0^\infty(\mathbb{R}^n)$, the following norms are equivalent ($m \in \mathbb{N}$):

a) $$|u|_m := \Big(\sum_{|\alpha|\leq m} |D^\alpha u|_0^2\Big)^{1/2}, \text{ where}$$

$$|u|_0^2 := \langle u,u\rangle_0 = \int_{\mathbb{R}^n} u(x)\overline{u(x)}dx$$

and

$$D^\alpha := (-i)^{|\alpha|} \frac{\partial^{\alpha_1}}{\partial x_1^{\alpha_1}} \cdots \frac{\partial^{\alpha_n}}{\partial x_n^{\alpha_n}}, \quad |\alpha| := \alpha_1 + \ldots + \alpha_n.$$

b) $$||u||_m := (2\pi)^{-n/2}|(1 + |\xi|^2)^{m/2}\hat{u}|_0$$

Hint: For a $\Longleftrightarrow$ b, start with the Fourier differentiation formula (see Chapter I.8), giving

$$|u|_m^2 = (2\pi)^{-n} \int_{\mathbb{R}^n} \Big(\sum_{|\alpha| \le m} \xi^{2\alpha} \Big) |\hat{u}(\xi)|^2 d\xi.$$

Then prove

$$(1 + |\xi|^2)^m \le \sum_{|\alpha| \le m} \xi^{2\alpha} \le c(1 + |\xi|^2)^m$$

and deduce

$$||u||_m \le |u|_m \le c^{1/2} ||u||_m. \quad \square$$

Exercise 2. For real non-negative s, define the *Sobolev space*

$$W^s(\mathbb{R}^n) := \{u \in L^2(\mathbb{R}^n) : (1 + |\xi|^2)^{s/2}\, \hat{u} \in L^2(\mathbb{R}^n)\}$$

and show:

a) For all whole numbers s, $W^s(\mathbb{R}^n)$ is the completion of $C_0^\infty(\mathbb{R}^n)$ relative to the (equivalent, by Exercise 1) s-norms.

b) By considering scalar products which induce the respective s-norms, $W^s(\mathbb{R}^n)$ becomes a Hilbert space.

c) The following inclusions are defined in a natural way, and are continuous and dense

$$C_0^\infty(\mathbb{R}^n) \subset W^\infty := \bigcap_{s=0}^{\infty} W^s \subset \ldots \subset W^{s+t} \subset \ldots \subset W^s \subset \ldots W^0 := L^2(\mathbb{R}^n).$$

Hint: For a: Investigate Cauchy sequences in $C_0^\infty(\mathbb{R}^n)$ relative to $|\cdot\cdot|_s$. For b: For whole numbers, this is clear by a. For arbitrary real s compare also [Hörmander 1963, 37 and 45 f.] or [Lions-Magenes 1968, 35-37]. For c: For the inclusions, note the monotonicity of $(1 + |\xi|^2)^{s/2}$ in s. For the proof that $C_0^\infty(\mathbb{R}^n)$ is dense in $W^s(\mathbb{R}^n)$ note that $C_0^\infty(\mathbb{R}^n)$ is dense in $C_\downarrow^\infty(\mathbb{R}^n)$, and then work with the Fourier transform; see also Theorem 1 below. □

Exercise 3. Show that for $t \in \mathbb{R}$, the formally self-adjoint pseudo-differential operator

$$(\Lambda^t u)(x) := (2\pi)^{-n} \int e^{i\langle x,\xi\rangle}(1 + |\xi|^2)^{t/2}\, \hat{u}(\xi) d\xi$$

defines an isomorphism of Hilbert spaces ($s \ge 0$ and $s+t \ge 0$) $W^{s+t}(\mathbb{R}^n) \to W^s(\mathbb{R}^n)$.

Hint: Note that the Parseval Formula (Chapter I.8) implies the equality

$$||u||_t = |\Lambda^t u|_0, \quad u \in W^t(\mathbb{R}^n),$$

and that the family $\{\Lambda^t : t \in \mathbb{R}\}$ forms a group since $\Lambda^t\Lambda^r = \Lambda^{r+t}$. □

Exercise 4. Let X be a compact, oriented, C^∞, Riemannian manifold (without boundary) of dimension n, and let E be a C^∞ Hermitian vector bundle over X of fiber dimension N.

a) For $s \in \mathbb{N} \cup \{0\}$, define the space

$$W^s(E) := \{u \in L^2(E) : \text{for each } P \in \mathrm{Diff}_s(E,E) \text{ there is a } v \in L^2(E) \text{ such that } \langle u,Pw\rangle_0 = \langle v,w\rangle_0 \text{ for all } w \in C^\infty(E)\}.$$

Show that a section $u \in L^2(E)$ lies in $W^s(E)$, exactly when, for each local representation of u in the form $u(x) = \Sigma u_i(x)e_i(x)$ relative to a local chart and local basis $e_1,\ldots,e_N$ of E, we have $\phi u_i \in W^s(\mathbb{R}^n)$, for all C^∞ functions ϕ with support in the domain of the chart.

b) Define $W^s(E)$ for $s \in \mathbb{R}^+$, using this local recipe.

Hint for a: Exercise 1a. Note that v is uniquely determined by P. One says: v arises by "weak application" of the formal adjoint operator P^* on u ("differentiation in the distributional sense").

For b: The crucial point is the independence of the choice of charts, the local trivializations, and the smoothing functions. Take care with the coordinate changes: It is not entirely trivial that each diffeomorphism $\kappa : U \to V$ between open subsets of $\mathbb{R}^n$ induced (via $v \mapsto v\circ\kappa$) an isomorphism $W^s_K(\mathbb{R}^n) \to W^s_{\kappa^{-1}(K)}(\mathbb{R}^n)$, where $K \subset V$ is compact and $W^s_K(\mathbb{R}^n) := \{v \in W^s(\mathbb{R}^n) : \text{support } v \subset K\}$. An elementary proof for this is found in [Hörmander 1963, 57-59]. In our context it is simpler to go back to Theorem 3.3, where we have shown the invariance of $\mathcal{P}\mathrm{Diff}_k$ under a coordinate change (but only for $k \in \mathbb{Z}$). However, the proof goes through smoothly for $k = s \in \mathbb{R}_+$. Then define $W^s(E)$ as in a), where $P \in \mathcal{P}\mathrm{Diff}_s(E,E)$. Instead of coordinate invariance, which is self-evident, one must show, as in a), that one obtains elements of $W^s(\mathbb{R}^n)$ locally. Details are found in [Hörmander 1966b, 169f] or [Nirenberg, 1970, 151ff.]. □

Remark 1. For a fixed choice of atlas, local trivializations of the bundle E, and an appropriate C^∞ partition of unity, one obtains a norm and scalar product which makes $W^s(E)$ a Hilbert space. Beginning with Exercise 1b, on manifolds one may define a Laplace operator Δ which is an elliptic, self-adjoint, positive-definite second order differential operator; for a natural s (and also for real s, via the spectral theorem), we then may explicitly set

$$||u||_s := (\langle\Delta^s u,u\rangle_0)^{1/2}, \quad u \in C^\infty(E).$$

By [Atiyah-Singer 1968a, 511]*, one can proceed in this way, if the vector bundle E is furnished with a "C^∞ connection" ∇ (i.e., a choice of rule for parallel translation of vectors of the bundle E along curves). Such a choice is always possible (as with the choice of a C^∞ partition of unity), and it is often "almost canonically" suggested by the particular construction of X and E. Then, one defines (as in the Newtonian idea of acceleration, where u would be regarded as a velocity vector field when $E = TX$) the "covariant derivative" of $u \in C^\infty(E)$ at the point $c(0)$ in the direction of $\dot{c}(0)$ ($c : I \to X$ a C^∞ curve) by

$$\lim_{t\to 0} \frac{\nabla_t u(c(t)) - u(c(0))}{t},$$

where ∇_t denotes parallel translation from $c(t)$ to $c(0)$ along c. Thus, we obtain the "covariant differentiation" operator

$$D_\nabla : C^\infty(E) \to C^\infty(E \otimes T^*X).$$

By composition with the differential operator D^*_∇ formally adjoint to D_∇, one obtains a "second derivative" in which the dependence on direction is eliminated. Then, set $\Delta := 1 + D^*_\nabla D_\nabla$. For these concepts, see Kobayashi, S. and K. Nomizu: Foundations of Differential Geometry II, Interscience Publishers, New York, 1969, pp. 337-343; or DeRham, G.: Varietés Differentiables, Herman, Paris 1955, pp. 127-132.

Remark 2. In the literature, much more general Sobolev spaces ("Bessel potentials", etc.) are treated. For these, one starts with L^p theory ($p \neq 2$) instead of the Hilbert spaces L^2, and works with "weights" other than our $(1 + |\xi|^2)^{s/2}$. It is interesting that in "the study of classes of differential equations with variable coefficients which are defined by conditions on their highest-order part" (Hörmander) only those W-spaces play a role which are distinguished in a way by their invariance on manifolds ("translation invariance" of L^2 and "diagonalizability" of the derivative by means of Fourier transformation). □

*The idea goes back to E. Magenes: Spazi di interpolazione ed equazioni a derivate parziali, Atti del VII Congresso dell' Unione Mat. Ital. (Genova 1963). Ediz. Cremonese, Rom 1964, 134-197. See also [Lions-Magenes 1968, 42].

Exercise 5. Let $m \in \mathbb{N} \cup \{0\}$.

a) For $\mathbb{R}^n_+ := \{x \in \mathbb{R}^n : x_n \geq 0\}$, define (as in Exercise 4a) the space

$$W^m(\mathbb{R}^n_+) := \{u \in L^2(\mathbb{R}^n_+) : \text{for each differential operator } P \text{ of order } m, \text{ there is } v \in L^2(\mathbb{R}^n_+) \text{ such that } \langle u,Pw\rangle_0 = \langle v,w\rangle_0 \text{ for all } w \in C_0^\infty(\mathbb{R}^n_+)\},$$

and prove that

$$W^m(\mathbb{R}^n_+) = \{u \in L^2(\mathbb{R}^n_+) : \text{there is } v \in W^m(\mathbb{R}^n) \text{ with } v|\mathbb{R}^n_+ = u\}.$$

Show that $W^m(\mathbb{R}^n_+)$ is a Hilbert space.

b) Define the space $W^m(X)$ for a compact, orientable manifold X *with boundary* via "localization", and carry over Exercise 2c.

Hint for a: Set $||u||_m := \inf\{||v||_m : v \in W^m(\mathbb{R}^n) \text{ and } v|\mathbb{R}^n_+ = u\}$. Be careful with restricting to the halfspace: For $m \neq 0$, one must distinguish between $W^m(\mathbb{R}^n_+)$ and $W^m_{\mathbb{R}^n_+}(\mathbb{R}^n)$, the space of W^m functions with support in $\mathbb{R}^n_+$; see [Hörmander 1963, 51-54].

For b: For invariance under diffeomorphism is trivial here, since (without any loss in the applications - see Chapter 8) we only consider whole numbers m. See also [Hörmander 1963, 60f.]. □

C. The Main Theorems on Sobolev Spaces

Here, we discuss briefly (without proofs, for which we refer to the literature) the three main theorems. The content of these results is illustrated in section D on "case studies".

We begin with a regularity theorem showing how one can pass to results in classical form from Hilbert space results given in the language of Sobolev spaces.

Theorem 1 (S. L. Sobolev, 1938): If X is a compact C^∞ manifold (with or without boundary) of dimension n, then $W^s(X) \subset C^k(X)$ for $s > \frac{n}{2} + k$, and the embedding is continuous.

More precisely: An element u of $W^s(X) \subset L^2(X)$ is a class of functions which agree almost everywhere. The theorem means this: In each class $u \in W^s(X)$, there is a representative in $C^k(X)$, and each sequence of elements in $W^s(X)$ which converges in the norm of $W^s(X)$ yields a sequence of representatives in $C^k(X)$ which converges in the norm of the Banach space $C^k(X)$. For $u \in W^s(X)$, we have the *Sobolev inequality*

$$||u||_{C^k} \leq \varepsilon||u||_{W^s} + C_\varepsilon||u||_{L^2},$$

where $\varepsilon > 0$ can be made arbitrarily small, if C_ε is sufficiently large.

Proof: See [Bers-John-Schechter 1964, 167] and [Palais 1965, 159f.] for $X = T^n := S^1 \times \ldots \times S^1$, and [Palais 1965, 169] for the transition to arbitrary X and to sections of vector bundles. For the Sobolev inequality, see [Adams 1975, 75-76/97 ff.], where X is a codimension 0 submanifold (with boundary) of $\mathbb{R}^n$.

In the treatment of boundary value problems with Hilbert space methods, the problem arises that in L^2-spaces a function is only uniquely defined modulo its values on sets of measure zero. Thus, restrictions of functions to the boundary make no sense. However, the following restriction theorem which also goes back to S. L. Sobolev, is helpful (for $Y = \partial X$ and $m = 1$):

Theorem 2. Let X be a compact, C^∞ manifold (possibly with boundary) with a compact submanifold Y of codimension m, and let E be a complex vector bundle over X. Then, for each integer $s > \frac{m}{2}$, the canonical restriction map $C^\infty(E) \to C^\infty(E|Y)$ extends to a continuous, linear, surjective map $\rho : W^s(E) \to W^{s-m/2}(E|Y)$.

Proof: For the hyperplane problem $W^s(\mathbb{R}^n_+) \to W^{s-1/2}(\mathbb{R}^{n-1})$, see [Hörmander 1963, 54f.] or [Lions-Magenes 1968, 38]; for $Y = \partial X$, see [Lions-Magenes 1968, 44-48]; for $X = T^n$ and $Y = T^{n-m}$ and the general case, see [Palais 1965, 161f]. □

Finally, the following lemma, named after Franz Rellich and proved by him in a different formulation, brings in compact operators (see Chapter I.3 ff.), and will furnish a further connection with the Fredholm theory of elliptic operators.

Theorem 3 (F. Rellich, 1930): If X is a compact, C^∞ manifold (possibly with boundary) and E is a complex vector bundle over X, then the inclusion $W^m(E) \to W^s(E)$ is compact for $m > s \geq 0$.

Proof: See [Adams 1975, 144], when X is a codimension 0 submanifold of $\mathbb{R}^n$. For $X = T^n := S^1 \times \ldots \times S^1$, see [Bers-John-Schechter 1964, 169 f.] or [Palais 1965, 158 f.], and [Palais 1965, 168] for the general case. □

D. Case Studies

To illustrate the preceding theorems, we consider some simple special cases:

Theorem 1A. $W^1(\mathbb{R}) \subset C^0(\mathbb{R}^n)$.

Proof: Let $u \in C_0^\infty(\mathbb{R}^n)$. Using the Fourier Inversion Formula (see Chapter I.8 above), we have for $x \in \mathbb{R}$:

$$|u(x)| = (2\pi)^{-1} \left|\int_{-\infty}^{\infty} e^{ix\xi}\hat{u}(\xi)d\xi\right| \leq (2\pi)^{-1}\int_{-\infty}^{\infty} |\hat{u}(\xi)|d\xi$$
$$\leq (2\pi)^{-1} \int_{-\infty}^{\infty} \{|\hat{u}(\xi)|(1 + |\xi|^2)^{1/2}\}(1 + |\xi|^2)^{-1/2}d\xi$$
$$\leq (2\pi)^{-1}\{\int_{-\infty}^{\infty} |\hat{u}(\xi)|^2(1 + |\xi|^2)^1 d\xi\}^{1/2}\{\int_{-\infty}^{\infty}(1 + |\xi|^2)^{-1}d\xi\}^{1/2},$$

where we may use the Schwarz Inequality $|\langle a,b\rangle| \leq (\langle a,a\rangle)^{1/2}(\langle b,b\rangle)^{1/2}$, since $C_0^\infty(\mathbb{R}^n) \subset W^1(\mathbb{R})$ and hence $\xi \mapsto \hat{u}(\xi)(1 + |\xi|^2)^{1/2}$ is in $L^2(\mathbb{R})$, and since $\xi \mapsto (1 + |\xi|^2)^{-1/2}$ also lies in $L^2(\mathbb{R})$ because $\int_{-\infty}^{\infty}(1 + |\xi|^2)^{-1}d\xi < \infty$. The definition in Exercise 1b then yields

$$\sup_{x\in \mathbb{R}} |u(x)| \leq K||u||_{W^1(\mathbb{R}^n)},$$

where the constant K does not depend on u. Since $C_0^\infty(\mathbb{R})$ is dense in $W^1(\mathbb{R})$ (see Exercise 2c), we are done. □

Theorem 2A. We write the n-dimensional torus T^n in the form $\mathbb{R}^n/2\pi\,\mathbb{Z}^n$. Then the restriction map $C^\infty(T^n) \to C^\infty(T^{n-1})$ (induced by the projection $(y,\theta) \mapsto y$ of T^n onto T^{n-1}) extends to a continuous linear map $W^s(T^n) \to W^{s-1/2}(T^{n-1})$, for $s \geq \frac{1}{2}$.

Proof: (after [Palais 1965, 143-162], while previously Peter Lax* had remarked that, in the periodic case, certain technical difficulties vanish):

Step 1: $C^\infty(T^n)$ consists of functions on $\mathbb{R}^n$ which are periodic of period 2π in each variable. The functions

$$e_\nu(x) := (2\pi)^{-n/2}e^{i\langle\nu,x\rangle}, \quad \nu \in \mathbb{Z}^n$$

form a complete orthonormal system for $L^2(T^n)$; see the theory of Fourier series (Chapter I.8 above). By definition (see Exercise 1b), we have $u \in W^s(T^n)$ exactly when $||u||_s^2 := \sum_{\nu\in \mathbb{Z}^n} |\hat{u}(\nu)|^2(1 + |\nu|^2)^s < \infty$, where

$$\hat{u}(\nu) := (2\pi)^{-n/2} \int_{T^n} e^{-i\langle\nu,x\rangle}u(x)dx$$

---

*P. Lax, Comm. Pure Appl. Math. 8 (1955), 615-633.

is the ν-th "Fourier coefficient", $\langle\nu,x\rangle := \nu_1x_1 + \ldots + \nu_nx_n$, and $|\nu|^2 := \nu_1^2 + \ldots + \nu_n^2$. Then the series $\Sigma\hat{u}_\nu e_\nu$ converges absolutely to u in the $W^s(T^n)$ topology.

Step 2: For $x = (y,\theta) \in T^{n-1} \times S^1 = T^n$ and $u \in C^\infty(T^n)$, we have $u(y,\theta) = \Sigma\hat{u}(\lambda,\mu)e_\lambda(y)e_\mu(\theta)$, where the sum is over all $(\lambda,\mu) \in \mathbb{Z}^{n-1} \times \mathbb{Z} = \mathbb{Z}^n$, and the convergence on T^n is uniform. Since $e_\mu(0) = (2\pi)^{-1/2}$, it follows that

$$u(y,0) = \sum_{\lambda\in\mathbb{Z}^{n-1}} e_\lambda(y)\Big(\sum_{\mu\in\mathbb{Z}} \hat{u}(\lambda,\mu)\Big),$$

where the series converges uniformly on T^n. Thus, we have

$$(u|T^{n-1})\hat{}\,(\lambda) = (2\pi)^{-1/2} \sum_{\mu\in\mathbb{Z}} \hat{u}(\lambda,\mu).$$

Step 3: From the Euler-Maclaurin Summation Formula, it follows (see [Palais 1965, 143 f.]) that for $s \geq \frac{1}{2}$, $b \geq 1$ and $a : \mathbb{Z} \to \mathbb{R}_+$, we have

$$\Big(\sum_{\mu\in\mathbb{Z}} a_\mu\Big)^2 b^{s-1/2} \leq C \sum_{\mu\in\mathbb{Z}} a_\mu^2(b + \mu^2)^s,$$

where C only depends on s. We set $a_\mu := \hat{u}(\lambda,\mu)$ and $b := (1 + |\lambda|^2)$, and then obtain

$$\Big(\sum_{\mu\in\mathbb{Z}} |\hat{u}(\lambda,\mu)|\Big)^2 (1 + |\lambda|^2)^{s-1/2} \leq C \sum_{\mu\in\mathbb{Z}} |\hat{u}(\lambda,\mu)|^2(1 + |\lambda|^2 + \mu^2)^s.$$

Thus, with the above Step 2, we have

$$||(u|T^{n-1})||_{s-1/2} \leq \frac{C}{2\pi} ||u||_s,$$

and we are done, since $C^\infty(T^n)$ is dense in $W^s(T^n)$. □

The following case study gives insight into the possible loss of differentiability under restrictions of Sobolev spaces, in contrast to the gain of differentiability in the C^k theory:

Theorem 2B. There is no continuous, linear map $W^s(\mathbb{R}^n) \to W^s(\mathbb{R}^{n-1})$ which extends the restriction map $C_0^\infty(\mathbb{R}^n) \to C_0^\infty(\mathbb{R}^{n-1})$ defined by $u \mapsto u(.,0)$.

Proof: On the ball $K^n := \{x \in \mathbb{R}^n : |x| \leq 1\}$, the function $|x|^\alpha$ is integrable, if $\alpha > -n$, since then in polar coordinates, we have

$$\int_{K^n} |x|^\alpha dx \le C \int_0^1 r^{\alpha+n-1} dr < \infty.$$

Now consider the function $u(x) := |x|^\alpha \chi(x)$, where χ is a C^∞ function with compact support and $\chi(x) = 1$ for all $x \in K^n$. For $\alpha = -\frac{1}{2}$ and $n = 2$, we have $u \in L^2(\mathbb{R}^2)$, but $u(.,0) \notin L^2(\mathbb{R})$, since $\int_0^1 x^{-1}dx = \infty$. Thus, the theorem is verified for $s = 0$, since there is a sequence $\{u_\nu\}_{\nu=1}^\infty$ with $u_\nu \in C_0^\infty(\mathbb{R}^2)$ which converges in $W^0(\mathbb{R}^2)$ $(= L^2(\mathbb{R}^2))$ to u, but $\{u_\nu(.,0)\}_{\nu=1}^\infty$ is not a Cauchy sequence in $W^0(\mathbb{R})$ $(= L^2(\mathbb{R}))$.

One can also easily construct counter-examples for $s > 0$, since by the above argument it follows that the function u (defined there) lies in $W^1(\mathbb{R}^n)$ exactly when $2\alpha > 2-n$ (see Exercise 1a). For example, $u(x) := |x|^{-1/4}\chi(x)$ is an element of $W^1(\mathbb{R}^3)$, but $u(.,.,0) \notin W^1(\mathbb{R}^2)$ and $u(.,0,0) \notin W^1(\mathbb{R})$. If a sequence $u_\nu \in C_0^\infty(\mathbb{R}^3)$ converges to $u \in W^1(\mathbb{R}^3)$ and the restrictions $u_\nu(.,0,0)$ were to converge in $W^1(\mathbb{R})$, then the limit in $W^1(\mathbb{R})$ would also be in $C^0(\mathbb{R})$ by Theorem 1A; this contradicts the form of u. □

Theorem 3A. Without the assumption that X is compact, Theorem 3 above is false.

Proof: For each $\nu \in \mathbb{N}$, one constructs a $u_\nu \in W^1(\mathbb{R})$ with $||u_\nu||_1 \le 3$, as in the picture. However, we have $||u_\nu - u_\mu||_0 \ge k|\nu-\mu|$, where k is independent of ν and μ. The u_ν lie in a bounded set of $W^1(\mathbb{R})$, but there is no subsequence convergent in $W^0(\mathbb{R}) = L^2(\mathbb{R})$. □

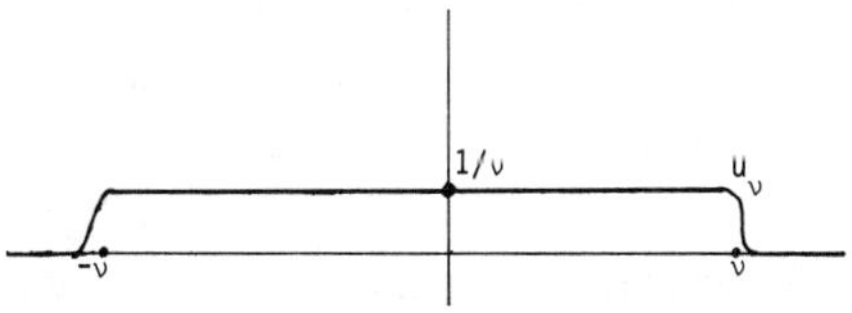

5. ELLIPTIC OPERATORS OVER CLOSED MANIFOLDS

In this chapter, we show that out of formal properties (invertibility) of symbols, a series of existence, regularity, and finiteness results for the associated pseudo-differential operators (Fredholm operators) can be obtained.

A. Continuity of Pseudo-Differential Operators

We denote by $OP_k(E,F)$ the set of *operators of order* k from E to F, where E and F are complex vector bundles over the C^∞ manifold X and $k \in \mathbb{Z}$: A linear operator $P : C_0^\infty(E) \to C^\infty(F)$ lies in $OP_k(E,F)$ exactly when it extends to a continuous map $P_s : W^s(E) \to W^{s-k}(F)$ for all $s \in \mathbb{R}$ with $s, s-k \geq 0$.

Convention: In what follows, the manifold X is "closed", i.e., compact, without boundary. We make this convention, in part for convenience (in order to make some proofs go easier), but also because otherwise some of the following theorems would be meaningless or false; see Exercise 10 below. Moreover, X continues to be oriented and is furnished with a fixed Riemannian metric; E and F are Hermitian vector bundles. Without loss of generality, we will occasionally assume that the "Hilbertable" Sobolev spaces are already furnished with a fixed norm or scalar product.

Theorem 1. $\mathcal{P}\mathrm{Diff}_k(E,F) \subset OP_k(E,F)$, $k \in \mathbb{Z}$.

Proof: Formally, this theorem says that the "analytical order" of a pseudo-differential operator (determined by the asymptotic behavior of its amplitude) coincides with its "functional analytic order" (which is expressed by its continuity relative to the norms of the Sobolev spaces). We need to prove the estimate $||Pu||_{s-k} \leq C||u||_s$, $u \in C^\infty(E)$, where C only depends on $P \in \mathcal{P}\mathrm{Diff}_k(E,F)$ and not on u. Since $C^\infty(E)$ is dense in $W^s(E)$, the theorem follows, because then P can be extended to a continuous linear operator $P_s : W^s(E) \to W^{s-k}(F)$.

Since the norms in $W^s(E)$ and $W^{s-k}(F)$ are locally defined by Exercise 4.4, it suffices to show the inequality for $u \in C_0^\infty(\mathbb{R}^n)$. By Exercise 4.3 we can further assume without loss of generality that $k = 0$ and $s = 0$. Let

$$(Pu)(x) = (2\pi)^{-n} \int e^{i\langle x,\xi\rangle} p(x,\xi)\hat{u}(\xi)d\xi, \quad u \in C_0^\infty(\mathbb{R}^n),$$

be a pseudo-differential operator, whose amplitude $p(x,\xi)$ vanishes for sufficiently large x and satisfies the estimate (see above)

$$|D_\xi^\alpha D_x^\beta p(x,\xi)| \le c(1 + |\xi|)^{k-|\alpha|}, \quad (x,\xi) \in \mathbb{R}^n \times \mathbb{R}^n,$$

for all multi-indices α and β. Then the Fourier transform $\hat{p}(\cdot,\xi)$ of the function $x \mapsto p(x,\xi)$ can be estimated by $|\hat{p}(z,\xi)| \le C_N(1 + |z|)^{-N}$ for all $N \in \mathbb{N}$, as was done for $\hat{u}$ in the proof of Theorem 3.1.

By the Parseval Formula (Chapter I.8), we have

$$||Pu||_0 = \langle Pu,Pu\rangle_0^{1/2} = (2\pi)^{-n}\langle \widehat{Pu},\widehat{Pu}\rangle_0^{1/2} \le (2\pi)^{-n}\tilde{c}||(1+|\xi|)^{-N}||_0||\hat{u}||_0$$
$$\le C||u||_0$$

for N sufficiently large; here, we have

$$\widehat{Pu}(\eta) = (2\pi)^{-n} \iint e^{-i\langle\eta,x\rangle} e^{i\langle x,\xi\rangle} p(x,\xi)dx\, \hat{u}(\xi)d\xi$$
$$= (2\pi)^{-n} \int \hat{p}(\eta-\xi,\xi)\hat{u}(\xi)d\xi,$$

whence $|\widehat{Pu}(\eta)| \le \tilde{\tilde{c}} \int (1+|\eta-\xi|)^{-N}|\hat{u}(\xi)|d\xi$. Apply the Schwarz Inequality to the integral $\int(1+|\eta-\xi|)^{-N/2}\{(1+|\eta-\xi|)^{-N/2}|\hat{u}(\xi)|\}d\xi$ and integrate with respect to η. □

<u>Exercise 1</u>. Interpret and prove:

$$P \in \mathcal{P}\mathrm{Diff}_0(E,F) \Rightarrow (P_0)^* = (P^*)_0. \quad □$$

<u>Exercise 2</u>. Let $\sigma_k : \mathcal{P}\mathrm{Diff}_k(E,F) \to \mathrm{Smbl}_k(E,F)$ be the well-defined (Exercise 3.3e), surjective (Theorem 3.4) symbol map. Show $\ker \sigma_k \subset OP_{k-1}(E,F)$.

<u>Hint</u>: This is trivial by axiom (***) (see Section 3.D above) and the preceding Theorem 1. □

<u>Exercise 3</u>. Show that the short exact sequence

$$\mathcal{P}\mathrm{Diff}_k(E,F) \xrightarrow{\sigma_k} \mathrm{Smbl}_k(E,F) \to 0$$

splits; i.e., there is a linear right inverse $\chi_k : \mathrm{Smbl}_k(E,F) \to \mathcal{P}\mathrm{Diff}_k(E,F)$ to σ_k which satisfies the continuity condition

$$\sup\{||\chi_k(\rho)(u)||_{s-k} : u \in C^\infty(X) \text{ and } ||u||_s = 1\}$$
$$\le C \sup\{|\rho(x,\xi)| : x \in X,\ |\xi| = 1\} \qquad (s,s-k > 0),$$

where C does not depend on $\rho \in \mathrm{Smbl}_k(E,F)$, and $|\rho(x,\xi)|$ is the usual matrix norm which may be defined via the Hermitian inner products on E_x and F_x.

Hint: Go through the proof of Theorem 3.4 again. □

Exercise 4. Show conversely, that for all s, $s-k > 0$ and all $P \in \mathcal{P}\mathrm{Diff}_k(E,F)$, we have the following inequality

$$\sup\{|\sigma_k(P)(x,\xi)| : x \in X, |\xi| = 1\} \leq \sup\{||Pu||_{s-k} : u \in C^\infty(X) \text{ and } ||u||_s = 1\}$$

Hint: By a theorem of Israil Gohberg (see also [Seeley 1965, 171]), one can find a sequence $\{\phi_\nu\}$ of functions in $C_0^\infty(\mathbb{R}^n)$ such that

(i) $\phi_\nu(x) = 0$ for $|x-x_0| > 1/\nu$

(ii) $||\phi_\nu||_0 = 1$ for all ν and

(iii) $||P\phi_\nu - \sigma(P)(x_0,\xi_0)\phi_\nu||_0 \to 0$ as $\nu \to \infty$,

where $(x_0,\xi_0) \in \mathbb{R}^n \times (\mathbb{R}^n \setminus \{0\})$ is any given point. Details are found in [Seeley 1965, 179] or [Cartan-Schwartz 1965, 22-05]. Caution: On $\mathcal{P}\mathrm{Diff}_k(E,F)$ itself there is a topology which is defined in a natural way by the condition that the map $P \mapsto P_s$ be continuous for all s. However, then σ_k is not continuous; see [Palais 1965, 175]. □

B. Elliptic Operators

As a generalization of our earlier definition for differential operators (see Section 2.D above), we call $P \in \mathcal{P}\mathrm{Diff}_k(E,F)$ *elliptic*, if $\sigma_k(P)(x,\xi)$ is an isomorphism from E_x to F_x for all $x \in X$ and $\xi \in (T^*X)_x$, $\xi \neq 0$. We write $P \in \mathrm{Ell}_k(E,F)$.

Theorem 2 (Main result). For each $P \in \mathrm{Ell}_k(E,F)$, there exists $Q \in \mathrm{Ell}_{-k}(F,E)$ such that $PQ - I_F \in OP_{-1}(F,F)$ and $QP - Id_E \in OP_{-1}(E,E)$.

Remark. This existence theorem forms the foundation of our theory of elliptic operators. Using the terminology introduced by David Hilbert, one calls Q a *parametrix* for P- although a "crude" one: The classical parametrix ("Green's function") inverts P not only modulo operators of order -1, but also of order $-\infty$ (i.e., modulo so-called "smoothing operators"; see [Hörmander 1971a].

Proof: Theorem 3.4 guarantees the existence of a $Q \in \mathcal{P}\mathrm{Diff}_{-k}(F,E)$ with $\sigma_{-k}(Q)(x,\xi) := (\sigma_k(P)(x,\xi))^{-1}$, whence $PQ \in \mathcal{P}\mathrm{Diff}_0(F,F)$ by Theorem 3.5a and $\sigma_0(PQ - Id_F) = 0$, and $PQ - Id_F \in OP_{-1}(F,F)$ by Exercise 2. □

Theorem 3. Let $P \in \mathrm{Ell}_k(E,F)$ and s, $s-k \geq 0$. Then we have:

a) (Finiteness): The extension $P_s : W^s(E) \to W^{s-k}(F)$ is a Fredholm operator with index independent of s.

b) (Existence): P^* is elliptic and $\mathrm{Coker}\, P_s \cong \mathrm{Ker}\,(P^*)_{s-k}$.

c) (Regularity): $\mathrm{Ker}\, P_s = \mathrm{Ker}\, P$.

d) (Homotopy-invariance): $\mathrm{index}\, P = \mathrm{index}\, P_s$ depends only on the homotopy class of $\sigma(P)$ in $\mathrm{Iso}^\infty_{SX}(E,F)$. Here $\mathrm{Iso}^\infty_{SX}(E,F)$ is the space of C^∞ bundle isomorphisms $\tau^*E \to \tau^*F$, where $\tau\colon SX \to X$ is the base-point map and $SX := \{(x,\xi) : x \in X,\ \xi \in T^*X, \text{ and } |\xi| = 1\}$ is the "co-sphere bundle"; $\mathrm{Iso}^\infty_{SX}(E,F)$ is equipped with a supremum norm as in Exercise 3.

Remark. In conjunction with c), existence says that the inhomogeneous equation $Pu = f$ has a solution exactly when $f \perp \mathrm{Ker}\, P^*$, and the solution is unique if constrained to be orthogonal to $\mathrm{Ker}\, P$ in $W^0(E)$. By regularity, all "classical solutions" (i.e., $u \in C^k(E)$) of homogeneous elliptic differential equations $Pu = 0$ (with C^∞ coefficients) lie in $C^\infty(E)$. In the context of distribution theory (see [Hörmander 1963]), one obtains the sharper result that every "weak solution" (in the distribution sense) is a "strong solution" (in the function sense); i.e., from the assumption $u \in W^0(E)$ and $\langle u,P^*f\rangle_0 = 0$ for all $f \in C^\infty_0(F)$, the conclusions $u \in C^\infty(E)$ and $Pu = 0$ follow. Such regularity results, which were first proved in 1940 by Herman Weyl in the case of the Laplace operator $P := \Delta$, are of special importance, when one is solving partial differential equations by variational methods (i.e., solving through extremal conditions); see [Hörmander 1963, 96] and [Lions-Magenes 1968, 214 ff.].

Proof of a: If $Q \in \mathrm{Ell}_{-k}(F,E)$ is a parametrix (see Thoerem 2) for P, then it follows from the Theorem of Franz Rellich (Theorem 4.3) that the composition

$$W^s(E) \xrightarrow[Q_{s-k}\circ P_s - \mathrm{Id}]{} W^{s+1}(E) \hookrightarrow W^s(E)$$

is a compact operator on $W^s(E)$, and correspondingly, $P_sQ_{s-k} - \mathrm{Id}$ is a compact operator on $W^{s-k}(F)$. Thus, $P_s : W^s(E) \to W^{s-k}(F)$ is a Fredholm operator by Theorem I.5.1. (There, actually the proof was explicitly given only for endomorphisms, but this is no restriction for separable Hilbert spaces, since they are all isomorphic.) By continuity considerations (Theorem I.5.2 and the preceding Exercises 3 and 4, or easy norm comparison for P_s and P_r by means of Λ^{s-r} of Exercise 4.3) it follows that index P_s = index P_r.

For b: Without loss of generality (Λ argument of Exercise 4.3), let $k = s = 0$. Then b) follows directly from Exercise 1 and Theorem I.3.1.

For c: By definition we have $\operatorname{Ker} P_{s+1} \subset \operatorname{Ker} P_s$, since $W^{s+1}(E) \subset W^s(E)$. Conversely, by Theorem 2 there is a bounded operator $K : W^s(E) \to W^{s+1}(E)$ such that $Q_{s-k}P_s u - \operatorname{Id} u = Ku$ for all $u \in W^s(E)$, where Q is a parametrix for P; thus $u \in W^{s+1}(E)$, if $P_s u = 0$. Thus, $\operatorname{Ker} P_s = \operatorname{Ker} P_{s+1} = \ldots = \operatorname{Ker} P$, since $C^\infty = \bigcap W^s$; see Exercise 4.2c and Theorem 4.1).

For d: For $Q \in \operatorname{Ell}_k(E,F)$ and $\sigma_k(Q) = \sigma_k(P)$, we obtain index Q = index P from Exercise 2 and the invariance of the index under perturbation by compact operators (Exercise I.5.7). In general, by Exercise 3, each continuous curve $\rho : I \to \operatorname{Smbl}_k(E,F)$ lifts to a corresponding continuous (in the operator norm) curve $\pi : I \to \mathcal{P}\operatorname{Diff}_k(E,F)$ with $\sigma_k \circ \pi = \rho$. Thus, if one can connect $\sigma_k(Q)$ and $\sigma_k(P)$ by a continuous curve in $\operatorname{Iso}^\infty_{SX}(E,F)$, then P and Q can be connected in $\operatorname{Ell}_k(E,F)$; and so for all s, the Fredholm operators Q_s and P_s lie in the same component and have the same index by Theorem I.5.2. Finally, if $Q \in \operatorname{Ell}_j(E,F)$ is an operator whose symbol $\sigma_j(Q)$ coincides with $\sigma_k(P)$ on SX, then $(\sigma_j(Q))^{-1} \circ \sigma_k(P)$ is the symbol of a self-adjoint operator $R \in \operatorname{Ell}_{k-j}(E,E)$. From $\sigma_k(P) = \sigma_j(Q) \circ \sigma_{k-j}(R) = \sigma_k(QR)$, it follows by preceding arguments that index P = index QR = index Q + index R = index Q, since index $R = 0$ by c) (see also the composition rule in Exercise I.2.4 and the following Exercise 5a). □

Convention: In the following, we write for short $\sigma(P)$ for the restriction of $\sigma_k(P)$ to SX.

Exercise 5. Let E, F, G, H be Hermitian vector bundles over the closed oriented Riemannian manifold X; $P \in \operatorname{Ell}_k(E,F)$, $Q \in \operatorname{Ell}_j(F,G)$, $R \in \operatorname{Ell}_{k'}(G,H)$. Show that the following expressions are defined, and prove the formulas:

a) index P^* = -index P

b) index QP = index P + index Q

c) index $P \oplus R$ = index P + index R

d) index $P = 0$, if $\sigma(P)(x,\xi)$ depends only on x and not on $\xi \in (SX)_x$.

Hint for d: A bundle isomorphism $E \to F$ and a multiplication operator $M_\psi \in Ell_0(E,F)$ are defined via $\psi(x) := \sigma(P)(x,\xi)$ for $\xi \in (SX)_x$. Apply Theorem 3d. □

Exercise 6. a) Form the closure $\overline{Smbl_k(E,F)}$ of $Smbl_k(E,F)$ in the supremum norm, and show that one then obtains all the continuous symbols. In particular, the space $Iso_{SX}(E,F)$ of all continuous isomorphisms from τ^*E to τ^*F, where $\tau : SX \to X$ is the projection, consists of the restrictions $p|SX$ for $p \in \overline{Smbl_k(E,F)}$

b) For each s, $s-k \geq 0$, form the closure (in the operator norm) of the set of all operators P_s with $P \in \mathcal{P}Diff_k(E,F)$, and show that σ_k can be continuously extended to a surjective map on this space if the target of σ_k is enlarged to $\overline{Smbl_k(E,F)}$. □

We now write $P \in \overline{\mathcal{P}Diff_k(E,F)}$, if P lies in the closure formed in Exercise 6, for all $s \geq 0$ (and $s-k \geq 0$). One easily sees that our results up to now (in particular on elliptic operators) remain valid in this larger class. The most important reason for passing to the closure arises from the multiplicative behavior of pseudo-differential operators:

Exercise 7. Consider two closed, oriented, Riemannian manifolds X and Y; and Hermitian vector bundles E and F over X, and G and H over Y. Moreover, let $P \in \mathcal{P}Diff_k(E,F)$ and $Q \in \mathcal{P}Diff_k(G,H)$, $k \in \mathbb{N}$.

a) Show that an operator $P \otimes Id_G \in \overline{\mathcal{P}Diff_k(E \otimes G,\ F \otimes G)}$ is defined by

$$(P \otimes Id_G)(u \otimes v) := Pu \otimes v, \quad u \in C^\infty(E), \quad v \in C^\infty(G),$$

which does not always lie in $\mathcal{P}Diff_k(E \otimes G,\ F \otimes G)$.

b) Over the manifold $X \times Y$, define the operator

$$P\#Q : C^\infty(E \otimes G) \oplus C^\infty(F \otimes H) \to C^\infty(F \otimes G) \oplus C^\infty(E \otimes H)$$

by the matrix

$$P\#Q := \begin{pmatrix} P \otimes Id_G & -Id_F \otimes Q^* \\ Id_E \otimes Q & P^* \otimes Id_H \end{pmatrix}.$$

Prove the Multiplication Theorem: If P and Q are elliptic, then $P\#Q$ is elliptic, and index $P\#Q$ = (index P)(index Q).

Hint for a: For the sake of simplicity, assume that all bundles are trivial line bundles (i.e., the "function case"). Then $P \otimes Id_{\mathbb{C}_Y} : C^\infty(X \times Y) \to C^\infty(X \times Y)$ is the operator obtained when P acts on the first variable while the second is held fixed. In local coordinates, for

$(x,y) \in X \times Y$ and $u \in C^\infty(X \times Y)$, we have:

$$(P \otimes Id)u(x,y) = (2\pi)^{-n} \iint e^{i<x-x',\xi>} p(x,\xi)u(x',y)dx'd\xi.$$

The amplitude $\tilde{p}$ of $P \otimes Id$ is then given by $\tilde{p}(x,y,\xi,\eta) = p(x,\xi)$ (up to a constant of integration). Show that the amplitude estimate

$$|D_\xi^\alpha D_x^\beta p(x,\xi)| \le c(1 + |\xi| + |\eta|)^{k-|\alpha|}$$

can only hold for large $|\alpha|$ when $D_\xi^\alpha D_x^\beta p(x,\xi)$ is identically zero (i.e., when p is a polynomial).

For the proof that $P \otimes Id \in \overline{\mathcal{P}Diff_k(X \times Y)}$, explicitly construct a family $\{R^t : t \in (0,1]\}$ with $R^t \in \mathcal{P}Diff_0(X \times Y)$ and $(P \otimes Id)R^t \in \mathcal{P}Diff_k(X \times Y)$ such that $(P \otimes Id)R^t$ converges in the operator norm to $P \otimes Id$ as $t \to 0$, as in [Atiyah-Singer 1968a, 513-516]. An alternative proof can be found in [Hörmander 1971b, 96f.], where the consideration of the difference variable z in x and y (and of ζ in ξ and η) is carried out in the framework of the theory of Fourier integral operators, with its more flexible methods; see also the theorem of Kuranishi, Theorem 3.2.

For b: For the origin of the somewhat strange form of $P\#Q$, compare the "golden rule" of tensoring chain complexes; see also Chapters III.1/2. Details may be found in [Seeley 1965, 190-193], [Cartan-Schwartz 1965, exposé 22], [Atiyah-Singer 1968a, 526-529].

Exercise 8. Let $P : C^\infty(E) \to C^\infty(F)$ be an elliptic operator (i.e., $\sigma(P) \in Iso_{SX}(E,F)$). Show that index $P = 0$, if $\sigma(P)$ can be extended to an isomorphism over all of $BX := \{\xi \in T^*X : |\xi| \le 1\}$.

Hint: Show that a homotopy between the symbol of P and the symbol of the multiplication operator $(Mu)(x) := \sigma(P)(x,0)(U(x))$ can be defined and apply Theorem 3d and Exercise 5d. □

Exercise 9. Now, let E and F be trivial line bundles over the closed manifold X. Show that index $P = 0$, if $\dim X > 2$.

Hint: Reduce this to Exercise 8 by a suitable deformation of $\sigma(P)(x,\xi)$; see [Nirenberg 1970, 160ff.]. Compare also with Section III.4.C below. □

Exercise 10. Carry the following theorems and exercises to the case of manifolds with boundary: Theorem 1, Exercise 1 (if the formal adjoint operator is defined by $\langle Pu,v\rangle = \langle u,P^*v\rangle$ for all u and v with support contained in the interior of X), Exercises 2-4, Theorem 2 (Why not Theorem 3?), Exercise 6, and Exercise 7a. □

6. ELLIPTIC BOUNDARY-VALUE SYSTEMS I (DIFFERENTIAL OPERATORS)

Below we will develop a systematic theory of elliptic boundary-value systems in a generality, simplicity, and apparent arbitrariness created for convenient proofs. First, we will discuss the "idea of ellipticity of boundary-value conditions", i.e., the connection between boundary value problems of partial differential equations and initial conditions of ordinary differential equations and thus finally the algebraic essence of the "idea of ellipticity."*

A. Differential Equations With Constant Coefficients

We begin with the local considerations which arise from the classical theory of homogeneous differential equations with constant coefficients in the half-space. Thus, for the moment, let p be a homogeneous polynomial of degree k in n variables and $p(D) = p(-i \frac{\partial}{\partial x_1}, \ldots, -i \frac{\partial}{\partial x_n})$ the associated homogeneous differential operator of order k with constant coefficients. In general, $\operatorname{Ker} p(D) = \{u : u \in C^\infty(R^n)$ and $p(D)u = 0\}$ will be an infinite dimensional function space. For example, for $n = 2$ let $p(x_1,x_2) = x_1^2 + x_2^2$, hence $p(D) = -\Delta$, whose kernel consists exactly of all harmonic functions. From numerical mathematics, which is concerned with the approximation of arbitrary solutions by solutions of particularly simple form, we know the importance - beyond the realm of ordinary linear differential equations - of the "exponential solutions." These are functions $u \in \operatorname{Ker} p(D)$ which can be written in the form $u(x) = f(x) \exp i(x_1\xi_1 + \ldots + x_n\xi_n)$ where f is a polynomial in n variables and $\xi_1,\ldots,\xi_n \in ¢$. In [Hörmander 1963, 76f.], we find the exact formulations and the proof that the exponential solutions are dense in $\operatorname{Ker} p(D)$ for a suitable topology. While systems of ordinary differential equations with constant coefficients can be solved completely by means of elementary functions, the same is true in approximation only, for the analogous partial differential equations.

The exponential solutions play a similarly significant role in the characterization of boundary value problems:

---

*Much of the material of this and the following two sections has not reached its final form. Rather, 30 years after the appearance of the programmatic paper of I. M. Gelfand, it seems to be in a state of flux. In fact, according to the most recent papers of M. F. Atiyah, P. B. Gilkey, and others on the "boundary integrands" in the formulas for the signature and Euler characteristic of a Riemannian manifold with boundary, some of our notions and theorems on elliptic boundary value problems appear in a new light. See also the Remarks 1 and 2 below in Section III.4.H.

Exercise 1. For each $(n-1)$-tuple $\eta = (\xi_1,\ldots,\xi_{n-1}) \in \mathbb{R}^{n-1}$, define in Ker $p(D)$ the subspace

$$M_\eta := \{u : u \in \text{Ker } p(D) \text{ and } u(x) = [\exp i(x_1\xi_1 + \ldots + x_{n-1}\xi_{n-1})]h(x_n) \text{ with } h \in C^\infty(\mathbb{R})\}$$

and show that

$$M_\eta \cong \{h : h \in C^\infty(\mathbb{R}) \text{ and } h \in \text{Ker } p(\xi_1,\ldots,\xi_{n-1}, -i\tfrac{d}{dt})\}.$$

Trick: Compute that

$$p(D)[(\exp i(x_1\xi_1 + \ldots + x_{n-1}\xi_{n-1}))h(x_n)]$$
$$= [\exp i(x_1\xi_1 + \ldots + x_{n-1}\xi_{n-1})]p(\xi_1,\ldots,\xi_{n-1},-i\tfrac{d}{dx_n})h(x_n)$$

and obtain the isomorphic representation of M_η by a mere separation of variables. □

In this fashion we associated with the partial differential operator $p(D)$ a family of vector spaces parametrized by $\eta \in \mathbb{R}^{n-1}$ whose elements are solutions of ordinary differential equations with constant coefficients. A number of results about the structure of the spaces M_η now follow painlessly by means of the most elementary stability considerations. From these, we will then develop the notion of ellipticity of boundary-value problems both in its analytic and its algebraic form.

Theorem 1. For a homogeneous polynomial p of degree k in n variables and for $\eta \in \mathbb{R}^{n-1}$, M_η consists of exponential solutions (i.e., the coefficient functions h are polynomials). If p is elliptic, then $\dim M_\eta = k$, and M_η decomposes in a natural way into two subspaces M_η^+ and M_η^- which are complementary for $\eta \neq 0$.

Proof: It is well-known (see [Pontryagin 1962, 45ff.] that the solutions h of the differential equations $q(-i\frac{d}{dt})h(t) = 0$ (where q is a polynomial, in one variable, of degree k) are completely determined by the zeros of q. Namely, when $\lambda_1,\ldots,\lambda_m$ are the roots of q with respective multiplicities $r_1,\ldots,r_m$, then Ker $q(-i\frac{d}{dt})$ is the linear span of the k exponential functions

$$t^p \exp i\lambda_\mu t, \quad \mu = 1,\ldots,m \text{ and } p = 0,\ldots,r_\mu-1.$$

For p elliptic, $q(t) = p(\xi_1,\ldots,\xi_{n-1},t)$ retains degree k, and M_η is a k-dimensional subspace of the vector space of exponential solutions of the equation $p(D)u = 0$. The asymptotic behavior as $t \to \infty$ (i.e., $x_n \to \infty$) is determined by the imaginary parts of the zeros. If $\eta = (\xi_1,\dots,\xi_{n-1}) \neq 0$, then $q(t)$ has no real zeros (at least for p elliptic), and M_η is the direct sum of M_η^+ and M_η^-, consisting of solutions which remain bounded (more precisely, go to zero) as $t \to \infty$, resp. as $t \to -\infty$. □

Remark 1. Conversely, the ellipticity of $p(D)$ can be defined by the condition

$$M_\eta^+ \cap M_\eta^- = \begin{cases} \{0\} & \text{for} \quad \eta \neq 0 \\ \{\text{const}\} & \text{for} \quad \eta = 0. \end{cases}$$

This means that $p(D)$ possesses no bounded exponential solutions other than constant functions.

Remark 2. In order to move on to boundary (or initial-value) problems, we consider the trajectories ("integral curves") $h(t)$ of solutions together with their derivatives up to order $k-1$ in "phase space" $\mathbb{C}^k$. Each trajectory corresponds uniquely to its initial values $(h(0), Dh_{|0}, \dots, D^{k-1}h_{|0})$, where $D := -i\frac{d}{dt}$. The trajectories, corresponding to points in M_η^+ $(\eta \neq 0)$ which tend to the origin, have initial values which lie in a subspace K_η^+ (the "plus-stable subspace") in phase space. The corresponding initial values for M_η^- lie in a subspace K_η^- (the "minus-stable subspace") complementary to K_η^+. The remaining integral curves, whose initial values lie in the open set $\mathbb{C}^k - (K_\eta^+ \cup K_\eta^-)$, have a positive distance from the origin and are unbounded as $t \to \infty$ and $t \to -\infty$. The isomorphic images K_η^+ of M_η^+ and K_η^- of M_η^- form the level set through a generalized saddle point at the origin of the phase space $\mathbb{C}^k$.* The necessary simple calculations may be found in [Nemitsky-Stepanov 1960, 187f.].

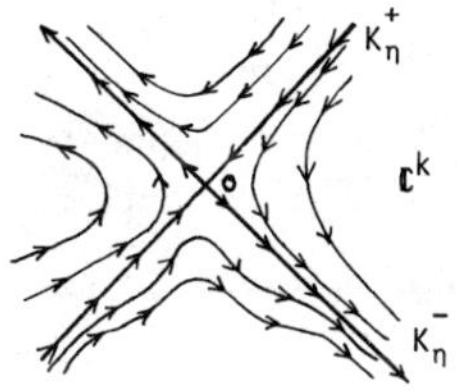

□

*Note that the spaces K_η^+ and M_η^+ are canonically isomorphic (for given p and η); thus we can identify them with each other. Therefore, in the following, we will only speak of M_η^+, and take M_η^+ as a space of functions or initial values, depending on context.

We come now to the definition of ellipticity of boundary-value problems in the halfspace $x_n \geq 0$ of $\mathbb{R}^n$ for a single elliptic differential equation

$$p(D)u = 0 \tag{*}$$

with the boundary conditions

$$p_j(D)u_{|x_n=0} = 0, \quad j = 1,\ldots,r. \tag{**}$$

Here $p, p_1,\ldots,p_r$ are homogeneous polynomials of degree k, and $k_1,\ldots,k_r < k$ respectively.

We call (*), (**) an *elliptic boundary-value problem*, if

$$k = 2r \tag{I}$$

and if

the system of equations (*), (**) has no nontrivial solutions in M_η^+ for each $\eta \neq 0$. (II)

Detailed motivation for this definition may be found in [Hörmander 1963, 242-246]. According to this reference, the ellipticity of a boundary-value problem is important primarily for the proof of regularity theorems and finiteness of the index (see below). In addition, L. Hörmander proved that from a regularity assumption, saying that all solutions are still C^∞ on the boundary, it already follows that system (*), (**) can possess no bounded exponential solutions in the x_n-direction and that $k = 2r$. (The ellipticity of $p(D)$ is not needed here.) Elliptic boundary-value conditions, first formulated in this sense by Jaroslav Boresovich Lopatinsky, are thus exactly those conditions which insure the smoothness of solutions also on the boundary.

In order to isolate the algebraic kernel of the ellipticity conditions for the boundary-value system (*), (**), we again translate the conditions (I) and (II) into the language of ordinary differential equations and their initial conditions. Thus, let $p,p_1,\ldots,p_r$ be homogeneous polynomials as above, with p elliptic of degree k, and let $\eta \in \mathbb{R}^{n-1} \setminus \{0\}$. As before, we assign, to the pair (p,η), the vector space M_η^+, which we can identify with the plus-stable subspace K_η^+. Now show:

<u>Exercise 2</u>. The data $p,p_1,\ldots,p_r$ of a boundary-value problem (p elliptic) defines a linear map $\beta_\eta^+ : M_\eta^+ \to \mathbb{C}^r$, for each point η.

<u>Trick:</u> Assign to each boundary operator $p_j(D)$, $j = 1,\ldots,r$, and each point $\eta = \{\xi_1,\ldots,\xi_{n-1}\} \in \mathbb{R}^{n-1} \setminus \{0\}$, the initial value problem

$$C^\infty(\mathbb{R}_+) \ni h \longmapsto p_j(\xi_1,\ldots,\xi_{n-1},-i\,\frac{d}{dt})h|_{t=0},$$

and thus a linear functional on the phase space $\mathbb{C}^k \cong M_\eta$.

The $p_1,\ldots,p_r$ together define (for each $\eta \neq 0$) a linear map β_η from $\mathbb{C}^k$ to $\mathbb{C}^r$; then consider the restriction to the plus-stable subspace M_η^+. □

Theorem 2. A boundary-value system (*), (**) is elliptic, exactly when the map $\beta_\eta^+ : M_\eta^+ \to \mathbb{C}^r$ is an isomorphism for each $\eta \neq 0$. Written out: For each $\eta = (\xi_1,\ldots,\xi_{n-1}) \in \mathbb{R}^{n-1} \setminus \{0\}$ and for each r-tuple of complex numbers $(g_1,\ldots,g_r) \in \mathbb{C}^r$, there is exactly one bounded (as $t \to \infty$) function $h \in \mathbb{C}^\infty(\mathbb{R})$ such that $p(\xi_1,\ldots,\xi_{n-1}, -i\,\frac{d}{dt})h = 0$ and $p_j(\xi_1,\ldots,\xi_{n-1},-i\,\frac{d}{dt})h|_{t=0} = g_j$; $j = 1,\ldots,r$.

Proof: First, suppose (*), (**) is elliptic. The condition (II) says that $M_\eta^+ \cap \{g \in \mathbb{C}^k : \beta(g) = 0\} = \{0\}$ for all $\eta \neq 0$; this means that β_η^+ is injective.

To prove that β_η^+ is surjective, we use the condition (I). Recall (see Remark 2 after Theorem 1 above) that M_η^+ and M_η^- together span the whole phase space $\mathbb{C}^k$ and are complementary to each other; in particular,

$$\dim M_\eta^+ + \dim M_\eta^- = k. \tag{I'}$$

Since β_η^+ is injective, we have $\dim M_\eta^+ \leq r$. However, $M_\eta^- \cong M_{-\eta}^+$, since $u \in \operatorname{Ker} p(D) \Longleftrightarrow \tilde{u} \in \operatorname{Ker} p(D)$, where $u(x_1,\ldots,x_n) = \{\exp i(x_1\xi_1 + \ldots + x_{n-1}\xi_{n-1})\}h(x_n)$ and $\tilde{u}(x_1,\ldots,x_n) = \{\exp i(-x_1\xi_1 - \ldots - x_{n-1}\xi_{n-1})\}\tilde{h}(x_n)$ with $\tilde{h}(t) := h(-t)$. Thus, we also have $\dim M_\eta^- \leq r$, whence $\dim M_\eta^+ = k - \dim M_\eta^- \geq k-r = r$, since $k = 2r$ by (I). Thus, $\dim M_\eta^+ = r$, and so β_η^+ is bijective.

Conversely, (II) follows easily from the bijectivity of β_η^+; moreover, $\dim M_\eta^+ = \dim M_\eta^- = r$ and because of (I'), we obtain condition (I). □

Remark 1. Since the boundary conditions $p_1,\ldots,p_r$ are supposed to define a linear map from phase space $\mathbb{C}^k$ to $\mathbb{C}^r$, we had to first assume that the degrees of the polynomials $p_1,\ldots,p_r$ are smaller than the degree k of the elliptic polynomial p. Actually, we can algebraically reduce every polynomial of degree greater than p (e.g., see [van der Waerden 1936/1960, §18]) and come to equivalent boundary (resp. initial) conditions.

Remark 2. For an elliptic differential equation with constant coefficients (homogeneous of degree $2r$), we have first the natural (initial-value) isomorphism $M_\eta \cong \mathbb{C}^{2r}$. Thus, elliptic boundary conditions cannot exist unless $\dim M_\eta^+ = r$, i.e., $M_\eta^+ \cong \mathbb{C}^r$, and this isomorphism is not canonically defined but depends on a choice of boundary conditions. In this sense, the algebraic-geometrical meaning of elliptic boundary-value problems lies in providing a fixed coordinate system (i.e., a fixed basis) for M_η^+. □

Exercise 3. a) For a boundary-value system of your choice, determine the families of vector spaces M_η, M_η^+, K_η^+ and the maps β_η^+. For example, show for the Laplace equation with $p(\xi_1,\xi_2) = \xi_1^2 + \xi_2^2$ that, at the point $\eta = \xi_1 > 0$,

$$M_\eta = \{c_1 e^{\eta(ix+y)} + c_2 e^{\eta(ix-y)} : c_1, c_2 \in \mathbb{C}\}$$
$$\cong \{c_1 e^{\eta t} + c_2 e^{-\eta t} : c_1, c_2 \in \mathbb{C}\} \cong \mathbb{C}^2$$

and

$$M_\eta^+ = \{c e^{\eta(ix-y)} : c \in \mathbb{C}\} \cong \{c e^{-\eta t} : c \in \mathbb{C}\} \cong \mathbb{C}.$$

Then determine the linear map β_η^+, e.g., for the Neumann boundary-value problem with $p_1(\xi_1,\xi_2) = \xi_2$, and show that it selects the basis vector $(0,-i/\eta)$ in M_η^+.

b) Show that, for the Cauchy-Riemann equation with $p(\xi_1,\xi_2) := \xi_1 + i\xi_2$, the vector spaces M_η^+ $(\eta \in \mathbb{R} \setminus \{0\})$ have different dimensions, depending on whether $\eta > 0$ or $\eta < 0$. □

B. Systems of Differential Equations with Constant Coefficients

In this section, p is a $N \times N$ matrix of homogeneous polynomials of degree k in n variables, and p is elliptic; i.e., $\det p(\xi) \neq 0$ for $\xi = (\xi_1,\ldots,\xi_n) \neq 0$.

Exercise 4. For $\eta = (\xi_1,\ldots,\xi_{n-1}) \in \mathbb{R}^{n-1} \setminus \{0\}$ define M_η, M_η^+ and M_η^-, as above. □

Theorem 3. The following statements hold for $\eta \neq 0$.

(i) $\dim M_\eta = Nk$

(ii) Each solution is uniquely determined by its initial value in phase space $\mathbb{C}^{Nk}$.

(iii) The vector spaces M_η^+ and M_η^-, interpreted as the plus-stable resp. minus-stable subspaces, are complementary in $\mathbb{C}^{Nk}$.

Proof: For fixed $\eta = (\xi_1,\ldots,\xi_{n-1})$, the elements of M_η are of the form

$$u(x) = \begin{pmatrix} u_1(x) = \{\exp i(x_1\xi_1 + \dots + x_{n-1}\xi_{n-1}\}h_1(x_n) \\ \vdots \\ u_N(x) = \{\exp i(x_1\xi_1 + \dots + x_{n-1}\xi_{n-1}\}h_N(x_n) \end{pmatrix},$$

where $x = (x_1,\dots,x_n) \in \mathbb{R}^n$; or more simply of the form

$$h(t) = \begin{pmatrix} h_1(t) \\ \vdots \\ h_N(t) \end{pmatrix}$$

where $t \in \mathbb{R}$. For such C^∞ functions on $\mathbb{R}$ with values in $\mathbb{C}^N$, we write $h \in C^\infty(\mathbb{C}^N_{\mathbb{R}})$; learnedly: "$h$ is a C^∞ section of the N-dimensional trivial bundle over $\mathbb{R}$."

For the proof of (i), we go through the canonical transformation of the $N \times N$ system

$$p(\xi_1,\dots,\xi_{n-1}, -i\,\frac{d}{dt})h = 0, \quad h \in C^\infty(C^N_{\mathbb{R}})$$

of order k to a $Nk \times Nk$ system

$$\dot{\tilde{h}} - A\tilde{h} = 0, \quad \tilde{h} \in C^\infty(\mathbb{C}^{Nk}_{\mathbb{R}})$$

of order 1 (as in [Pontryagin 1962, 89f.]). The solutions $\tilde{h}$ can be written ([Pontryagin 1962, 97]) explicitly in the form $\overset{(\nu)}{q}(t)\exp i\lambda t$, where λ is an eigenvalue of A, $\nu \in \{1,\dots,r\}$, r is the multiplicity of λ,

$$\overset{(\nu)}{q}(t) := \frac{t^{\nu-1}}{(\nu-1)!}\,e_1 + \frac{t^{\nu-2}}{(\nu-2)!}\,e_2 + \dots + e_\nu,$$

and $e_1,e_2,\dots,e_r$ is a system of vectors in $\mathbb{C}^{Nk}$ given through the Jordan normal form (see [Pontryagin 1962, 277-295]) so that

$$\begin{aligned} Ae_1 &= \lambda e_1 \\ Ae_2 &= \lambda e_2 - ie_1 \\ &\ \vdots \\ Ae_r &= \lambda e_r - ie_{r-1}. \end{aligned}$$

In this way, we establish (ii) and (iii) along with (i).

In [Hörmander 1963, 269], we find a direct proof of (i) by directly working out the algebraic essence of the statement - without the conceptual, but perhaps somewhat diverting, discussion of solution curves:

Set $P(t) := p(\xi_1,\ldots,\xi_{n-1},t)$. One knows from the theory of elementary divisors (e.g., [Albert 1941]) that invertible $N \times N$ matrices $A(t)$ and $B(t)$ exist, whose elements are polynomials as are the entries of the inverse matrices, such that $A(t)P(t)B(t) = Q(t)$, where $Q(t)$ is a diagonal matrix. For $D := -i\frac{d}{dt}$, the operators $A(D)$ and $B(D)$ are bijective on $C^\infty(\mathbb{C}^N_{\mathbb{R}})$, and we have $P(D) = A^{-1}(D)Q(D)B^{-1}(D)$. Thus,

$$M_\eta \cong \{h \in C^\infty(\mathbb{C}^N_{\mathbb{R}}) : P(D)h = 0\}$$

$$= \{B(D)g : g \in C^\infty(\mathbb{C}^N_{\mathbb{R}}) \text{ and } A^{-1}(D)Q(D)g = 0\}$$

$$= \{B(D)g : Q(D)g = 0\} \cong \{g : Q(D)g = 0\}.$$

Because of the polynomial form of $A(t)$ <u>and</u> $A(t)^{-1}$, $\det A(t)$ does not depend on t. The same holds for $\det B(t)$. Thus, there is a constant $C \neq 0$ such that $\det Q(t) = C \det P(t)$. The sum of the degrees of the diagonal elements of $Q(t)$ (and hence, $\dim M_\eta$) is the same as the degree of $\det P(t)$, namely Nk if p is elliptic of degree k.

Note that the transformation $B(D)$ does not alter the asymptotic behavior of g. Thus, the direct decomposition of M_η into M_η^+ and M_η^- is directly given via $Q(D)$ (or through $\det P(t)$) and its zeros with positive imaginary part). □

For the differential operator $p(D_1,\ldots,D_n)$, a system of boundary conditions is given by $N \times M_j$ matrices p_j $(j = 1,\ldots,r)$, whose elements are homogeneous polynomials of degree k_j. The *boundary-value problem* is written exactly as in (*), (**) above. For *ellipticity*, the condition (I) must be correspondingly generalized to $Nk = 2\,\Sigma_{j=1}^r M_j$, while condition (II) needs no further modification.

<u>Exercise 5</u>. As above, define the initial-value map $\beta_\eta : \mathbb{C}^{Nk} \to \oplus\, \mathbb{C}^{M_j}$, and show that the boundary-value system is elliptic exactly when the restriction $\beta_\eta^+ : M_\eta^+ \to \bigoplus_{j=1}^{k} \mathbb{C}^{M_j}$ is an isomorphism for all $\eta \neq 0$. □

C. <u>Variable Coefficients</u>

The preceding developments easily carry over to the case where X is a n-dimensional C^∞ manifold (with boundary Y) equipped with a Riemannian metric, E and F are N-dimensional C^∞ complex vector bundles over X, and $P : C^\infty(E) \to C^\infty(F)$ is an elliptic differential operator of order k. Here, the geometrical character of our definitions (their invariance under coordinate changes) is manifest, so that we may pass to global concepts. For this, first show:

<u>Exercise 6</u>. For each $y \in Y$, the symbol $\sigma(P)(y,\ldots)$ of P defines a homogeneous partial differential operator with constant coefficients $P_y : \mathcal{E}_y \otimes E_y \to \mathcal{E}_y \otimes F_y$, where $\mathcal{E}_y$ is the space (isomorphic to $C^\infty(\mathbb{R}^n)$) of C^∞ functions on the tangent space $(TX)_y$; thus $\mathcal{E}_y \otimes E_y$ is the space of C^∞ functions on $(TX)_y$ with values in E_y (isomorphic to the n-fold product $C^\infty(\mathbb{R}^n) \times \ldots \times C^\infty(\mathbb{R}^n)$).

<u>Warning</u>: Here it does not matter whether P is actually elliptic or not, and this somewhat contrived construction of P_y applies at each $y \in X$, not only at boundary points. □

As usual, let $T'Y$ be the bundle of nonvanishing covectors on Y; i.e., T^*Y without the zero section. We obtain, as the first main result of the heuristic considerations of this paragraph, the following

<u>Theorem 4</u>. Each elliptic differential operator P over the manifold X with boundary Y defines C^∞ vector bundles M, M^+, and M^- (with fiber dimension not necessarily constant; see Exercise 3b) over $T'Y$.

<u>Proof</u>: By selecting an inner normal $\nu \in (T^*X)_y$ by means of the Riemannian metric, we obtain a half-space problem, which defines the spaces M_η, M_η^+, M_η^- as above, where now $\eta \in (T^*Y)_y$, $\eta \neq 0$. Namely, set

$$M_\eta := \{f \in \operatorname{Ker} P_y : f = \{\exp i\langle ..,\eta\rangle\}h, \text{ where } h \text{ is a } C^\infty \text{ function on } (TX)_y/(TY)_y \text{ with values in } E_y\},$$

and correspondingly define, with the appropriate boundedness conditions, the spaces M_η^+ and M_η^-, where $(TX)_y/(TY)_y$ is identified with $\mathbb{R}$ in an oriented way by means of the Riemannian metric so that we know what it means to go to $+\infty$ or $-\infty$. Passing to the equivalent system of ordinary differential equations, we obtain the polynomial $\sigma(P)(y,\eta+t\nu) = \Sigma_{\kappa=0}^{k} c_\kappa t^\kappa$ of degree k with coefficients in $\operatorname{Hom}(E_y,F_y)$ and the linear ordinary differential operator $\sigma(P)(y,\eta+D\nu)$, which in local coordinates $u = (u_1,\ldots,u_n)$ about y (which respect the Riemannian metric at y and map the neighborhood of the boundary Y near y to $\mathbb{R}^{n-1}$) takes the familiar form $p(\xi_1,\ldots,\xi_{n-1},-i\,\frac{d}{dt})$, where $\eta = \xi_1 du + \ldots + \xi_{n-1}du_{n-1}$ and $\nu = du_n$.

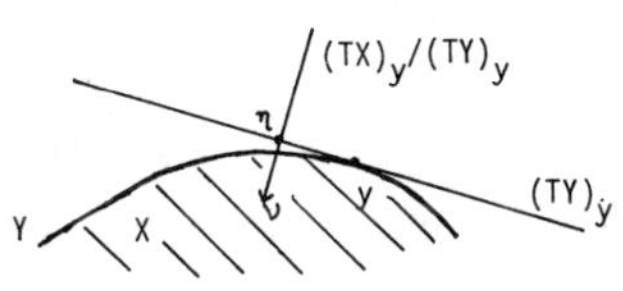

In this way, the families M, M^+ and M^- of vector spaces are defined and parametrized by $T'Y$. Considerations in local coordinates show that M, M^+, and M^- are actually C^∞ vector bundles, since the respective solutions (resp., their initial values in phase space) depend smoothly on the coefficients $c_0,\ldots,c_k$, or more generally, on (y,η): This follows easily from the explicit form of the solutions given in the proof of Theorem 3 (see also the alternative proof there). □

Now let $G = \bigoplus_{j=1}^{r} G_i$ be a vector bundle on the boundary Y of X, where $G_j \in \mathrm{Vect}(Y)$ has fiber dimension $M_j \in \mathbb{N}$. Let $R = (R_1,\ldots,R_r)$ be differential boundary operators, $R_j : C^\infty(E) \to C^\infty(G_j)$ of order ℓ_j. Our second main result introduces a further global invariant. For this, let $\pi_Y : T'Y \to Y$ be the natural projection (base-point map).

Theorem 5. Each boundary-value system $(P,R) : C^\infty(E) \to C^\infty(F) \oplus C^\infty(G)$ with P elliptic defines, in a natural way, a vector bundle homomorphism $\beta^+ : M^+ \to \pi_Y^*(G)$.

Proof: M^+ is the bundle defined from $\sigma(P)$ in Theorem 4, and β^+ is the initial-value map given, at the point (y,η), by

$$M_\eta^+ \xrightarrow{R_y|M_\eta^+} E_y \otimes G_y \xrightarrow{\delta} G_y, \qquad \beta_\eta^+ : M_\eta^+ \to G_y$$

Here, $R_y : E_y \otimes E_y \to E_y \otimes G_y$ is the differential operator with constant coefficients defined from the symbol $\sigma(R) = (\sigma(R_1),\ldots,\sigma(R_r))$ taken at the point y as in Exercise 6, and δ is the map defined by $\delta(f) := f(0)$. □

The boundary-value system (P,R) is defined to be *elliptic* when β^+ is an isomorphism.

Exercise 7. By introducing a coordinate system, show that this definition of ellipticity is just a summary of the local ellipticity defined above. □

7. ELLIPTIC DIFFERENTIAL OPERATORS OF FIRST ORDER WITH BOUNDARY CONDITIONS

In II.5 we saw that a topological object $\sigma(P) \in \mathrm{Iso}_{SX}(E,F)$ is associated with each elliptic operator $P : C^\infty(E) \to C^\infty(F)$, whereby index P depends only on the homotopy type of $\sigma(P)$. Here SX is the covariant sphere bundle for a Riemannian metric for X, and E and F are Hermitian vector bundles over X.

A. The Topological Interpretation of Boundary Value Conditions (Case Study)

We consider, on a manifold X with boundary the elliptic boundary-value system $(P,R) : C^\infty(E) \to C^\infty(F) \oplus C^\infty(G)$ where E, F and G are vector bundles on X and on the boundary Y of X respectively. The object $\sigma(P) \in \mathrm{Iso}_{SX}(E,F)$ is well defined, but it does not contain the necessary information on the index of the boundary value problem (P,R) which may depend on the specific choice of the boundary conditions R, by the Theorem of I. N. Vekua (Theorem 1.1). In the case of a differential operator P of 1st order (but see Exercise 1) we will show, roughly, how the given boundary conditions canonically determine a continuation of $\sigma(P)$ beyond SX to the closed manifold $SX \cup BX|Y$. We obtain a topologically more significant object (conceptually: a closed line packs more topological information than an open one) whose homotopy type does in fact determine index (P,R), as we will see in Chapter 8 and in Section III.4.H. Contrary to the technical explanations of the algebraic meaning of ellipticity of boundary-value problems in the preceding chapter, we are concerned in the following case study with the geometric-topological interpretation.

Theorem 1. An elliptic system (P,R) of N partial differential equations of first order over the compact, oriented, Riemannian manifold X with N/2 boundary conditions over the boundary Y of X defines (uniquely, up to homotopy) a continuous map of the closed manifold $SX \cup BX|Y$ into $GL(2N,\mathbb{C})$ which coincides with $\sigma(P) \oplus \mathrm{Id}_N$ on SX. Here, $GL(2N,\mathbb{C})$ is the group of complex, invertible $2N \times 2N$ matrices and $BX := \{\xi \in T^*X : |\xi| \leq 1\}$.

Proof (a verbal communication of I. M. Singer*): Let

$$(P,R) : C^\infty(E) \to C^\infty(F) \oplus C^\infty(G)$$

* See also [Atiyah-Bott 1964a, 180-184], [Cartan-Schwartz 1965, 25/05-25/07] and [Palais 1965, 346-350].

be an elliptic boundary-value system with $P \in \mathrm{Diff}_1(E,F)$ and $E = F = \mathbb{C}_X^N$ and $G = \mathbb{C}_Y^{N/2}$, N even. Let $\nu \in (T^*X)_y$ be the inner normal at the point $y \in Y$. Each covector $\xi \in (T^*X)_y$ can be written in the form $z\nu + \eta$ with $z \in \mathbb{R}$ and $\eta \in (T^*Y)_y$, where $(T^*Y)_y$ can be taken to be a proper subspace of $(T^*X)_y$ by means of the Riemannian metric (see Exercise 2.10 above). We write $\sigma(\xi) := \sigma_1(P)(y,\xi)$ and obtain (since $\sigma(\xi)$ is a homogeneous polynomial in $\xi = (z\nu,\eta)$):

$$\sigma(z\nu+\eta) = \sigma(z\nu) + \sigma(\eta) = z\sigma(\nu) + \sigma(\eta) : E_y \to F_y \quad \text{linear.}$$

By a corresponding choice of basis for F_y, we may assume (without loss of generality) that $\sigma(\nu) = \mathrm{Id}$, whence

$$\sigma(z\nu+\eta) = z\mathrm{Id} + \sigma(\eta). \tag{1}$$

Consider the space M_η^+, defined as in Theorem 6.4, consisting of the C^∞-functions $h : \mathbb{R} \to E_y$ with $\sigma(D\nu+\eta)h = Dh + \sigma(\eta)h = 0$ which remain bounded as $t \to +\infty$; here $D = -i\frac{d}{dt}$. These are the functions of the form $h(t) = h_0 e^{i\lambda t}$, where λ is an eigenvalue of the endomorphism $\sigma(\eta)$: $E_y \to E_y$ with $\mathrm{Im}\,\lambda \geq 0$ and $h_0 \in E_y$ is an element of the associated eigenspace. Corresponding remarks hold for M_η^-. Thus, the spaces $M_\eta^\pm$ are naturally isomorphic to the (+) (resp. (-)) - eigenspaces of $\sigma(\eta)$. For $\eta \neq 0$, the homomorphism $\lambda\mathrm{Id} + \sigma(\eta) = \sigma(\lambda\nu+\eta)$ is regular for all $\lambda \in \mathbb{R}$ by the ellipticity of P; i.e., $\sigma(\eta)$ has no real eigenvalues, and E_y can be represented as the direct sum

$$E_y \cong M_\eta^+ \oplus M_\eta^-. \tag{2}$$

Now let $G_1,\ldots,G_r$ be vector bundles over Y with $\oplus G_j = G$ and

$$R_j : C^\infty(E) \to C^\infty(G_j), \quad j = 1,\ldots,r,$$

be boundary conditions given by differential expressions such that the associated initial value map

$$\beta_\eta^+ : M_\eta^+ \to G_y$$

is an isomorphism for all $y \in Y$ and $\eta \in (T^*Y)_y \setminus \{0\}$. By Theorem 6.5 and Exercise 6.7, this is the condition of ellipticity for boundary-value systems. We now show that $\sigma(p)(y,\cdot) : (SX)_y \to \mathrm{Iso}(E_y,E_y)$ is (stably-) homotopic to a constant map, and that the

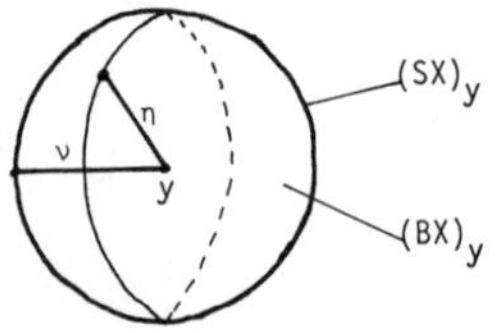

homotopy is defined in a natural way by using $\{\beta_\eta^+ : \eta \in (T^*Y)_y \setminus \{0\}\}$. Thus, $\sigma(P)$ (more precisely $\sigma(P) \oplus \mathrm{Id}_{\mathbb{C}^N}$; see the Homotopy 2. below) can be extended to $(BX)|Y$.

<u>Homotopy 1</u>: For $\eta \in (T^*Y)_y \setminus \{0\}$, let $\pi_\eta^\pm : E_y \to M_\eta^\pm \subset E_y$ be the projections defined by (2). By means of

$$s\sigma(\eta) + (1-s)(i\pi_\eta^+ - i\pi_\eta^-), \quad s \in I, \tag{4}$$

we obtain a homotopy of $\sigma(\eta)$ to the map $i\pi_\eta^+ - i\pi_\eta^-$; the geometric meaning of this is that one may concentrate the eigenvalues of $\sigma(\eta)$ on

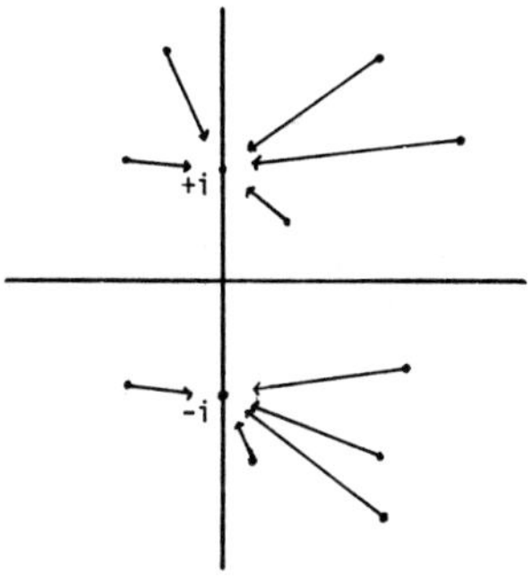

the eigenvalues $\{+i\}$, $\{-i\}$ which are independent of η, while the eigenspaces still depend on η. By using $h_0 = h_0^+ = h_0^- \in E_y$ with $h_0^\pm \in M_\eta^\pm$ one calculates that the eigenvalues of the endomorphism defined in (4) always remain non-real; thus, $z\mathrm{Id}$ + (4) is nonsingular for $z \in \mathbb{R}$. Hence, we have a homotopy in the space of elliptic symbols (i.e., in $(SX)|Y \to GL(N,\mathbb{C})$ here) from σ to σ_1 with $\sigma_1(z\nu+\eta) := z\mathrm{Id} + i\pi_\eta^+ - i\pi_\eta^-$.

<u>Homotopy 2:</u> By means of $e^{-i\phi}i\pi_\eta^+ - e^{i\phi}i\pi_\eta^-$, $\phi \in [0,\pi/2]$, each $\sigma_1(\eta)$ in $\mathrm{Iso}(E_y,E_y)$ can be connected with the identity, but this deformation depends on the choice of η and does not go through uniformly for all $\eta \in (T^*Y)_y - 0$. $GL(N,\mathbb{C})$ is too small to implement the further homotopy. Hence, we enlarge σ_1 by direct sum with $\mathrm{Id}_G \oplus \mathrm{Id}_G$ to a map $(SX)|Y \to \mathrm{Iso}(E \oplus G \oplus G,\ E \oplus G \oplus G)$, which we also denote by σ_1. Because of the splitting $E_y \cong M_\eta^- \oplus M_\eta^+$, $\eta \in (T^*Y)_y \setminus \{0\}$, we have

$$\sigma_1(\eta) = \begin{pmatrix} -i & & & 0 \\ & i & & \\ & & 1 & \\ 0 & & & 1 \end{pmatrix},$$

where the diagonal elements mean respectively the identities on M_η^- or M_η^+ or G multiplied by the coefficients $-i$ or $+i$ or 1. By the deformation $e^{-i(\pi/2)s}\mathrm{Id}_G \oplus e^{i(\pi/2)s}\mathrm{Id}_G$, $s \in I$, we can uniformly deform $\sigma_1(\eta)$ to

$$\sigma_2(\eta) := \begin{pmatrix} -i & & & 0 \\ & i & & \\ & & -i & \\ 0 & & & i \end{pmatrix}$$

on $M_\eta^- \oplus M_\eta^+ \oplus G_y \oplus G_y$.

Homotopy 3: Now it remains to deform $\sigma_2(\eta)$ to a constant ($\sigma_2(\eta)$ still depends on the positions of the eigenspaces $M_\eta^\pm$), in such a way that no real eigenvalues appear and $\sigma_2(z\nu+\eta) := z\mathrm{Id} + \sigma_2(\eta)$ does not become singular for $z \in \mathbb{R}$ under the deformation. This we achieve with the help of the boundary isomorphism $\beta_\eta^+ : M_\eta^+ \cong G_y$ given by the ellipticity of the boundary-value problem in (3). Indeed, there is a homotopy (which switches the second and third diagonal members in $\sigma_2(\eta)$) of $\sigma_2(\eta)$ to a constant map (on $(SY)_y$)

$$\eta \longmapsto \sigma_3(\eta) := \begin{pmatrix} 1 & & & \\ & 0 & (\beta_\eta^+)^{-1} & \\ & \beta_\eta^+ & 0 & \\ & & & 1 \end{pmatrix} \circ \sigma_2(\eta) = \begin{pmatrix} -i & & & \\ & -i & & \\ & & i & \\ & & & i \end{pmatrix},$$

which is multiplication by $-i$ on E_y and by i on $G_y \oplus G_y$; the homotopy

$$\begin{pmatrix} 0 & (\beta_\eta^+)^{-1} \\ \beta_\eta^+ & 0 \end{pmatrix} \sim \begin{pmatrix} \mathrm{Id} & 0 \\ 0 & \mathrm{Id} \end{pmatrix}$$

follows from the Homotopy Lemma which says that

$$\begin{pmatrix} 0 & B \\ A & 0 \end{pmatrix} \sim \begin{pmatrix} \mathrm{Id} & 0 \\ 0 & AB \end{pmatrix}$$

in the space of automorphisms of the vector space $V \times W$, if V and W are complex vector spaces and $A : V \to W$ and $B : W \to V$ are linear with $AB \in \mathrm{Iso}(W,W)$. One proves the Homotopy Lemma by composing two homotopies: First connect

$$\begin{pmatrix} 0 & B \\ A & 0 \end{pmatrix} \text{ and } \begin{pmatrix} i\mathrm{Id} & 0 \\ 0 & iAB \end{pmatrix} \text{ by } \begin{pmatrix} (i \sin \phi)\mathrm{Id} & (\cos \phi)B \\ (\cos \phi)A & (i \sin \phi)AB \end{pmatrix},$$

$\phi \in [0,\pi/2]$, and then multiplication by $e^{-i\psi}$, $\psi \in [0,\pi/2]$, provides the final homotopy.

Homotopy 4: We have now deformed $\sigma \oplus \mathrm{Id}$ to the constant map σ_3 on $(SY)_y$; we extend this map to a map on all of $(SX)_y$ which is homotopic to $\sigma_1 \oplus \mathrm{Id}$, by means of the parametrization

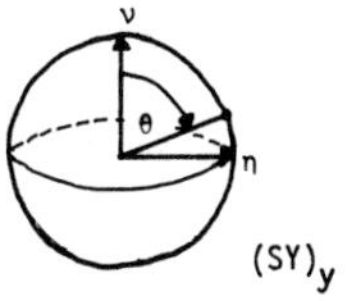

$$(SX)_y = \{(\cos \theta)\nu + (\sin \theta)\eta : \eta \in (SY)_y \text{ and } \theta \in [0,\pi]\}.$$

Namely, define on $E_y \oplus (G_y \oplus G_y)$ the automorphism

$$\sigma_3((\cos \theta)\nu + (\sin \theta)\eta) := \begin{pmatrix} \cos \theta - i \sin \theta & 0 \\ 0 & \cos \theta + i \sin \theta \end{pmatrix}$$

which by definition is homotopic to the constant map $\mathrm{Id} \oplus \mathrm{Id}$. □

B. Generalizations (Heuristic)

Remark 1. In the preceding proof, we have explicitly shown with a sequence of homotopies how one can continuously extend the map $\sigma \oplus \mathrm{Id}_N$: $SX \to GL(2N,\mathbb{C})$ to a map $\tilde{\sigma} : SX \cup (BX)|Y \to GL(2N,\mathbb{C})$ with $\tilde{\sigma}(y,0) = \mathrm{Id}_{2N}$ for all $y \in Y :=$ boundary of X. In that proof, we have chosen formulations that make the generalization for arbitrary bundles clear. To be sure, we must make precise what we understand by "stable homotopy"; i.e., why we are content with an extension of $\sigma \oplus I_{2G} : (SX)|Y \to \mathrm{Iso}(E \oplus G \oplus G, F \oplus G \oplus G)$ on $(BX)|Y$, even though this object does not immediately extend over all of SX, since G is defined only over Y. See also under Section III.1.C.

To bring about further generalizations of the proof, we remark that the special form of the boundary conditions (which were given by differential operators) played no role, since only the isomorphism β^+_η was needed; β_η^+ possibly could be defined through pseudo-differential boundary conditions. The definition of the spaces $M_\eta^\pm$ and the construction of the symbol homotopies are made very easy by the polynomial form of $\sigma(P)$, i.e., its derivation from a differential operator. This is why we have devoted so much space to this case study. The construction of an extension of $\sigma(P) \oplus \mathrm{Id}_N$ as an isomorphism over $(BX)|Y$ goes through more generally for elliptic pseudo-differential operators with "transmission properties" which allow elliptic boundary problems; see Chapter 8 and Section III.4.H. □

Remark 2. To each elliptic operator $P \in \mathcal{P}\mathrm{Diff}_k(E,F)$ over the n-dimensional Riemannian manifold X, one can assign a *local index*, namely the homotopy class of

$$\begin{array}{ccc} \sigma(P)(x,\cdot) : (SX)_x & \to & \mathrm{Iso}(E_x,F_x) \\ \wr\| & & \wr\| \\ S^{n-1} & \to & GL(N,\mathbb{C}) \end{array},$$

where N is the fiber dimension of E. By the Bott Periodicity Theorem, whereby (see Exercise III.1.8) for N sufficiently large (which we may achieve here by adding the identity)

$$\pi_{n-1}(GL(N,\mathbb{C})) = \begin{cases} \mathbb{Z} & \text{for } n \text{ even} \\ 0 & \text{for } n \text{ odd} \end{cases},$$

we obtain an integer degree(P) for the local index; by continuity, it is independent of the choice of x. If X is a manifold with boundary Y, then by II.6 we can determine the vector spaces $M_\eta^\pm$ and the integer $\dim M_\eta^+ - \dim M_\eta^- = \mu(P)$, which is independent of the choice of $\eta \in T'Y$ and does not automatically vanish for $n = 2$.

If the operator P admits elliptic boundary conditions, then the condition $\mu(P) = 0$ follows from Theorem II.6.2 and Theorem II.6.5, and by the preceding Theorem 1, the condition degree(P) = 0 holds. These two conditions are closely connected. By a communication from M. F. Atiyah, in the special case $N = 1$, $n = 2$ (where degree(P) is the classical winding number of $\sigma(P)(x,\cdot)$ in $\mathbb{C} \setminus \{0\}$ about the point 0), the equation

$$\mathrm{degree}(P) = \pm\mu(P) \tag{*}$$

holds. Also, in the general case, we have

$$\mathrm{degree}(P) = \pm\ \mathrm{degree}\ M^+ \tag{**}$$

where degree M^+ is the integer degree of a map $f_y : S^{n-3} \to GL(M,\mathbb{C})$, $y \in Y$, which is used to join together the trivial bundles over the upper and lower hemispheres of S^{n-2} along the equator S^{n-3} (see Appendix, before Exercise 4) to obtain the bundle $M_y^+ := \{M_\eta^+ : \eta \in (T^*Y)_y \setminus \{0\}\}$ over $(SY)_y \cong S^{n-2}$. Again, by continuity arguments, it is clear that degree M^+ does not depend on the choice of y of f_y. Equations (*) and (**) then represent a reformulation of the Bott Periodicity Theorem. For this, see also [Atiyah-Bott 1964a, 178], [Cartan-Schwartz 1965, 25-05], and in particular [Palais 1965, 351], where a K-theoretic formulation of (*) and (**) is given; see further, the Remark after Exercise 1.

As a result of Theorem 1, we have obtained (in degree(P) $\neq$ 0) a topological obstruction to there being elliptic boundary conditions for the elliptic differential operator P. In the classical theory $E = F = \mathbb{C}_X$, the obstruction can arise only in the case $n = 2$, since for $n \geq 3$ every homogeneous elliptic polynomial is of even order $2k$ and possesses an equal number k of zeros in the upper and lower half-planes [Hörmander 1963, 246]. The situation is different for $n = 2$, where degree$(\frac{\partial}{\partial \bar{z}}) = 1$. However, for all even $n \geq 4$ there are elliptic differential operators P with degree(P) $\neq$ 0; e.g., the Dirac operators (see [Palais 1965, 91f.] or [Booss 1972, 26 ff.]), defined using the Clifford module $\mathbb{R}^{2^{m-1}}$ over $X = \mathbb{R}^{2m}$, have local index 1.

The example is comparable to the pecularity shown by the Cauchy-Riemann operator $\partial/\partial\bar{z}$ in the classical theory for $n = 2$. Actually, there are many elliptic operators which arise in Riemannian geometry, that have a nonvanishing local index, and, for closed manifolds, characterize important topological invariants such as the "Euler number" and "signature" by their global indices; see [Booss 1972, 30] and [Atiyah-Patodi-Singer 1975, 46]. In the last work, as an expedient for the calculation involving the corresponding topological invariants of manifolds with boundary, a nonlocal theory of boundary-value problems is applied, with which one obtains a Fredholm theory in which the above "obstructions" become irrelevant.

<u>Exercise 1</u>. For each elliptic system

$$(P,R) : C^\infty(E) \to C^\infty(F) \oplus C^\infty(G)$$

of partial differential equations of order $k > 1$ with boundary conditions on the bounded domain X of $\mathbb{R}^n$, where E, F, G are trivial bundles, construct an "equivalent" (in which sense?) elliptic system

$$(\tilde{P},\tilde{B}) : C^\infty(\tilde{E}) \to C^\infty(\tilde{E}) \oplus C^\infty(\tilde{G})$$

of differential equations of order 1 with boundary conditions.

Hint: For the exact formulations, see [Agmon-Nirenberg 1964]. The operator $\tilde{P}$ which we define here is only elliptic in an extended sense, which one calls Douglis-Nirenberg elliptic. For reduction to the classical case (i.e., by splitting $\tilde{E}$ and applying Λ-type operators componentwise), see also [Hörmander 1966a, 134-136]. Here, we only give the rough idea (note also the Remark after Exercise 2). Let $P = \sum_{|\alpha|\le k} a_\alpha D^\alpha$, where $a_\alpha \in C^\infty(\mathrm{Hom}(E,F))$ (i.e., $a_\alpha(x)$ is a linear map from $\mathbb{C}^N$ to $\mathbb{C}^N$, if $E = F = \mathbb{C}^N_X$ and $x \in X$. Then begin with the "order lowering" procedure for ordinary differential equations of higher order, constructing a differential operator $\tilde{P} \in \mathrm{Diff}_1(\tilde{E},\tilde{E})$, where we set $\tilde{E} = \tilde{F} = \bigoplus_{0\le|\alpha|\le k} E \times\{\alpha\}$ and $\tilde{P}$ is given by the following matrix. (We consider only the case $N = 1$, $n = k = 2$, whence $P = a_{00} + a_{10}\partial_1 + a_{01}\partial_2 + a_{20}\partial_1^2 + a_{11}\partial_1\partial_2 + a_{02}\partial_2^2$, where $\partial_j = -i\frac{\partial}{\partial x_j}$.)

$$\begin{pmatrix} a_{00} & a_{10} & a_{01} & a_{20} & a_{11} & a_{02} \\ -\partial_1 & 1 & 0 & 0 & 0 & 0 \\ -\partial_2 & 0 & 1 & 0 & 0 & 0 \\ 0 & -\partial_1 & 0 & 1 & 0 & 0 \\ 0 & 0 & -\partial_1 & 0 & 1 & 0 \\ 0 & 0 & -\partial_2 & 0 & 0 & 1 \end{pmatrix}.$$

Write the matrix in greater generality! How does the corresponding $\tilde{R}$ appear, if $R = (R_1,\ldots,R_r)$ are differential boundary operators with $R_j : C^\infty(E) \to C^\infty(G_j)$ of order k_j where $G_j = \partial X \times C^{N_j}$ and $G = \bigoplus_{j=1}^r G_j$. Compare $\mathrm{Ker}(P,R)$ and $\mathrm{Ker}(\tilde{P},\tilde{R})$ and correspondingly, $\mathrm{Coker}(P,R)$ and $\mathrm{Coker}(\tilde{P},\tilde{R})$; show $\mathrm{index}(P,R) = \mathrm{index}(\tilde{P},\tilde{R})$. □

Exercise 2. Consider the disk $X := \{z \in \mathbb{C} : |z| \le 1\}$ with the circle $Y := \{z \in \mathbb{C} : |z| = 1\}$ as boundary.

a) Go through the construction of Theorem 1 for the "transmission operator" (see Exercise 1.9)

$$T : (u,v) \longmapsto \left(\frac{\partial u}{\partial \bar{z}}, \frac{\partial v}{\partial z}, (u-v)|Y\right)$$

b) Change the Laplace operator Δ on X as in Exercise 1 to a system $\tilde{\Delta}$ of partial differential operators of order 1, and check that $\tilde{\Delta}$ is elliptic. What are the new boundary conditions that are obtained for the Dirichlet boundary-value problem $u|Y = 0$? For $y \in Y$ and $\eta \in (SY)_y = \{\pm 1\}$, determine the spaces $\tilde{M}^{\pm}_{\eta}$ and the isomorphism $\tilde{\beta}^{+}_{\eta}$. As in Theorem 1, extend $\sigma(\tilde{\Delta})$ by means of $\tilde{\beta}^{+}$ from $(SX)|Y = S^1 \times S^1$ to a map $\sigma(\tilde{\Delta}, \tilde{\beta}^{+}_{\eta})$ on the solid torus $(BX)|Y$.

c) Go through the corresponding construction for the Neumann boundary-value problem $\frac{\partial u}{\partial \nu}|Y = 0$, and compare $\sigma(\tilde{\Delta}, \tilde{\beta}^{+})$ with $\sigma(\tilde{\Delta}, \tilde{\gamma}^{+})$, where $\tilde{\gamma}^{+}_{\eta}$ denotes the boundary isomorphism obtained from the Neumann boundary-value problem relative to $\tilde{\Delta}$.

d) Does the map $\sigma(\Delta)$, which is the constant $-i$ on SX, have different extensions to $SX \cup (BX)|Y$? Can one apply the boundary isomorphisms β^{+}_{η} or γ^{+}_{η} here, and relate them to the results of a), b), and c)?

Remark. For the practical calculation of the index of an elliptic boundary-value system, the explicit determination of $\tilde{P}$ is often lengthy and in fact superfluous, since the index is completely determined by the symbol and boundary isomorphisms; see Theorem 8.3d below. In particular, it suffices to reduce elliptic boundary-value systems of order k to elliptic systems of order 1 on the "symbol level", just as in the previous Exercise where $\sigma(P) \oplus \mathrm{Id}_{\tilde{E}}$ is homotopic in $\mathrm{Iso}_{SX}(E \oplus \tilde{E}, F \oplus \tilde{E})$ to $\sigma(\tilde{P})$. This deformation procedure on the "symbol level" was introduced in [Atiyah-Bott 1964a, 180 ff], and is entirely analogous to the "linearization of polynomial clutching functions", which plays a decisive role in one of the proofs of the Bott Periodicity Theorem (see [Atiyah-Bott 1964b, 241 f.]). It has the merit of also carrying over to the case of arbitrary manifolds with boundary and nontrivial bundles E, F, and C. This requires some standard tricks: One must identify the bundles E and F in a neighborhood of the boundary Y of X, by means of the isomorphism $\sigma(P)(y,\nu) : E_y \to F_y$, where ν is the inner normal at the boundary point y, etc.

In such general cases it is possibly advantageous to go yet a step higher: One does not specify a homotopy-theoretic extension of $\sigma(P) \in \mathrm{Iso}_{SX}$ to an isomorphism over $SX \cup BX|Y$ by means of the boundary isomorphism β, but rather one directly constructs, from $\sigma(P)$ and β, a difference vector bundle in $K(BX, SX \cup BX|Y)$, as sketched in [Palais 1965, 346-351].

Beside these two ways of finding the "correct" topological object while preserving the information about the index of an elliptic boundary value problem, viz.

- linearization and continuation of the symbol by means of homotopies,

and

- K-theoretic axiomatic characterization of a difference vector bundle,

there is a third way oriented more strongly towards functional analysis, viz.

- assignment of families of Wiener-Hopf operators to elliptic boundary value problems.

All three approaches depend essentially on the Bott Periodicity Theorem (see Chapter III.1) and not only on its result, but on essential elements of the different proofs - just as, conversely, the Periodicity Theorem can be derived via an investigation of a simple boundary-value problem on the disc (see Exercise III.1.9 below).

In Theorem I.9.1, we already provided the Index Theorem for Wiener-Hopf operators. We will prove the Periodicity Theorem, following [Atiyah 1969, 116-120], as a generalization of this theorem. Hence, the third approach via Hilbert space theory is most convenient for us and offers some formal advantages for the demonstration of the analytical main theorems on the index of elliptic boundary-value problems. Admittedly, this approach may be less conceptual than the first and less elegant than the second.

8. ELLIPTIC BOUNDARY-VALUE SYSTEMS II (SURVEY)

Before coming to the definition and treatment of a class of "boundary value systems", chosen very large for our purposes, we elucidate by means of simple examples the technique of reducing a boundary-value problem to a problem on the boundary. This "Poisson principle" provides the main idea for the solution of elliptic boundary-value problems, particularly for the computation of the index.

A. The Poisson Principle

Siméon-Denis Poisson and other mathematicians of the beginning 19th century occupied themselves with the observation that in "equilibrium" the temperature and heat distribution in a three dimensional body is completely determined by the temperature distribution on its surface.

Similar statements are true for magnetic and electrostatic fields (e.g., the "potential" of an electric field is determined by the charges/charge densities on the surface). The mathematical content of such laws of nature can be expressed variously depending on the context:

Exercise 1 (Poisson Integral Formula): Show that every harmonic function on the disk $\{z \in \mathbb{C} : |z| < \rho\}$, which has a continuous extension to the boundary, can be expressed in terms of its values u_0 on the boundary curve $\{z : |z| = \rho\}$.

Hint: For $0 \leq r < \rho$ and $0 \leq \psi \leq 2\pi$, prove the formula $u = Lu_0$, where L is the "Poisson operator" given by

$$(Lu_0)(re^{i\psi}) := \frac{1}{2\pi}\int_0^{2\pi} \frac{\rho^2 - r^2}{\rho^2+r^2-2r\rho\cos(\psi-\phi)}\, u_0(\rho e^{i\phi})\,d\phi,$$

by going back to the Cauchy Integral Formula

$$f(z) = \frac{1}{2\pi i}\int_\gamma \frac{f(\zeta)}{\zeta - z}\,d\zeta, \qquad |z| < \rho, \text{ where } \gamma : [0,2\pi] \to \mathbb{C}$$

is the curve $\gamma(\phi) := \rho e^{i\phi}$ and f is holomorphic (in the disk of radius ρ). Observe that u is harmonic ($\Delta u = 0$) exactly when u is the real part of a holomorphic function. Details are found in [Ahlfors 1953, 175 ff], but see also [Dym-McKean 1972, 164]. □

Exercise 2. Let X be a codimension - 0, bounded, submanifold of $\mathbb{R}^n$ with C^∞ boundary Y, and let Δ be the Laplace operator and R_0, R_1 be differential operators on Y. Set $u_0 := u|Y$ and $u_1 := \frac{\partial u}{\partial \nu}|Y$ for $u \in C^\infty(X)$, where $\partial/\partial\nu$ is a normal field on the boundary Y of X. Show: The solution of the boundary-value problem

$$\Delta u = 0 \text{ in } X \text{ and } R_0u_0 + R_1u_1 = f \text{ on } Y \qquad (*)$$

is "equivalent" to the solution of a system of pseudo-differential equations

$$(\mathrm{Id}-Q_0)u_0 - Q_1u_1 = 0 \text{ and } R_0u_0 + R_1u_1 = f \qquad (**)$$

on Y, where $Q_0, Q_1 \in \mathcal{P}\mathrm{Diff}_0(Y)$ can be given explicitly.

Hint: Begin with Green's Formula,

$$\int_X (u\Delta v - v\Delta u)\,dx = \int_{\partial X} (u_0v_1 - u_1v_0)\,dy, \qquad u,v \in C^\infty(X),$$

where dy is the volume element on $\partial X = Y$. (More generally, for every differential operator $P \in \mathrm{Diff}_k(X)$, the difference $\langle u,Pv\rangle - \langle P^*u,v\rangle$ can be estimated, using Stokes' Theorem, by a form on ∂X in which only the derivatives on ∂X of u and v of order $\leq k-1$ enter; see the Appendix, Exercise 8, and [Palais 1965, 73-75]. Now determine (e.g., by elementary distribution theory) a "fundamental solution" E of the Laplace operator (i.e., $\Delta E = \delta$); in classical terminology E is the "Newtonian kernel" of Δ. For $u \in \operatorname{Ker}\Delta$ and $v := E$, derive the "Poisson Integral Formula"

$$u(x) = L(u_0,u_1)(x) := \int_{\partial X} u_0(y)\frac{\partial E}{\partial \nu}(x-y)\,dy - \int_{\partial X} u_1(y)E(x-y)\,dy,$$

$x \in X \setminus Y$. Now show (by letting $x \rightsquigarrow \partial X$) that pseudo-differential operators Q_0 and Q_1 are defined such that for each solution $u \in C^\infty(X)$ of the boundary-value problem (*), we have the equation $u_0 = Q_0u_0 + Q_1u_1$, and that conversely, each solution pair (u_0,u_1) of the pseudo-differential system (**) furnishes a solution $u := L(u_0,u_1)$ of (*), by means of the Poisson integral - and indeed, such that $u_1 = \frac{\partial u}{\partial \nu}$. Details are found in [Hörmander 1966a, 187] or [Jörgens 1970/1982, §§9.3-9.5]. □

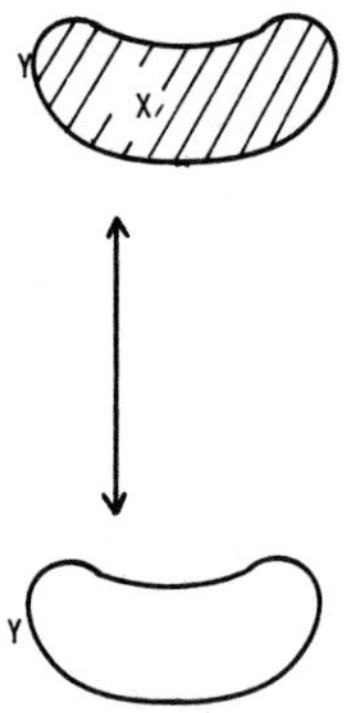

(*)

(**)

<u>Remark</u>. Pseudo-differential operators thus turn up very naturally when one goes from a boundary-value system, which may only contain differential operators, to a problem on the boundary.

B. <u>The Green Algebra</u>

The function analytic treatment of boundary-value problems is often troublesome. One reason for this lies in the natural "asymmetry" of the operators

$$(P,R) : C^\infty(E) \to C^\infty(F) \oplus C^\infty(G)$$

$$u \longmapsto (Pu,Ru),$$

where E and F are vector bundles over X, and G is a vector bundle over ∂X; so, a formal adjoint operator to (P,R) would be of an entirely different form and no longer defines a boundary-value problem. Also, one has no clear composition rules. Here, the following algebra which extends the class of boundary-value systems, is helpful; the construction goes back, in particular, to the work of Mark Josifovich Vishik and Grigory Ilyich Eskin. However, for the most part, we follow the specific approach of Louis Boutet de Monvel and employ his definition and method of proof; see [Boutet de Monvel 1971] and [Grubb-Geymonat 1977], where there is a non-technical introduction to the Boutet-de-Monvelian calculus in the appendix.

We consider systems of equations of the form

$$\begin{aligned}(P + G)e + Lg &= f\\ Re \qquad + Qg &= h\end{aligned}$$

or - in other words - operators of the form

$$\begin{pmatrix} P+G & L \\ R & Q \end{pmatrix} : \begin{matrix} C^\infty(E) \\ \oplus \\ C^\infty(G) \end{matrix} \to \begin{matrix} C^\infty(F) \\ \oplus \\ C^\infty(H) \end{matrix} ,$$

where E and F are Hermitian vector bundles over the compact, oriented, C^∞ Riemannian manifold X with boundary $Y := \partial X$, and G and H are Hermitian vector bundles over Y; e, f, g, h are C^∞ sections of the corresponding vector bundles. Here, $P \in \mathcal{P}\mathrm{Diff}_k(E,F)$ is a pseudo-differential operator over X, $Q \in \mathcal{P}\mathrm{Diff}_{k'}(G,H)$ is a pseudo-differential operator over the boundary Y of X, $R := \Sigma Q_i(P_i(..)|Y)$ is a "trace operator" which is composed of pseudo-differential operators Q_i over the boundary and P_i over all of X, and $L := \{P_0((..)\delta(Y))\}|(X \setminus Y)$ is a "Poisson operator". Here $P_0 \in \mathcal{P}\mathrm{Diff}_{k-1}(\tilde{G},F)$ where $\tilde{G}$ is a vector bundle over X with $\tilde{G}|Y = G$ and $\delta(Y)$ is the Lebesgue measure on the boundary Y of X, so that $g\delta(Y)$ can be regarded as a "generalized section" of the vector bundle $\tilde{G}$ for $g \in C^\infty(E)$. Hence, the operator L is defined in the sense of distribution theory. For a direct definition of these "Poisson operators" ("potentials") by means of Fourier representation - without distributions - see [Boutet de Monvel 1971, 26-28].

By a certain symmetry condition (the "transmission condition") on the behavior of the amplitudes of the pseudo-differential operators on the boundary, it follows that the images of $C^\infty(E)$ under P and $C^\infty(G)$ under L lie in $C^\infty(F)$; this symmetry condition holds in particular for differential operators and more generally for rational symbols, and represents no real restriction for our considerations. More precisely, one shows that, for each $g \in C^\infty(G)$, Lg lies in $C^\infty(F|(X\setminus Y))$ and extends to a C^∞ section of F (over all of X) which we also denote by Lg.

One has to add to P a "singular Green operator" G which is a finite sum of our "Poisson operators" and "trace operators" if one wants to describe (for example) the changes in the solution of an elliptic boundary-value problem resulting from a modification of the boundary conditions, or, more generally, if one wishes to secure the existence of a "parametrix" of this form (see Theorem 3a below) for arbitrary or only differential boundary-value systems.

We have the following theorems; we refer to [Boutet de Monvel 1971] and [Vishik-Eskin 1967] for their proofs.

Theorem 1. Let P, G, L, R, Q be defined as above, and let m be the order (the homogeneity of the symbol) of P, ℓ the order of L, r the order of R and $-m+\ell+r+1$ the order of Q. Then the operator

$$A = \begin{pmatrix} P+G & L \\ R & Q \end{pmatrix} : \begin{matrix} C^\infty(E) \\ \oplus \\ C^\infty(G) \end{matrix} \to \begin{matrix} C^\infty(F) \\ \oplus \\ C^\infty(H) \end{matrix}$$

possesses (for s sufficiently large) a continuous extension

$$A_s : \begin{matrix} W^s(E) \\ \oplus \\ W^{t+\ell-1/2}(G) \end{matrix} \to \begin{matrix} W^t(F) \\ \oplus \\ W^{s-r-1/2}(H) \end{matrix}, \quad t := s-m$$

to Sobolev spaces. (See Chapter 4 above.) □

Theorem 2. The operators of the stated type form an "algebra" (called the "Green algebra" by Boutet de Monvel) which is closed under sums A+B and composites $A \circ C$ - so far as they are defined, and in particular, their bundles or domains and codomains fit. In certain cases*, it is also closed under the formation of formal adjoints. Moreover, we have

While the adjoint L of a Poisson operator L always is a trace operator, the adjoint of a restriction operator $C^\infty(E) \to C^\infty(E|Y)$ (which arises in the Dirichlet problem, for example) is not a Poisson operator of the form $C^\infty(E|Y) \to C^\infty(E)$. One can place necessary and sufficient conditions on the corresponding symbol classes so that R* is a Poisson operator (namely, if R is of "class 0"; see [Grubb-Geymonat 1977]). Similar restrictions hold for G.

$$\sigma_X(A+B) = \sigma_X(A) + \sigma_X(B) \quad \text{and} \quad \sigma_X(A\circ C) = \sigma_X(A)\circ\sigma_X(C),$$

where $\sigma_X(A) := \sigma(P) \in \mathrm{Hom}_{S(X)}(E,F)$ is the *inner symbol* of A; hence, $\sigma(P)(x,\xi) \in \mathrm{Hom}(E_x,F_x)$ for $x \in X$ and $\xi \in (T^*X)_x$ with $|\xi| = 1$. (For the notion of "boundary symbol", that may be also introduced in greater generality, see the following discussion of the elliptic case.) □

C. The Elliptic Case

We first recall the approach chosen for deriving the main analytic theorems for elliptic operators on *closed* manifolds in Chapter 5 above: There, we were successful in deriving the various results (Fredholm properties) of the operators from formal properties of the associated symbols. Roughly speaking, the symbol map was the complete "linearization" of the operator. The latter was carried out from point to point and for each tangent vector individually so that this type of "snapshot", of pointwise "freezing", produced from an operator of functional analysis a family of matrices, the symbol, which could be investigated with the tools of linear algebra. (E.g., characterization of the ellipticity of the operator P by the condition that the matrices $\sigma(P)(x,\xi)$ are invertible for all non-vanishing covectors.)

For the analytical treatment of elliptic boundary-value systems, we need a refinement of this symbol calculus. For example, let us take an element

$$A = \begin{pmatrix} P+G & L \\ & \\ R & Q \end{pmatrix} : \begin{matrix} C^\infty(E) \\ \oplus \\ C^\infty(G) \end{matrix} \to \begin{matrix} C^\infty(F) \\ \oplus \\ C^\infty(H) \end{matrix}$$

of the "Green algebra" over the manifold X with boundary Y. For elliptic P, by definition, $\sigma(P)(y,\eta+t\nu) : E_y \to F_y$ is an isomorphism of vector spaces, where $y \in Y$, $\eta \in (SY)_y$, $\nu \in (SX)_y$ is the inwardly-directed normal relative to a Riemannian metric, and $t \in \mathbb{R}$. We denote by $p(y,\eta)$ the function defined on $\mathbb{R}$ by $t \mapsto \sigma(P)(y,\eta+t\nu)$ with values in $\mathrm{Aut}(E_y)$, if we identify F_y with E_y by means of $\sigma(P)(y,\eta+0)$. If P is an elliptic operator of order 0 and $\lim_{t\to+\infty} p(y,\eta+t\nu) = \lim_{t\to-\infty} p(y,\eta+t\nu)$,

then $p(y,\eta)$ defines (e.g., by means of the Cayley transformation; see Section I.9.G above) a continuous map $S^1 \to \mathrm{Aut}(E_y)$ and hence a "discrete Wiener-Hopf operator"

$$T_{p(y,\eta)} : H_0(S^1) \otimes E_y \to H_0(S^1) \otimes E_y$$

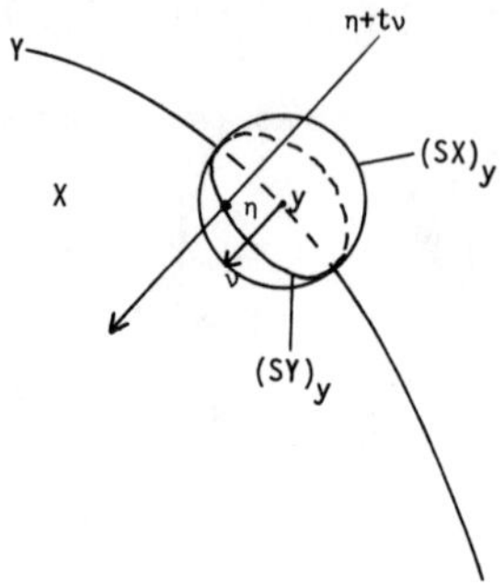

which is a Fredholm operator with

$$\text{index } T_{p(y,\eta)} = -W(\det(p(y,\eta)),0);$$

here (as in Chapter I.9), $H_0(S^1) \otimes E_y$ is the Hilbert space of square integrable functions on the circle S^1, with values in the vector space E_y, which can be analytically continued to the disk $\{z \in \mathbb{C} : |z| < 1\}$, and T_f (for $f : S^1 \to \text{Aut}(E_y)$) is the operator defined by pointwise multiplication by f and subsequent orthogonal projection onto $H_0(S^1) \otimes E_y$. See Chapter I.9 for details, and in particular, Exercise 6 there.

These constructions go through with certain modifications, if $p(y,\eta)$ does not define a continuous map $S^1 \to \text{Aut}(E_y)$: One then works, as in Exercise I.9.9 and Theorem I.9.2, on the real line and forms the corresponding multiplication operators with projection, the "continuous Wiener-Hopf operators" $W_{p(y,\eta)}$ which are Fredholm operators on suitably chosen "Hardy spaces". (If the order $k \neq 0$, we must develop the "Hardy space" $H_0(\mathbb{R})$ introduced in Section I.9.G into a Sobolev chain, as we have done with the Hilbert space $L^2(X)$ in Chapter II.4.) Details may be found in [Vishik-Eskin 1967, 304-311, in particular Lemma 3.1] and [Boutet de Monvel 1971, 14-20]; see also [Grubb-Geymonat 1976, Appendix].

In each case, from the ellipticity of P there follows the invertibility of $p(y,\eta)$ - regarded as a homomorphism $\sigma(P)(y,\eta+t\nu) : E_y \to F_y$ for all $t \in \mathbb{R}$ - but not the invertibility of $p(y,\eta)$ - regarded as a Wiener-Hopf operator. To each elliptic operator P on a manifold X with boundary Y, we can assign a family of Fredholm operators $W_{p(y,\eta)}$ which depends continuously on the unit covector $\eta \in SY$ whose basepoint y lies in Y. By Theorem I.7.1, an "index bundle" in $K(SY)$ can be constructed in a canonical way from such a family; we denote this bundle by $j(P)$, the *indicator bundle* of P. □

Exercise 3. Compare the construction of $j(P)$ with the definition of the bundle M^+ in Chapter II.6 for an elliptic differential operator P. In particular, investigate $P \in \{\Delta, \partial/\partial\bar{z}, \partial/\partial\bar{z} \oplus \partial/\partial z\}$.

Hint: [Boutet de Monvel 1971, 35], [Vishik-Eskin 1967, 328-331]. □

Exercise 4. Show that each operator $A = \begin{pmatrix} P+G & L \\ R & Q \end{pmatrix}$ of the Green algebra defines a family $\sigma_Y(A)$ of Wiener-Hopf operators $\sigma_Y(A)(y,\eta)$ parametrized by $\eta \in S(Y)$, and that the usual addition and composition rules for these "boundary symbols" hold.

Hint: [Grubb-Geymonat 1977, Appendix]. □

Exercise 5. Show that, for $y \in Y$ and $\eta \in (SY)_y$, $\sigma_Y(A)(y,\eta)$ is a Fredholm operator, if P is elliptic. Then define the indicator bundle $j(A) \in K(SY)$, as the index bundle of this family of Fredholm operators, and show $j(A) = j(P) + [\pi^*(G)] - [\pi^*(H)]$, where $\pi : SY \to Y$ is the projection.

Hint: The boundary symbol $\sigma_Y(\tilde{A})(y,\eta)$ with $\tilde{A} = \begin{pmatrix} 0 & L \\ R & Q \end{pmatrix}$ is an operator of finite rank for each $y \in Y$ and $\eta \in (SY)_y$. □

We can now define $A = \begin{pmatrix} P+G & L \\ Q & R \end{pmatrix}$ to be *elliptic* when P is elliptic and the Wiener-Hopf operator $\sigma_Y(A)(y,\eta)$ is invertible, for all $y \in Y$ and $\eta \in (SY)_y$. The purpose of the boundary conditions R and the potential L is, by this definition, exactly to "use up" the kernel and cokernel of the Wiener-Hopf operators $W_{p(y,\eta)}$. We then have $j(A) = 0$, and hence $j(P) = [\pi^*H] - [\pi^*G]$. □

Exercise 6. For a differential operator P with (differential) boundary operator R, show that (P,R) is an elliptic boundary-value system in the sense of Chapter II.6, exactly when $\begin{pmatrix} P & 0 \\ R & 0 \end{pmatrix}$ is an elliptic operator in the Green algebra.

Hint: Compare the condition of invertibility of the Wiener-Hopf operators $\sigma_Y(A)(y,\eta)$ with the Lopatinsky condition (II) in Section 6.A above. See also Exercise 3. How can $\mathcal{B}^+$ be interpreted? [Boutet de Monvel 1971, 45f.].

Theorem 3. If $\mathrm{Ell}_k(X;Y)$ is the class of elliptic operators in the Green algebra over the compact, oriented, Riemannian C^∞ manifold X with boundary $\partial X = Y$, then we have

a) Each $A \in \mathrm{Ell}_k(X;Y)$ possesses a "parametrix" $B \in \mathrm{Ell}_{-k}(X;Y)$;

i.e., $AB - Id$ and $BA - Id$ are operators of order -1.

b) If $A \in Ell_k(X;Y)$, then the extensions A_s, defined above in Theorem 1 for $s >> 0$, are Fredholm operators on the corresponding Sobolev spaces.

c) If A, B, and C are elliptic, then

$$\operatorname{index}(A \oplus B) = \operatorname{index} A + \operatorname{index} B.$$

Also, if $A \circ C$ is well defined, then

$$\operatorname{index}(A \circ C) = \operatorname{index} A + \operatorname{index} B.$$

d) Two elliptic operators A and B are called *stably equivalent* ($A \sim B$), if the interior and boundary symbols of $A \oplus Id_N$ and $B \oplus Id_M$ can be continuously deformed into each other while maintaining ellipticity. Then, we have

$$\operatorname{index} A = \operatorname{index} B.$$

e) If R, R' are boundary operators, and L, L' are Poisson operators, and Q is a pseudo-differential operator on the boundary, then

$$\begin{pmatrix} Id-L'R' & L \\ R & Q \end{pmatrix} \sim \begin{pmatrix} Id-R'L' & -R'L \\ -RL' & Q-RL \end{pmatrix},$$

where the second operator is a pseudo-differential operator on the boundary.

Arguments: a) follows easily from the definition of ellipticity, if one knows that the inverse of the boundary symbol $\sigma_Y(A)$ is again the boundary symbol of an operator of the Green algebra; see [Boutet de Monvel 1971, 19f. and 34f]. Incidentally, for an elliptic boundary system $A = \binom{P}{R}$ in the sense of Chapter II.6, one can find a parametrix of the form $B = (\tilde{P}+G \quad L)$. One obtains b), c), and d) from Theorem 1 and the general theory of Fredholm operators in Hilbert space as the corresponding statements in Chapter II.5. e) is the consequential (see Exercise 7 below) result of a topological exercise. [Boutet de Monvel 1971, 44f]. □

Exercise 7 (M. S. Agranovich and A. S. Dynkin, 1962): If $A_1 = \binom{P}{R_1}$ and $A_2 = \binom{P}{R_2}$ are two elliptic boundary-value systems for the same elliptic operator P on the manifold X with boundary Y, then $\operatorname{index} A_1 - \operatorname{index} A_2 = \operatorname{index} Q$, where Q is a pseudo-differential operator on Y defined in a canonical way by means of A_1 and A_2.

Hint: Set $Q := R_2L_1$, if $B_1 := \begin{pmatrix} P' & L_1 \\ 0 & 0 \end{pmatrix}$ is a parametrix for A_1, and show that $A_2B_1 \sim Q$ with Theorem 3e.

Part III. The Atiyah-Singer Index Formula

"In perhaps most cases when we fail to find the answer to a question, the failure is caused by unsolved or insufficiently solved simpler and easier problems. Thus all depends on finding the easier problem and solving it with tools that are as perfect as possible and with notions that are capable of generalization."
(D. Hilbert, 1900)

1. INTRODUCTION TO ALGEBRAIC TOPOLOGY (K-THEORY)

It is the goal of this part to develop a larger portion of algebraic topology by means of a theorem of Raoul Bott concerning the topology of the general linear group $GL(N,\mathbb{C})$ on the basis of linear algebra, rather than the theory of "simplicial complexes" and their "homology" and "cohomology". There are several reasons for doing so. First of all, it is of course a matter of taste and familiarity as to which approach to "codifying qualitative information in algebraic form" (Atiyah) one prefers. In addition, there are objective criteria such as simplicity, accessibility and transparence, which speak for this path to algebraic topology. Finally, it turns out that this part of topology is most relevant for the investigation of the index problem.

Before developing the necessary machinery, it seems advisable to explain some basic facts on winding numbers and the topology of the general linear group $GL(N,\mathbb{C})$. Note that the group $GL(N,\mathbb{C})$ moved to the fore in Part II already in connection with the symbol of an elliptic operator, and that the group $\mathbb{Z}$ of integers was in a certain sense the topic of Part I, the Fredholm theory. In the following, Part III, the concern is - roughly - the deeper connection between the previous parts. Thereby we will be guided by the search for the "correct" and "promising" generalizations of the theorem of Israil Gohberg and Mark Krein on the index of Wiener-Hopf operators (see Chapter I.9, Theorem 1, Exercise 8 and Theorem 2). □

A. Winding Numbers

"How can numerical invariants be extracted from the raw material of geometry and analysis?" (Hirsch). A good example is the concept of winding number, surely the best known item of algebraic topology: In his studies of celestial mechanics the French physicist and mathematician Henri Poincaré turned to stability questions of planetary orbits. Many of the related problems are not completely solved even today (e.g., the "three body problem" of describing all possible motions of three points which are interrelated by way of gravitation. However, this problem is solved for practical purposes, as is shown by the successful landing of the lunar module Luna 1.)

As a tool for the qualitative investigation of non-linear (ordinary) differential equations, Poincaré introduced in 1881 the notion of the "index" $I(P_0)$ of a "singular point" P_0 for a system of two ordinary differential equations

$$\dot{x} = F(x,y)$$

$$\dot{y} = G(x,y).$$

To do this, surround P_0 by a closed curve C in the phase portrait (see the following examples) and measure on it the angle of the rotation

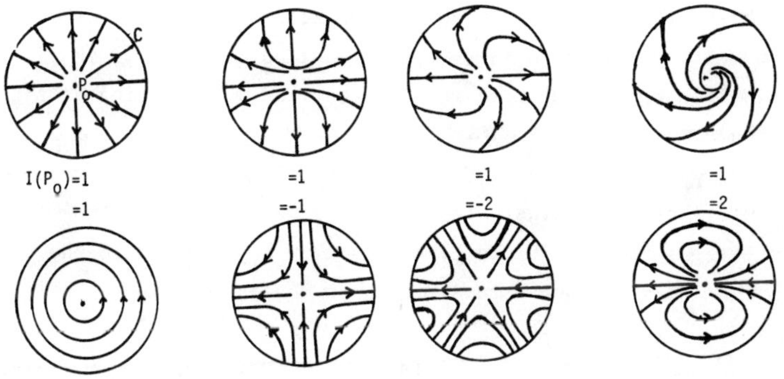

performed by the vector field $(F(x,y),G(x,y))$ when (x,y) once passes along C counterclockwise. The angle is an integral multiple of 2π, and this integer is $I(P_0)$. In case of a magnetic field one can actually see $I(P_0)$ in the rotation of the needle, when a compass is moved along C. Among other things, one has the theorem (see [Abraham-Marsden 1978,

75-76]): If the equilibrium position P_0 is stable, then $I(P_0) = 1$. We will come back to this in Section III.4.E.

Poincaré returned to this topological argument in 1895, when he considered all closed curves in an arbitrary "space" and classified them according to their deformation properties.* His simplest result can be expressed in today's terminology (we follow [Atiyah 1967b, 237-241]) as follows:

Theorem 1. Let $f : S^1 \to \mathbb{C}^\times$ be a continuous mapping of the circle S^1 to the punctured plane of non-zero complex numbers $C^\times$. In other words, we have a closed path in the plane not passing through the origin. Then the following hold:

(i) f possesses a "winding number" which states how many times the path rounds the origin; we write $W(f,0)$ or degree(f).

(ii) This degree is invariant under continuous deformations.

(iii) degree(f) is the only such invariant, i.e., f can be deformed to g, if and only if degree(f) = degree(g).

(iv) For each integer m, there is a mapping f with degree(f) = m.

Arguments: Instead of a formal proof, we briefly assemble the different ways of defining or computing degree(f).

Geometrically: Replace f by $g := f/|f|$. This is a mapping on S^1 to S^1. Approximate g by a differentiable map h, and count (algebraically, i.e. with a sign convention according to the derivative of h) the number of points in the preimage of a point which is in general position. This method can also be characterized as "counting of the intersection numbers": Draw an arbitrary ray emanating from the origin which does not pass through a self-intersection point of the path. Now count the intersections of the path with the ray according to the "traffic rule of the right of way" (H. Weyl) - thus with a plus sign if the path has the right of way, and a minus sign when the ray has the right of way.

*The "fundamental group" was introduced in Poincaré's work Analysis Situs (Oeuvres 6, 193-288), whose theme is purely topological-geometric-algebraic: an "analysis situs in more than three dimensions." P. expected the abstract formalism to "do in certain cases the service usually expected of the figures of geometry." He mentioned three areas of application: In addition to the Riemann-Picard problem of classifying algebraic curves, and the Klein-Jordan problem of determining all subgroups of finite order in an arbitrary continuous group, he particularly stressed its relevance for analysis and physics: "one easily recognizes that the generalized analysis situs would allow treating the equations of higher order and specifically those of celestial mechanics (the same way as H. P. had done it before with simpler types of differential equations; B.B.)... I also believe that I did not produce a useless work, when I wrote this treatise."

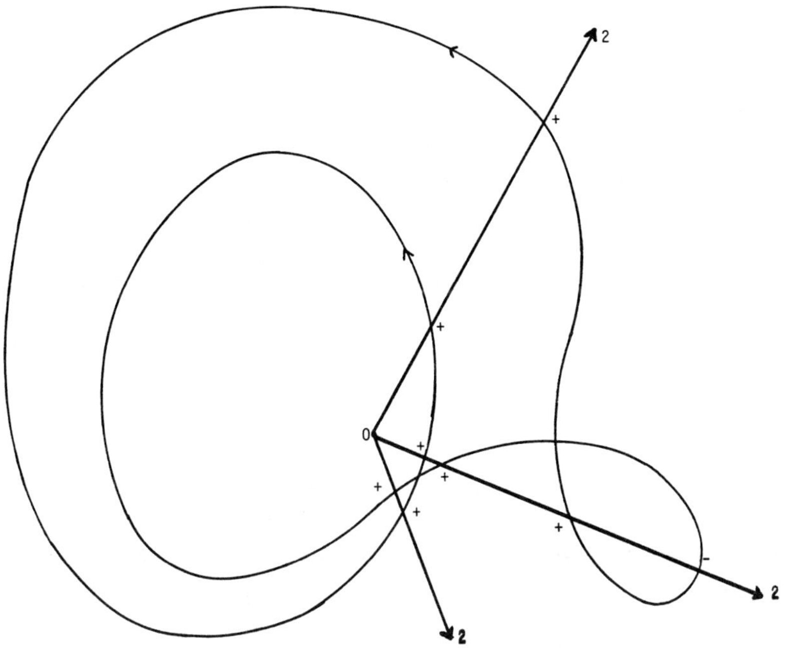

Combinatorially: We approximate with a piecewise linear path g, and then use combinatorial methods; i.e., we permit deleting and adding of those edges of our polygonal path which are boundaries of 2-simplices (these are triangles whose interior is completely contained in $\mathbb{C}^{\times}$, and thus do not contain the origin.

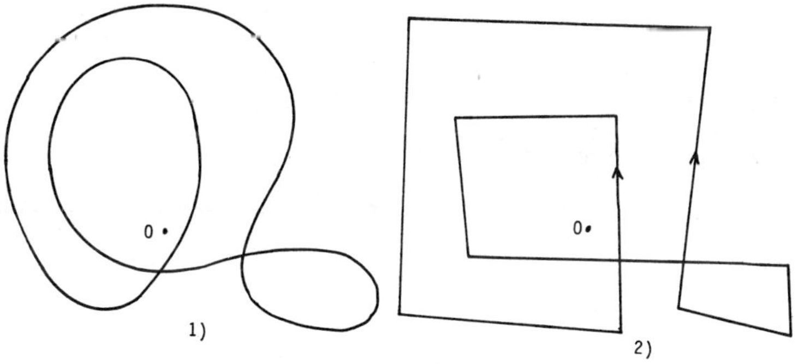

*(Continuation of footnote)
The complexity and limited understanding of the topological problems did not however permit Poincaré to carry out his program completely: "Each time I tried to limit myself I slipped into darkness."

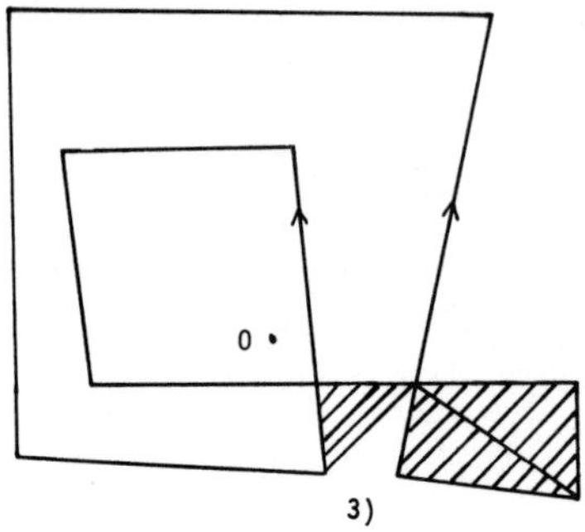

3)

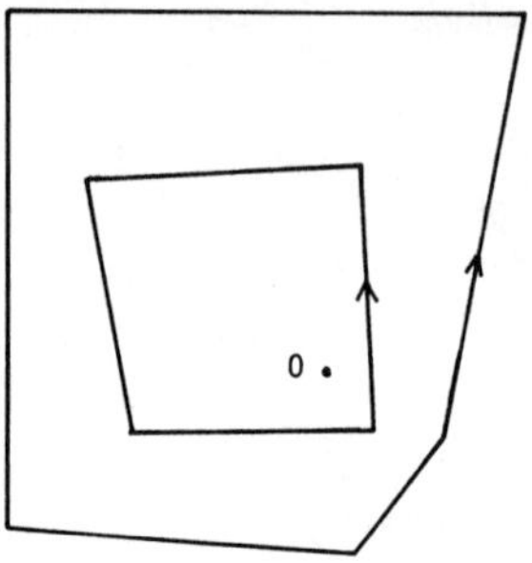

4)

Differential: We approximate by a differentiable g, and then set

$$\text{degree } f := \frac{1}{2\pi i}\int \frac{dg}{g}.$$

Here, we have regarded g as a map $[0,2\pi] \to \mathbb{C}^{\times}$, and then the integral is defined as $\int_0^{2\pi} \frac{\dot{g}(\tau)}{g(\tau)}\,d\tau$. From the Cauchy Integral Formula, it follows that the integral is a multiple of $2\pi i$, and hence degree f is an integer.

Algebraic: Approximate by a finite Fourier series

$$g(\phi) = \sum_{\nu=-k}^{k} a_\nu e^{i\nu\phi}.$$

We regard g as a map $S^1 \to \mathbb{C}^{\times}$ with $S^1 = \{z \in \mathbb{C} : |z| = 1\}$ and write $g(z) = \sum_{\nu=-k}^{k} a_\nu z^\nu$. Consider now the extension of g to the disk $|z| < 1$, where it is a finite Laurent series. We then obtain a meromorphic function h, and set

$$\text{degree } f := N(h) - P(h)$$

where N and P denote the number of zeros and poles in $|z| < 1$.

Function Analytic: Set degree $f := -\text{index } T_f$, where T_f is the Wiener-Hopf operator, assigned to f, on the space $H_0(S^1) \subset L^2(S^1)$ spanned by the functions $z^0, z^1, z^2, \ldots$. T_f is defined by

$$(\widehat{T_f u})(n) := \begin{cases} \sum_{k=0}^{\infty} \hat{f}(n-k)\hat{u}(k) & \text{for } n \geq 0 \\ 0 & \phantom{\text{for }} n < 0 \end{cases} \quad ; \quad u \in H_0(S^1),$$

where $\hat{f}(m) := \langle f, z^m\rangle$ is the m-th Fourier coefficient of f. For the details of this, see Theorem I.9.1 above - and Theorem I.9.2 for the analogous representation degree $f = \text{index}(I + K_\phi)$ via the (continuous) Wiener-Hopf operator

$$(K_\phi u)(x) := \int_0^\infty \phi(x-y)u(y)dy; \quad x \in \mathbb{R}_+, \quad u \in L^2(\mathbb{R}^+),$$

where $\phi \in L^1(\mathbb{R})$ with $\hat{\phi} = f \circ \kappa - 1$, and $\kappa t := \frac{t-i}{t+i}$ is the Cayley transformation.

As a first example, one may recall the standard map $a : S^1 \to \mathbb{C}^\times$ which is given by $a(z) = z$, $z \in \mathbb{C}$, $|z| = 1$. Here the equivalence of the various definitions is clear. We omit the proof for complicated cases and refer to [Ahlfors 1953, 151]. Compare also Theorem 2 and Theorem 7 below and - in another context - [Hirsch 1976, 120-131] or [Bröcker-Jänich 1973, 161f.]. □

In Theorem 1 all closed curves in the punctured plane are compared and the "essentially different" ones separated. The different definitions of winding number listed there reflect the main branches of topology with their different techniques, goals and connections. Accordingly, depending on the point of view chosen, very many generalizations of Theorem 1 to higher dimensions are possible (see also Exercise 10 below):

If one sticks with the classification of systems of ordinary differential equations - which was the point of departure for Poincaré's topological papers - one would first try to distinguish the different possibilities of bending the real line into a closed curve in space or other higher dimensional spaces. In this fashion H. Poincaré (but also see the footnote above) conceived - among other things - the "fundamental group" $\pi_1(X,x_0)$ of a space X which arises from the homotopic classification of closed paths $S^1 \to X$ (which pass through the point $x_0 \in X$). Here only the embedding question (which depends on the structure of X) is of interest, while the embedded images themselves of the compactified line are topologically identical. The reason is that there is just one way of bending the real line into a closed manifold, namely the form of a circle which may be traversed several times and may wind so many times around one or the other hole, but still remains topologically a circle.

The situation is different, when we pass from ordinary to partial differential equations. Here the classification essentially requires a differentiation between the various possibilities of bending the plane or higher dimensional Euclidean spaces into closed manifolds (see also

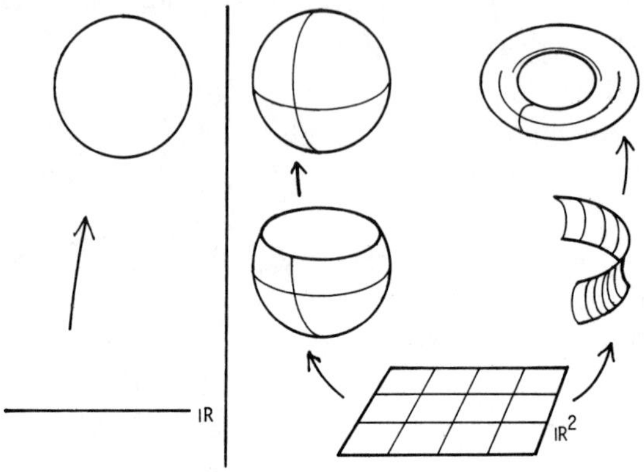

[Atiyah 1976b].) Now "genuine" global difficulties arise, since - as the above diagrams illustrate - there are already for the plane $\mathbb{R}^2$ very different ways of "bending it together." For $\mathbb{R}^2$ it is still possible to survey completely the different forms which can be classified according to the "genus" of the surface, the number of its "handles" (see for instance Hirsch 1976, 204f.). The corresponding problem of the "bending" of $\mathbb{R}^3$ has not been solved, although it is of special importance for the analysis of space-time processes of the real world by means of partial differential equations. For example it is as yet unknown whether the "Poincaré conjecture" (according to which every simply-connected, three dimensional, closed manifold is homeomorphic to the 3-sphere S^3) is true or not.

The following generalization of Theorem 1 is due to Raoul Bott. It is a true achievement of topology, and in addition, touches on the heart of the index problem for systems of elliptic differential equations. □

B. The Topology of the General Linear Group

We consider continuous maps $f : S^{n-1} \to GL(N,\mathbb{C})$, $2N \geq n$, where S^{n-1} is the unit sphere in $\mathbb{R}^n$ and $GL(N,\mathbb{C})$ is the general linear group of all invertible linear maps from $\mathbb{C}^N$ to $\mathbb{C}^N$.

<u>Theorem 2</u> (<u>R. Bott, 1958</u>). If n is odd, each such f can be deformed to a constant map. If n is even, one can define an integer $\deg(f)$ such that f can be deformed to another map g exactly when $\deg(f) = \deg(g)$. Moreover, there exist maps having arbitrarily prescribed integer degree.

Arguments. (*We will completely prove Theorem 2 in a different form below*):

1. Theorem 1 is the special case of Theorem 2 for $n = 2$ and $N = 1$. For $n = 1$ and N arbitrary, we recover the well-known fact that $GL(N,\mathbb{C})$ is pathwise connected; see above, before Exercise I.6.3.

2. As we have formulated it here, the Bott Theorem is expressed in terms of deformations (i.e., "homotopies"), a central theme of topology. In the formalism of homotopy theory, the Bott Theorem says that for all n, N with $2N \geq n$ the homotopy groups $\pi_{n-1}(GL(N,\mathbb{C}))$ (i.e., the group of homotopy classes of maps $S^{n-1} \to GL(N,\mathbb{C})$) are given as follows:

$$\pi_{n-1}(GL(N,\mathbb{C})) \cong \begin{cases} 0 & \text{if } n \text{ is odd} \\ \mathbb{Z} & \text{if } n \text{ is even.} \end{cases}$$

In the second case, the isomorphism is given by

$$\text{degree} : \pi_{n-1}(GL(N,\mathbb{C})) \overset{\cong}{\to} \mathbb{Z}.$$

Using these concepts the classically expressed result of our Theorem 1 above means that the first homotopy group (the "fundamental group") of $\mathbb{C}^\times$ is isomorphic to $\mathbb{Z}$.

Since Theorem 2 yields an isomorphism of $\pi_{n+1}(GL(N,\mathbb{C}))$ with $\pi_{n-1}(GL(N,\mathbb{C}))$, it is also known as a "periodicity theorem." Incidentally, there is a corresponding theorem (with period 8) for $GL(N,\mathbb{R})$. It has close connections with the theory of real elliptic skew-adjoint operators; see Section III.4.J below.

3. As with Theorem 1 above, there are various ways in which the degree (for even n) can be defined:

First, a *differential* definition of degree(f) is possible with the help of a known, explicitly defined, invariant differential form ω on the C^∞ manifold $GL(N,\mathbb{C})$. For the not entirely simple definition of this "Weltkonstante" (F. Hirzebruch), we refer to [Hirzebruch 1966, 587f]. One then sets

$$\text{degree}(f) := \int_{S^{n-1}} f^*(\omega),$$

where $f^*(\omega)$ is the pull-back form over S^{n-1}, and shows (!) that the invariant, so defined, is an integer.

As an alternative to this direct, somewhat computationally cumbersome definition, one can also define degree(f) *geometrically* - by means of a stepwise reduction to the more intuitive, but topologically no less demanding, notion of "mapping degree" of a continuous mapping of the (n-1)-sphere into itself. (In Theorem 1 the two concepts still coincided.) First one shows that, without loss of generality, one can take $2N = n$, since in the case $2N > 2$, f can be deformed into a map of the form $g(x) = \begin{pmatrix} h(x) & 0 \\ 0 & Id \end{pmatrix}$, where $h: S^{n-1} \to GL(n/2,\mathbb{C})$. All further constructions do not depend on the choice of g, since this gives, more precisely,

$$\pi_{n-1}(GL(N,\mathbb{C})) \cong \pi_{n-1}(GL(n/2,\mathbb{C})) \quad \text{for} \quad N \geq n/2.$$

{This follows from the exact homotopy sequence

$$\pi_{i+1}(U(m),U(n)) \to \pi_i(U(n)) \to \pi_i(U(m)) \to \pi_i(U(m),U(n))$$

induced by $U(n) \hookrightarrow U(m)$ for $n \leq m$ (where $U(n) := \{A \in GL(n,\mathbb{C}) : AA^* = I\}$), and the exact sequence

$$\pi_{i+1}(S^{2n+1}) \to \pi_i(U(n)) \to \pi_i(U(n+1)) \to \pi_i(S^{2n+1})$$

from the "fibration" $U(n) \to U(n+1) \to S^{2n+1}$. For $i < 2n$, we have $\pi_{i+1}(S^{2n+1}) = \pi_i(S^{2n+1}) = 0$; namely, show with the Sard Theorem (after differentiable approximation) that on account of different dimensions, not every map in such homotopy classes can be surjective, whence one always has a point in the complement of the image in S^{2n+1} from which one can contract. Thus, it follows that $\pi_{i-1}(U(m)) \cong \pi_{i-1}(U(n))$ for $1 \leq i \leq 2n \leq 2m$. Since the unitary group $U(n)$ is a deformation retract of $GL(n,\mathbb{C})$ (prove!), the statement follows; see [Steenrod 1951, §5.6 and §19.5].}

Hence, let $2N = n$. Then the first column of the matrix f defines a map $f_1 : S^{n-1} \to \mathbb{C}^N - \{0\}$. Since $\mathbb{C}^N - \{0\} = \mathbb{R}^n - \{0\}$, we have a map $g := f_1/|f_1| : S^{n-1} \to S^{n-1}$, for which a degree (the "mapping degree", the "most" natural generalization of Theorem 1) may be easily defined: One approximates g by a differentiable map h, chooses a point $y \in S^{n-1}$ "in general position" (i.e., so that the differential $h_{*|x}$ - see Section II.2.C above - has maximal rank for all $x \in h^{-1}(y)$, meaning that it is an isomorphism of tangent spaces), and counts the algebraic number of points in $h^{-1}(y)$ (i.e., the points x where h_{*x} reverses orientation are counted negatively). Details for this are found in [Hirsch 1976, 121-131]. For other definitions of the "mapping degree" of g, see [Bröcker-Jänich 1973/1982, 14.9.6-10] (on "intersection numbers" - thus, again a geometrical definition, but in a more general setting), and [Eilenberg-Steenrod 1952, 304.ff] (on the "homology" of S^{n-1}; i.e., fundamentally a combinatorial definition); furthermore, see Exercise 10 below ("K-theoretical" definition).

It turns out that the mapping degree of g (a) contains all the "essential" information on the qualitative behavior of f and (b) is always divisible by $(N-1)!$; we then define

$$\operatorname{degree}(f) := \frac{(-1)^{N-1}\operatorname{degree}(g)}{(N-1)!},$$

where the sign $(-1)^{N-1}$ is there for secondary technical reasons.

One can perhaps best visualize (a) (concerning the term "visualize" see 4 below. Strictly speaking it is more a comprehension aid) as follows: Topologically, the linear independence of the column vectors of a matrix means "approximately" the same as their being perpendicular. Thus, if such is the case, each function $S^{n-1} \to GL(N,\mathbb{C})$ is so "rigid" that it can be completely classified topologically by means of a single column.

In fact - and this now concerns (b) - the maps $S^{n-1} \to GL(N,\mathbb{C})$ are so "rigid" that in general, by far not every function $S^{n-1} \to \mathbb{C}^N \setminus \{0\}$ can appear as the first column. What is the reason? Isn't it possible to make an invertible matrix out of any non-zero vector by adding orthogonal vectors? Yes and no: It is possible to do so at each single point, but not if the additions are to be made uniformly in continuous dependence on the points of S^{n-1}. One may think of a sphere which as the 2-sphere has no unit tangential vector field (see Exercise 11 below), and thus eliminates the identity map as a possible first column.

For an evaluation of the "difficulty" of the divisibility theorem (b), see [Hirzebruch 1966, 182], where the theorem was proven in somewhat different form "pretty much at the end of the study (on the Theorem of Riemann-Roch, B.B.) as a corollary of cobordism theory, but which on the other hand, belongs "not at the end, but at the beginning", namely with "the Bott periodicity theory which is the basis of the newer proofs of the Riemann-Roch Theorem".*

---

*In the language of Section III.4.A, the facts are the following: The Chern character $ch_N : \tilde{K}(S^{2N}) \to H^{2N}(S^{2N};\mathbb{Q})$ is given by $(-1)^{N-1}c_N/(N-1)!$, since all other terms in the formula for ch_N cancel because of Bott periodicity which here takes the form $\tilde{K}(S^{2N}) \cong H^{2N}(S^{2N};\mathbb{Z})$. If E is the vector bundle over S^{2N} of complex dimension N, which is constructed by gluing with $f : S^{2N-1} \to GL(N,\mathbb{C})$, then (regarding the sign see above)

4. What can be said so far on the *substance* of the Bott Periodicity Theorem? We stay in the case n even, $2N \geq n$. Then, we have three statements:

(i) degree : $\pi_{n-1}(GL(N,\mathbb{C})) \to \mathbb{Z}$ is well defined,
(ii) the map is surjective, and
(iii) the map is injective.

As we have seen, for (i) two approaches are available, a "differential" one and a "geometrical" one. In this way, the degree was defined one-time as an integral of a differential form (hence it is, a priori, a real number), and the other time as a quotient of two integers (a priori, as a rational number). Either way, statement (i) causes no difficulties except for the integrability and divisibility theorems needed, since the homotopy invariance of degree is rather clear from the definition. One might simply say: (i) concerns the definition of an integer homotopy invariant - this is homology and comparatively simple.

Also, (ii) is comparatively simple: Namely, one can form (as when tensoring elliptic operators in Exercise II.5.7 above) the tensor product of $f : S^{n-1} \to GL(N,\mathbb{C})$ and $g : S^{m-1} \to GL(M,\mathbb{C})$,

$$f\#g : S^{m+n-1} \to GL(2MN,\mathbb{C}),$$

$$(x,y) \longmapsto \begin{pmatrix} f(x) \otimes \mathrm{Id}_M & -\mathrm{Id}_N \otimes g^*(y) \\ \mathrm{Id}_N \otimes g(y) & f^*(x) \otimes \mathrm{Id}_M \end{pmatrix};$$

here f and g are extended homogeneously to all of $\mathbb{R}^n$ resp. $\mathbb{R}^m$. From the simple multiplication formula degree$(f\#g)$ = (degree f)(degree g), it then follows that degree$(a_k) = 1$, where $a : S^1 \to GL(1,\mathbb{C})$ is the standard map $(a(z) := z)$ of degree 1 and the map

$$a_k := a \,\#\, \overset{k\text{-times}}{.\ \ .\ \ .} \,\#\, a : S^{2k-1} \to GL(2^{k-1},\mathbb{C})$$

is its k-fold power. Thus, product theory furnishes a generating map, meaning a generating element of the infinite cyclic group of homotopy classes of maps $S^{n-1} \to GL(N,\mathbb{C})$ for $n = 2k$, whence the surjectivity of "degree" follows.

---

*(footnote cont.)
degree $f = ch_N([E] - [\mathbb{C}^N_{S^{2N}}])$, where $[E]$ is the class of E in $K(S^{2N})$; furthermore the Nth Chern class $C_N(E)$ is mapped to degree(g) under the isomorphism $H^{2N}(S^{2N};\mathbb{Z}) \cong \mathbb{Z}$, where $g : S^{2N-1} \to S^{2N-1}$ is defined by means of f as before.

The statement (iii) is in contrast highly nontrival: While in (i) and (ii) one had to construct particular objects (a homotopy-invariant "number" and a suitable homotopically nontrivial map), one must now show that mappings of the same degree may be deformed into each other, in particular, that each map of degree 0 is homotopic to a constant map. Having defined the invariants, one wants to verify their relevance and determine their value. This is exactly "homotopy" and extremely non-conceptual. It can be seen from the fact that, although the homotopy classes of maps $S^{n-1} \to GL(N,\mathbb{C})$ (for $2N < n$) as well as those from spheres to spheres are in a fashion "closer" to our space-time comprehension, they are largely unknown, because of their extreme complexity.

5. Without commenting on the very different proofs of the Periodicity Theorem which exist to date, we would like to point out that *all* proofs proceed by induction on n, more precisely by induction from n to n+2. In the language of the product theory presented above, it must be shown that $f \longmapsto f \# a$ is an isomorphism of the homotopy group of dimension n-1 with the homotopy group of dimension n+1.

6. Interestingly, the "*algebraic*" definition of winding number (see Theorem 1) was initially not susceptible to generalization to the present situation. However, the discovery of the topological significance of elliptic boundary-value problems (see above II.6 - II.8 and below, Section III.4 H) lead Raoul Bott and Michael F. Atiyah to a new and elementary* proof of the Periodicity Theorem. In a very deep sense, which we will explain below in a particularly suitable form when presenting the proof, this proof generalizes and unifies the algebraic and function analytic definitions and furthermore "gets to the heart of the problem" (Atiyah). □

The "Lie group" $GL(N,\mathbb{C})$ and its real analogue play an important role everywhere in mathematics. Accordingly the Periodicity Theorem and its off-spring "K-theory" (see below) not only have immediate applications to the index problem of elliptic oscillations (see particularly Chapters 2-4 below) but also proved to be a useful tool for a number of deep geometric problems. An example is the computation of the number of linearly independent vector fields on a sphere which the British mathematician John Frank Adams carried out in 1962 with just these tools. A detailed exposition of this and some other applications can be found in [Husemoller 1975, Chapter 15].

---

* In comparison with the original proof which employed essentially the tools of the modern calculus of variations due to Marston Morse; see, e.g. [Milnor 1963, particularly 124-132].

The deep relation between the topology of the general linear group and the geometry of differentiable manifolds, which becomes manifest in these successes, can be described intuitively as follows: Among the simplest global topological invariants of a compact oriented n-dimensional manifold X is the "Euler characteristic" $e(X)$. A famous theorem, proven in 1895 by Henri Poincaré for $n = 2$, and in general by Heinz Hopf in 1925, says that $e(X)$ can be found by means of a differentiable structure on X as the number of "singularities". Here a singularity is an isolated zero x of a tangent vector field v on X, and the counting must be done with the proper multiplicity, namely the (local) "index" of v at x (i.e., the degree of the mapping $S^{n-1} \to S^{n-1}$) which is given by v on the surface of a ball about x. For a conceptually very plausible proof for $n = 2$ see [Brieskorn 1976, 166-171], see also Section III.4.D/E. below.

Today very many global topological invariants are known which are defined on X by means of a classical (Riemannian or complex) structure; see [Hirzebruch 1956/1966]. These "characteristic classes" are without exception generalizations of the Euler characteristic since "roughly, one considers the cycles where a given number of vector fields become "dependent", as [Atiyah 1968b, 59] remarks. The question, of which way a system of linearly independent vectors can become dependent, forms the link between topology and $GL(N,\mathbb{C})$ (and $GL(N,\mathbb{R})$). □

C. The Ring of Vector Bundles

Let X be a compact topological space. We consider the abelian semigroup $\mathrm{Vect}(X)$ (defined in the Appendix) of isomorphism classes of complex vector bundles over X. If X consists of a single point, then $\mathrm{Vect}(X) \cong \mathbb{Z}_+$. Now we generalize the construction which one uses to go from the semi-group $\mathbb{Z}_+$ to the group $\mathbb{Z}$, in such a way that we can assign a group $K(X)$ to the semigroup $\mathrm{Vect}(X)$.

Theorem 3. Each abelian semi-group A (with zero element) yields in a canonical way an abelian group $B := A \times A/\sim$ and a semi-group homomorphism $\phi : A \to B$ induced by $a \to (a,0)$. Here $\sim$ is the equivalence relation on $A \times A$ defined by

$$(a_1,a_2) \sim (a_1',a_2') \iff (a_1 \oplus a, a_2 \oplus a) = (a_1' \oplus a', a_2' \oplus a'), \quad a,a' \in A.$$

Proof: Let $\Delta: A \to A \times A$ be the diagonal homomorphism $a \mapsto (a,a)$ of semi-groups. Then B consists of the cosets of $\Delta(A)$ in $A \times A$; i.e., $B = \{(a_1,a_2) + \Delta(A) : a_i \in A,\ i = 1,2\}$, where $(a_1,a_2) + \Delta(A) :=$

$\{(a_1 \oplus a, a_2 \oplus a) : a \in A\}$. B is a quotient semi-group in which there is an inverse for each element given by

$$-(a_1,a_2) + \Delta(A) = (a_2,a_1) + \Delta(A).$$

Thus, B is a group. In this notation, the semi-group homomorphism is given by

$$\phi(a) := (a,0) + \Delta(a). \quad \square$$

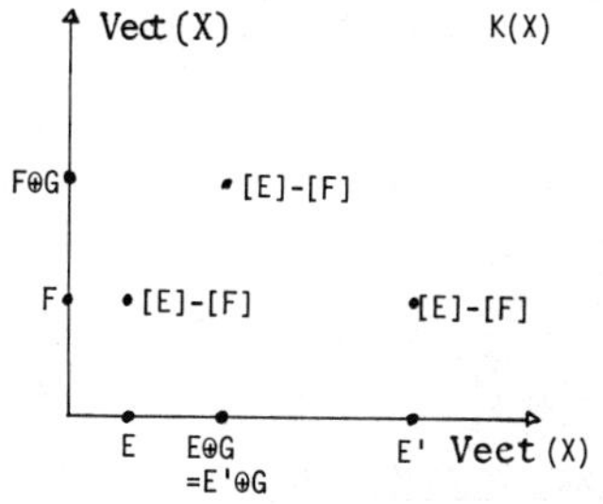

Remarks. 1. It is advisable to go through the definition of B carefully as we did in the proof, since the intuition one gains in the transition from $\mathbb{Z}_+$ to $\mathbb{Z}$ is partly deceptive: Namely, a semi-group is *not* always embedded in a group. (The cancellation rule must already hold in the semi-group). The natural homomorphism $\phi: A \to B$ is not necessarily injective; see Exercise 2.

2. In a certain sense, B is the "best possible" group that can be made from the semi-group A. More precisely, the following *universal property* of ϕ holds: Every semi-group homomorphism $h : A \to G$ from A to an arbitrary group G can be "factored" through ϕ in exactly one way; i.e., there is exactly one group homomorphism $h' : B \to G$ such that the adjacent diagram is commutative. The universal property is a generalization of the trivial observation that ϕ becomes an isomorphism, if A is already a group, and, conversely, the latter follows from the functorial property of the assignment $A \mapsto (B,\phi)$ on the category of semi-groups; see [Atiyah 1967a, 42f.].

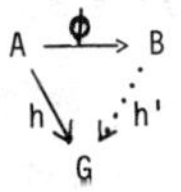

Alternatively, B can be defined as a quotient group FA/RA, where FA is the "free abelian group" on A (which consists of all finite linear combinations of elements of A with coefficients in $\mathbb{Z}$), and RA is the subgroup of FA generated by the subset $\{1(a_1 \oplus a_2) + (-1)a_1 + (-1)a_2 : a_i \in A\}$. The homomorphism $\phi: A \to FA/RA$ is defined in the natural way and fulfills the homomorphism condition $\phi(a_1 \oplus a_2) = \phi(a_1) +$

$\phi(a_2)$, since $\phi(a_1 \oplus a_2) - \phi(a_1) - \phi(a_2)$ is represented by $1(a_1 \oplus a_2) + (-1)a_1 + (-1)a_2$. That ϕ is the "universal solution of the factorization problem" is shown as follows: The uniqueness of h' is clear, since "$h'(\phi(a)) = h(a)$ for $a \in A$" implies that h' is already given on a set of generators of FA/RA, whence h' itself is uniquely defined. From $h'(\phi(a_1 \oplus a_2)) - h'(\phi(a_1)) - h'(\phi(a_2)) = 0$, it follows that h' is a group homomorphism. From the universal property, it follows easily that the two methods of group construction are equivalent, and in particular that B and FA/RA are isomorphic. □

Now let $A := \mathrm{Vect}(X)$, X compact. We denote the associated abelian group B by $K(X)$. Then, for each vector bundle E over X, we obtain (by means of ϕ) an element $[E] \in K(X)$, and every element of $K(X)$ can be written as a linear combination of such elements; see also Exercise 4a. For $[\mathbb{C}_X^N]$, we also simply write N.

Exercise 1. In the formalism developed here, describe the canonical extension of subtraction $\delta: \mathbb{Z}_+ \times \mathbb{Z}_+ \to \mathbb{Z}$ to the "difference-bundle construction" $\mathrm{Vect}(X) \times \mathrm{Vect}(X) \to K(X)$. First show $K(X) \cong \mathbb{Z}$, if X is a point.

Exercise 2. Show that the cancellation rule does not always hold in $\mathrm{Vect}(X)$.

Hint: First illustrate with the two real bundles TS^2 and $\mathbb{R}^2_{S^2}$ over the 2-sphere. These are not isomorphic (see Exercise 11 below), but forming the direct sum of each with the trivial line bundle $\mathbb{R}_{S^2}$ we arrive at isomorphic bundles (with the tangent bundle TS^2 one thinks of the direct sum with the normal bundle NS^2 of the canonical embedding of S^2 in $\mathbb{R}^3$). In general search for a nontrivial vector bundle E that becomes trivial when a trivial bundle is added to it. A detailed discussion of special "cancellation type" rules can be found in [Husemoller 1975, Chapter 8]; for example, the "Uniqueness Theorem for Vector Bundles" says that trivial bundles over manifolds of dimension n may be cancelled when the other summand has fiber dimension $\geq n/2$. □

Exercise 3. a) Show that each element of $K(X)$ can be written in the form $[E] - N$, where $E \in \mathrm{Vect}(X)$ and $N \in \mathbb{N}$.

b) Show that two vector bundles E and F define the same element of $K(X)$ (i.e., $[E] = [F]$) exactly when $E \oplus \mathbb{C}_X^N = F \oplus \mathbb{C}_X^N$, for some N.

c) One says that the bundles E and F are *stably-equivalent*, when there are natural numbers M and N such that

$$E \oplus ¢_X^N \cong F \oplus ¢_X^M.$$

Show that the set $I(X)$ of stable equivalence classes form a group relative to the operation of direct sum.

Hint for c): See Appendix, Exercise 6. □

Exercise 4. a) Show that, by means of the tensor product $\otimes$ for vector bundles (Appendix, Exercise 3), a multiplicative structure for $K(X)$ is furnished, making it a commutative ring with unit $[¢_X]$.

b) Show that each continuous map $f: Y \to X$ induces a ring homomorphism $f^* : K(X) \to K(Y)$ which only depends on the homotopy class of f.

c) Let $i : Y \hookrightarrow X$ be the inclusion of a closed subset Y of X. Set $K(X,Y) := \operatorname{Ker}(K(X/Y) \to K(Y/Y))$, where X/Y is the space obtained from X when Y is "collapsed" to a point $\{Y/Y\}$.

(i) Let Y consist only of a single point x_0. Show that the group $K(X,Y)$ forms an ideal in $K(X)$, and that $K(X)$ splits into a direct sum

$$\begin{aligned} K(X) &\cong K(X,x_0) \oplus K(x_0) \\ &\cong K(X,x_0) \oplus \mathbb{Z} = \tilde{K}(X) \oplus \mathbb{Z}, \end{aligned}$$

where $\tilde{K}(X) := K(X,x_0)$, the "essential" part of $K(X)$, is isomorphic to $I(X)$.

(ii) In general, define a natural map $j^* : K(X,Y) \to K(X)$, and show that the short sequence $K(X,Y) \overset{j^*}{\to} K(X) \overset{i^*}{\to} K(Y)$ is exact.

Hint for a): Use the "universal property" (Remark 2) to factorize $\operatorname{Vect}(X) \times \operatorname{Vect}(X) \to K(X)$ through $K(X) \times K(X)$. For b): Appendix, Theorem 2; in particular, $K(X) \cong K(Y)$ when X and Y are homotopy equivalent. For c): Work with the retraction $r : X \to \{x_0\}$, for the splitting in (i). See Exercise 3 for $I(X)$. First define j^* in (ii) more generally, when $j : (X',Y') \to (X,Y)$ is a map of pairs of spaces (i.e., $j: X' \to X$ continuous with $j(Y') \subset Y$). Then set $X' = X$ and $Y' = \phi$. To check that $\operatorname{Im}(j^*) \subset \operatorname{Ker}(i^*)$, factor

$$\begin{array}{ccc} (Y,\phi) & \xrightarrow{ji} & (X,Y) \\ & \searrow \quad \nearrow & \\ & (Y,Y) & \end{array},$$

and note that $K(Y,Y) = 0$. To prove the other direction, work with the representation as in Exercise 3a. See also [Atiyah 1967a, 69f.]. □

Theorem 4. Let X be a compact space and $[X,F]$ the set of homotopy classes of continuous maps $T : X \to F$, where F is the space of Fredholm operators in a Hilbert space H. The construction of index bundles (see Chapter I.7 above) induces a bijective map index: $[X,F] \to K(X)$ under which composition in F and addition in $K(X)$ correspond, as do adjoints in F and negatives in $K(X)$.

Proof: See Theorem I.7.3 above. □

1. K-Theory and Functional Analysis. In the proof of Theorem 3 and the subsequent remark, we learned two different constructions of the group $K(X)$ and of these the first is probably the most natural in connection with functional analysis (as in Theorem 4). On the other hand, the second construction immediately yields the "universal property" which historically motivated this formal group construction in the papers of Claude Chevalley and Alexandre Grothendieck concerned with algebraic geometry. It arose as a tool for the study of problems in which functions are involved which are additive on a semigroup with integral values. This may also explain the relevance of $K(X)$ for our index problem of elliptic operators:

In Part II, we associated with each elliptic pseudo-differential operator $P : C^\infty(E) \to C^\infty(F)$ (E and F C^∞-bundles over the closed Riemannian C^∞-manifold X) its symbol $\sigma(P) \in \mathrm{Iso}_{SX}(E,F)$, and we proved that index P only depends on the homotopy type of $\sigma(P)$. Now let $S'X := B^+X \cup_{SX} B^-X$ be the n-sphere bundle over X which arises by glueing two copies B^+X and B^-X of the covariant unit-ball bundle $BX := \{(x,\xi) : x \in X$ and $\xi \in (T^*X)_x$ with $|\xi| \leq 1\}$ along their common boundary SX. We lift E over B^+X and F over B^-X and glue them (Appendix, Exercise 4) over SX by means of $\sigma(P)$. This way we obtain a vector bundle on $S'X$ whose isomorphism class $[E \xrightarrow{\sigma(P)} F]$ only depends on the homotopy type of $\sigma(P)$. Conversely, the space of pseudo-differential operators is so rich that $\sigma(P)$ has any desired homotopy type for suitable P. Because of the special form of $S'X$, we therefore obtain all isomorphism classes of vector bundles over $S'X$ in this fashion. Thus the theory of elliptic equations yields a semigroup homomorphism index : $\mathrm{Vect}(S'X) \to \mathbb{Z}$ which fits into the following diagram:

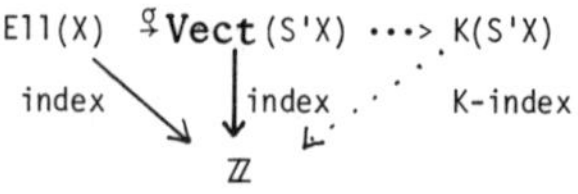
Ell(X)
g
Vect(S'X)
K(S'X)
index
index
K-index
ℤ

Here Ell(X) is the class of elliptic pseudo-differential operators on the closed manifold X. Thus the universal property of K(S'X) guarantees the existence and uniqueness of a K-index, which makes the diagram commutative and which being a group homomorphism is easier to analyze especially after more is known about the group K(S'X).*

2. K-theory and cohomology. Exercise 3 says that K is a contravariant functor on the category of compact topological spaces and continuous maps into the category of commutative rings (with identity) and ring homomorphisms, and this functor bears great resemblance (also in aspects not explained here) with the cohomology functor H^*; see [Eilenberg-Steenrod 1952, 13f]. The "characteristic classes" of vector bundles (see [Hirzebruch 1966]) yield a variety of interesting operations $K \to H^*$, and one can show that in fact $K(X) \otimes_{\mathbb{Z}} \mathbb{R} \cong H^{even}(X; \mathbb{R})$; see [Atiyah 1967a].

---

*Presently, three different ways are known for associating with an elliptic operator P a K-theoretic object via its symbol. The construction presented here which ends up in $K(B^+X \cup B^-X)$ is the most conceptual, since $[E \xrightarrow{\sigma(P)} F] \in K(B^+X \cup B^-X)$ can be represented directly by a vector bundle on $B^+X \cup B^-X$. The other constructions yield objects in K(BX,SX) or - equivalently, see the following section - in K(TX) and basically amount to forming the difference class $[E \xrightarrow{\sigma(P)} F] - [F]$. The advantage of this "non-conceptual" construction of a difference bundle (for details see Chapters 2/3 below) rests on the fact that the object $[E \xrightarrow{\sigma(P)} F]$ contains too much useless information on the specific form of the vector bundles E and F which is completely irrelevant to the index problem. For example, one can make E (or F) trivial by adding a vector bundle V (Appendix, Exercise 6) while the index of $P \oplus Id_V : C^\infty(E \oplus V) \to C^\infty(F \oplus V)$ does not change. More precisely: By evaluating the row-exact commutative diagram (↙•↗ is the identity on K(X)) one finds that $K(B^+X \cup B^-X) \cong K(X) \oplus K(BX,SX)$, whereby the first

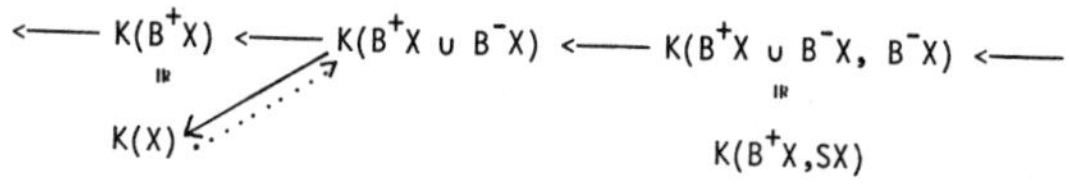
K(B+X)
K(B+X ∪ B-X)
K(B+X ∪ B-X, B-X)
K(X)
K(B+X,SX)

summand is irrelevant for the index problem and the second one is best dealt with in the non-relative form K(TX) in the framework of "K-theory with compact support."

D. K-Theory with Compact Support

Until now, we have only defined $K(X)$ for compact X. For locally compact X, we now set $K(X) := K(X^+,+) = \mathrm{Ker}(K(X^+) \xrightarrow{i^*} K(+))$, where $X^+ = X \cup \{+\}$ is the 1-point compactification of X (by the addition of the point $+$, where $i : \{+\} \to X^+$ is the inclusion of this point). For compact X, this definition brings nothing new. Alternatively, $K(X)$ can be expressed in terms of "complexes" of vector bundles - see [Atiyah-Singer 1968a, 489 ff.]; e.g., the elements of $K(Y \setminus Z) = \underset{\sim}{K}(Y,Z)$ can be taken to be equivalence classes of isomorphisms $\sigma: E|Z \overset{\cong}{\to} F|Z$, where E and F are (complex) vector bundles over the compact set Y with closed subset Z. (One thinks of the symbol of an elliptic operator over the manifold X and sets $Y := BX$ and $Z := SX$, where $Y \setminus Z$ is then diffeomorphic to the full cotangent bundle T^*X.)

<u>Exercise 5</u>. a) Verify that $K(X)$ is a ring (without unit element) when X is non-compact. b) Show functoriality for *proper* maps $f : Y \to X$; these are the maps which can be continuously extended to Y^+.

<u>Hint for b)</u>: One may also define f to be proper exactly when $f^{-1}(K)$ is compact for all compact subsets $K \subset X$. From this comes the notion of "K-Theory with compact support" - see also [Eilenberg-Steenrod 1952, 5 and 269 ff.]. In particular, each homeomorphism is proper, and we have $K(X) \cong K(Y)$ for homeomorphic X and Y, and $f^* = \mathrm{Id}$, if $Y = X$ and f is homotopic (within the class of homeomorphisms) to the identity. On the other hand, the mere homotopy type of X does not determine $K(X)$. (Example: $K(\mathbb{R}) \ncong K(\text{point})$). □

<u>Theorem 5</u>. If X and Y are locally compact spaces, then (in addition to the ring structures of $K(X)$ and $K(Y)$) there is an *outer product*

$$\boxtimes : K(X) \otimes K(Y) \to K(X \times Y).$$

<u>Remark</u>: The outer product admits a particularly simple and natural definition, if one adopts the above introduction of $K(X)$ via complexes, and forms the tensor product of complexes; see [Atiyah-Singer 1968a, 490] and also the outer tensor product for elliptic operators in Exercise II.5.7b and for matrix-valued functions in Section B of this chapter.

<u>Proof</u>: 1) If X and Y are compact, then $\boxtimes$ is well defined since one can form the vector bundle $E \boxtimes F$ over $X \boxtimes Y$, where E is a vector bundle over X, and F is over Y, and $E \times F$ has fiber $E_x \otimes F_y$ over (x,y).

2) In order to carry this definition over to locally compact X and Y, we prove the exactness of the short sequence

$$0 \to K(X \times Y) \to K(X^+ \times Y^+) \to K(X^+) \oplus K(Y^+), \tag{*}$$

whereby $K(X \times Y)$ is identified with the subgroup of $K(X^+ \times Y^+)$ which vanishes on the "axes" X^+ and Y^+. For this, we begin with the short exact sequence

$$K(A,B) \overset{j^*}{\to} K(A) \overset{i^*}{\to} K(B) \tag{**}$$

from Exercise 4c (above) for compact topological spaces A and B with $i : B \to A$ and for the case where B is a "retract" of A; i.e., there is a continuous map $r : A \to B$ which is the identity on B. Then $ri = \mathrm{Id}$ on B and $i^*r^* = \mathrm{Id}$ on $K(B)$, where $r^* : K(B) \to K(A)$ is naturally defined; thus, we see that i^* is surjective and r^* is injective. Furthermore, we obtain (prove!) a $g^* : K(A) \to K(A,B)$ with $g^*j^* = \mathrm{Id}$, whence j^* is injective. One says: The sequence "splits" (see [Eilenberg-Steenrod 1952, 229f.]), and we obtain a decomposition

$$K(A) \cong K(A,B) \oplus K(B). \tag{***}$$

From (***), we get (*): one time we set

$$A := X^+ \times Y^+ \quad \text{and} \quad B := X^+ \times \{+\}, \tag{1}$$

and then

$$A := (X^+ \times Y^+)/X^+ \quad \text{and} \quad B := Y^+.$$

Since B is a retract of A in both cases, (**) splits and we obtain the formulas

$$K(X^+ \times Y^+) \cong K(X^+) \oplus K(X^+ \times Y^+, X^+) \tag{1'}$$

and

$$K(X^+ \times Y^+/X^+) \cong K(Y^+) \oplus K(X^+ \times Y^+/X^+, Y^+), \tag{2'}$$

from which the desired splitting (*)

$$K(X^+ \times Y^+) \cong K(X^+) \oplus K(Y^+) \oplus K(X \times Y)$$

follows because $K(X^+ \times Y^+/X^+, Y^+) \cong K(X \times Y)^+, +) = K(X \times Y)$. One can check that the splitting is compatible with the naturally defined arrows in (*).

3) Now let $x \in K(X) \subset K(X^+)$ and $y \in K(Y) \subset K(Y^+)$. Then $x \boxtimes y \in K(X^+ \times Y^+)$ is well defined by 1). Actually, $x \boxtimes y$ can be

regarded also as an element of $K(X \times Y)$ by (*), since $i^*(x \boxtimes y) = 0$ where $i : X^+ \hookrightarrow X^+ \times Y^+$ is the canonical inclusion (and correspondingly for $Y^+ \hookrightarrow X^+ \times Y^+$). For a proof of this, we write (Exercise 3a) $x = [E] - N$ and $y = [F] - M$, where $E \in \mathrm{Vect}_N(X^+)$ and $F \in \mathrm{Vect}_M(Y^+)$, $N,M \in \mathbb{Z}_+$; we then have $x \boxtimes y = [E \boxtimes F] - [N \boxtimes F] - [E \boxtimes M] + [N \boxtimes M]$, and so $i^*(x \boxtimes y) = [E \otimes F_+] - [E \otimes M] - [N \otimes F_+] + [N \otimes M] = 0$, since the fiber $F^+ \cong \mathbb{C}^M$. (Beware: M denotes the trivial M-dimensional bundle over Y^+ in the first formula, but in the second, it is only the vector space $\mathbb{C}^M$.) □

An important example of a locally compact space is furnished by Euclidean space $\mathbb{R}^n$, whose 1-point compactification is the n-sphere S^n; by definition $K(S^n) \cong K(\mathbb{R}^n) \oplus \mathbb{Z}$ holds, where the second summand is $K(\{+\}) \cong \mathbb{Z}$, given by the dimension of the vector bundle. This shows that $K(\mathbb{R}^n)$ is actually the interesting part of $K(S^n)$.

Exercise 6. For an arbitrary paracompact X, go through the splitting $K(S^n \times X) \cong K(\mathbb{R}^n \times X) \oplus K(X)$.

Hint: Work with the sequence (**) from step 2 of the preceding proof, where $B := X^+$ is a retract of $A := S^n \times X^+/S^n$. Note that $A = (S^n \times X)^+$ and $A/B = (\mathbb{R}^n \times X)^+$, homeomorphically. □

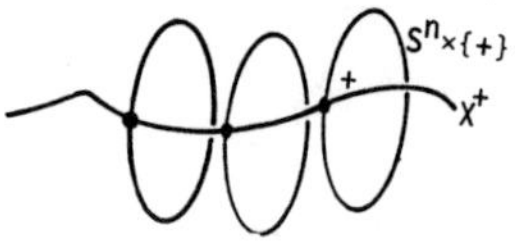

The importance of K-theory with compact support stems for one thing from the fact that applications frequently involve non-compact, but locally compact, spaces such as Euclidean space or tangent spaces. Of course it is possible, without undue difficulties, to avoid non-compact spaces altogether (as with the passage from the tangent bundle TX to the double ball bundle $B^+X \cup_{SX} B^-X$ in the comment following Section C above) - as artificial as this construction may appear. However, the splitting of Exercise 6 (also see the footnote at the end of Section C) makes the locally compact formalism genuinely simpler and perhaps conceptually clearer. In the following proof of the Bott Periodicity Theorem which we adopt from [Atiyah 1969], we will therefore stay in the category of locally compact spaces: We already know that $K(\mathbb{R}^0) = K(\text{point}) \cong \mathbb{Z}$ and we obtain $K(\mathbb{R}^1) = 0$, since all complex vector bundles on the circle are trivial ($GL(N,\mathbb{C})$ is connected..., see the glueing classification in the Appendix, Theorem 2). With some pains (as well as some projective geometry, see Appendix, Exercise 2a, and [Atiyah 1967a, 46 f.]), we could still compute $K(\mathbb{R}^2) \cong \mathbb{Z}$. How does it go?

The theorem which we will prove says that the sequence of these K-groups can be continued alternatingly, thus $K(\mathbb{R}^3) = 0$, $K(\mathbb{R}^4) \cong \mathbb{Z}$, $K(\mathbb{R}^5) = 0$ etc., and that, more generally, for each locally compact X, there is a natural isomorphism $K(\mathbb{R}^2 \times X) \cong K(X)$.

E. Proof of the Periodicity Theorem of R. Bott

In Chapter I.9, we became acquainted with Wiener-Hopf operators, and as a generalization of the discrete (and there rather trivial) index theorem of Israil Gohberg and Mark Krein (Theorem I.9.1), we gave a construction $(V,f) \longmapsto F \longmapsto \text{index } F$. In the pair (V,f), V is a vector bundle over a compact parameter space X, and f is an automorphism of π^*V, where $\pi: S^1 \times X \to X$ is the projection. For $z \in S^1$ and $x \in X$, $f(z,x)$ is then an automorphism of the fiber V_x, and it depends continuously on x and z. Now, $F: X \to \mathcal{F}$ is the associated family of Fredholm operators (after the Wiener-Hopf recipe, formed on certain Hilbert spaces of "half-space functions") and $\text{index } F \in K(X)$ is the "index bundle" of F, which is defined (in the special case that the kernels of all of the Fredholm operators F_x have constant dimension) as $[\text{Ker } F] - [\text{Coker } F]$. Moreover, we have seen that $\text{index } F$ only depends on the homotopy class of f.

Exercise 7. a) From the pair (V,f), how can one construct a vector bundle over $S^2 \times X$ that only depends on V and the homotopy class of f?

b) Show that every $E \in \text{Vect}(S^2 \times X)$ can be obtained in this way.

Hint for a): Decompose S^2 into the two hemispheres B^+ and B^- with $B^+ \cap B^- = S^1$, and form the bundle $(\pi^+)^*V \cup_f (\pi^-)^*V$ by means of the clutching construction (Appendix, Exercise 4), where $\pi^\pm: B^\pm \times X \to X$ is the projection.

For b): Argue as in the proof of Theorem 2 of the Appendix, where X consists only of a single point. The parameter space plays only a subordinate role, and so the proof actually carries over. It is convenient to normalize the map f that one obtains so that $f(1,x)$ is the identity on V_x. □

Theorem 6. Let X be locally compact. Then a homomorphism α: $K(\mathbb{R}^2 \times X) \to K(X)$ can be defined (here, by the index of a Wiener-Hopf family of Fredholm operators; see also Exercise 9) with the following properties:

(i) α is functorial in X.

(ii) If Y is another locally compact space, then we have the following multiplication rule, expressed by the commutative diagram:

$$\begin{array}{ccc} K(\mathbb{R}^2 \times X) \otimes K(Y) & \xrightarrow{t'} & K(\mathbb{R}^2 \times X \times Y) \\ \downarrow \alpha_X \otimes \mathrm{Id} & & \downarrow \alpha_{X\times Y} \\ K(X) \otimes K(Y) & \xrightarrow{t} & K(X \times Y), \end{array}$$

where t and t' are the outer tensor products $\boxtimes$ defined in Theorem 5.

(iii) $\alpha(b) = 1$, where b is the Bott class, a kind of basis element in $K(\mathbb{R}^2)$ that we define by $b := [E_{-1}] - [E_0] \in K(S^2)$, where E_m is the line bundle defined over S^2 by means of the clutching function $f(z) = z^m$; see Appendix, Theorem 2 or the preceding Exercise 7a. Since E_{-1} and E_0 have the same dimension (one), b lies in $K(S^2,\{+\}) = K(\mathbb{R}^2)$. Thus, here we set $X = \{\text{point}\}$, and identify $K(\text{point})$ with $\mathbb{Z}$, as in Exercise 1.

Proof: 1) We start with X compact. By Theorem 1 and Exercise 8 of Chapter I.9, in conjunction with the preceding Exercise 7, we can (for each vector bundle E over $S^2 \times X$) go through a construction $E \longmapsto (V,f) \longmapsto F \longmapsto \text{Index } F$. In this way, a semi-group homomorphism $\text{Vect}(S^2 \times X) \to K(X)$ is defined, which can be extended (see the Remark after Theorem 3) uniquely to a group homomorphism α': $K(S^2 \times X) \to K(X)$. The restriction of α' to $K(\mathbb{R}^2 \times V)$ (which may be regarded as a subgroup of $K(S^2 \times X)$ by Exercise 6) then provides a homomorphism α: $K(\mathbb{R}^2 \times X) \to K(X)$. The "functoriality" of α means that, for each element $u \in K(\mathbb{R}^2 \times X)$ and each continuous map $g : X' \to X$ (where X' is another compact space), we have

$$\alpha_{X'}((g \times \mathrm{Id}_{\mathbb{R}^2})^* u) = g^* \alpha_X(u),$$

which is clear from the functorial nature of index bundles (Exercise I.7.5).

2) We apply this definition for $X = \{\text{point}\}$ and calculate $\alpha(b) = \alpha'[E_{-1}] - \alpha'[E_0]$. By construction, we have $\alpha'[E_m] = -m$ $(m \in \mathbb{Z})$, since

by Theorem I.9.1 we have the assignments

$$E_m \overset{\text{clutch}}{\longmapsto} (\mathbb{C}, z^m) \overset{\text{Wiener-Hopf}}{\longmapsto} T_{z^m} \overset{\text{index}}{\longmapsto} -m.$$

Hence $\alpha(b) = 1$, and so (iii) is fulfilled.

3) To prove the multiplication rule (ii) - first for X,Y compact - we must consider the difference $t(\alpha_X \otimes \mathrm{Id})(u \otimes v) - \alpha_{X\times Y}(t(u \otimes v))$ for $u \in K(\mathbb{R}^2 \times X)$ and $v \in K(Y)$; without loss of generality, we may assume $v = 1$ (the class of the trivial line bundle $\mathbb{C}_Y$ over Y), since all of the maps arising here are $K(Y)$-module homomorphisms.

By the functoriality of α, the difference $\pi^*\alpha_X(u) - \alpha_{X\times Y}(\pi^*u)$ vanishes, as needed. (Here $\pi: X \times Y \to X$ is the projection.)

4) The definition of $\alpha: K(\mathbb{R}^2 \times X) \to K(X)$ for locally compact X carries over just as in the proof of (ii) in the compact case via the one-point compactification with the decomposition above in the proof of Theorem 5 and in Exercise 6. □

Theorem 7 (Periodicity Theorem). For each locally compact space X, we have that $\alpha: K(\mathbb{R}^2 \times X) \to K(X)$ is an isomorphism, whose inverse β: $K(X) \to K(\mathbb{R}^2 \times X)$ is given via outer multiplication $x \mapsto \beta(x) := b \boxtimes x$ by the Bott class b.

Proof: The proof follows by repeated use of the multiplicative property of α expressed in Theorem 6(ii). In the following, for short we write only xy for the outer product $x \boxtimes y$ or $t(x \otimes y)$.

$\alpha\beta = \mathrm{Id}$: For the proof here, in (ii) substitute $\{\text{point}\}$ for X and X for Y; then we have $\alpha\beta(x) = (\alpha(b))x$ for each $x \in K(X)$ by (ii), whence $\alpha\beta = \mathrm{Id}$, since $\alpha(b) = 1$ by (iii).

$\beta\alpha = \mathrm{Id}$: Let $u \in K(\mathbb{R}^2 \times X)$. We want to show that $\beta\alpha u := b(\alpha(u)) = u$, or equivalently (if multiplying by b from the right) $(\alpha(u))b = \tilde{u}$, where $\tilde{u} := \rho^*u$ and $\rho: X \times \mathbb{R}^2 \to \mathbb{R}^2 \times X$ switches factors. From (ii) with $\mathbb{R}^2$ for Y, it follows that $(\alpha(u))(b)$ equals $\alpha(ub)$. Now comes the trick, through which the proof of $\beta\alpha = \mathrm{Id}$ can be reduced to the rather banal fact $\alpha\beta = \mathrm{Id}$ already proven: On $K(\mathbb{R}^2 \times X \times \mathbb{R}^2)$, where the element ub lies, the map τ^*, which is the "lifting" along the switching map

$$\tau: \mathbb{R}^2 \times X \times \mathbb{R}^2 \to \mathbb{R}^2 \times X \times \mathbb{R}^2$$

$$(a,b,c) \longmapsto (c,b,a),$$

is the identity, since τ is homotopic to the identity on $\mathbb{R}^2 \times X \times \mathbb{R}^2$

through homeomorphisms. (See the hint for Exercise 5b.) Namely, on $\mathbb{R}^4 = \mathbb{R}^2 \times \mathbb{R}^2$, τ is given by the matrix

$$\begin{pmatrix} 0 & 0 & 1 & 0 \\ 0 & 0 & 0 & 1 \\ 1 & 0 & 0 & 0 \\ 0 & 1 & 0 & 0 \end{pmatrix}$$

which has determinant +1 and hence - one thinks of the transition to Jordan normal form - lies in the same connected component of $GL(4, \mathbb{R})$ as the identity.

Hence, we have $(\alpha(u))b = \alpha(ub) = \alpha(\tau^*(ub)) = \alpha(b\tilde{u}) = \alpha\beta\tilde{u} = \tilde{u}$. □

Exercise 8. Show the following consequences of Theorem 7:

a) $K(X \times S^2) \cong K(X) \otimes K(S^2)$ for X compact.

b) $K(\mathbb{R}^n) = \begin{cases} \mathbb{Z} & \text{for } n \text{ even} \\ 0 & \text{for } n \text{ odd} \end{cases}$

c) $K(S^n) = \begin{cases} \mathbb{Z} \oplus \mathbb{Z} & \text{for } n \text{ even} \\ 0 & \text{for } n \text{ odd} \end{cases}$

d) $N \geq n/2 \Rightarrow \pi_{n-1}(GL(N,\mathbb{C})) = \begin{cases} \mathbb{Z} & \text{for } n \text{ even} \\ 0 & \text{for } n \text{ odd} \end{cases}$

Hint: While a), b) and c) follow directly from Theorem 7 with Exercise 6, the derivation of d) requires two further considerations. First, for a compact manifold X of dimension $n-1$, we have that for each $E \in \mathrm{Vect}_N(X)$ with $N \geq m := [n/2 - 1]$, there is an $F \in \mathrm{Vect}_m(X)$ such that $E = F \oplus \mathbb{C}_X^{N-m}$; i.e., each vector bundle over X is stably equivalent (see Exercise 3c above) to a vector bundle of fiber dimension m. This is the "basis theorem" for vector bundles. The "uniqueness theorem" then says that (under the same assumptions) stably-equivalent vector bundles of fiber dimension $N \geq m+1$ are isomorphic. For the proofs of these two lemmas (e.g., see [Husemoller 1975, Chapter 8]) one needs some homotopy theory. The rest is trivial, since we can now represent the group $I(X)$ of stable-equivalence classes of bundles for $N \geq n/2$ by $\mathrm{Vect}_N(X)$; hence, in particular (see Exercise 4c), $\mathrm{Vect}_N(S^n) \cong I(S^n) \cong K(\mathbb{R}^n)$. The "classical" form of the periodicity theorem now follows, since $\pi_{n-1}(GL(N,\mathbb{C})) \cong \mathrm{Vect}_N(S^n)$ by Theorem 2 of the Appendix. □

Exercise 9. As an alternative to Theorem 6, give a construction of the isomorphism $\alpha\colon K(\mathbb{R}^2 \times X) \to K(X)$ by means of a family of elliptic boundary-value problems.

Hint: For $X = \{\text{point}\}$ and $f : S^1 \to GL(N,\mathbb{C})$, consider (over the disk $|z| \leq 1$) the "transmission operator" (see Exercise II.1.9 above)

$$A_f : (u,v) \longmapsto \left(\frac{\partial u}{\partial \bar{z}}, \frac{\partial v}{\partial z}, fu|S^1 - v|S^1\right),$$

where u,v are N-tuples of complex-valued functions. By the same recipe, one can also treat families of such boundary-value problems which are parametrized over a space X; see [Atiyah 1968a, 118-122]. □

Remark: The connection with the construction of α via Wiener-Hopf operators lies, roughly speaking, in the "Poisson principle" (i.e., in the Agranovich-Dynin formula of Chapter II.8 above), by which boundary-value problems can be translated into problems on the boundary. Namely, extend (as in Exercise II.3.5) the discrete Wiener-Hopf operator T_f to a pseudo-differential operator of order 0 on the circle S^1 (via the identity on the basis elements of the form z^m with $m < 0$), which we still denote by T_f. Then, we have that

$$S_f : (u,v) \longmapsto \left\{\frac{\partial u}{\partial \bar{z}}, \frac{\partial v}{\partial z}, T_f(zu|S^1 - v|S^1)\right\}$$

is an elliptic problem with pseudo-differential boundary conditions; it has the same kernel as the transmission problem A_{zf} and isomorphic cokernel. On the other hand, the operator S_f is formed directly by the composition of A_{zId} with the "primitive boundary-value problem" $(u,v,w) \longmapsto (u,v,T_f w)$. Here Id is the constant function that assigns the identity in $GL(N,\mathbb{C})$ to each $z \in S^1$. While Exercise II.1.9 says that index $A_{Id} = 1$, one finds index $A_{zId} = 0$, whence index A_{zf} = index S_f = index T_f. Within the Green algebra (see Chapter II.8 below), S_f then induces the connection

$$\left(\begin{array}{cc:c} \frac{\partial}{\partial \bar{z}} & 0 & 0 \\ & \frac{\partial}{\partial z} & 0 \\ \hdashline M_{zf}\circ r & -\mathrm{Id}\circ r & \end{array}\right) \overset{S_f}{\longleftrightarrow} \left(\begin{array}{cc:c} & & 0 \\ & \mathrm{Id} & 0 \\ \hdashline 0 & 0 & T_f \end{array}\right)$$

between a conventional elliptic system of partial differential equations of the first order over the disk B^2 with "ideally simple" boundary-value conditions which are formed via the restriction $r: C^\infty(B^2) \to C^\infty(S^1)$ and a trivial multiplication operator, and a "primitive boundary-value problem" that consists of a (somewhat complex) elliptic pseudo-differential operator only on the boundary.

Actually, the "change of planes" from differential boundary value problems in the plane to pseudo-differential operators on the boundary line can be pushed further, and using a polynomial approximation to f, a purely algebraic definition of the homomorphism α can be given. In doing so, no Hilbert space theory and Fredholm operators are needed just as in the wholly elementary, but in parts quite tedious, proof [Atiyah-Bott 1964b]. A detailed discussion of the advantages and disadvantages of the different methods for the construction of α is in [Atiyah 1968a, 131-136]. □

As a simple corollary of the periodicity theorem, we prove the classical fixed point theorem of topology.

<u>Theorem 8</u> (<u>L. E. J. Brouwer, 1911</u>): Each continuous map f of the closed n-dimensional ball B^n into itself has a fixed point.

<u>Proof</u>: If $f(x) \neq x$ for all $x \in B^n$, a continuous map $g: B^n \to S^{n-1}$ is defined by

$$g(x) := (1-\alpha(x))f(x) + \alpha(x)x,$$

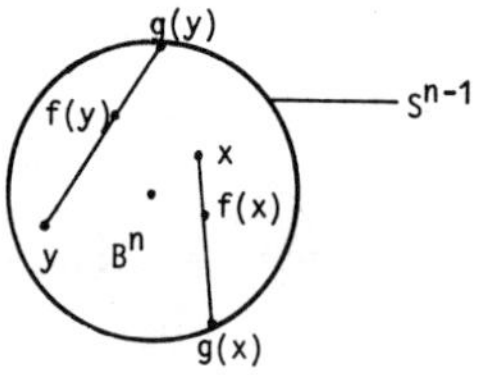

where $\alpha(x) \geq 0$ is chosen such that $||g(x)|| = 1$; g is the identity on S^{n-1}, and therefore a retraction of B^n to S^{n-1} (i.e., $g \circ i = \mathrm{Id}$, where $i : S^{n-1} \to B^n$ is inclusion).

For odd n, we note that the composition

$$\begin{array}{ccccc} K(S^{n-1}) & \xrightarrow{g^*} & K(B^n) & \xrightarrow{i^*} & K(S^{n-1}) \\ \cong & & \cong & & \cong \\ \mathbb{Z} \oplus \mathbb{Z} & & \mathbb{Z} & & \mathbb{Z} \oplus \mathbb{Z} \end{array}$$

cannot be the identity, since $\mathbb{Z}$ and $\mathbb{Z} \oplus \mathbb{Z}$ are not isomorphic. For even n, note that the corresponding composition of "suspensions" (see Appendix, Exercise 4):

$$\begin{array}{ccccc} K(S^{n}) & \xrightarrow{(Sg)^*} & K(SB^n) & \xrightarrow{(Si)^*} & K(S^{n}) \\ \cong & & \cong & & \cong \\ \mathbb{Z} \oplus \mathbb{Z} & & \mathbb{Z} & & \mathbb{Z} \oplus \mathbb{Z}, \end{array}$$

would also be the identity, yet not the identity. □

Exercise 10. Define the *mapping degree*, deg $f \in \mathbb{Z}$, of an arbitrary continuous map $f: S^n \to S^n$, and show that homotopic maps have the same mapping degree.

Hint: Each homomorphism $h: \mathbb{Z} \to \mathbb{Z}$ clearly has a "degree", namely d, such that $h(m) = dm$ for all $m \in \mathbb{Z}$. Thus, for even n work with the subgroup $K(\mathbb{R}^n) \cong \mathbb{Z}$ of $K(S^n)$, which is carried into itself by f^*. For odd n, consider the suspension $Sf: S^{n+1} \to S^{n+1}$. □

Exercise 11. Show that the tangent bundle $T(S^n)$ is nontrivial, for even $n \geq 2$.

Hint: More generally, each nowhere vanishing vector field v on S^n $(v \in C^\infty(TS^n))$ provides a homotopy between the identity and the antipodal map $A: S^n \to S^n$, while degree Id $\neq$ degree A for even $n \geq 2$, contradicting Exercise 10. □

The preceding examples show that important notions and results of classical algebraic topology can be developed as well, and perhaps more quickly and more easily, on the basis of linear algebra via K-theory than they can by establishing, say, homology theory by means of simplicial theory.

A number of sharper results, such as the well-known converse by Hopf of Exercise 10 (equality of mapping degrees implies homotopy) in case $n \geq 2$, cannot be obtained by K-theoretic means but require deeper geometric considerations. At least according to Atiyah, the Periodicity Theorem is not only simpler than most major theorems of classical algebraic topology, but also more relevant for the index problem and, more generally, for many investigations of manifolds. Hereby, the "philosophy" is that the usual algebraic topology destroys the structures too much while K-theory "comparable to molecular biology" (Atiyah) searches for the essential macromolecules which make up the manifold. It is then clear that for a "comparison" of the topology of the intricate manifold with the "building blocks" one needs to first know the topology of the general linear group which comprises the transition functions and, more generally, the Periodicity Theorem in its K-theoretic form.

In the following chapters we will apply the Periodicity Theorem in various ways for computing the index of elliptic problems. Conversely, the index of special elliptic differential equations served to prove the Periodicity Theorem; see especially Theorem 6 and Exercise 9. This is no contradiction but an indication of how closely K-theory and index theory of linear elliptic oscillation equations are related - both being linked by the catchwords "linear", "finite dimensional" and "deformation invariant".

2. THE INDEX FORMULA IN THE EUCLIDEAN CASE

A. Index Formula and Bott Periodicity

In the preceding chapter, we used analytic tools (the Gohberg-Krein Index Theorem for Wiener-Hopf operators) to prove the Bott Periodicity Theorem. We will now use it to derive an index theorem for elliptic integral operators in $\mathbb{R}^n$. The basic idea is perhaps best described via homotopy theory:

An elliptic pseudo-differential operator of order k in $\mathbb{R}^n$ is given in the form

$$(Pu)(x) = (2\pi)^{-n} \int_{\mathbb{R}^n} e^{i<x,\xi>} p(x,\xi)\hat{u}(\xi)d\xi$$

where u is a C^∞-function on $\mathbb{R}^n$ with compact support and values in $\mathbb{C}^N$, and the "amplitude" (see above Chapter II.3) p is an $N \times N$ matrix-valued function with

$$\sigma(P)(x,\xi) := \lim_{\lambda\to\infty} \frac{p(x,\lambda\xi)}{\lambda^k} \in GL(N,\mathbb{C}), \quad \xi \neq 0.$$

Thus for fixed x, we have a continuous map

$$\sigma(P)(x,\ldots) : S^{n-1} \to GL(N,C),$$

that has a well-defined degree for n even and N sufficiently large which does not depend on x because of continuity ($\mathbb{R}^n$ is connected) and which was denoted by degree(P) in Remark 2 following Theorem II.7.1.

In order to get interesting global problems, P is usually combined with $Nk/2$ boundary conditions to form an elliptic system in the sense of Chapter II.6. However, this is only possible in the stated fashion, if the "local index" degree(P) vanishes. A sort of extremely simple boundary condition - lacking the topological and analytical difficulties discussed in Chapters II.7/8 - arises when we put $k = 0$ and $p(x,\xi) = Id$ for x outside a compact subset K of $\mathbb{R}^n$. The class of elliptic pseudo-differential operators of order 0 in $\mathbb{R}^n$ with this property of "being equal to identity at infinity" was first investigated by Robert T. Seeley and will be denoted by $\mathrm{Ell}_c(\mathbb{R}^n)$. Obviously, see Exercise 1 below, every $P \in \mathrm{Ell}_c(\mathbb{R}^n)$ has a finite dimensional kernel and cokernel; therefore index P is well-defined. If furthermore $p(x,\xi) = \mathrm{Id}_N$ for $|x| \geq r$, r real, then $\sigma(P)(x,\xi) \in GL(N,\mathbb{C})$ for $|x| + |\xi| \geq r$. In this way P defines a continuous mapping $S^{2n-1} \to GL(N,\mathbb{C})$ where S^{2n-1} is the sphere of radius r in $\mathbb{R}^{2n}$ (the (x,ξ)-space). Since index P = index P + Id (Id the identity operator on functions), we may assume without loss of generality that $N \geq n$. By the homotopy theoretic form of the Bott Periodicity Theorem (Theorem 1.2 or Exercise 1.8d), we have $\pi_{2n-1}(GL(N,\mathbb{C})) \cong \mathbb{Z}$. Thus we have three integral invariants:

index P	the analytic index (defined in the sense of analytic function theory),
degree(σ(P)(..,..))	the topological index (defined via homotopy theory by the global behavior of σ(P)),
degree(P)	the local index (defined via homotopy theory for even n by the pointwise behavior of $\sigma(P)(x,..)$.

We have degree(P) = 0 - which is trivial - and index P = ±degree(σ(P)(..,..)) (see Theorem 1 below; be careful with the sign). The second formula is not trivial. Just as with the Gohberg-Krein Index Formula for Wiener-Hopf operators on the circle and the straight line (Theorem I.9.1, Exercise I.9.6, and Theorem I.9.2), its significance derives from the fact that on the left side the analytic index defined globally by the operator P is an object of the analysis of infinite-dimensional function spaces, while on the right side the topological index is given by the symbol, i.e., by locally defined data of the linear algebra of finite dimensional vector spaces (which are suitably "integrated").

The proof of this index formula (see Theorem 1 below) roughly rests on the fact that $\mathrm{Ell}_c(\mathbb{R}^n)$ is so rich that the analytic index (which as in II.5 only depends on the symbol and does not change under "small" deformations of the symbol) can be considered an additive function on $\pi_{2n-1}(GL(N,\mathbb{C}))$, and thus as a multiple of the topological index. Comparing the topological and the analytic index for the generators of the homotopy groups, we obtain equality. □

B. The Difference Bundle of an Elliptic Operator

We will no longer pursue these homotopy-theoretic arguments, but rather we carry out the details of the proofs in the more convenient formalism of K-theory, which also permits an easier transition to the more general situation of the following section.

Exercise 1. Let X be a (not necessarily compact) oriented C^∞ Riemannian manifold, and let $\mathrm{Ell}_c(X)$ be the class of elliptic pseudo-differential operators of order 0 on X which are "equal to the identity at infinity"; i.e., for each $P \in \mathrm{Ell}_c(X)$, there is a compact subset $K \subset X$ such that $P\phi = \phi$ for all C^∞ sections ϕ (in the domain of definition of P) with support $\phi \cap K = \emptyset$. The same condition should hold for the formal adjoint operator P^*.

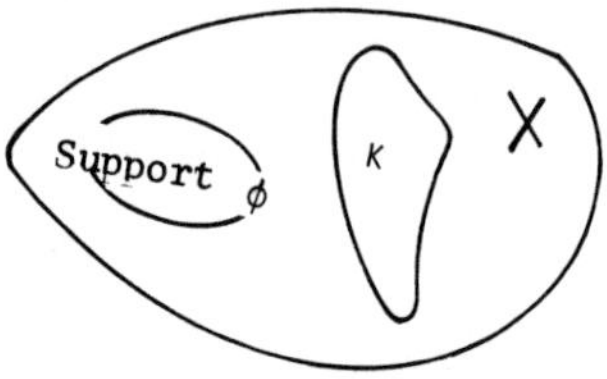

a) Show that this definition coincides with the definition given in A for $X = \mathbb{R}^n$.

b) Show that index P is well defined, depends only on $\sigma(P)$, and remains constant under a C^∞ homotopy of the symbol within the space of elliptic symbols which are the identity at infinity.

Hint for b): Instead of repeating the proofs of II.5, one can also reduce the present case to the results of II.5 directly. Indeed, one can embed K in a bounded, compact, codimension-zero submanifold Y of X, and then investigate the "doubled" operator $\tilde{P}$ on the closed manifold $\tilde{X} := Y \cup_{\partial Y} Y$. Show that index $\tilde{P} = 2$ index P. □

Exercise 2. Show that the symbol $\sigma(P)$ of an elliptic operator $P \in \mathrm{Ell}_c(X)$ defines an element $[\sigma(P)] \in K(TX)$ in a natural way, and that each element $a \in K(TX)$ can be represented in this way; here, X is as in Exercise 1 and TX is the tangent bundle of X, which can be identified with the cotangent bundle T^*X by means of the Riemannian metric on X.

Remark: Much more generally one can define, for locally compact Y, the group $K(Y)$ through "*complexes of vector bundles with compact support*"; these are short sequences $0 \to E^0 \overset{\alpha}{\to} E^1 \to 0$, where E^0 and E^1 are complex vector bundles over Y and α is a vector bundle isomorphism outside a compact subset of Y. Two complexes $E^\cdot = 0 \to E^0 \overset{\alpha}{\to} E^1 \to 0$ and $F^\cdot = 0 \to F^0 \overset{\beta}{\to} F^1 \to 0$ are called *equivalent*, if there is a complex $G^\cdot = 0 \to G^0 \overset{\gamma}{\to} G^1 \to 0$ with compact support over $Y \times I$ such that

$E^\cdot = G^\cdot|Y \times \{0\}$ and $F^\cdot = G^\cdot|Y \times \{1\}$. The equivalence classes form a semi-group $C(Y)$ with sub-semi-group $C_\emptyset(Y)$ of *elementary complexes* with empty "support" (i.e., the bundle maps over Y are isomorphisms). Then the sequence

$$0 \to C(Y)/C_\emptyset(Y) \xrightarrow{d} K(Y^+) \xrightarrow{i^*} K(+) \to 0 \tag{*}$$

is exact and splits; hence $C(Y)/C_\emptyset(Y)$ proves to be isomorphic to $K(Y)$.

The construction (which goes back to Michael Atiyah and Friedrich Hirzebruch) of "difference bundles" $d(E^\cdot)$ for complexes $E^\cdot = 0 \to E^0 \xrightarrow{\alpha} E^1 \to 0$ over Y with compact support K looks roughly as follows: We choose a compact neighborhood L of K such that K is contained in the interior L^0 of L. In order to extend the complex $E^\cdot$ to all

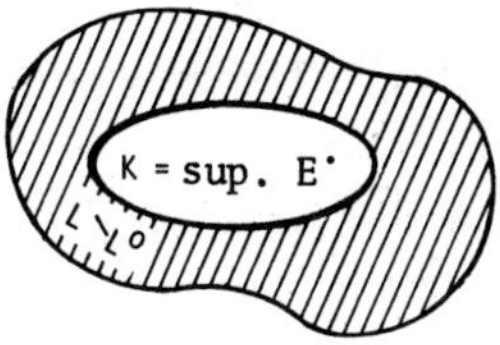

of Y^+, we replace it by an equivalent complex whose bundles are trivial over $L \setminus L^0$:

$$0 \to E^0|L \oplus F \xrightarrow{\alpha \oplus \mathrm{Id}} E^1|L \oplus F \to 0$$

where $F \in \mathrm{Vect}(L)$ is chosen (by means of Appendix, Exercise 6) so that $E^1|L \oplus F$ is trivial. Since α is an isomorphism on $L \setminus L^0$, we must have that $E^0|L \oplus F$ is trivial at least on $L \setminus L^0$.

Let $\tau_i \colon E^i|L\setminus L^0 \oplus F|L\setminus L^0 \to (L\setminus L^0) \times \mathbb{C}^N$, $i = 0,1$, be trivializations with τ_1 arbitrary and $\tau_0 := \tau_1 \circ (\alpha \oplus \mathrm{Id})$. Then the clutched bundle (Appendix, Theorem 2)

$$G^i := (E^i|L \oplus F) \cup_{\tau_i} ((Y^+ \setminus L^0) \times \mathbb{C}^N) \in \mathrm{Vect}(Y^+),$$

and thus $d(E^\cdot) := [G^0] - [G^1] \in K(Y^+)$ are well defined. Since the fiber dimensions of G^0 and G^1 coincide, $d(E^\cdot)$ lies in $K(Y)$. Incidentally, one calculates easily that $d(E^\cdot) = d(E^\cdot \oplus H^\cdot)$, if $H^\cdot$ is an elementary complex, and that $d(E^\cdot)$ does not depend on the choice of F.

Because of the universal property (see Remark 2 after Theorem 1.3) of K, it suffices to define the *splitting homomorphism* $b: K(Y^+) \to C(Y)/C_\emptyset(Y)$ additively on $\mathrm{Vect}(Y^+)$. For $E \in \mathrm{Vect}(Y^+)$, one sets $b(E) := E^\cdot|Y$, where $E^\cdot$ is the complex $0 \to E \overset{\beta}{\to} p^*i^*E \to 0$, $p: Y^+ \to \{+\}$ is the retraction, and β is an arbitrary extension of $\beta_+ := \mathrm{Id}$ to an isomorphism on a neighborhood of $+$. The support of $E^\cdot$ is then compact and contained in Y, and hence $b(E) \in C(Y)$. As an element of $C(X)/C_\emptyset(X)$, $b(E)$ is independent of the choice of the extension β. One sees immediately that $db \oplus p^*i^* = \mathrm{Id}$, whence in particular $K(+)$ is the cokernel of d; and with suitable homotopies for the vector bundle homomorphisms, we have $bd = \mathrm{Id}$, whence d is injective.

Further details of this construction are found in [Atiyah-Singer 1968a, 489 ff.] and in [Segal 1968, 139-151], where complexes

$$0 \to E^0 \xrightarrow{\alpha_1} E^1 \xrightarrow{\alpha_2} \cdots \xrightarrow{\alpha_n} E_n \to 0$$

of length n with $\alpha_{i+1} \circ \alpha_i = 0$ are considered, which are exact outside a compact subset of Y. Incidentally, by means of tensor products of complexes of arbitrary length (e.g., see [Eilenberg-Steenrod 1952, 140 ff.] a *ring structure* on $K(Y)$ may be introduced in a natural way.*

If the locally compact space Y can be represented in the form $Z - A$ (where Z is compact and A is closed in Z), then often in the literature for this special case one sees

---

*For complexes $E^\cdot = 0 \to E^0 \overset{\alpha}{\to} E^1 \to 0$ and $F^\cdot = 0 \to F^0 \overset{\beta}{\to} F^1 \to 0$ of length 1, one obtains as an outer product the complex

$$E^\cdot \boxtimes F^\cdot := 0 \to E^0 \boxtimes F^0 \overset{\phi}{\to} E^1 \boxtimes F^0 \oplus E^0 \boxtimes F^1 \overset{\psi}{\to} E^1 \boxtimes F^1 \to 0$$

of length 2 where ("golden rule" of multilinear algebra)

$$\phi := \alpha \boxtimes \mathrm{Id} + \mathrm{Id} \boxtimes \beta \quad \text{and} \quad \psi := -\mathrm{Id} \boxtimes \beta + \alpha \boxtimes \mathrm{Id}.$$

With the help of Hermitian metrics on the vector bundles one can rewrite $E^\cdot \boxtimes F^\cdot$ as a complex of length 1

$$0 \to E^0 \boxtimes F^0 \oplus E^1 \boxtimes F^1 \overset{\theta}{\to} E^1 \boxtimes F^0 \oplus E^0 \boxtimes F^1 \to 0,$$

where

$$\theta := \begin{pmatrix} \alpha \boxtimes \mathrm{Id} & -\mathrm{Id} \boxtimes \beta^* \\ \mathrm{Id} \boxtimes \beta & \alpha^* \boxtimes \mathrm{Id} \end{pmatrix}$$

and α^* and β^* are the adjoint homomorphisms. Details are in [Atiyah 1967a, 93f.].

$$K(Y) = K(Z \setminus A) = K(Z,A) \overset{(*)}{=} C_{Z \setminus A}(Z)/C_{\emptyset}(Z).$$

Hence, an element of $K(Y)$ is written as an equivalence class of an isomorphism $\sigma\colon E^0|A \to E^1|A$, where E^0 and E^1 are complex vector bundles over Z. In our applications (where Y is the tangent bundle TX) one prefers $K(BX,SX)$ to $K(TX)$, since $B(X)$ and $S(X)$ are compact for X compact. In fact, the construction of the difference bundles in K-theory began with compact base space, although the proof (see e.g., [Atiyah 1967a, 88-94] in its basic idea is not as simple as that for the more general locally compact space.

Hint for Exercise 2: Construct the "difference bundle" $[\sigma(P)] \in K(TX)$ as in the preceding remark. In addition, show

$$[\sigma(P)] = [\mathbb{C}^N_{B_0} \cup_{\sigma(P)} \mathbb{C}^N_{B_\infty}] - [N]$$

in the special case $X = \mathbb{R}^n$ where $TX^+ = (\mathbb{R}^{2n})^+ = S^{2n} = B_0 \cup B_\infty$ with $B_0 \cap B_\infty = S^{2n-1}$, and

$$\sigma(P)(\ldots,\ldots) : S^{2n-1} = \{(x,\xi) : |x| + |\xi| = r\} \to GL(N,\mathbb{C}),$$

where r is so large that $P\phi = \phi$ for all N-tuples ϕ of complex-valued functions such that $\operatorname{support}(\phi) \cap \{x\colon |x| \le r\} = \emptyset$.

For the reverse direction set $V := TX$ and represent $a \in K(V)$ (as with the "splitting homomorphism" in the preceding remark) by a complex $0 \to F^0 \overset{\phi}{\to} F^1 \to 0$, where the bundles F^i are restrictions to V of bundles of the same fiber dimension N over V^+; i.e., outside a compact subset $L \subset V$, we have isomorphisms $\tau_i\colon F^i|V \setminus L \overset{\cong}{\to} (V \setminus L) \times \mathbb{C}^N$ such that $\phi := (\tau_1)^{-1}\tau_0$ is a bundle isomorphism over $V \setminus L$.

If $\pi\colon V \to X$ is the base point map, then replace the bundles F^i on $V \setminus L$ (where they are trivial) by π^*E^i, where E^i is the restriction of F^i to the zero section of V. On L this cannot be done in general. However, the following artifice (after [Atiyah-Singer 1968a, 492 f.]) is helpful: Choose an open, relatively compact subset Y of X that includes $\pi(L)$ and a real number $\rho > 0$ such that the compact set L is

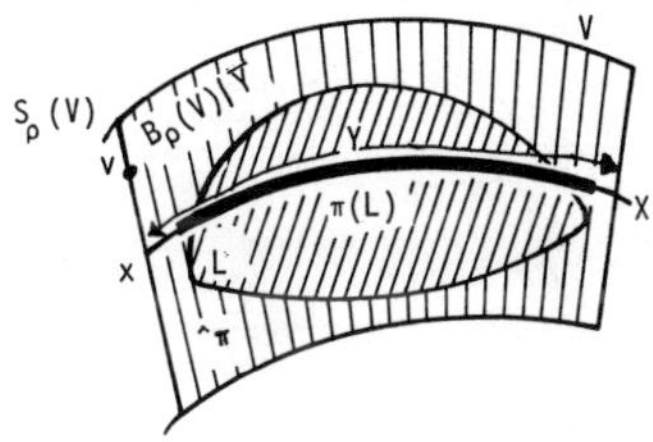

contained in the "ball bundle" $B_\rho(V)|\overline{Y}$. Now show that $\overline{Y}$ is a deformation retract of $B_\rho(V)|\overline{Y}$, and conclude (with Appendix, Theorem 1) that there are isomorphisms, $\theta_i : F^i|(B_\rho(V)|\overline{Y}) \to \pi^*E^i(B_\rho(V)|\overline{Y})$, which are extensions of the trivialization isomorphisms given above over $\overline{Y}\setminus Y$. Thus, one must show that θ_i can be chosen so that the homomorphisms

$$\theta_i(v) : F^i_v \to (\pi^*E^i)_v = F^i_x, \quad x \in \overline{Y}\setminus Y \text{ and } \pi(v) = x,$$

coincide with the composition $(\tau_i(x))^{-1} \circ \tau_i(v)$. (Furthermore, if we require that θ_i is the identity on the zero section then it is uniquely determined up to homotopy.) Now define $\alpha := \theta_1 \circ \phi \circ \theta_0^{-1}$ over $\partial(B_\rho(V)|\overline{Y}) = (S_\rho(V)|\overline{Y}) \cup (B_\rho(V)|\overline{Y}\setminus Y)$, and on $V|\overline{Y}$ (modulo the zero section), extend it to be homogeneous of degree 0, and the given trivialization on $\pi^{-1}(Y\setminus\overline{Y})$.

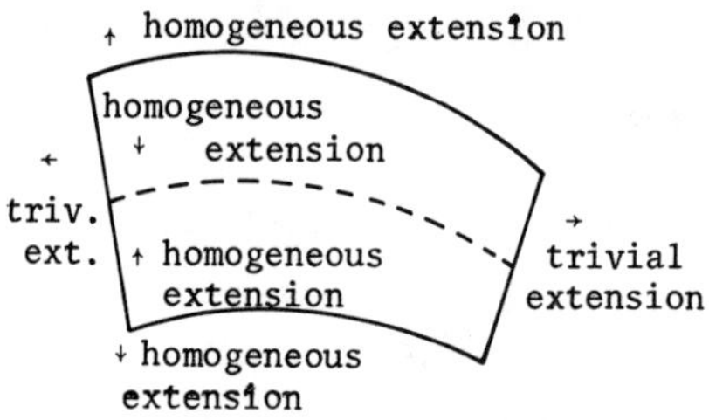

Finally, approximate the complex $0 \to \pi^*E^0 \overset{\alpha}{\to} \pi^*E^1 \to 0$ so obtained (where α is homogeneous of degree 0, and is induced by an isomorphism $E^0 \to E^1$ outside a compact subset of the base X) by a C^∞ mapping with the same properties, where (without loss of generality - see Appendix, Exercise 7b) the E^i can be taken to be C^∞ vector bundles. Incidentally, what simplifications can be made for $X = \mathbb{R}^n$? □

C. The Index Formula

Theorem 1. For all $P \in \mathrm{Ell}_c(\mathbb{R}^n)$, we have the index formula

$$\text{index } P = (-1)^n \alpha^n([\sigma(P)]).$$

Here $\alpha^n : K(\mathbb{R}^{2n}) \overset{\cong}{\to} K(\mathbb{R}^0) \cong \mathbb{Z}$ is the "periodicity homomorphism" produced by iteration of

$$\alpha_X : K(\mathbb{R}^2 \times X) \to K(X) \quad \text{for } X = \mathbb{R}^{2(n-1)}, \mathbb{R}^{2(n-2)}, \ldots ;$$

(see Theorem 1.7 above).

Proof: 1) For locally compact X, we have the following commutative diagram:

$$\begin{array}{ccc} \mathrm{Ell}_c(X) & \xrightarrow{[\sigma(..)]} & K(TX) \\ & \underset{\text{index}}{\searrow} \quad \underset{\text{index}}{\swarrow} & \\ & \mathbb{Z} & \end{array}$$

Here the "analytical index" is defined on $\mathrm{Ell}_c(X)$ by Exercise 1b, and by Exercise 2 (surjectivity of the difference bundle construction $[\sigma(\cdot\cdot)]$) it is well defined on $K(TX)$ and trivially additive (Exercise I.1.2). Hence, for $X = \mathbb{R}^n$ (where $K(TX)$ is isomorphic to $\mathbb{Z}$ by the Bott Periodicity Theorem above), the index is a multiple of this isomorphism, whence

$$\text{index } P = C_n \alpha^n [\sigma(P)],$$

where the constant does not depend on P, but indeed may depend on n.

2) We now want to show that $C_n = (-1)^n$. For this we must find a $P \in \mathrm{Ell}_c(\mathbb{R}^n)$ with $[\sigma(P)] = b \boxtimes \ldots \boxtimes b \in K(\mathbb{R}^{2n})$ ($b \in K(\mathbb{R}^2)$ the Bott class of Theorem 1.6 above) and index $P = (-1)^n$. The main problem consists in finding a sufficiently simple operator P, so that one can compute its analytical index. We already know that P can not have constant coefficients, since P is the identity at infinity; also, P is not a differential operator, since it has vanishing order.

These days, there are various ways to solve this problem; see [Atiyah 1967b, 243 f.], [Atiyah 1970a, 110 ff.], and [Hörmander 1971, 141-146]. For us, it is most convenient to first show that we can restrict ourselves to the case $n = 1$. As in Exercise II.5.7b, we have (with analogous proof) the following multiplicativity properties: $P = Q \# R$, $P \in \mathrm{Ell}_c(\mathbb{R}^n)$, $Q \in \mathrm{Ell}_c(\mathbb{R}^m)$, $R \in \mathrm{Ell}_c(\mathbb{R}^k)$, and $m+k = n$ imply $\sigma(P) = \sigma(Q) \# \sigma(R)$ and $[\sigma(P)] = [\sigma(Q)] \boxtimes [\sigma(R)]$ and index $P = (\text{index } Q)(\text{index } R)$. Since α^n is multiplicative by construction, we have $C^n = (C_1)^n$.

3) Thus, let $n = 1$ (i.e., $X = \mathbb{R}$ and $TX = \mathbb{R}^2 = \mathbb{C} = \{x+i\xi\}$. Then (by definition) b is represented by the complex

$$\begin{array}{ccccccc} 0 \to & \mathbb{C}_{\mathbb{C}} & \xrightarrow{\ .(x+i\xi)^{-1}\ } & \mathbb{C}_{\mathbb{C}} & \to 0 \\ & \downarrow & & \downarrow & \\ & TX & & TX & \end{array} ,$$

which, as it stands, still does not represent any pseudo-differential operator. As in Exercise 2, we can deform the bundle map ϕ, which at the point (x,ξ) is defined on the fiber $\mathbb{C}$ by

$$\phi(x,\xi) : z \longmapsto z(x+i\xi)^{-1},$$

to a map α with

$$\alpha(x,\xi) = 1 \qquad |x| \geq 1$$

$$\alpha(x,\lambda\xi) = \alpha(x,\xi) \quad \text{for} \quad \lambda > 0$$

$$\alpha(x,\xi) \neq 0 \qquad \xi \neq 0.$$

For $|x| \leq 1$, we explicitly set

$$\alpha(x,\xi) = \begin{cases} e^{i\pi(x-1)} & \\ & \text{for} \\ 1 & \end{cases} \begin{array}{l} \xi > 0 \\ \\ \xi < 0 . \end{array}$$

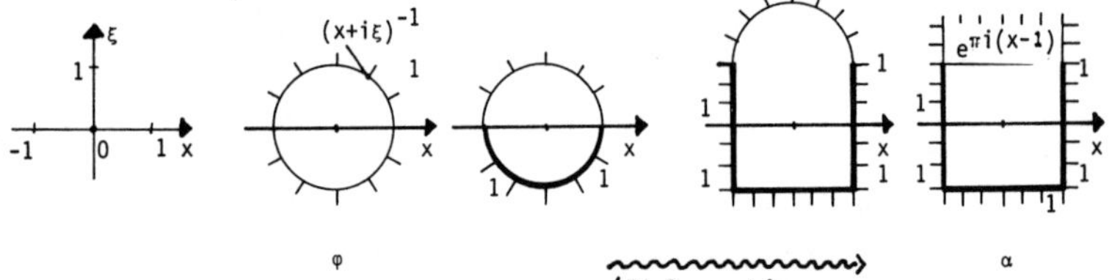

φ

(Deformation: dialation of one)

α

Thus, after smoothing, we can represent α as the symbol of an elliptic pseudo-differential operator T of order 0 on $\mathbb{R}$, which is the identity outside of the interval $[-1,1]$ and inside it equals the "Toeplitz operator" $\tilde{T} := e^{\pi i(x-1)}P + (\mathrm{Id} - P)$ on the circle $S^1 \cong [-1,1]$, where $P : \sum_{-\infty}^{\infty} a_\nu z^\nu \longmapsto \sum_{\nu=0}^{\infty} a_\nu z^\nu$ is the projection operator. By construction, index T = index $\tilde{T}$, and according to Exercise II.3.5c, we obtain index $\tilde{T} = W(e^{\pi i(x-1)},0) = -1$. Thus, index (b) = -1, and $C_1 = -1$ then follows. □

<u>Exercise 3</u>. How can one directly prove $C_2 = 1$, without using induction from step 2 of the preceding proof?

<u>Hint</u>: Consider, on the disk $X := B^2$, the "transmission operator"

$$A : (u,v) \longmapsto \left(\frac{\partial u}{\partial \bar{z}}, \frac{\partial v}{\partial z}, (u-v)|S^1\right); \quad u,v \in C^\infty(X)$$

with index $A = 1$ (Exercise II.1.9), and construct an operator A' (in the Green algebra $\mathrm{Ell}(X,\partial X)$) which is stably equivalent to A, and is equal to the identity in a neighborhood of ∂X; use the deformation procedure of Exercise II.7.2a or Theorem II.7.1. Show index A' = index A with Theorem II.8.3d, whence index A" = 1 if $A'' \in Ell_c(\mathbb{R}^2)$ is the extension of A' to all of $\mathbb{R}^2$ with A" = Id outside B^2.

Then it only remains to show that $[\sigma(A'')]$ is actually $b \boxtimes b$. For this, represent b as in Theorem 1 by the complex $0 \to ¢_{¢} \xrightarrow{.\zeta^{-1}} ¢_{¢} \to 0$ and derive (using the recipe given in the above footnote to Exercise 2) the representation of $b \boxtimes b$ by the complex $0 \to ¢_{TZ} \oplus ¢_{TZ} \overset{\theta}{\to} ¢_{TZ} \oplus ¢_{TZ} \to 0$ over the tangent bundle $TZ = ¢^2 = \{(z,\zeta)\}$ of the space $Z = \mathbb{R}^2 = ¢ = \{z\}$. Then

$$\theta(z,\zeta) := \begin{pmatrix} z^{-1} & -(\overline{\zeta})^{-1} \\ \zeta^{-1} & (\overline{z})^{-1} \end{pmatrix}$$

can be deformed into $\sigma(A')(z,\zeta)$. □

Theorem 1 is a beautiful result of a purposeful application of modern topological methods to questions of analysis. Its theoretical ramifications are manifold, and we mention briefly:

(i) Generalizations of the Gohberg-Krein Index Formula for Wiener-Hopf operators on the circle or halfline (see above Ch. I.9) to elliptic pseudo-differential operators of order 0 on n-dimensional Euclidean space. See also [Prössdorf 1972] for a systematic comparison of these two interesting operator classes.

(ii) Analytic definition of the degree of a (2n-1)-dimensional homotopy class of GL(N,¢): In its homotopy theoretic form (see Section A), Theorem 1 supplies an explicit formula for the index, if one uses either one of the definitions of "degree" given explicitly in Section 1.B above. Conversely, one can use Theorem 1 to define the degree (by analytic means) and generalize in this fashion the "algebraic" definition, which in Section 1.A was available only for $n = 1$, to the case $n > 1$. Although "the index of an operator is usually a less computable quantity than an integral, say, the actual computation is for many theoretical goals unimportant, while the analytic definition entails numerous theoretical advantages." ([Atiyah 1967b, 244], where a number of "advantages" - a priori integrality, connections with analytic function theory and Lie groups - are discussed in detail.)

(iii) Applications to Boundary-Value Problems. In Chapter II.7, we learned methods for "trivializing" the symbol of an elliptic boundary-value problem along the boundary. Thereby (more precise discussion in

[Boutet de Monvel 1971, 40]) one usually obtains an operator of order 0 which equals the identity near the boundary and therefore can be continued to an operator of $\mathrm{Ell}_c(\mathbb{R}^n)$, if the manifold with boundary is a region with boundary in $\mathbb{R}^n$. As was sketched in the hint to Exercise 3, the index does not change under these manipulations. Therefore, the index of a boundary-value problem can be computed as the case may be via Theorem 1; however, the derivation of a closed formula (see Ch. 4 below) requires a reduction to the Agranovich-Dynin Formula (Theorem II.8.3e or Exercise II.8.7) and thus to the study of pseudo-differential operators also on the boundary which is a closed manifold.

(iv) In the following chapter, finally, we will derive from Theorem 1 an index formula for elliptic operators on closed (= compact without boundary) manifolds, provided the latter can be imbedded in Euclidean space "trivially" (i.e. with trivial normal bundle). □

3. THE INDEX FORMULA FOR CLOSED MANIFOLDS

A. The Index Formula

Let X be a closed (i.e., compact, without boundary), oriented, C^∞ Riemannian manifold of dimension n, which is "trivially" embedded (i.e., with trivial normal bundle) in the Euclidean space $\mathbb{R}^{n+m}$. Let E and F be Hermitian vector bundles over X and $P \in \mathrm{Ell}_k(E,F)$, $k \in \mathbb{Z}$. Then we have the following formula:

Theorem 1 (M. F. Atiyah, I. M. Singer 1963).

$$\text{index } P = (-1)^n \alpha^{n+m}([\sigma(P)] \boxtimes b^m),$$

where $[\sigma(P)] \in K(TX)$ is the "symbol class" of P, $b \in K(\mathbb{R}^2)$ is the "Bott class", and $\alpha^{n+m}\colon K(\mathbb{R}^{2(n+m)}) \to \mathbb{Z}$ is the iteration of the Bott isomorphism.

Remark. The preceding situation of "trivial" embedding arises in many applications; e.g., when X is a hypersurface, in particular the boundary of a bounded domain in $\mathbb{R}^{n+1}$. For the more general case of "non-trivial" embedding, for different modes of expressing the "topological index" (the right side of the formula), and for a comparison of the various proofs, see the commentary below.

Proof: 1) First, we want to visualize the contents of the formula. The left side is a well-defined (by Chapter II.5) integer which depends only on the homotopy type of the symbol $\sigma(P) \in \mathrm{Iso}_{SX}(E,F)$. However, how is the right side defined? The construction of $[\sigma(P)] \in K(TX)$ was carried out in Exercise 2.2 for $k = 0$; the case $k \neq 0$ adds nothing new. (One can reduce it to the case $k = 0$ directly via composition with Λ^{-k}.)

Now, consider the following figure:

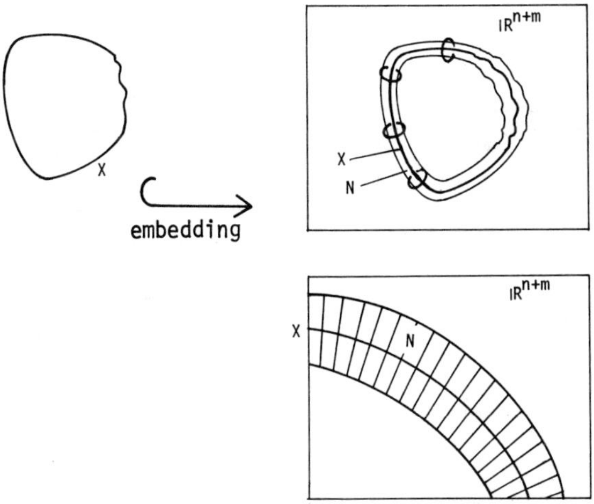

Here N is a "tubular neighborhood" of X in $\mathbb{R}^{n+m}$; i.e., a neighborhood of X which locally (and also globally, because of the "triviality" of the embedding) has the form $X \times \mathbb{R}^m$.

The index formula then says that the following diagram is commutative:*

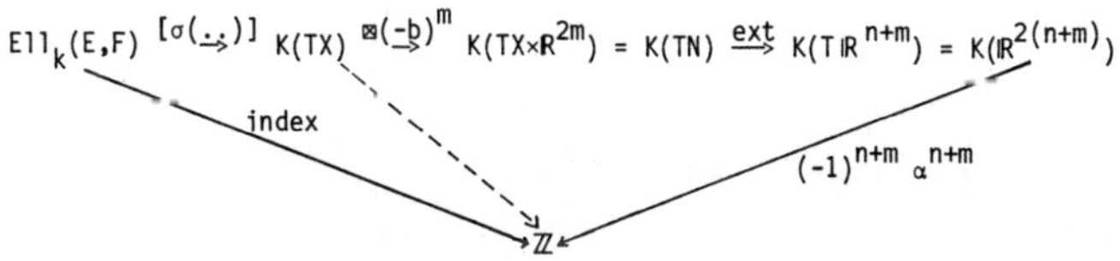

---

*We prefer working with the finitely generated abelian groups $K(TX)$, rather than with single elliptic operators or classes of such: It is precisely the advantage of topological methods that complicated objects of analysis, whose structure is only partially explored, can be replaced purposely by simple "quantities" (in the case before us, by the rank r of the group $K(TX)$). Incidentally, it is known that $r = \sum_{k=0}^{n} \operatorname{rank} H^{2k}_{\mathrm{comp}}(TX)$. Therefore, justified by Exercise 2.2, we will henceforth consider the "analytic index" not on $\mathrm{Ell}_k(E,F)$ but directly on $K(TX)$ as indicated by the broken line in the diagram.

Here, ext is induced by the map $(T\mathbb{R}^{n+m})^+ \to (TN)^+$, which maps the complement of the open set TN in $T\mathbb{R}^{n+m}$ to the point at infinity $+ \in (TN)^+$.

2) We choose a very direct way for the proof of the index formula, by systematically replacing the horizontal homomorphisms (K-theoretic operations) by operations associated with corresponding elliptic operators: Thus, for each elliptic operator P over X, we construct an elliptic operator P' over N, which is the identity at infinity and satisfies index $P' = (-1)^m$ index P and $[\sigma(P')] = [\sigma(P)] \boxtimes b^m$.

We take P' to be $P \# T^m$, where $T \in \mathrm{Ell}_c(\mathbb{R})$ is the standard operator with index $T = -1$ given in the proof of Theorem 2.1; further, note that in Exercise II.5.7 the tensor product # was only defined for operators of order $k > 0$. Hence, more precisely, we take $P' \in \mathrm{Ell}_c(N)$ to be an operator whose symbol on the unit cosphere bundle coincides with $\sigma(P) \# \sigma(T) \# \ldots \# \sigma(T)$; e.g., for $m = 1$:

$$(\sigma(P) \# \sigma(T))(x,t;\xi,\tau) = \begin{pmatrix} \sigma(P)(x,\xi) & -\mathrm{Id}_{F_x} \otimes \sigma(T^*)(t,\tau) \\ \mathrm{Id}_{E_x} \otimes \sigma(T)(t,\tau) & \sigma(P^*)(x,\xi) \end{pmatrix},$$

$$x \in X, \quad t \in \mathbb{R}, \quad (\xi,\tau) \in S(X \times \mathbb{R}).$$

As in Exercise 2.2 (since $\sigma(T)(t,\tau) = 1$ for t sufficiently large and for τ negative), this symbol can be conveniently deformed to a symbol σ' with $\sigma'(x,t;\xi,\tau) = \mathrm{Id}$ for t sufficiently large, if we identify $E' \oplus F'$ and $F' \oplus E'$ by means of switching the summands. Here $E' := p^*E$ where $p: N = X \times \mathbb{R} \to X$ is the projection, whence $E'_{x,t} = E_x$; F' is defined similarly.

3) Without loss of generality, we can take $E' \oplus F'$ to be a trivial bundle, since otherwise we can form $P' \oplus \mathrm{Id}_G$, where the bundle G over N is chosen so that $E' \oplus F' \oplus G$ is trivial. Hence, $E' \oplus F'$ is extended to a trivial bundle over all of $\mathbb{R}^{n+m}$. Then extend the operator P' to all of $\mathbb{R}^{n+m}$, by the identity outside N, to an operator $P'' \in \mathrm{Ell}_c(\mathbb{R}^{n+m})$. Since P' is the identity near the "boundary" $\overline{N} \setminus N$ of N, it follows that if $P''u = 0$ (on $\mathbb{R}^{n+m}$), then the support of u lies entirely in the interior of N, whence $\mathrm{Ker}\, P'' = \mathrm{Ker}\, P'$. By the same argument for the formal adjoint operators, it follows that index P'' = index P'; $[\sigma(P'')] = \mathrm{ext}[\sigma(P')]$ by construction.

4) The formula given for index P then follows from the index formula in the Euclidean case (Theorem 2.1). □

Remarks. 1. Instead of tensoring with the standard operator $T \in \mathrm{Ell}_c(\mathbb{R})$, we can also (for $m = 2$) tensor with the standard transmission operator $T \in \mathrm{Ell}(B^2, S^1)$, thereby constructing an *elliptic boundary-value problem over the bounded manifold* $\overline{N}$, whose inner symbol can be deformed by the well-known procedure (using the boundary symbols) such that it becomes the identity near the boundary $\overline{N} \setminus N$. The desired operators P' and P'' are then provided.

Correspondingly, for arbitrary m, one can find a boundary-value problem whose index is 1 and whose symbol induces the bundle $b \boxtimes \ldots \boxtimes b$ (m-times): For this, one takes the differential operator $d + d^*$ (in the exterior calculus of differential forms; see Appendix, Exercise 8) from the forms of even order to those with odd order, with a suitable elliptic boundary-value problem in the sense of Chapter II.6.

2. The advantage of the construction of P' via boundary-value problems is best exhibited, when *the embedding of* X *in* $\mathbb{R}^{n+m}$ *is not "trivial"* i.e., when N is no longer $X \times \mathbb{R}^m$. In [Atiyah-Singer 1968a] and [Atiyah 1970a] devices of "equivariant K-theory" (with transformation groups) are employed for explicitly stating or axiomatically characterizing the desired operator P' with the help of symmetry properties of standard operators over the sphere. The use of "equivariant K-theory" is avoided (also in the case $N \neq X \times \mathbb{R}^m$ not treated by us) by passing to boundary-value problems and by another method proposed by [Hörmander 1971c] using "hypo-elliptic" operators and stronger analytical tools.

3. Just as we learned (in Sections 1.A/B) different ways for defining "degree", the index, computed here in K-theoretic terms, can be determined in *cohomological or integral form*; see Section 4.A. This is possible without new or modified proofs, but simply by routine exercises in algebraic topology, the transition from K-theory to cohomology, whereby simply "one set of topological invariants is translated into another... Which formula provides the "best answer" is largely a matter of taste. It depends on which invariants are most familiar or can be computed most easily (M. F. Atiyah, I. M. Singer). □

B. Comparison of the Proofs: The Cobordism Proof

Michael Atiyah and Isadore Singer gave two more proofs of the Index Formula, in addition to the "embedding proof" given above. These are the original "cobordism proof" and the newer "heat equation proof". We cannot summarize them here, but we will comment briefly (see the tabulated crude survey at the end of this section). Since all three of these proofs appear to be somewhat complicated, several authors (among others, [Bojarski 1963], [Calderon 1967] and [Seeley 1965/1967]) tried to give simpler or more elementary proofs for the Euclidean case. In the judgement of [Atiyah 1967b, 245] "these different proofs differ only in their use and presentation of algebraic topology" (instead and at times together with the Bott Periodicity Theorem "older but not at all elementary parts of topology" are employed) - "the analysis is essentially the same in origin".

The "Cobordism proof" is sketched in [Atiyah-Singer 1963] and worked out in detail in [Brieskorn et al. 1963], [Cartan-Schwartz 1965] and [Palais 1965]. It was the first proof: It begins with a compact oriented Riemannian manifold (without boundary) of dimension 2ℓ and defines $d\colon \Omega^j \to \Omega^{j+1}$ and $d^*\colon \Omega^{j+1} \to \Omega^j$ as the exterior ("Cartan"-) derivative of forms and its adjoint. (These are considered in Exercise 8 of the Appendix. More precisely, we have here $\Omega^j := C^\infty(\Lambda^j(T^*X) \otimes \mathbb{C})$.) Then $d+d^*\colon \Omega \to \Omega$ is a self-adjoint elliptic differential operator of first order whose square is the Laplace operator Δ of Hodge theory; here $\Omega = \sum_{j=0}^{2\ell} \Omega^j$. The formula

$$\tau(v) := i^{p(p-1)+\ell} * v, \quad v \in \Omega^p,$$

defines on Ω an "involution" (i.e., $\tau\circ\tau = \mathrm{Id}$), where $*\colon \Omega^p \to \Omega^{2\ell-p}$ is the duality operator. If $\Omega^\pm$ denote the ± 1-eigenspaces of τ, we define the signature operator (see Section 4.D below and [Atiyah-Singer 1968b, 575]) $D^+\colon \Omega^+ \to \Omega^-$ to be the restriction of $d+d^*$ to Ω^+. Then D^+ is an elliptic operator and its index is the "signature" of the manifold X, and is frequently named after Friedrich Hirzebruch.

For sufficiently many special cases (spefically for $X = \mathbb{P}(\mathbb{C}^{\ell+1})$ and $X = S^{2\ell}$) one can now compute the signature index D^+ using some cohomology theory and derive an index formula for manifolds of even dimension. "Sufficiently many" here means four things:

(i) By a deep result of cobordism theory by René Thom, every even-dimensional manifold Y is in a certain sense "cobordant" to the special manifolds; in other words, there is a bounded manifold Z whose boundary - roughly - is built up from X and Y.

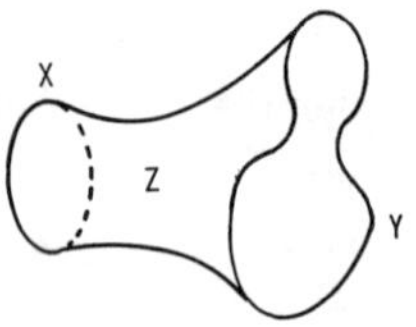

(ii) Furthermore, René Thom proved the vanishing of the signature for bounded manifolds. Hence the index of the signature operator on an arbitrary 2ℓ-dimensional manifold X can be computed from the indices of the special signature operators [Hirzebruch 1956/1966, §8].

(iii) For a Hermitian C^∞-vector bundle E over X, let

$$\Omega_E^j := C^\infty(\Lambda^j(T^*X) \otimes E)$$

be the space of j-forms "with coefficients in E". By means of a "covariant derivative" ∇_E (inducing parallel translation along paths), one can define the operator

$$d_E(u \otimes v) := du \otimes v + (-1)^j u \wedge \nabla_E v; \quad u \in \Omega_E^j, \quad v \in C^\infty(E),$$

on Ω_E^j and its adjoint d_E^*, and one can establish the index formula for the generalized signature operator D_E^+ of the vector bundle E which is defined by means of an involution on Ω_E.

This concludes the proof since every elliptic operator on a closed even-dimensional oriented manifold X is "equivalent" in the sense of K-theory to a generalized signature operator. More precisely: $K(TX)$ is a ring over $K(X)$ and the subgroup $K(X)\cdot[\sigma(D^+)]$ generated by generalized signature operators via $[\sigma(D_E^+)] = [E]\cdot[\sigma(D^+)]$ is so large, namely a "subgroup of finite index", that "practically" all of $K(TX)$ is generated. Here "practically" means up to the image of $K(X)$ in $K(TX)$ and up to 2-torsion, where the index must vanish as an additive function with values in $\mathbb{Z}$.

At this place, the theory of pseudo-differential operators enters in order to achieve that the symbols are arbitrary bundle isomorphisms over SX, and to allow reduction of the index computation from $Ell(X)$ to $K(TX)$. Further, the Bott Periodicity Theorem is used in somewhat generalized form in representing $K(TX)$ approximately by $K(X)$; see [Atiyah-Bott-Patodi 1973, 321f.].

(iv) The general index formula can be extended to an odd-dimensional manifold X, using the multiplicative property of the index by tensoring with the standard operator T with index 1 on S^1 and by passing to the even-dimensional manifold $X \times S^1$. If one is not interested in the sign in the index formula, one can avoid the explicit definition of T and simply pass to the "squared" (relative to the tensor product) operator on the even-dimensional manifold $X \times X$. □

C. Comparison of the Proofs: The Imbedding Proof

The "Imbedding Proof", which was given in the present monograph, following [Atiyah 1967b], [Atiyah-Singer 1968a] and [Atiyah 1970a]. In this proof, the methods remain topological with the consideration of K(TX) instead of the operator space Ell(X). The idea goes back to the proof of the Riemann-Roch Theorem by Alexander Grothendieck, Section 4.G below. One shows first using the Bott Periodicity Theorem that every elliptic operator on the sphere or Euclidean space is equivalent, in the sense of K-theory, to one of infinite many (more precisely $|\mathbb{Z}|$-many) standard operators, and then the case of an arbitrary elliptic operator on an arbitrary closed manifold is reduced to the standard case by imbedding.

The advantage as well as weakness of this proof lies in its perhaps somewhat forced directness. It succeeds on the one hand in eliminating cohomology and cobordism theory completely, bringing out the function analytic and topological pillars (Theorems of Gohberg-Krein and Bott) plainly and in their most elementary form, and in achieving through this simplicity of tools the greatest susceptibility to generalization (see Section 4.L). On the other hand, under the imbedding (except for particularly smooth ones, e.g. holomorphic embeddings of algebraic manifolds in a complex projective space) the special structure of "classical" operators is completely destroyed. For example, the signature operator does not at all become another signature operator under the imbedding, and to affect the proof of the Riemann-Roch Theorem for arbitrary compact complex manifolds (to mention another problem defined by "classical" operators; see also Section 4.G below), one has to leave this category, whereby many of the interesting and sometimes open problems of modern differential topology become less transparent. □

D. Comparison of the Proofs: The Heat Equation Proof

In comparison with the first two proofs, which argue more topologically, the "heat equation proof" offers a completely different and initially purely analytic approach to the index problem. The germinal idea goes back to papers of Marcel Riesz on spectral theory of positive self-adjoint operators, and was presented by M. F. Atiyah as early as 1966 at the International Congress of Mathematicians in Moscow, and then published, also in connection with applications of the Index Formula to fixed point problems, in [Atiyah-Bott 1967], [Atiyah 1968b] and in related form in [Calderon 1967] and [Seeley 1967]:

1. One starts with an operator $P \in \mathrm{Ell}_k(E,F)$, $k > 0$, where E and F are Hermitian C^∞-vector bundles on the n-dimensional, closed, oriented, Riemannian manifold X. Then the operator P^*P is a non-negative self-adjoint operator of order $2k$ with a discrete spectrum (see Ch. I.5 above) of non-negative eigenvalues $0 \leq \lambda_1 \leq \lambda_2 \leq \ldots$ (the multiplicity may be larger than 1, hence "$\leq$"); and the series

$$\theta_{P^*P}(t) := \sum_{m=1}^{\infty} e^{-t\lambda_m} \tag{0}$$

converges for all $t > 0$. By the way, for $X = S^1$ and $P = -i\frac{d}{dx}$ we do obtain the "theta function" $\theta(t) = \Sigma_{m=0}^{\infty} e^{-tm^2}$ of analytic number theory since the square numbers are exactly the eigenvalues of $P^*P = \Delta = -d^2/dx^2$.

Correspondingly, one forms the function θ_{PP^*}. The operators P^*P and PP^* have the same non-zero eigenvalues, and only the eigenvalue 0 has in general different multiplicities, namely $\dim \operatorname{Ker} P$ and $\dim \operatorname{Ker} P^*$ (see the remark following Theorem I.3.2 above). This way, one has a new index formula

$$\operatorname{index} P = \theta_{P^*P}(t) - \theta_{PP^*}(t), \quad t > 0. \tag{1}$$

Trivially, one may choose an arbitrary function ϕ on $\mathbb{R}$ with $\phi(0) = 1$ instead of the function $m \mapsto e^{-tm}$ and thus obtain for each ϕ a further index formula

$$\operatorname{index} P = \sum_{\lambda \in \operatorname{Spec}(P^*P)} \phi(\lambda) - \sum_{\lambda \in \operatorname{Spec}(PP^*)} \phi(\lambda).$$

Now, the theta function is distinguished by permitting near $t = 0$ an asymptotic development

$$\theta_{P^*P}(t) \sim \sum_{m \geq -n} t^{m/2k} \int_X \mu_m(P^*P) \tag{2}$$

where $\mu_m(P^*P)$ is, for each $m \in \mathbb{Z}$, a certain density on X which can be formed canonically with the coefficients of the operator P^*P. Using (1) we get from (2) the explicit integral representation

$$\operatorname{index} P = \int_X \mu_0(P^*P) - \mu_0(PP^*). \tag{3}$$

2. The convergence of the series in (0) has implications for the construction of solutions of the heat conduction equation [Gilkey 1974, §3].

$$\frac{\partial u}{\partial t}(x,t) + \Box u(x,t) = 0, \quad x \in X, \quad t \in [0,\infty)$$

with the initial condition $u(\ldots,0) = u_0 \in L^2(E)$. Here u is the unknown function on $X \times [0,\infty)$ with values in the bundle E (the "heat distribution"), and $\square = P^*P$ is a generalized Laplace operator on X. Then

$$H_t := e^{-\square t} = \mathrm{Id} - t\square + \frac{t^2}{2!}\square^2 - \frac{t^3}{3!}\square^3 + - \ldots, \quad t \geq 0,$$

is a well-defined family of bounded operators on the Hilbert space $L^2(E)$, which satisfies the heat equation

$$\frac{\partial}{\partial t} H_t + \square H_t = 0$$

with initial value $H_0 = \mathrm{Id}$. Thus H_t yields for each initial distribution u_0 the heat distribution at time t by the formula $u(\ldots,t) = H_t u_0$.

Since the eigenfunctions $\{v_m\}$ of $\square$ form a complete orthogonal system for $L^2(E)$, the formula

$$\theta_\square(t) = \operatorname{trace} e^{-\square t} \tag{4}$$

is meaningful. The convergence of the series in (0) means that the solution operators H_t of the parabolic heat conduction equation belongs to the "trace class" for $t > 0$. By means of the theory of pseudo-differential operators it follows more precisely that H_t is a "smoothing operator", i.e., an operator of order $-\infty$ which is representable as integral operator

$$(H_t v)(x) = \int_X K_t(x,.)v\omega, \quad v \in L^2(E), \quad x \in X,$$

with C^∞-weight function $(x,y) \longrightarrow K_t(x,y) \in L(E_y,E_x)$ and the volume element ω. Then $\theta_\square(t) = \operatorname{trace} H_t = \int_X \mu_t$, $t > 0$, where

$$\mu_t := \operatorname{trace} K_t(.,.)\omega = \sum_{m=1}^{\infty} e^{-\lambda_m t} |v_m(\cdot)|^2 \omega, \quad t > 0,$$

is a density on X which at each $x \in X$ is the pointwise trace of the operator K_t. While μ_t can be expressed in terms of the coefficients of the operator P only very indirectly (via the eigenvalues and eigenfunctions of $\square = P^*P$), one has at each single point $x \in X$ an asymptotic expansion

$$\mu_t(x) \sim \sum_{m \geq -n} t^{m/2k} \mu_m(x) \quad \text{for} \quad t \to 0,$$

where the μ_m are purely local invariants of P^*P which then implies (2).

4. A proof of (2) with a recipe for the computation of μ_m extracted from the theory of pseudo-differential operators is due to [Seeley 1967]. It shows that the μ_m depend rationally on the coefficients of P and their derivatives of orders $\leq n$. A more intuitive and heuristic description of the μ_m for the special case $X = \pi^n := \mathbb{R}^n/2\pi\mathbb{Z}^n$, to which we paid special attention in our Sobolev case studies (Chapter II.4), can be found in [Atiyah-Bott-Patodi 1973, 300f.]. The idea of the proof goes back to the mathematicians Subbaramiah Minakshisundaram and Åke Pleijel, who in 1949 (long before an effective machinery for pseudo-differential operators was established) computed the μ_m for the case $P^*P = \Delta$, where Δ is the invariantly defined Laplace-Beltrami operator which depends only on the Riemannian metric on X, i.e., $k = 1$ and $E = \mathbb{C}_X$. Precisely, S. Minakshisundaram and Å. Pleijel (and later R. T. Seeley, when generalizing their results) studied in place of P^*P the positive self-adjoint operator $\square = \mathrm{Id} + P^*P$ and in place of the theta function the zeta function $\zeta(z) := \Sigma_{m=1}^{\infty} (\lambda_m)^{-z}$ summed over all (discrete positive) eigenvalues of $\square$. As is well-known, the zeta function is well-defined for $\mathrm{Re}(z) > \dim X$ and can be continued to a meromorphic function in the z-plane with finitely many real poles of order 1 with behavior at poles known in principle. Among other things it is found that $z = 0$ is not a pole and that the value $\zeta(0)$ can be expressed explicitly in terms of $\square$; in fact,

$$\zeta(0) = \int_X \rho_0(\square), \tag{5}$$

where the right hand side is fairly complicated but can be computed in principle. On the other hand, $\zeta(z)$ can be interpreted within spectral theory as $\mathrm{trace}(\square^{-z})$, and finally $\zeta(0)$ appears as the constant term in the asymptotic expansion of $\theta_\square(t)$ as $t \to \infty$. This establishes the connection with the heat conduction approach. In particular, the measure $\mu_0(\square)$ sought there is identical with the measure $\rho_0(\square)$ in equation (5), and (2) follows from (5) and similar computations of the residues of ζ at its poles.

5. Thus, a general algorithm is available that is capable of producing the right hand side of the index formula (3) in finitely many steps by means of a computer for example. In contrast to the index formulas of the cobordism and imbedding proofs (into which enter the derivatives of the coefficients of P up to order 2 only) this formula is in the general case complicated numerically and algebraically mainly by the appearance of derivatives up to order n (= dim X). While for algebraic curves of

complex dimension 1 (= Riemannian surfaces, $n = 2$), the formula can be handled well computationally, the general situation requires so much effort that M. F. Atiyah and R. Bott by their own admission had initially "little hope for interpreting these integrals directly in terms of the characteristic classes of E, X", and therefore "the beautiful formula appeared to be useless in this context." Only a series of more recent papers on curvature tensors revealed that in the special case when X is even-dimensional and P is the signature operator D^+, all higher order derivatives "cancel out" in Seeley's formula for the measure $\mu_0(P^*P)$, and only the derivatives up to order 3 remain. Details of such computations appeared first in [McKean-Singer 1967], in the case (not all that fortunate for this aspect) that P is the operator $d+d^*: \Omega^{even} \to \Omega^{odd}$ with the Euler characteristic as index (see Section 4.D). This was generalized in 1971 by V. K. Patodi - again by means of symmetry considerations - and extended to Riemann-Roch operators (see Section 4.G) in particular. P. Gilkey succeeded shortly thereafter in replacing Patodi's complicated group theoretic cancellation procedure for the higher derivatives by an axiomatic argument which was drastically simplified in [Atiyah-Bott-Patodi 1973] through the use of stronger tools of Riemannian geometry. It says roughly that in each integrand with the general qualitative properties of $\mu_0(P^*P)$ the annoying higher derivatives can be disregarded and $\mu_0(P^*P)$ can be identified with the (normalized) Gaussian curvature.

In this fashion, a new purely analytic proof of the Hirzebruch Signature Theorem (the index formula for "classical" operators) is achieved which implies the general index formula, as in the cobordism proof (see above, items (iii) and (iv) in the sketch B), and with the same topological arguments.

6. The significance of the heat equation proof, which cannot be extended (just like the cobordism proof) to families of elliptic operators and operators with group action, is at present difficult to estimate. Its authors, who - as an aside - acknowledge that their "whole thinking on these questions has been stimulated and influenced very strongly by the recent paper of Gelfand on Lie algebra cohomology" [Gelfand 1970], point out that their proof is hardly shorter than the imbedding proof since it "uses more analysis, more differential geometry and no less topology. On the other hand, it is more direct and explicit for the classical operators associated with Riemannian structures: In particular, the local form of the Signature Theorem and its generalizations are of considerable interest in itself and should lead to further developments" [Atiyah-Bott-Patodi 1973, 281].

This prediction appears to materialize even beyond the realm of differential geometry: The approach via the zeta function of the Laplace-Beltrami operator Δ on X (whose values yield real-valued invariants of the Riemannian metric ρ of X - "spectral invariants" - at any point where the zeta function does not have a pole) has been extended to the systems case, where the Laplace equation is replaced by the system of partial differential equations of the "total" Laplace operator of Hodge theory which can be represented as the square of a self-adjoint operator A ("Dirac operator"). In analogy to the zeta function, one considers for an operator A, which is not positive, the function $\eta(z) := \Sigma_{\lambda\neq 0}(\text{sign }\lambda)|\lambda|^{-z}$, where summation is over the eigenvalues of A with proper multiplicity. Again η is a holomorphic function for large $|z|$, which can be continued meromorphically to the whole z-plane. Corresponding to the asymptotic expansion above in equation (2) for the theta and zeta functions, one can look for an integral formula for $\eta(0)$. Such a formula

$$\eta(0) = \int_Y \alpha(\tilde{\rho}) - \text{integer} \tag{6}$$

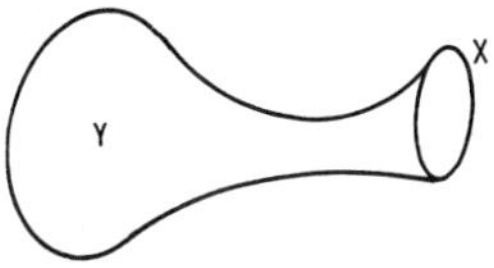

is proven in [Atiyah-Patodi-Singer 1975, Theorem 4.14], when X can be obtained as the boundary of a 4-dimensional manifold Y with Riemannian metric $\tilde{\rho}$. Here $\tilde{\rho}$ is assumed to induce on X the metric ρ and to render a neighborhood of X in Y isometric to $X \times [0,1]$. The integrand $\alpha(\tilde{\rho})$ is explicitly known (the "ℓ-th Hirzebruch L-polynomial in Pontryagin forms of the Riemannian metric $\tilde{\rho}$"), as well as the integral correction term (the "signature of Y" defined by the topology of Y). In this way, a formula is obtained which relates the spectral invariant $\eta(0)$, measuring the asymmetry of the spectrum of A, with a differential geometric and a purely topological invariant.

This formula lies deeper than (2) or (3), where the integrand was of local type, and has numerous relations to the cobordism as well as the imbedding (particularly via boundary-value problems) proof. It is not as esoteric as it may appear to someone not so much interested in differential-topological problems, since it yields - roughly - a direct geometric interpretation for the peculiarities of the distribution of the eigenvalues of A: Newer results in this direction by Victor W. Guillemin and others, say for example, that a Riemannian manifold X is isometric with S^n if the spectra of the Laplace operators coincide. One cannot always expect that two manifolds with equal spectra of their Laplace operators are isometric (the 16-dimensional tori* yield a counterexample), but it appears that at least "extreme" distributions of the eigenvalues (when they are not randomly distributed, but lumped together near integers, for example) carry with them "extreme" geometric situations (in our example, the closedness of the geodesics). According to an announcement in [Singer 1976], this answers in principle the classical question "Can you hear the shape of the drum?" (Mark Kac) which had already motivated [McKean-Singer 1967]. Further, a gate opened to the "inverse problem"** of "mathematical modelling" of real phenomena: Theoreticians frequently and

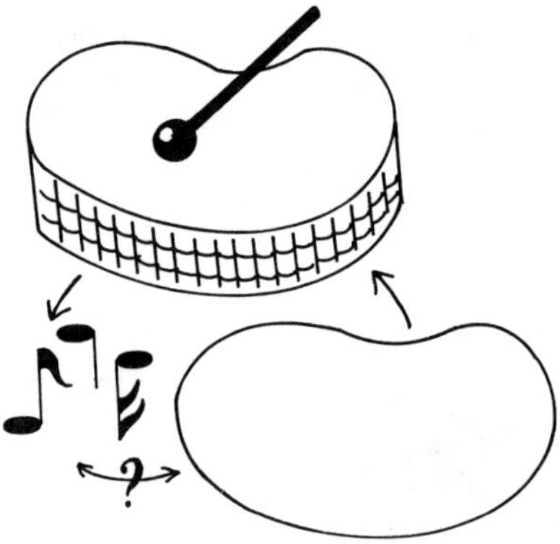

and perforce only "apply" theory, i.e., they try, similar to the axiomatic method within mathematics, to draw conclusions as far-reaching as possible

---

*J. Milnor: Eigenvalues of the Laplacian operator on certain manifolds. Proc. Nat. Acad. of Sciences U.S.A. 51 (1964), 542.

**According to a communication by Richard Bellmann the "inverse problem" in its most general formulation goes back to Carl Gustav Jacob Jacobi (1804-1851). Today, the Inverse Problem is studied frequently in very different contexts, reaching from algebraic problems [Ulam 1960, 32] to "structure identification and parameter estimation" in control theory. A special version of the Inverse Problem is the inversion of spectral analysis, "spectral synthesis", which goes far beyond the newer results of Riemannian geometry presented here [Benedetto 1975].

about the concrete behavior from relatively modest assumptions about the existence of certain laws. Conversely, the practitioner needs in general the "inverse" of the theory, namely the exposure of regularities in the observations at hand. (Somewhat overstated, the practitioner desires to fit a curve to given measurements, while the theoretician sees his strength in a detailed discussion of the properties of a given curve.) In this sense, the novelty consists in the attempt to estimate the parameters of a differential equation, when information about special solutions (eigenfunctions and eigenvalues, for example) is available.

The following last chapter contains a survey on some reformulations, applications and generalizations of the Atiyah-Singer Index Formula.

4. APPLICATIONS (SURVEY)

With the Atiyah-Singer Index Formula, we proved "one of the deepest and hardest results of mathematics" which "is probably enmeshed more widely with topology and analysis than any other single result" [Hirzebruch-Zagier 1974, VIII]. In this book, we are mainly interested in the varied sources and parts which flow together or are put together in the Index Formula. But the formula itself is of great interest also. It can express important and far-reaching ideas in various areas of application, through numerous corollaries, specializations, reformulations and generalizations.

The appraisals of the role of the Atiyah-Singer Index Formula in these applications are contradictory: On the one hand, it permits us to attack complicated topological questions with relatively simple analytic methods; on the other hand, this formula frequently serves only "to derive a number of wholly elementary identities, which could have been proved much more easily by direct means." This is the judgement of Friedrich Hirzebruch and Don Zagier about the relationship between the Index Formula and topics from elementary number theory. Some of the following principal theorems, which at first could be proved only in the framework of the Atiyah-Singer theory, have been proved directly in the meantime. This is true for the general Riemann-Roch Theorem (see Section G below) in [Toledo-Tong 1976], its consequences for the classification of certain algebraic surfaces drawn by Kunihiko Kodaira in [Hirzebruch 1973], and for some of the vector field computations of Atiyah-Dupont (see Section E) in [Koschorke 1975].

There is much that is unclear in the relationship between the Atiyah-Singer Index and its applications. It is important and interesting that

Schematic Comparison	Cobordism Proof	Imbedding Proof	Heat Equation Proof
Where to start:	Signature operator D_E^+ on $\mathbb{P}(\mathbb{C}^{\ell+1})$ and $S^{2\ell}$ (with coefficients in an arbitrary bundle E). *Cohomology.*	Standard operators T_1 of index 1 on the n-sphere (or Euclidean space). *Pseudo-differential operators.*	Heat equation on $X \times \mathbb{R}$, X arbitrary Riemannian manifold. *Pseudo-differential operators.*
How to continue:	Signature for "cobordant" X. *Coholomology* Signature for arbitrary even-dimensional differentiable manifolds. *Cobordism.*	Standard operators $\{T_m : m \in \mathbb{Z}\}$ on the sphere with index $T_m = m$. Proof that every elliptic operator on the sphere can be deformed to some T_m in the class of elliptic differential operators. *K-Theory (Bott Periodicity).*	Laplace operator on 4n-dimensional Riemannian manifold X. Signature operator D^+ on X. Signature operator D_E^+ on arbitrary closed oriented Riemannian manifolds of even dimension. *Riemannian geometry.*
Intermediate results:	General signature formula. *Topology.*	General index formula in the Euclidean case. *Analysis and topology.*	General signature formula. *Analysis.*
How to generalize:	Generalize index formula for even-dimensional manifolds, by showing that every elliptic operator can be deformed to a signature operator. *K-theory (Bott).* Generalize index formula for odd-dimensional manifolds by tensoring with the standard operator T_1 on the 1-sphere. *Tensor product.*	General index formula for arbitrary manifolds through imbedding. *Tensoring.*	Same as for the Cobordism Proof.

Schematic Comparison	Cobordism Proof	Imbedding Proof	Heat Equation Proof
Advantages/ special features:	Derives directly and continuously from the classical papers on "analysis on manifolds."	Very direct proof. Can be generalized widely. Stresses the unity of analysis and topology.	Most direct and explicit for the "classical operators". Exceptionally rich interrelations (promising).
Difficulties/ Weaknesses:	Cohomology, cobordism theory	Destruction of the special structure of the classical operators in the imbedding process.	Varied and sophisticated tools.
Traditions:	F. HIRZEBRUCH	I. GOHBERG, M. KREIN, A. GROTHENDIECK	M. RIESZ, I. GELFAND, D. GILKEY

these many relationships exist, although considerable future efforts will be required to bare the real reasons for these relationships, and to - in the end - better understand the unity of mathematics and the specifics and interrelationships of its parts. Once again, we quote [Hirzebruch-Zagier 1974, VII]: "That a connection exists, a number of people realized essentially at the same time.... And since neither we nor anybody knows why there must be such a connection, this seemed like an ideal topic for a book in order to confront other mathematicians with a puzzle for their embarrassement or their entertainment - as the case may be."

Since there is not space to truly introduce the following areas of applications, we chose to give this section the form of a literature survey.

A. Cohomological Formulation of the Index Formula

In Theorem 3.1, the Index Formula is phrased in the language of K-theory: Its right side only involves vector bundles and operations with vector bundles. The conversion to cohomological form is carried out in [Atiyah-Singer 1968b, 546-559]. While the individual calculations are somewhat complicated (if more or less routine topological exercises nowadays), the underlying method (namely, the construction of a "functor" that assigns to each vector bundle a "characteristic" cohomology class of the base X of E with coefficients in $\mathbb{Z}$, $\mathbb{Q}$, or $\mathbb{R}$) is rather clear. Here, we follow [J. W. Milnor, 1974].

Let E be a complex vector bundle of fiber dimension N over the paracompact space X, with projection $\pi: E \to X$, and let E_0 be the subspace of E obtained by removing the zero section. As a real vector bundle of fiber dimension $2N$, E is oriented, since all complex bases $e_1,\dots,e_N$ of the fiber E_x, $x \in X$, yield real bases $e_1, ie_1, e_2, ie_2, \dots, e_N, ie_N$ of the same orientation. In the language of cohomology, an orientation for E_x is the choice of a generating element $\mathcal{O}_x$ for $H^{2N}(E_x, E_x \setminus \{0\}; \mathbb{Z})$. Now, the following observations go back to R. Thom:

(i) $H^i(E,E_0;\mathbb{Z}) = 0$ for all $i < 2N$.

(ii) The orientation of E defines a "total orientation class" $U \in H^{2N}(E,E_0;\mathbb{Z})$ by the condition $(j_x)^*U = \mathcal{O}_x$ for all $x \in X$, where $j_x: (E_x, E_x \setminus \{0\}) \to (E,E_0)$ is the embedding.

(iii) Via the cup product $u \mapsto \pi^*(u) \cup U$, $u \in H^i(X;\mathbb{Z})$, with the orientation class U, an isomorphism $\phi: H^*(X;\mathbb{Z}) \xrightarrow{\cong} H^*(E,E_0;\mathbb{Z})$ is defined which raises the dimension of the cohomology classes by $2N$. This is the "*Thom isomorphism*" that can be given more generally for all real oriented vector bundles of arbitrary fiber dimension. For a comparison with the Bott isomorphism of K-Theory, see [Atiyah-Singer loc. cit.]. □

The N-th *Chern class* $c_N(E)$ of E is the class $\phi^{-1}(U \cup U) \in H^{2N}(X; \mathbb{Z})$; the *total* Chern class

$$c(E) = 1 + c_1(E) + \ldots + c_N(E); \quad c_i(E) \in H^{2i}(X, \mathbb{Z})$$

is obtained (for example) by the axiomatic conditions of functoriality (i.e., $f^*c(E) = c(f^*E)$ for $f: Y \to X$) and homomorphy (i.e., $c(E \oplus F) = c(E) \cup c(F)$. Because of this homomorphism property, the Chern classes are not only defined on $\mathrm{Vect}(X)$, but also on $K(X)$, since $c(E) = 1$ if E is a trivial bundle.

By means of the formal factorization $c(E) = (1+y_1) \cup \ldots \cup (1+y_N)$ with $y_i \in H^2(X, \mathbb{Z})$, where the $-y_i$ are the hypothetical "zeros" of the polynomial $1 + c_1(E)t + \ldots + c_N(E)t^N$, one obtains the *Chern character*

$$\mathrm{ch}(E) := \sum_{i=1}^{N} e^{y_i} = N + \sum_{i=1}^{N} y_i + \frac{1}{2!} \sum_{i=1}^{N} y_i^2 + \ldots,$$

which extends to a ring homomorphism $\mathrm{ch}: K(X) \to H_c^*(X; \mathbb{Q})$, thereby providing a natural transformation from K-Theory to singular cohomology theory with rational coefficients and compact support. □

Results. a) If P is an elliptic operator on a closed oriented manifold X of dimension n, then we have

$$\text{index } P = (-1)^{n(n+1)/2}\{\phi^{-1}\mathrm{ch}[\sigma(P)] \cup \tau(TX \otimes \mathbb{C})\}[X].$$

Here $[\sigma(P)] \in K(X)$ is the difference bundle of the symbol $\sigma(P)$ of P, $\mathrm{ch}[\sigma(P)] \in H^*(TX; \mathbb{Q})$ is its Chern character, $\phi.\ H^*(X; \mathbb{Q}) \to H^*(TX, (TX)_0; \mathbb{Q}) = H_c^*(TX; \mathbb{Q})$ is the Thom isomorphism, $[X] \in H_n(X; \mathbb{Q})$ is the fundamental cycle of the orientation of X, and

$$\tau(E) := \frac{y_1}{1-e^{-y_1}} \cdot \ldots \cdot \frac{y_N}{1-e^{-y_N}} \in H^*(X, \mathbb{Q})$$

is the Todd class of the complex vector bundle E of fiber dimension N, whose Chern class is factored as above. (Here, take E to be the complexification $TX \otimes_{\mathbb{R}} \mathbb{C}$ with fiber dimension n.) Incidentally, by means of Riemannian geometry one can express $\mathrm{ch}(E)$ and $\tau(E)$ by the "curvature matrix" of the vector bundle E which one equips with a Hermitian metric; e.g., see [Atiyah-Singer 1968b, 551] and [Atiyah-Bott-Patodi 1973, 310], or section 2.G of Part IV.

b) More generally, one can drop the orientability of X and obtain the formula

$$\text{index } P = (-1)^n\{\text{ch}[\sigma(P)] \cup \pi^*\tau(TX \otimes_{\mathbb{R}} \mathbb{C})\}[TX].$$

Here [TX] is the fundamental cycle of the tangent bundle TX which admits an orientation as an "almost-complex manifold" (namely, divide the tangent space of TX into a "horizontal" = "real" and a "vertical" = "imaginary" part); and $\pi: TX \to X$ is the projection. The calculation of the right sides in a) and b), naturally involves only the evaluation of the highest dimensional components of the cup product on the respective fundamental cycles. □

B. The Case of Systems (Trivial Bundles)

In [Atiyah-Singer 1968b, 600-602], a drastic simplication of the Index Formula is proved for the case of trivial vector bundles (i.e, when the elliptic operator P is applied to a system of N complex-valued functions). The symbol of such an operator is then a continuous map $\sigma(P): SX \to GL(N,\mathbb{C})$, where SX is the unit sphere bundle of X. Looking at the induced cohomology homomorphism $(\sigma(P))^*: H^*(GL(N,\mathbb{C})) \to H^*(SX)$ (coefficients arbitrary), it becomes evident that in order to obtain a useful formula in this case, one must know something about the cohomology of the Lie group $GL(N,\mathbb{C})$.

Since the unitary group U(N) is a deformation retract of $GL(N,\mathbb{C})$, we have $H^*(GL(N,\mathbb{C})) \cong H^*(U(N))$. Moreover, since we have a natural map $\rho_N: U(N) \to U(N)/U(N-1) = S^{2N-1}$, we may obtain all essential information on $H^*(GL(N,\mathbb{C}))$ from the well-known cohomology of S^{2N-1}. More precisely: Let $u_i \in H^{2i-1}(S^{2i-1})$ be the natural generating element, where S^{2i-1} is oriented as the boundary of the ball in $\mathbb{C}^i$. Then set $h_i^i := (\rho_i)^*(u_i) \in H^{2i-1}(U(i))$; e.g., $h_N^N \in H^{2N-1}(U(N))$. By the normalization condition $j^*h_i^N = h_i^i$, where $j: U(i) \to U(N)$ is the canonical embedding for $i \leq N$, one obtains the additional elements $h_i^N \in H^{2i-1}(U(N))$, $i \leq N$, which together form a system of generators for the algebra $H^*(U(N))$. □

Results. For an elliptic system P of N pseudo-differential equations for N complex-valued functions on a closed manifold X of dimension n, we have

$$\text{a) index } P = (-1)^n \left\{ \sum_{i=1}^{N} \frac{(-1)^{i-1}(\sigma(P))^*h_i^N}{(i-1)!} \cup \tau(X)\right\}[SX].$$

Here [SX] is the fundamental cycle of the canonical orientation of SX and $\tau(X)$ is the lift to SX of the Todd class of the complexification of TX; see Section A above.

b) If $n \leq 3$ or if X is a hypersurface in $\mathbb{R}^{n+1}$, then $\tau(X) = 1$, and hence

$$\text{index }(P) = \begin{cases} (-1)^{N+n-1} \dfrac{(\sigma(P))^* h_n^N}{(n-1)!} & \text{for } N > n \\ -\dfrac{\text{mapping degree }(\rho \circ \sigma(P))}{(n-1)!} & \text{for } N = n \\ 0 & \text{for } N < n \end{cases}$$

For the determination of the mapping degree of the composition $\rho \circ \sigma(P)$: $SX \to S^{2n-1}$, see [Bröcker-Jänich 1973/1982, 14.9.6-10].

c) In the framework of Hodge theory (which provides a canonical isomorphism between the "harmonic differential forms" of degree p on a manifold and the p-th cohomology of the manifold with coefficients in $\mathbb{C}$ - see also Section D below), there are explicit differential forms $\omega_i \in \Omega^{2i-1}(U(N))$ (so-called "bi-invariant forms"), which represent the generators h_i^N. Hence, one obtains the integral formula

$$\text{index } P = (-1)^n \int_{SX} \sigma(P)^* \omega \wedge \pi^* \tau,$$

where $\omega := \sum_{i=1}^{N} \dfrac{(-1)^{i-1}\omega_i}{(i-1)!} \in \Omega^*(U(N))$ is a "global invariant"; $\pi\colon SX \to X$ is the projection and $\tau \in \Omega^*(X)$ is the (total) differential form corresponding to the Todd class, which as we have already remarked in A is explicitly given by a general expression involving the curvature of the Riemannian manifold X. □

C. Examples of Vanishing Index

In a series of special cases one can conclude that the index of an operator vanishes by using the index formula in Theorem 3.1 or its alternative formulations in Sections A and B above, without having to go through all of the somewhat complicated topological computations. For some of these results (e.g., for a) and the special case $N = 1$ and $n > 2$ in e), one does not need the full index formula, but rather only the simpler theorem (see Exercise 2.2 above) that the index is a homomorphism $K(TX) \to \mathbb{Z}$. □

Results. Let X be a closed manifold of dimension n, $E,F \in \mathrm{Vect}_N(X)$, and $P \in \mathrm{Ell}_k(E,F)$. Then we have $\mathrm{index}(P) = 0$ in the following cases:

a) n odd and P a differential operator.

b) $N < n$ and (X a hypersurface in $\mathbb{R}^{n+1}$ or $n \leq 3$)

c) $N = n/2$ and Euler number $e(X) \neq 0$

d) $N = n/2$ and n not divisible by four

e) $N < n/2$.

(b)–e): and E and F trivial.)

Arguments: b) was already gone over in Result B.b). The derivation of c)-e) from the Result B.a) can be found in [Atiyah-Singer 1968b, 602f.]. a) follows very nicely from Result A.b): Let $\alpha: \xi \longmapsto -\xi$ be the antipodal map on the tangent bundle TX. Since $\sigma(P)$ at the point x is written in terms of a matrix of homogeneous polynomials of the k-th degree with coefficients in $\mathbb{C}$ and coordinates in $(T^*X)_x$ as variables, we have the symmetry condition

$$\sigma(P)(\alpha(\xi)) = (-1)^k \sigma(P)(\xi), \quad \xi \in (TX)_x. \tag{1}$$

Here we have identified TX and T^*X by means of a Riemannian metric on X. Via multiplication by $e^{it\pi}$, $t \in [0,1]$, one obtains a homotopy in $\mathrm{Iso}_{SX}(E,F)$ from $\sigma(P)$ to $-\sigma(P)$, hence $[\sigma(P)]$ and $[-\sigma(P)]$ are equal in $K(TX)$. We can then neglect the sign in (1) and obtain

$$\alpha^*[\sigma(P)] = [\sigma(P)], \tag{2}$$

if P is a differential operator.

We now apply Result A.b):

$$\begin{aligned}\mathrm{index}\, P &= (-1)^n\{ch[\sigma(P)] \cup \tau(X)\}[Tx] \\ &= (-1)^n\}\alpha^* ch[\sigma(P)] \cup \tau(X)\}(\alpha_*[TX]) \\ &= (-1)^n\{ch[\sigma(P)] \cup \tau(X)\}(-1)^n[TX] \\ &= -\mathrm{index}\, P, \text{ whence } \mathrm{index}\, P = 0.\end{aligned}$$

Here we have used (2) in the third equality, to obtain $\alpha^* ch[\sigma(P)] = ch[\sigma(P)]$ in $H^*(TX;\mathbb{Q})$. Note also that α inverts only the vertical part of the tangent space TX, leaving the horizontal part unchanged; in local coordinates $(x^1,\ldots,x^n)$, with $\xi \in TX$ represented by $(x^1,\ldots,x^n, \xi_1,\ldots,\xi_n)$ where $\xi = \Sigma\xi_i dx^i$, we have $\alpha(\xi)$ represented by $(x^1,\ldots,x^n, -\xi^1,\ldots,-\xi^n)$. Thus, the orientation of TX is reversed by α, precisely when n is odd.

Incidentally, with somewhat more topology, (see [Atiyah-Singer 1968b, 600]) one can show directly for odd n and P a differential operator that $\sigma(P)(\alpha(\cdot\cdot))$ and $(\sigma(P)(\cdot\cdot))^{-1}$ are stably homotpic, whence $[\sigma(P)] + [\sigma(P)] = 0$ and $[\sigma(P)]$ is of finite order. Then index $P \in \mathbb{Z}$ must also be of finite order and hence zero, since index: $K(TX) \to \mathbb{Z}$ is a homomorphism. In this way, one obtains a) without recourse to the explicit index formula.

One can also directly prove e) for the special case $N = 1$ and $n > 2$ without the full index theorem (see also Exercise II.5.9 and the literature given there, where the same result is derived topologically in a "pedestrian" way). For trivial line bundles the space of elliptic symbols can be expressed very simply: Since $GL(1,\mathbb{C}) = \mathbb{C}^{\times}$ can be contracted to the circle S^1, the index is defined on the set of homotopy classes $[SX,S^1] \cong H^1(SX;\mathbb{Z})$. Since (by Exercise II.5.8) the index of an elliptic operator P is zero when its symbol $\sigma(P)$ depends only on x (and not on $\xi \in (SX)_x$), it follows that the index vanishes on the image of π^* in the following long exact cohomology sequence:

$$\begin{array}{ccccccccc} \cdots \to & H^1(BX) & \to & H^1(SX) & \to & H^2(BX,SX) & \to & H^2(BX) & \to \cdots \\ & & \overset{\pi^*}{\nearrow} & \Big\downarrow{\scriptstyle \text{index}} & & & & & \\ & H^1(X) & & \mathbb{Z} & & & & & \end{array}$$

(We have omitted the coefficient ring $\mathbb{Z}$ from the cohomology groups here.) As we already reported in A(i) above, by a classical result of R. Thom, $H^2(BX,SX;\mathbb{Z}) = 0$ for $n > 2$, whence π^* is surjective. Thus, we have proved that index $P = 0$ for each elliptic operator P defined on the space of complex-valued functions on a manifold of dimension ≥ 3. Compare with [Atiyah 1970a, 103f.], where similar topological arguments are needed in certain cases for $n = 2$ and for nontrivial bundles. □

D. Euler Number and Signature

We have seen already how (e.g., in our proof of the Bott Periodicity Theorem) analytic methods are utilized in topology and, conversely, how the Index Formula expresses the analytic index by topological means. This deep inner relationship between the topology of manifolds and the analysis of linear elliptic operators is further revealed by the fact that certain invariants of manifolds can be realized as indices of "classical" elliptic operators which can be defined quite naturally on these manifolds. A detailed presentation is contained in [Mayer 1965]; for the case of bordered manifolds, which remained obscure for a long time (since the "classical" operators do not all admit elliptic boundary value systems in the sense of II.6 - II.8; see [Booss 1972]), we refer to [Atiyah-Patodi-Singer 1975].

The invariant which we consider here is the Euler number $e(X)$. It is defined by the observation (a matter of solid geometry and probably known long ago to Greek mathematicians) that, for every "triangulization" of a closed oriented surface X, the alternating sum $e(X) := \alpha_0 - \alpha_1 + \alpha_2$ of the number of vertices (α_0), edges (α_1) and faces (α_2) is always the same and depends only on the number of "handles", the "genus" g of the surface: $e(X) = 2 - 2g$. (The picture shows the beginning of a triangulization and the surface has genus 4.) The Euler number can be

interpreted as the alternating sum $e(X) = \beta_0 - \beta_1 + \beta_2$ of the 0-, 1- and 2-dimensional "holes" whereby the interior of X consists of a 0-dimensional and g 1-dimensional holes and the exterior of further g 1-dimensional holes and the 2-dimensional total space, thus $\beta_0 = \beta_2 = 1$ and $\beta_1 = 2g$.

In the language of singular homology β_i is the rank of the ith group of homology $H_i(X; \mathbb{Z})$, the ith "Betti number", and in this form the definition of the Euler number can be extended to a topological manifold X of dimension $n > 2$, and if the α_i are again defined by a triangulization of X then we obtain

$$e(X) = \alpha_0-\alpha_1 + \ldots + (-1)^n\alpha_n = \beta_0-\beta_1 +- \ldots + (-1)^n\beta_n.$$

The Euler number is among the best understood topological invariants. For example one knows (see e.g., [Alexandroff-Hopf 1935, 309 and 358], [Seifert-Threlfall 1934, 246] and [Greenberg 1967, 99-103]):

(i) $e(X \times Y) = e(X) \cdot e(Y)$,

(ii) $\dim X$ odd $\Rightarrow e(X) = 0$,

(iii) $e(S^{2m}) = 2$ and $e(\mathbb{P}\mathbb{C}^m) = m$. □

An invariant which is sharper in several aspects (see Section E) is the signature of an oriented topological manifold X of dimension $4q$ defined as follows. Let $Q : (a,b) \longmapsto (a \cup b)[X]$, $a,b \in H^{2q}(X; \mathbb{R})$, be the real-valued symmetric bilinear form obtained by valuation of the cup product on the fundamental cycle $[X]$ of the orientation of X. Then $\operatorname{sign}(X) := \operatorname{sign}(Q) := p^+ - p^-$ where p^+ (resp. p^-) denotes the maximal dimension of subspaces of $H^{2k}(X; \mathbb{R})$ on which the quadratic form is positive (resp. negative).

Q is non-degenerate, whence $p^+ + p^- = \beta_{2q}$ (clearly $\beta_i = \dim H_i(X; \mathbb{R}) = \dim H^i(X; \mathbb{R})$). Furthermore, the Poincaré duality $H_i(X; \mathbb{R}) \cong H^{4q-i}(X; \mathbb{R})$ implies $e(X) \equiv \beta_{2q} \bmod 2$, and we obtain the formula

(iv) $e(X) \equiv \operatorname{sign}(X) \bmod 2$. □

We now want to describe these two invariants analytically, when the oriented closed n-dimensional manifold X is equipped with a differentiable structure and a Riemannian metric. Referring to Exercise 8 of the Appendix and Section 3.B above, let $\Omega = \sum_{j=1}^{n} \Omega^j$ be the space of (complexified) exterior differential forms with exterior derivative $d: \Omega \to \Omega$ and its adjoint $d^*: \Omega \to \Omega$.

<u>Results</u>. a) The operator $d+d^*: \Omega \to \Omega$ is an elliptic self-adjoint differential operator of order 1, whence $\operatorname{index}(d+d^*) = 0$.

b) The Laplace operator $\Delta := (d+d^*)^2 = dd^* + d^*d: \Omega \to \Omega$ is an elliptic self-adjoint operator of order 2, which is "homogeneous of degree 0" (i.e., it sends Ω^j to Ω^j, $0 \leq j \leq n$). For $\Delta_j := \Delta|\Omega_j$, we have the "main theorem of Hodge theory" $\operatorname{Ker} \Delta_j \cong H^j(X; \mathbb{C})$.

c) By restricting $d+d^*$ to the even forms, an elliptic differential operator $D: \Omega^{\text{even}} \to \Omega^{\text{odd}}$ of order 1 with $\operatorname{index} D = e(X)$ is obtained.

d) If $n \equiv 0 \bmod 4$ and $\tau: \Omega \to \Omega$ is the involution (defined in Section 3.B) with the ± 1-eigenspaces $\Omega^{\pm}$, then $d+d^*(\Omega^{\pm}) \subset \Omega^{\mp}$, and an elliptic operator $D^+: \Omega^+ \to \Omega^-$ of order 1 with $\operatorname{index} D^+ = \operatorname{sign}(X)$ results.

e) There are formulas $e(X) = \chi(TX)[X] = \frac{1}{2\pi} \int_X K$ (C. F. Gauss/O. Bonnet), and (if $n = 4q$) $\operatorname{sign}(X) = L_q(p_1, \ldots, p_q)[X]$ (F. Hirzebruch). Here $\chi(TX) = \phi^{-1}(U \cup U) \in H^n(X; \mathbb{Z})$ is a certain "characteristic class" defined as in Section A which one can apply to the fundamental cycle, and

K is a form expressible in terms of the curvature tensor of X. $L_q(p_1,\dots,p_q)$ is a polynomial in the "Pontryagin classes" $p_j := (-1)^j c_{2j}(TX \otimes \mathbb{C})$, which is given by

$$L_q := \frac{x_1}{\tanh x_1} \cdot \dots \cdot \frac{x_q}{\tanh x_q},$$

if we have the formal factorization $\sum_{j=1}^{q} p_j = (1+x_1^2)\dots(1+x_q^2)$. In particular, we have $L_1 = \frac{1}{3} p_1$, $L_2 = \frac{1}{45}(7p_2-p_1^2)$, $L_3 = \dots$, whence for $\dim X = 4$

$$\operatorname{sign}(X) = \frac{1}{3}\int_X \tilde{p}_1,$$

where $\tilde{p}_1 \in \Omega^4(X)$ is a differential form representing $p_1 \in H^4(X;\mathbb{C})$.

Arguments (from [Atiyah-Singer 1968b, 573-577]): For a) and b), one checks that $\sigma(d)(x,\xi)v = i\xi \wedge v$ and $\sigma(d^*)(x,\xi)v = -i{*}(\xi \wedge {*v})$, whence $\sigma(\Delta)(x,\xi)v = ||\xi||^2 v$ for $x \in X$, $\xi \in (T^*X)_x$ and $v \in \Lambda^*(T^*X)_x$; thus, the ellipticity of $d+d^*$ and Δ follows.

Since we have the splitting $\Omega^j = \operatorname{Ker}\Delta_j \oplus \operatorname{Im}\Delta_j = \operatorname{Ker}\Delta_j \oplus \operatorname{Im} d_{j-1} \oplus \operatorname{Im}(d_j)^*$ from Chapter II.5 and trivially $\Omega^j = \operatorname{Ker} d_j \oplus \operatorname{Im}(d_j)^*$, we obtain $H^j(X,\mathbb{C}) \cong \operatorname{Ker} d_j/\operatorname{Im} d_{j-1} \cong \operatorname{Ker}\Delta_j$ (the "space of harmonic j-forms"), where the first isomorphism is a classical result of Georges de Rham. Note that the right vector space is defined via the Riemannian structure, the middle vector space via the differentiable structure, and the left purely via the topology of X.

Since $\Delta = (d+d^*)^2 = (d+d^*)^*(d+d^*)$, it follows that $\operatorname{Ker}\Delta = \operatorname{Ker}(d+d^*)$, whence $\operatorname{index} D = e(X)$. Incidentally, the above facts (i) and (ii) then follow immediately.

For the proof of d), note that $(D^+)^* = D^-: \Omega^- \to \Omega^+$, and so $\operatorname{index} D^+ = \dim\operatorname{Ker} D^+ - \dim\operatorname{Ker} D^-$, where $\operatorname{Ker} D^\pm = \{w \in \operatorname{Ker}\Delta : \tau w = \pm w\}$. Now, a pair (a,b) (with $a \in \operatorname{Ker}\Delta_j$, $j \neq 2q$, and $b \in \operatorname{Ker}\Delta_{4q-j}$) belongs to $\operatorname{Ker} D^+$ (i.e., $\tau a = b$) exactly when $(a,-b)$ lies in $\operatorname{Ker} D^-$. Thus, we have $\operatorname{index} D^+ = \dim H^+ - \dim H^-$, where $H^\pm = \{w \in \operatorname{Ker}\Delta_{2q} : {*w} = \pm w\}$ and $\dim_{\mathbb{C}} H^\pm = \dim_{\mathbb{R}}\{w \in \operatorname{Ker}\Delta_{2q} : w \text{ real and } {*w} = \pm w\}$. The definition of $*$ implies that $\int {*w} \wedge w = \langle w,w\rangle$, where $\langle\cdot,\cdot\rangle$ is the scalar product in Ω^{2q}. Thus, H^+ is a maximal subspace on which the quadratic form $w \longmapsto \int w \wedge w$ is positive, and H^- is a maximal subspace on which it is negative. From Hodge-de Rham theory (see b)), it then follows that $\dim H^\pm = p^\pm$.

The explicit formulas in e) follow from the formulas given above in A, which one must somewhat generalize for the derivation of the signature formula (via operators with transformation groups; see also under Section L).

E. Vector Fields on Manifolds

Among the basic concepts of the analysis of a dynamical system, as well as of the geometry of a differential manifold X, is the notion of a "vector field". It is a C^∞ section v of the tangent bundle (in classical terminology a "variety of line elements"), thus $v \in C^\infty(TX)$.

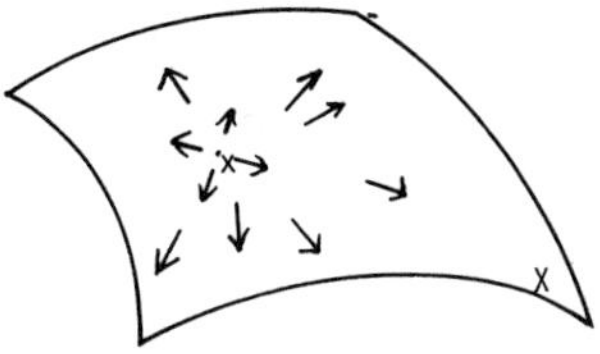

Each v defines an ordinary differential equation $\dot{c}(t) = v(c(t))$ for differentiable paths ("trajectories") $c: \mathbb{R} \to X$. According to the classical existence and uniqueness theorems, the equation has a unique solution for any given "initial value" $c(0) = x_0 \in X$; for details see [Bröcker-Jänich 1982, 74-87]. In every theory of flows the "singularities" $\Sigma := \{x \in X$ and $v(x) = 0\}$ of v are of special interest, as they describe the "stagnation points" of dynamics, the equilibrium points, or the so-called "stationary solutions" (the constant paths $c(t) = x_0$). For applications, one may think intuitively of examples from oceanography, of a magnetic field, or of dynamic laws of economics. We saw above in Section 1.A that every singularity x of a vector field v defines locally a map $S^{n-1} \to S^{n-1}$, $n = \dim X$, whose mapping degree we denote by $I_v(x)$.

The geometric interest in vector fields usually stems from the question of "parallelizability" of the manifold: Is it possible to assign to a tangent vector $v \in (TX)_x$ at a point x a tangent vector $v' \in (TX)_{x'}$ at x' parallel to v in a way which is independent of the path from x to x' used in the process? This question is important for the physical notion of space and finally for the analysis of motion (since the concept of acceleration depends on parallel displacement). It is equivalent to the question of whether an n-dimensional manifold X possesses n vector fields that are linearly independent at each point $x \in X$; i.e., whether the tangent space TX is trivial. (For example, by a famous theorem of J. F. Adams S^{n-1} is paralellizable if and only if $\mathbb{R}^n$ is a "division algebra", i.e., for $n = 2m$ and $m = 1,2,4$). □

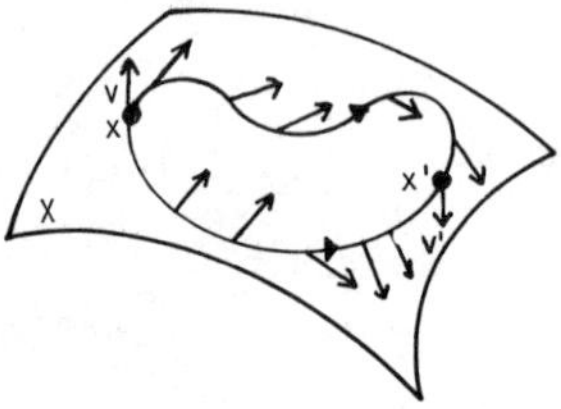

The geometrical idea and also the historical origin (with Eduard Stiefel, 1935, and almost simultaneously and in a similar connection with Hassler Whitney) of the topological invariants of Riemannian or complex manifolds (nowadays called "characteristic classes" - see Section A above) lies in the general investigation of r vector fields $v_1,\ldots,v_r$ on a manifold and their singular set Σ consisting of those points x where $v_1(x),\ldots,v_r(x)$ are linearly dependent. One knows, for example, that the set Σ generically has dimension $r-1$, and that the "cycle" Σ defines a homology class (with suitable coefficient group) which is a "characteristic class" of X and is independent of $v_1,\ldots,v_r$; see also the large survey [Thomas 1969].

The index formula (see Section A above) says that the symbol of an elliptic operator is a kind of characteristic class, and its index is a "characteristic" integer. From this arose the program to illuminate the connection between elliptic operators and vector fields on manifolds. Specifically, the existence of a certain number of vector fields implies certain symmetry properties for "classical operators" and corresponding results for their indices such as Euler number, signature and "characteristic numbers". □

Results. Let X be a closed, oriented, Riemannian manifold.

a) Each vector field v "generically" has only finitely many zeros (i.e., by an arbitrarily small perturbation a vector field can be put in this form), and we have $e(X) = \sum_{x\in X} I_v(x)$, where $I_v(x)$ is the local index of v at x defined above; thus, the right side of the formula consists of the finite weighted sum of zeros of v.

b) If $\dim X = 4q$ and X has a tangent field of 2-dimensional planes (i.e., an oriented 2-dimensional subbundle of the tangent bundle), then $e(X) \equiv 0 \bmod 2$ and $\operatorname{sign}(X) \equiv e(X) \bmod 4$.

c) If the 4q-dimensional manifold X has at least r vector fields that are independent everywhere, then $\operatorname{sign}(X) \equiv 0 \bmod b_r$, where the values of b_r are given in the following table ($b_{r+8} = 16\ b_r$):

r	1	2	3	4	5	6	7	8
b_r	2	4	8	16	16	16	16	32

Arguments: For $\dim X = 2$, a) is a classical result of H. Poincaré, for which one can find a very clear sketch of the proof in [Brieskorn 1976, 166-171]. The generalization for $\dim X > 2$ originated from H. Hopf, who also showed that $e(X) = 0$, if and only if, there is a nowhere zero vector field on X.

For b) and c), we refer to [Mayer 1965], [Atiyah 1970c] and [Atiyah-DuPont 1972], where a series of related results are proved by combining results of analysis, topology, and algebra, and in some cases using the theory of real elliptic operators (see Section J below). To illustrate the methods, we will only treat the theorem

$$\exists v \in C^\infty(TX) \quad \text{with} \quad v(x) \neq 0 \quad \forall x \in X \Rightarrow e(X) = 0,$$

which one can obtain from the explicit Atiyah-Singer Index Formula or the Gauss-Bonnet Formula which says that $e(X) = \chi(TX)[X]$ - see D.e. By definition, $\chi(TX) = \phi^{-1}(U \cup U) = \tilde{v}^* i^* U$, where $U \in H^n(BX,SX;\mathbb{Z})$ is the orientation class of TX, $\phi\colon \Sigma H^j(X;\mathbb{Z}) \to \Sigma H^{j+n}(BX,SX;\mathbb{Z})$ is the Thom isomorphism, $i\colon (BX,\emptyset) \to (BX,SX)$ is the trivial embedding, and $\tilde{v}\colon X \to BX$ is the normalized vector field $\tilde{v} = v/|v|$, where $|v| \neq 0$ by assumption. Since $i\tilde{v}(X) \subset SX$, we have $\tilde{v}^* i^* = 0$.

To get the same result, without recourse to the explicit index formula, one can apply the general theory of elliptic operators (see Chapter II.5 above) and the results D.c) and D.d) of Hodge theory. The key issue for this is the fact that $\sigma(d+d^*)(x,\xi)w = i\xi \times w$ for $x \in X$, $\xi \in (T^*X)_x$ and $w \in \Lambda^*(T^*X)_x$. Here $\times\colon \Lambda^*V \times \Lambda^*V \to \Lambda^*V$, $V := (T^*X)_x$, is the "interior" ("Clifford") product, in contrast to the "exterior" ("Cartan") multiplication. For each real Euclidean vector space V of dimension n, the exterior algebra Λ^*V, as a vector space of dimension 2^n, is canonically isomorphic to the Clifford algebra $\mathrm{Cliff}(V)$ which is generated by $N \times N$ matrices $A_1,\ldots,A_n$ ($N = 2^{n/2}$) with $A_i^2 = -\mathrm{Id}$ and $A_iA_j = -A_jA_i$ for $i \neq j$. The "interior" product $\times$ in $\Lambda^*(V)$ is then defined via the matrix multiplication in $\mathrm{Cliff}(V)$.

By means of the Riemannian metric on X, a vector field v can be regarded as a 1-form which yields a 0-th order differential operator via $u \longmapsto u \times v$, $u \in \Omega(X)$ (i.e., via pointwise Clifford multiplication by v).

Since $u \times v \times v = -|v|^2 u$, in the case where v is nowhere zero, we have that this differential operator is an automorphism on $\Omega(X)$, which maps Ω^{even} and Ω^{odd} onto Ω^{odd} and Ω^{even} respectively. Since $(i\xi \times w) \times v(x) = i\xi \times (w \times v(x))$, we know that $\sigma(d+d^*)$ commutes with the differential operator. We denote by T the restriction of the operator to Ω^{odd}. Then $T^{-1}DT - D^* \in OP_0$ by Chapter II.5, and hence $\text{index } D = \text{index } T^{-1}DT = \text{index } D^* = -\text{index } D$, whence $e(X) = \text{index } D = 0$. □

F. Abelian Integrals and Riemann Surfaces

Perhaps one of the first (in our topological sense) "quantitative" results of analysis is contained in the major work of the Norwegian mathematician Niels Henrik Abel entitled "Mémoire sur une propriété générale d'une classe très étendue de fonctions transcendantes". It was written in 1826 but published only posthumously in 1841. In it Abel takes up a dispute of the numerical analysis of the 18th century about "rectifiability", the possibility of solving integrals by means of elementary functions (algebraic functions, circular functions, logarithm and exponential functions). Since Jakob Bernoulli and Gottfried Wilhelm Leibniz* there was an interest particularly in the integration of irrational functions which turn up in many problems of science and technology. The efforts (already of the 17th century) to rectify the ellipse, whose arc length is important for astronomy, leads to the computation of the integral

$$I(x) = a \int_0^x \frac{1 - k^2 t^2}{\sqrt{(1-t^2)(1-k^2t^2)}}\, dt, \quad k = (a^2-b^2)/a^2.$$

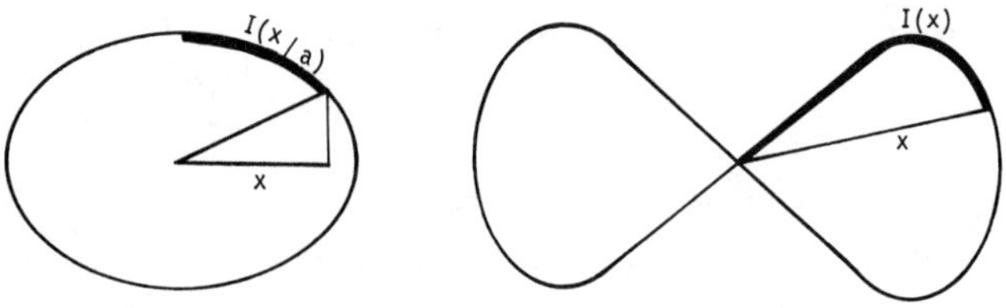

*Who drew the attention of the mathematicians to constructions and curves "quas natura ipsa simplici et expedito motu producere potest" (which nature itself can produce by simple and complete motions), quoted after [Brill-Noether 1894, 124]. See also [Kline 1972, 411 f.].

While investigating the deformation of an elastic rod under the influence of forces acting at its extremities, Jakob Bernoulli (1694) came across further irrational integrands. In this connection he also introduced the "lemniscate" $\{(\pm\sqrt{(x^2+x^4)/2}, \pm\sqrt{(x^2-x^4)/2}): 0 \le x \le 1$ whose arclength is given by

$$I(x) = \int_0^x \frac{dt}{\sqrt{1-t^4}}.$$

(For this and the following, see [Siegel I 1965/1969, 1.1-1.2].) For the "lemniscatic integral" Leonhard Euler proved (1753) the addition theorem

$$I(u) + I(v) = I(w) \quad \text{where} \quad w = \frac{u\sqrt{1-v^4} + v\sqrt{1-u^4}}{1 + u^2v^2},$$

It generalized a result of the Italian mathematician Giulio Carlo de'Toschi di Fagnano (1714) on the doubling of the circumference of the lemniscate with compass and ruler alone:

$$2I(u) = I(w) \quad \text{for} \quad w^2 = \frac{4u^2(1-u^4)}{(1+u^4)^2}.$$

Prior to this, Johann Bernoulli (1698) "accidentally" discovered that the difference of two arcs of the cubic parabola ($y = x^3$) is integrable by elementary functions and added to the problem of rectifying curves* the new problem of finding arcs of parabolas, ellipses, hyperbolas etc. whose sum or difference is an elementary quantity - "just as arcs of a circle could be compared with one another through the expressions for $\sin(\alpha+\beta)$, $\sin 2\alpha$, etc." [Brill-Noether 1894, 206].

A little later L. Euler succeeded in extending his addition theorem for the leminiscatic integral to "elliptic integrals of the first kind" i.e., in proving that

$$I(u) + I(v) = I(w) \quad \text{with} \quad w = \frac{u\sqrt{P(v)} + v\sqrt{P(u)}}{1 + u^2v^2} \tag{E}$$

where

---

*It was suspected already that this was impossible. But only in 1833, did J. Liouville prove rigorously that the "elliptic integrals" could not be solved "elementarily", i.e., expressed by a finite combination of algebraic, circular, logarithmic and exponential functions. (As an aside, it is now known that the problem of the elementary integration of arbitrary functions is recursively undecidable.)

$$I(u) := \int_0^u \frac{dt}{\sqrt{P(t)}} \quad \text{and} \quad P(t) = 1 + at^2 - t^4, \quad a \in \mathbb{R}.$$

Based on a comparison of elliptic arcs also due to Fagnano, Euler finally found another generalization to "elliptic integrals of higher kind". These are integrals of the form

$$I(u) := \int_0^u \frac{r(t)}{\sqrt{P(t)}}\, dt,$$

where r is a rational function in one variable and P is a polynomial of third or fourth degree with simple zeros. Here the addition theorem takes the form

$$I(u) + I(v) = I(w) + W(u,v) \tag{E'}$$

where w is, as above, an algebraic function of the arbitrary upper integration limits u and v and $W(u,v) = S_1(u,v) + \log S_2(u,v)$ with two rational functions S_1, S_2.

Euler already noticed that his methods cannot be used, e.g., for the treatment of "hyperelliptic integrals" (with a polynomial P of fifth or higher degree). But only N. H. Abel found the explanation for "the difficulties which Euler's formulation necessarily encountered when dealing with hyperelliptic integrals: The constant of integration ($I(w)$ in (E) or (E')), appearing in the transcendental equation, could not be replaced, as in the elliptic case, by a single integral but only by two or more hyperelliptic integrals - a remarkable circumstance which was in no way predictable... The question about the minimal number of integrals to which a given sum of integrals could be reduced, remained as the cardinal question; it caused Abel to produce the elaborate and laborious counts which constitute the main results of his great Paris paper and which brought him into the possession of the notion of "genus" of an algebraic structure long before Riemann" [Brill-Noether 1894, 211 f.]. □

Abelian Addition Theorem. Let R be a rational function and F a polynomial in two variables. For $a, x \in \mathbb{R}$ (a fixed), consider the "Abelian integral" $I(x) := \int_a^x R(t,y)dt$, where y satisfies the equation $F(t,y) = 0$ (For $F(t,y) = y^2 - P(t)$ with P as above and $R(t,y) = 1/y$, we obtain an elliptic integral of the first kind; and for $R(t,y) = r(t)/y$ with r a rational function, we get an elliptic integral of a higher kind.) For a given value of t, there may be several corresponding solutions ("roots") of the equation $F(t,y) = 0$. Thus, one must specify which root will be substituted for y in $R(t,y)$. Hence, one selects an integration path $\gamma: I \to \mathbb{C}$ in the plane (with $\mathrm{Re}\,\gamma(0) = a$, $\mathrm{Re}\,\gamma(1) = x$, and $F(\mathrm{Re}\,\gamma, \mathrm{Im}\,\gamma) = 0$), and regards $I(x)$ as a line integral. Then the sum of m (m sufficiently large) arbitrary Abelian integrals $I(x_i)$ with respective fixed integration paths ($x_i \in \mathbb{R}$, $i = 1,\ldots,m$) can be written as the sum of only g Abelian integrals $I(\hat{x}_j)$ ($j = 1,\ldots,g$) and a remainder term $W(x_1,\ldots,x_m)$ which is the sum of a rational function S_1 and the log of another rational function S_2 of the limits of integration $x_1,\ldots,x_m$:

$$\sum_{i=1}^{m} I(x_i) = \sum_{j=1}^{g} I(\hat{x}_j) + W(x_1,\ldots,x_m),$$

where the $\hat{x}_j = \hat{x}_j(x_1,\ldots,x_m)$ are algebraic functions and the number g only depends on the specific form of the polynomial F. □

According to C. G. J. Jacobi, Abel attempted to solve two problems with his addition theorem, "the representability of an integral by closed expressions", and "the investigation of general properties of integrals of algebraic functions" [Brill-Noether 1894, 205]. In fact, for one thing, the addition theorem says something about the rectification question: If $g = 0$ then $I(x) = W(x)$ for $m = 1$ where W is constructed from rational functions and the logarithm. Thus, in this case, the integral $I(x)$ can be solved by elementary functions. If $g > 0$ then in general at least g additional higher transcendental functions are needed, namely the $I(\hat{x}_i)$. Most of all, the theorem is an addition theorem just like the sum formulas for trigonometric functions, and can be used in the back-up files of computers as interpolation formulas for realizing standard functions of science and technology.*

We quoted Abel's theorem, in order to point out one of the earliest occurrences of the fundamental invariant g which, as the index, possess the dual character of being both analytic and algebraic. Without entering a discussion of the deep function theoretic aspects and geometric interpretations of Abel's addition theorem, we will note some results exhibiting the significance of the quantity g. These essentially are due to Bernhard Riemann (except c)).

*See Hart, J. F. & E. W. Cheney: Computer approximations. SIAM Series in Applied Mathematics, Wiley, New York 1968, §§1.5, 4.2.

Results. a) Each polynomial $F(t,y)$ defines a compact *Riemann surface*, an oriented surface with a complex-analytic structure (i.e., a topological manifold with a distinguished atlas whose coordinate changes are holomorphic functions). Conversely, one can define a complex-analytic structure on each compact, oriented topological surface X such that X can be regarded as the Riemann surface of a polynomial $F(t,y)$.

b) Topologically, a compact Riemann surface X is characterized by its "*genus*" g, the number of handles that must be fastened to the sphere in order to obtain X. Twice g is the number of closed curves needed to generate the first homology of X; i.e., there are closed curves $\gamma_1,\ldots,\gamma_{2g}$ (namely, along the lengths and girths of the handles) such that every closed curve γ in X is "homologous" to a unique integral linear combination $\Sigma n_i\gamma_i$.

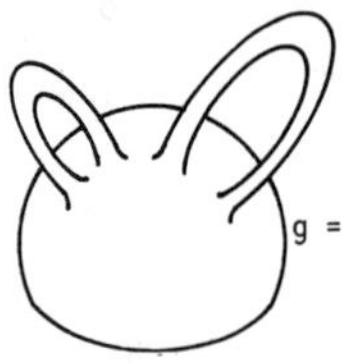

Analytically (see the Abelian Addition Theorem) X has another invariant, the maximal number g_1 of linearly independent holomorphic differential forms $\alpha_1,\ldots,\alpha_{g_1}$ of degree 1 on X. Actually, $g = g_1$; i.e., the "numerical complexity"* of the Abelian integral $\int R(t,y)dt$ with $F(t,y) = 0$ is equal to the genus of the Riemann surface F. In particular, for the elliptic integrals, one obtains an "*elliptic curve*", namely the torus of genus 1 with two generating cycles γ_1 and γ_2. (For the connection with the classical notion of "double periodicity" in elliptic integrals, see [Mumford 1976, 149-155], for example.) More generally, one can define a "periodicity matrix" $\omega_{ij} := \int_{\gamma_i} \alpha_j$; $i = 1,\ldots,2g$; $j = 1,\ldots,g$, which determines the complex-analytic structure of X by a theorem of R. Torelli.

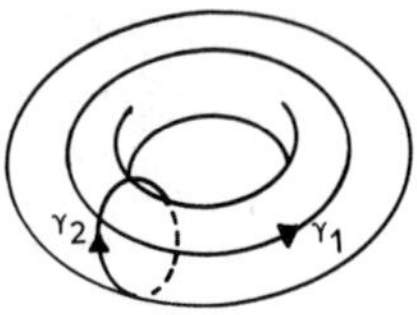

c) On each Riemann surface X of genus g there is a "*Cauchy-Riemann operator*" $\bar\partial: f \longmapsto \frac{\partial f}{\partial \bar z} d\bar z$ which assigns to each complex-valued C^∞ function f on X a complex differential form of degree 1; $\bar\partial$ is an elliptic operator and index $\bar\partial = 1-g$.

*Riemann (1857) calls this quanity the "Klassenzahl" (class number). The term "genus" originated with Alfred Clebsch (1864).

Arguments: a) and b) follow from the classical theory of Riemann surfaces. For c), note that $\dim \operatorname{Ker} \bar{\partial} = 1$, since $\operatorname{Ker} \bar{\partial}$ consists of the global holomorphic functions and these must be constant by the maximum principle. $\operatorname{Coker} \bar{\partial} = \operatorname{Ker} \bar{\partial}^*$ consists of the "(anti-) holomorphic differential forms" which constitute a vector space isomorphic to the space of holomorphic 1-forms (see b)).

One can also directly obtain c) from the Atiyah-Singer Index Formula, since the Euler class $\chi(TX)$ is the only Chern class of a Riemann surface, whence $\operatorname{index} \bar{\partial} = C\chi(TX)[X] = Ce(X) = C(1-2g+1) = C(2-2g)$. It is then not difficult to derive that $C = 1/2$; see also the following Section G, Results a)/c)/d), which include c) as a special case; and [Mumford 1976, 132-141] where c) is proven, with these generalizations in mind, in the form "arithmetic genus = geometrical genus" by elementary geometrical and algebraic tools of classical projective geometry. □

G. The Theorem of Riemann-Roch-Hirzebruch

We deal next with a class of theorems for which the Atiyah-Singer Index Formula yields new proofs or generalizations (see d)). According to [Brill-Noether 1894, 280 f.], who introduced the term "Riemann-Roch Theorem", it deals with the "counting of the constants of an algebraic function", and more generally, with establishing relations between the "constants" (number of singularities, degree, order, genus etc.) of an algebraic curve, algebraic surface or complex manifold. Thus history and motivation of "Riemann-Roch" - detailed in [Brill-Noether 1894]* - are closely tied to the unsolved problem of "completely" classifying algebraic varieties, see [Hirzebruch 1973].

We cannot convey the abundance of results in this subject. They are still too scattered and the diversity of approaches too uncertain**. We therefore restrict ourselves to a few stages which are essential for us, and where, out of the complexity of the problems, developed little by little, some unifying, very rich, and consequential aspects:

---

*For a function theoretic interpretation of Riemann-Roch, see also [Siegel II 1965/1971, 4.6-4.7], where Riemann-Roch is used to prove Abel's theorem and is expressly called "algebraic" in contrast to the "transcendental nature" of Abel's Theorem.

**Alfred Clebsch, "who surely did not lack knowledge and versatility" wrote very openly in a letter of August 1864 to Gustav Roch that he himself "understood very little of Riemann's treatise even after greatest efforts, and that Roch's dissertation remained for the most part incomprehensible to him." (Quoted after [Brill-Noether 1894, 320].) The historical process of "correctly understanding" the theorem apparently has not reached its conclusion at least as far as algebraic functions are concerned.

- Bernhard Riemann's "transcendental" idea of the analysis on Riemannian manifolds, i.e., his attempt to consider the totality of integrals of a fixed algebraic function field (just the Abelian integrals).

- The function theoretic treatment of the problems by Karl Weierstrass.

- The interpretation from the point of view of differential geometry in the language of Hodge theory due to Kunihiko Kodaira which was the basis for Friedrich Hirzebruch's generalization of Riemann-Roch to higher dimensional algebraic varieties.

Results: a) On a compact Riemann surface X of genus g, we consider meromorphic functions $w: X \to \mathbb{C}$ which have poles at the points $x_i \in X$, $i = 1,\ldots,r$ of order at most $m_i \in \mathbb{N}$ and zeros at the points $x_j \in X$, $j = r+1,\ldots,s$, of order at least $-m_j \in \mathbb{N}$. These form a complex vector space $L(\mathcal{D})$ whose dimension $\ell(\mathcal{D})$ is given by the following formula:

$$\ell(\mathcal{D}) - \ell'(\mathcal{D}) = \deg(\mathcal{D}) - g + 1,$$

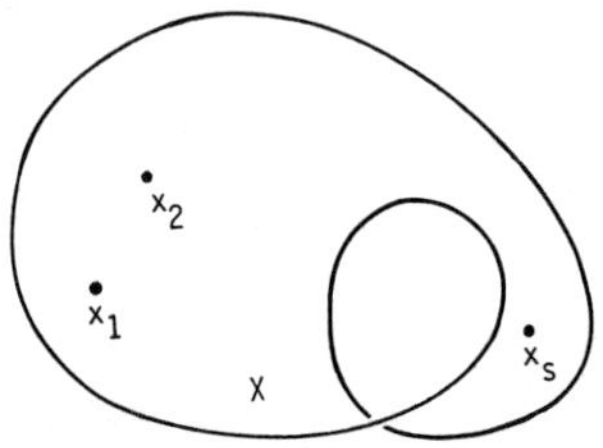

where $\mathcal{D} := \Sigma m_j x_j$, $\deg(\mathcal{D}) := \Sigma m_j$ and $\ell'(\mathcal{D})$ is the dimension of the vector space $L'(\mathcal{D})$ of meromorphic differential forms on X with the corresponding behavior on the zeros and poles.

b) For $\deg(\mathcal{D}) \geq 2g-1$, $\ell'(D)$ vanishes, whence one obtains a proper formula for $\ell(\mathcal{D})$ in this case.

c) For each formal integral linear combination $\mathcal{D}$ of points of X (called a *"divisor"*), there is a holomorphic vector bundle $\{\mathcal{D}\}$ of complex fiber dimension 1 such that $L(\mathcal{D})$ is isomorphic to the vector space $\operatorname{Ker} \bar{\partial}_{\{\mathcal{D}\}}$ of holomorphic sections of $\{\mathcal{D}\}$, and $L'(\mathcal{D})$ is isomorphic to the vector space $\operatorname{Coker} \bar{\partial}_{\{\mathcal{D}\}} \cong \operatorname{Ker} \bar{\partial}^*_{\{\mathcal{D}\}}$ of (anti-)holomorphic 1-forms "with coefficients in $\{\mathcal{D}\}$". Here, $\bar{\partial}_{\{\mathcal{D}\}} : \Omega^0(\{\mathcal{D}\}) \to \Omega^{0,1}(\{\mathcal{D}\})$ is the elliptic differential operator obtained from $\bar{\partial}$ by "tensoring" (see d) below). With this construction, we can write result a) in the form $\operatorname{index} \bar{\partial}_{\{\mathcal{D}\}} = \deg(\mathcal{D}) - g+1$.

d) More generally, one can consider the Dolbeaut complex $0 \to \Omega^0 \xrightarrow{\bar\partial} \Omega^{0,1} \xrightarrow{\bar\partial} \cdots \xrightarrow{\bar\partial} \Omega^{0,n} \to 0$ for a *Kähler manifold* X of dimension n (i.e., a complex manifold with Hermitian metric $\Sigma g_{ik} dz_i d\bar{z}_k$ whose associated 2-form $\alpha = \Sigma \bar{g}_{ik} dz_i \wedge d\bar{z}_k$ is closed, i.e., $d\alpha = 0$). Here $\Omega^{0,p}$ denotes the space of complex exterior differential forms of degree p, which can be written in the form $a_{i_1 \cdots i_p} d\bar{z}_{i_1} \wedge \cdots \wedge d\bar{z}_{i_p}$ relative to local coordinates $z_1, \ldots, z_n$. As in Exercise 8b of the Appendix, $\bar\partial$ is the exterior derivative. If V is a holomorphic vector bundle over X, then via tensoring (as above with the signature operator) one can construct the generalized Dolbeaut complex

$$0 \to \Omega^0(V) \xrightarrow{\bar\partial_V} \Omega^{0,1}(V) \xrightarrow{\bar\partial_V} \cdots \xrightarrow{\bar\partial_V} \Omega^{0,n}(V) \to 0,$$

where $\Omega^0(V) = C^\infty(V)$. In analogy with the above derivation of the Euler number, one defines the *Euler characteristic*

$$\chi(X,V) := \operatorname{index} \bar{D}_V = \sum_{p=0}^{n} (-1)^p \dim H^p(X;\mathcal{O}(V^*)),$$

where $\bar{D}_V := (\bar\partial_V + \bar\partial_V^*) | \Sigma\Omega^{0,\mathrm{even}}(V)$ is an elliptic operator of order 1, and the following equations hold: $H^{n-p}(X;\mathcal{O}(V^*))$:= (n-p)-th cohomology of X "with coefficients in the sheaf of germs of holomorphic sections in V^*" $\cong \operatorname{Ker} \bar\partial_V^{0,p} / \operatorname{Im} \bar\partial_V^{0,p-1} \cong \operatorname{Ker}(\Box_V : \Omega^{0,p}(V) \to \Omega^{0,p}(V))$ (where $\Box_V := \bar\partial_V \bar\partial_V^* + \bar\partial_V^* \bar\partial_V$) $\cong$ space of global holomorphic differential forms of degree p on X with coefficients in V.

We then have the "Riemann-Roch-Hirzebruch Theorem"

$$\chi(X,V) = (\operatorname{ch}(V) \cup \tau(TX))[X].$$

Here $\operatorname{ch}(V)$ is the Chern character of V and $\tau(TX)$ is the Todd class of X; see Section A above.

If V is the trivial line bundle $\mathbb{C}_X$, then $\chi(X) := \chi(X,V)$ is called the *arithmetic genus* of X and we have

$$\chi(X) = \begin{cases} \frac{1}{2} c_2(X)[X] \\ \frac{1}{12}(c_1^2(X) + c_2(X))[X] \\ \frac{1}{24}(c_1(X) \cup c_2(X))[X] \\ \text{etc.} \end{cases} \quad \text{for } \dim_{\mathbb{C}} X = \begin{cases} 1 \\ 2 \\ 3 \\ \text{etc.,} \end{cases}$$

where $c_i(X) \in H^{2i}(X)$ is the i-th Chern class of TX.

Arguments: a) follows using c) and d) from the Atiyah-Singer Index Formula; a direct proof can be found in [Weyl 1913/1955, 117-119]. If $\mathcal{D} = 0$ (in particular, $\deg(\mathcal{D}) = 0$), then one recovers results F b)/c). Notice that $\ell'(\mathcal{D}) = \ell(K_X - \mathcal{D})$, where K_X like $\mathcal{D}$ is a divisor which is canonically (up to a certain "linear equivalence") defined by the zeros and poles of an arbitrary nontrivial form on X. One has $\deg(K_X) = 2g-2$. For details of the definition of "canonical divisors", and for an elementary proof of $\ell(\mathcal{D}) - \ell(K_X-\mathcal{D}) = \deg(\mathcal{D}) - g+1$, see [Mumford 1976, 104-107 and 145-147]; the duality $\ell'(\mathcal{D}) = \ell(K_X-\mathcal{D})$ originated with Richard Dedekind and Heinrich Weber, and is a special case of a very general duality theorem of Jean-Pierre Serre (cf. [Hirzebruch 1956/1966, §15.4]).

For b): A non-trivial mermorphic function on a compact Riemann surface has as many poles as zeros (counting orders). Thus, $\ell(\mathcal{D}) = 0$, if $\deg(\mathcal{D}) < 0$. Since $\deg(K_X) = 2g-2$, it follows that $\deg(K_X-\mathcal{D}) < 0$, for $\deg(\mathcal{D}) \geq 2g-1$.

For c): For the construction of the line bundle $\{\mathcal{D}\}$, we cover X with a finite collection of open sets $\{U_j\}_{j\in J}$ such that on each U_j there is defined a meromorphic function f_j which has the zero and pole behavior that is the opposite of the portion of $\mathcal{D}$ involving points of U_j; so that f_i/f_j is a nowhere-zero holomorphic function on $U_i \cap U_j$. For example, one can choose U_j so small that at most one point x_k (of the finitely many points $x_1,\ldots,x_s$ found in $\mathcal{D}$) lies in U_j, and then take f_j to be a locally defined rational monomial which has a zero of order m_k if $m_k > 0$, and a pole of order $-m_k$ if $m_k < 0$. The clutching construction (Exercise 4 of Appendix) then yields the bundle $\{\mathcal{D}\}$ for which $\{f_i\}$ defines a global meromorphic section f with zero and pole behavior opposite $\mathcal{D}$. Via $g \longmapsto g/f$, an isomorphism is defined from the global holomorphic sections of $\{\mathcal{D}\}$ to the vector space of meromorphic functions on X with the zero and pole behavior prescribed by the divisor $\mathcal{D}$. Details of this construction, which is rather typical for the topological approach to analytic problems of function theory and algebraic geometry, can be found in [Hirzebruch 1956/1966, §15.2].

For d): The derivation from the Atiyah-Singer Index Formula can be found in [Palais 1965, 324 ff.] and [Schwarzenberger 1966], and in a somewhat more general context (see L below) also in [Atiyah-Singer 1968b, 563-565]. This transition from the classical Riemann-Roch Theorem to complex manifolds of arbitrary dimension was initiated by Friedrich Hirzebruch (1953). His proof depended essentially on the additional condition that X can be holomorphically embedded in a complex projective space of a suitable dimension. By going back to the index formula for elliptic operators, one can drop this restriction. Moreover, recently a rather differential geometric proof using the heat equation has been given for d); for the rather easily calculable case of an elliptic curve ($n = 1$, $g = 1$) refer to [Atiyah-Bott-Patodi 1973, 311 f.], and for a sketch of the general case see [loc. cit., 317 f.]. □

The Riemann-Roch-Hirzebruch Theorem appears here as a special application of the Atiyah-Singer Index Formula for elliptic operators of which one, namely $\overline{D}_V$, was constructed taking advantage of the complex structure. Even so, d) and already a) are the quintessential models for the structure of the Index Formula: On the left side we have the difference of globally defined quantities each of which can change even under small variations of the initial data (as the dimension $\ell(\mathcal{D})$ in a)), while on the right we have an expression in terms of topological invariants of the problem (in a) it is the genus g of the Riemannian surface X and the degree of the divisor $\mathcal{D}$).

In the end, for complex manifolds, the general index formula and the Riemann-Roch-Hirzebruch Theorem (the special index formula for elliptic operators $\overline{D}_V$) are equivalent: Let $\overline{D} := \overline{\partial} + \overline{\partial}^* : \Sigma\Omega^{0,2p} \to \Sigma\Omega^{0,2p+1}$ be the "Riemann-Roch operator" in $\mathbb{C}^n$. Then $\sigma(\overline{D})$ produces a generator of the homotopy group $\pi_{2n-1}(GL(N,\mathbb{C}))$, $N = 2^{n-1}$, and all of $K(TX)$ is generated by the $[\sigma(\overline{D}_V)]$ modulo the image of $K(X)$; see [Atiyah-Bott-Patodi 1973, 321 f.].

A survey of function theoretic and geometric applications (e.g., estimates for the dimensions of systems of curves or differential forms and determination of Betti numbers of complicated manifolds) can be found for the case of curves (a), $\dim_{\mathbb{C}} X = 1$) in [Mumford 1976, 147 ff.], for the classification of surfaces ($\dim_{\mathbb{C}} X = 2$) in [Kodaira 1964] including a multitude of very concrete geometric facts which are derived directly from the general Riemann-Roch-Hirzebruch Theorem (d) for line bundles), and for the general case in [Schwarzenberger 1966]. □

H. The Index of Elliptic Boundary-Value Problems

We will now have a look at the index problem for boundary-value problems which was solved in the forties and fifties for a number of special cases in [Vekua 1959/1962, 316-330], [Bojarski 1959, 15-18] and the sources stated there (see also Theorem II.1.1 above). In the programmatic article of [Gelfand 1960], it was called "description of linear elliptic equations and their boundary problems in topological terms", and this initiated the search for the index formula for elliptic operators on closed manifolds [Atiyah-Singer 1963]. In the center of the rapid development of "elliptic topology" (1963-1975), with alternate forms and proofs of the index formula and their widespread applications, remained manifolds without boundary, although [Atiyah-Bott 1964a] demonstrated the topological significance of elliptic boundary conditions, and showed how - in principle - an index formula for elliptic boundary value problems may be obtained by reducing it to the unbordered case ("Poisson principle", see Chapter II.7 and Section II.8.A). In connection with Section C, one can find in this fashion very quick proofs (and supplements) for the boundary-value problems with vanishing index listed in [Agranovich 1965].

In working out the details of the Atiyah-Bott concept, one encounters two main difficulties:

1. Let $\sigma \in \mathrm{Iso}_{SX}(E,F)$ be the symbol of an elliptic (differential) operator $P \in \mathrm{Ell}(E,F)$ over the bordered Riemannian manifold X with boundary Y, where E and F are vector bundles over X. If N is a sufficiently large natural number, it is relatively easy to extend (as exercised in Chapter II.7) $\sigma + \mathrm{Id}_{\mathbb{C}^N}$ on the "symbol level" to an elliptic symbol $\sigma' \in \mathrm{Iso}_{SX \cup BX|Y}(E',F')$, using elements of the proof of the Bott Periodicity Theorem in such a way that $\sigma'(y,\xi) = \mathrm{Id}_{E'_y}$ for all $y \in Y$, and $\xi \in (T^*X)_y$ with $|\xi| \leq 1$. Hereby E_y was identified with F_y by means of $\sigma(y,\nu)$, ν was an inner normal, and $E' := E \oplus \mathbb{C}^N_X$. The only hypothesis needed is that P admits elliptic boundary-value problems. The particular choice of boundary-value problem determines more exactly the way in which $\sigma \oplus \mathrm{Id}_{C^N}$ is extended.

Doing the corresponding deformation on the "operator level", i.e., deforming $P \oplus \mathrm{Id}^N$ to P' "stably equivalent" to P and $P' = \mathrm{Id}'$ near Y, poses some essential difficulties. It is, for example, necessary to pass from differential operators to the class of pseudo-differential operators, and for the transition, the boundary behavior must be suitably restricted (see above before Theorem II.8.1); specifically, in [Boutet de Monvel 1971, 39 f.] it is shown that if the "transmission condition" $\sigma(P)(y,-\nu) = (-1)^k \sigma(P)(y,\nu)$ is postulated, then the desired operator deformation is possible if and only if the indicator bundle $j(P) \in K(SY)$ (see above before Exercise II.8.3) vanishes. On the other hand, the "free" extension of $\sigma \oplus \mathrm{Id}_{\mathbb{C}^N}$ characterized above exists (see below result b)), if only the difference bundle $j(P)$ lies in the image of the "lifting" $\pi_y^* : K(Y) \to K(SY)$.

It was mainly L. Boutet de Monvel, M. I. Vishik and G. I. Eskin who dealt in long series of papers with this predominantly analytic problem of finding the "correct" operator class for the formation of the "parametrix" and the necessary operator deformations. On the positive side, these difficulties demonstrate the effectiveness of topological methods in analysis: Once the index formula (see result d) below) is established, the computation of the index of a boundary-value problem with, say, a partial differential operator, does not require the deformation of the operators with all its difficulties, but only that of the symbol according to the simple rules given in Chapter II.7.

2. Another basic difficulty is the following. Operators (such as the signature operator D^+) which play such an important role for closed manifolds in applications of the index formula as well as in the "cobordism" and "heat equation" proofs (see above Sections D and III.3.B/D) do not admit elliptic boundary value problems in the sense of III.6-8 on a bordered manifold X. See result b) below and [Atiyah-Patodi-Singer 1975, 46]. Accordingly $\sigma(D^+)$ cannot be deformed over the boundary Y of X to the identity (or any map which only depends on the base point $y \in Y$ but not on the covectors $\xi \in (T^*X)_y$). However, M. F. Atiyah, V. K. Patodi and I. M. Singer (loc. cit.) discovered that D^+ admits in an extended sense certain "globally elliptic" boundary conditions R^+ which canonically belong to D^+ and for which $\operatorname{index}\begin{pmatrix} D^+ \\ R^+ \end{pmatrix}$ is well-defined and equal to $\operatorname{sign}(X)$. Using the Gauss-Bonnet Formula $e(X) = (2\pi)^{-1} (\int_X K + \int_Y s)$ for the Euler number $e(X)$ as a model (X a bordered surface with "Gaussian curvature" K on X and with "geodesic curvature" s of Y in X), they (loc. cit., 54-57) could derive an index formula for a certain class of boundary value problems $\begin{pmatrix} P \\ R \end{pmatrix}$. The right hand side of this formula is the sum of two terms which are given explicitly in the "heat equation proof" and of which one only depends on the behavior of P in the interior resp. on the doubling of X, and the other only depends on the boundary behavior of $\begin{pmatrix} P \\ R \end{pmatrix}$. In particular, they obtained for $P := D^+$ the new formula

$$\operatorname{sign}(X) = \int_X L_q(p_1,\ldots,p_q) - \eta(0)$$

(already mentioned in III.3.D(6)) for the signature of a 4q-dimensional compact, oriented, bordered Riemannian manifold. See also [Gilkey 1975].

Efforts to obtain similar explicit "analytic" formulas for arbitrary elliptic boundary-value problems apparently did not achieve full success; see especially [Calderon 1967], [Seeley 1966] and [Fedosov 1974]. Thus - for the time being - we depend on our "mad" topological methods of symbol formation (see below result a)/d)), in spite of the associated computationally unfortunate destruction of the special problem structure which may be simpler initially, and despite the delicate technical questions of lifting homotopies to the "operator levels" in the proofs. (The role of the topology chosen for the symbol deformation is not clear. While we deformed in the C^0-topology (see page 184), there is also the C^1-topology which - according to a recent theorem of F. Waldhausen - has possibly less room for movement but a richer homotopy type.) □

Results: Let X be an n-dimensional, compact, oriented Riemannian manifold with boundary Y.

a) There is a unique construction, whereby one can assign to each elliptic element A of the Green algebra over X (see Chapter II.8 above) a difference bundle $[A] \in K(T(X\setminus Y))$ such that the following conditions hold:

(i) $[A] = [B]$, if A and B are stably equivalent,

(ii) $[A \oplus B] = [A] + [B]$ and $[A \bullet B] = [A] + [B]$, when the composition is defined,

(iii) $[A] = [\sigma(P)] + t[\sigma(Q)]$, for $A = \begin{pmatrix} P & 0 \\ 0 & Q \end{pmatrix}$ and $P = \mathrm{Id}$ near Y so that $[\sigma(P)] \in K(T(X\setminus Y))$ is well defined, Here t is the composition already investigated above in Chapter III.3, namely

$$K(TY) \underset{\cong}{\overset{\boxtimes b}{\longrightarrow}} K(TY \times \mathbb{R}^2) \cong K(TN) \overset{\text{ext}}{\longrightarrow} K(T\hat{X}) \cong K(T(X\setminus Y))$$

for the "trivial" embedding of Y into the open manifold $\hat{X} = X \cup (Y \times [0,\infty))$ with trivial normal bundle N; $\hat{X}$ and $X\setminus Y$ are diffeomorphic,

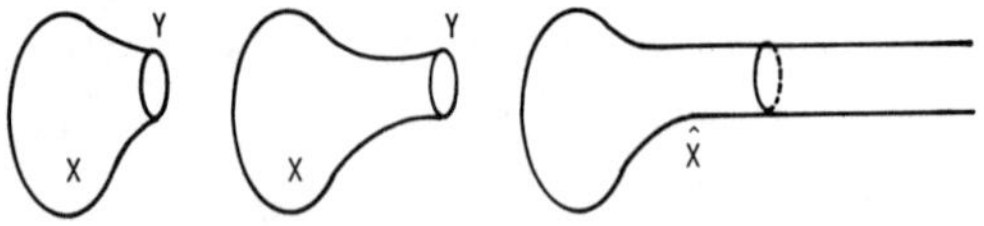

whence we have the isomorphism on the right.

(iv) $[A] = [\bar{\sigma}(P,\beta^+)]$, if $A = \begin{pmatrix} P \\ R \end{pmatrix}$ is an elliptic boundary-value system of differential operators and $\bar{\sigma}(P,\beta^+)$ is the extension of

$\sigma(P) \oplus \mathrm{Id}_{\mathbb{C}^N} \in \mathrm{Iso}_{SX}$ (defined by the boundary isomorphism β^+) to an object in $\mathrm{Iso}_{SX \cup BX|Y}$; see Chapter II.6/7.

b) For a differential operator P, at each point $y \in Y$, there is a vector bundle M_y^+ over the base $(SY)_y = S^{n-2}$, and a difference bundle $[\sigma_y(P)] \in K(\mathbb{R}^n)$ arising from $\sigma_y(P) := \sigma(P)(y,\cdot\cdot)$. Now, $M_y^+ - \dim M_y^+$ defines an element of $K(\mathbb{R}^{n-2})$, and we have (because of an unfortunate sign choice, we use P^* instead of P) $[\sigma_y(P^*)] = -b \times (M_y^+ - \dim M_y^+)$. The tensoring with the Bott element on the right side extends to a homomorphism $\gamma\colon K(SY) \to K(BX|Y,\ SX|Y)$, and we have the global formula $r^*[\sigma(P^*)] = -\gamma(M^+)$, which can be written in the form $r^*[\sigma(P)] = \gamma(j(P))$ in the case that P is a pseudo-differential operator. Here r^* : $K(BX,SX) \to K(BX|Y,\ SX|Y)$ is the restriction homomorphism, and M^+ resp. $j(P)$ are the "indicator bundles" defined via ordinary differential equations (Chapter II.6) or Wiener-Hopf operators (Chapter II.8); these indicator bundles coincide (up to a sign convention) for differential operators.

c) An elliptic pseudo-differential operator P over X can then be completed to produce an elliptic element $\begin{pmatrix} P & * \\ * & * \end{pmatrix}$ of the Green algebra over X, if and only if the following two equivalent conditions are met:

(i) $j(P) \in \mathrm{Ker}\,\gamma$

(ii) $\sigma(P) \oplus \mathrm{Id}_{\mathbb{C}^N}$ (N suitable) extends to an isomorphism over $SX \cup BX|Y$.

If P is a differential operator, then $\deg(P) = \nu(P)$, where $\deg(P)$ is defined as the "degree" of $\sigma(P)(y,\cdot\cdot)\colon (SY)_y \to GL(N,\mathbb{C})$, $y \in Y$ (i.e., the local degree of P), and $\nu(P)$ is the difference $\dim M_\eta^+$ $\dim M_\eta^-$, $\eta \in (SY)_y$.

d) Each embedding of the bounded manifold X in a Euclidean space $\mathbb{R}^m$ defines (as in Chapter III.3 and also in a(iii) above) a homomorphism $t'\colon K(T(X \setminus Y)) \to K(\mathbb{R}^{2m})$, and we then have the formula $\mathrm{index}\,A = \alpha^m t'[A]$ for each elliptic element A of the Green algebra. Here $[A]$ is the difference bundle of A (well defined by a), and $\alpha^m\colon K(\mathbb{R}^{2m}) \to K(\{0\}) \cong \mathbb{Z}$ is the iteration of the Bott isomorphism.

The cohomological version (see Section A above) of the index formula is

$$\mathrm{index}\,A = (-1)^{n(n+1)/2}\{\phi^{-1}\mathrm{ch}[A] \cup \tau(TX \otimes \mathbb{C})\}[X].$$

In this formula, $[X] \in H_n(X,Y;\mathbb{Q})$ denotes the "fundamental cycle" of the orientation of X, $\tau(TX \otimes \mathbb{C}) \in H^*(X,\mathbb{Q})$ is the "Todd class" of the complexification of TX, $ch\colon K(T(X\setminus Y)) \to H^*(BX, SX \cup BX|Y,\mathbb{Q}) = H_c^*(T(X\setminus Y);\mathbb{Q})$ is the "Chern character ring homomorphism", $\phi\colon H^*(X,Y;\mathbb{Q}) \to H^*(BX, SX \cup BX|Y;\mathbb{Q})$ is the "relative Thom isomorphism", and $\cup\colon H^*(X,Y;\mathbb{Q}) \times H^*(X;\mathbb{Q}) \to H^*(X,Y;\mathbb{Q})$ is the "relative cup product".

In the case of systems (see Section B above), where A is given by a system of equations, $[A]$ is defined by a continuous matrix-valued map $\sigma(A)\colon SX \cup BX|Y \to GL$, and we have the integral formula

$$\text{index } A = \int_{\partial(BX)} \sigma(A)^*\omega \wedge \pi^*\tau,$$

where ω and τ are the explicit differential operators (given in Section B above) on U resp. GL and on X, and $\pi\colon \partial(BX) \to X$ is the projection.

In the classical case, where X is a bounded domain in $\mathbb{R}^n$, one finally obtains the simple formula index $A = \sigma^n[A]$, which coincides, in the special case $A = \begin{pmatrix} P & 0 \\ 0 & Q \end{pmatrix}$ with $P = \text{Id}$ near Y, with Theorem 2.1/ Theorem 3.1.

<u>Arguments</u>: a) was announced in [Palais 1965, 349 f.], and proved in [Boutet de Monvel 1971, 47-50]. There, (iv) is replaced by the much weaker requirement that $[A] = 0$, if A belongs to a certain list of canonical operators of "Dirichlet type".

For b) and c), note that r^* and γ fit into the short exact sequences which are horizontal and vertical in the following diagram:

$$\begin{array}{ccccc}
 & & & & K(Y) \\
 & & & & \downarrow \pi_Y^* \\
 & & & & K(SY) \\
 & & & & \downarrow \gamma \\
K(BX, SX \cup BX|Y) & \xrightarrow{q^*} & K(BX,SX) & \xrightarrow{r^*} & K(BX|Y, SX|Y) \\
\wr\wr & & \wr\wr & & \wr\wr \\
K(T(X\setminus Y)) & & K(TX) & & K(\mathbb{R}\times TY)
\end{array}$$

Here, $\pi_Y\colon SY \to Y$ is the projection, and r and q are both embeddings; γ is the composition $K(SY) \overset{\cong}{\to} K(\mathbb{R} \times TY) \cong K(BX|Y, SX|Y)$. See [Boutet de Monvel 1971, 39 f.], and - also for the local version - [Palais 1965, 351].

c) is a trivial consequence of Exercise II.8.5: By definition,

$j(A) = j(P) + [\pi^*G] - [\pi^*H]$ vanishes for each elliptic element $A : C^\infty E \oplus C^\infty G \to C^\infty F \oplus C^\infty H$ of the Green algebra (G,H vector bundles over Y). Hence, $j(P) \in K(SY)$ lies in the image of π_Y^*, and hence in the kernel of γ. Because of b) and the exactness of the horizontal sequence, this means that it is possible to "lift" $[\sigma(P)] \in K(BX,SX)$ to $K(BX,SX \cup BX|Y)$, obtaining the difference bundle $[A] \in K(T(X \setminus Y))$. Incidentally, a) goes much further than c), since it is shown in a) how a special completion of the elliptic operator P to an element A of the Green algebra, generates the "lifting". In [Boutet de Monvel 1971, 35 f.] it is shown in addition that each bundle $\pi_Y^*(G)$ with arbitrary $G \in K(Y)$ is the "indicator bundle" $j(P)$ of an elliptic pseudo-differential operator P on X.

With d), concrete computations are made easy for some classical boundary-value problems. Indeed, one recovers without difficulty the Theorem of Vekua (Theorem II.1.1) with Exercise II.7.2. In view of the Atiyah-Patodi-Singer Signature Formula (see Difficulty 2 above), it seems probable (in order to remain in the language of integral formulas) that the index integral over the "integration domain" $\partial(BX) = SX \cup BX|Y$ should be replaced by the sum of two integrals with different integration domains. In fact, the proof of d) (in which one lifts the K-theoretic resp. homotopy-theoretic construction in a) to the level of operators) already runs in this direction. Roughly speaking, the proof of d) goes as follows: We start with $A \in \mathrm{Ell}_k(X;Y)$. Via composition with certain standard operators (construction modeled on operators like the Riesz Λ-operators; see Exercise II.4.3 above) of vanishing index, we obtain an operator $A' \in \mathrm{Ell}_0(X;Y)$ with $\mathrm{index}\, A' = \mathrm{index}\, A$. To A', we can now add another elementary boundary-value system B with $\mathrm{index}\, B = 0$, such that the indicator bundle $j(A' \oplus B)$ vanishes. With a topological argument, it follows that $A' \oplus B$ is stably equivalent to a $A'' = \begin{pmatrix} P'' & L'' \\ R'' & Q'' \end{pmatrix} \in \mathrm{Ell}_0(X;Y)$, where P'' equals the identity near Y. With Theorem II.8.3e, this allows a deformation of A'' to $A''' := \begin{pmatrix} P'' & 0 \\ 0 & Q'' \end{pmatrix}$, whence $\mathrm{index}\, A = \mathrm{index}\, A''' = \mathrm{index}\, P'' + \mathrm{index}\, Q''$. Thus, in principle, the index problem for boundary-value systems is reduced to the index problem over closed manifolds, since $\mathrm{index}\, P'' = \mathrm{index}\, \overline{P''}$, where $\overline{P''}$ is the operator extended to $X \cup_Y X$ by means of the identity. Details are in [Boutet de Monvel 1971, 47 f.], where a) and d) are proved simultaneously. For a heuristic treatment of the situation, when all the operators arising are differential operators at the outset, see also [Cartan-Schwartz 1965, 25-08 f.] and [Schulze 1976 a/b]. □

J. Real Operators

Up to now, we have considered operators between spaces of sections of *complex* vector bundles. One can also consider operators with *real* coefficients which operate only on sections of real bundles. If P is such a real elliptic and skew-adjoint operator, then trivially $\text{index } P = 0$. This is uninteresting, but now $\dim(\text{Ker } P)$ is a homotopy-invariant mod 2. The reason for this stems from the fact that the nonzero eigenvalues of P all come in complex conjugate pairs $(\lambda, \overline{\lambda})$; if one deforms the operator P so that λ goes to zero, then $\overline{\lambda}$ goes to zero and consequently $\dim(\text{Ker } P)$ increases by two. Already in 1959, R. Bott had discovered a real analogue to his "periodicity theorem" (see Chapter III.1 above), and proved that the homotopy groups $\pi_i(GL(N, \mathbb{R}))$, for large N, are periodic in i with period 8, and for $i \equiv 0$ or $i \equiv 1 \bmod 8$ are isomorphic to $\mathbb{Z}_2$.

In [Atiyah-Singer 1969] the connection between these two analytical and topological mod 2 invariants was determined, and the index theorem was carried over to the real case. Together, these considerations provide a new and topologically much simpler proof of the real Bott Periodicity Theorem. Two details are particularly noteworthy: First, in order to connect the two invariants of the real theory, one must go outside of the real theory, since the amplitude p of a real skew self-adjoint pseudo-differential operator P is defined via the Fourier transform, and is thus not real in general, but rather complex with the condition $p(x,-\xi) = \overline{p(x,\xi)}$. Thus, the symbol of a real operator does not immediately yield a suitable element of $\pi_i(GL(N, \mathbb{R}))$, but rather, must be interpreted first as a mapping $f: S^{2n-1} \to GL(N,\mathbb{C})$ (as in Chapter III.2) with the added condition $f(-\xi) = \overline{f(\xi)}$. Another peculiarity lies in the fact that these mod 2 invariants (although having only the values 0 or 1) are in a certain sense more complicated topologically, or in any case, are of a different type than the usual homology or cohomology classes (also, if one takes $\mathbb{Z}_2$ coefficients). Thus, in concrete situations they can provide decisive additional information. This program was carried out for vector fields (see Section E above) in [Atiyah 1970c] and [Atiyah-Dupont 1972]. □

K. The Lefschetz Fixed-Point Formula

Let $f: X \to X$ be a continuous map with X compact, and let $\text{Fix}(f) = \{x \in X: f(x) = x\}$ be the fixed-point set of X. Salomon Lefschetz (1926) introduced the formula $L(f) = \Sigma\nu(x)$, where the sum is over all fixed-points of f. For details of the definition of the integer $\nu(P)$ (which is 1 for an isolated fixed point and equals 0 for a point where $f = Id$ in a neighborhood) and for the proof, see [Alexandroff-Hopf 1935, 531-542] or [Greenberg 1967, 222-224]. The "*Lefschetz number*" $L(f)$ is defined as the alternating sum $\Sigma(-1)^i \operatorname{trace}(H^i f)$, where $H^i f$: $H^i(X,\mathbb{C}) \to H^i(X,\mathbb{C})$ is the cohomology endomorphism of the complex vector space $H^i(X,\mathbb{C})$ induced by f. M. F. Atiyah and R. Bott refined this beautiful formula in the mid 60's. Furthermore, the formula is simplicially defined and in general hardly computable - or in other words, only the topology of X enters into the formula, while additional structures are ignored. Atiyah and Bott removed this weakness by bringing the additional structures into play.

Results. Let X be a compact C^∞ manifold without boundary, and let P be an elliptic differential operator. Let $f\colon X \to X$ be differentiable and commute with P; here we need to assume that f "lifts" to a bundle mapping, say $\tilde{f}$, so that f acts on sections via $f \cdot s(x) = \tilde{f}(s(f^{-1}(x)))$. Then f yields a well defined endomorphism of the finite-dimensional vector spaces Ker P and Coker P. We define the *Atiyah-Bott-Lefschetz number* as

$$L(f,P) := \operatorname{trace}(f|\operatorname{Ker} P) - \operatorname{trace}(f|\operatorname{Coker} P).$$

If f is the identity, then by definition $L(f,P) = \operatorname{index} P$, and one can apply the Atiyah-Singer Index Formula. In the other "extreme case", where f has only isolated fixed points with multiplicity ± 1, one obtains the formula

$$L(f,P) = \Sigma\nu(x) \qquad \text{(ABL)}$$

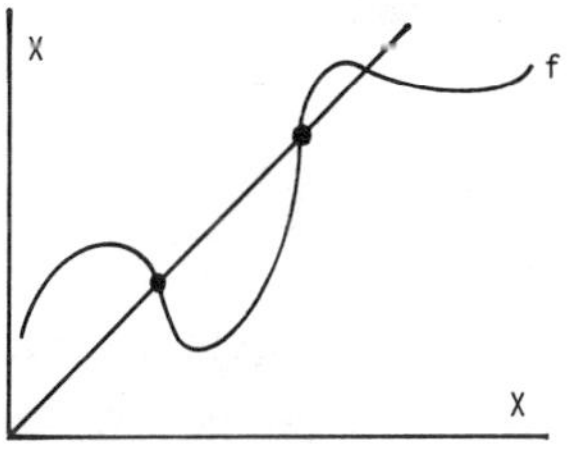

where x runs over the fixed-points of f and the complex number $\nu(x)$ depends only on the differential $f_*(x)\colon (TX)_x \to (TX)_x$. (The "simplicity" or "transversality" of the fixed-point x means that the endomorphism $Id - f_*(x)$ is invertible, and the "multiplicity" ± 1 is understood to be the sign of $\det(Id - f_*(x))$.)

In the following list, there are some applications of the Atiyah-Bott-Lefschetz formula (ABL) with the expressions for the respective values of $\nu(x)$:

No.	Situation: X,P,f	$\nu(x)$; $x \in \mathrm{Fix}(f)$	Meaning of Formula
1.	X oriented, Riemannian manifold, $P := D = d+d^*\|\Omega^{\mathrm{even}}$ (see Section D above) $f\colon X \to X$ smooth	$\nu(x) = \dfrac{\det(\mathrm{Id}-f_*(x))}{\|\det(\mathrm{Id}-f_*(x))\|}$ $= \mathrm{sign}\det(\cdots)$ $= \pm 1$	$L(f,P) = L(f)$ $\Sigma\nu(x)$ = "number" of fixed-points ABL $\subset$ Lefschetz
2.	X complex manifold $P := \overline{D}$ (see Section G above) $f\colon X \to X$ holomorphic	$\nu(x) = \det(\mathrm{Id}-f_{*\mathbb{C}}(x))^{-1}$ $f_{*\mathbb{C}}(x)$ is the complex differential ($\mathbb{C}$-linear approx. to f at x); we have $\det(\mathrm{Id}-f_*(x))$ $= \|\det(\mathrm{Id}-f_{*\mathbb{C}}(x))\|^2$	In particular, ABL contains well-known theorems on algebraic functions for the case of Riemann surfaces: [Hirzebruch 1966, 595 ff.]
3.	X complex manifold, $P := \overline{D}_V$; V holomorphic bundle over X (see Section G above), $f\colon X \to X$ holomorphic, $\phi\colon f^*V \to V$ holomorphic	$\nu(x) = \dfrac{\mathrm{trace}\,\phi(x)}{\det(\mathrm{Id}-f_{*\mathbb{C}}(x))}$	ABL has corollaries in algebraic geometry
4.	X oriented, Riemannian manifold of dimension 2q, $P := D^+$ signature operator (see Section D above), $f\colon X \to X$ orientation preserving isometry.	$\nu(x) = i^q \prod_{j=1}^{q} \mathrm{Cot}(r_{j,x})$ where $(TX)_x$ is decomposed into a direct sum of 2-dimensional planes $V_{j,x}$ on which $f_*(x)$ is a rotation by $r_{j,x}$.	ABL is a discrete analog of the Hirzebruch Signature formula (f=Id). We obtain theorems of the kind: If f is of order p^k with p a prime $\neq 2$ (i.e., $f^{(p^k)} = \mathrm{Id}$), then f cannot have exactly one fixed-point, since then $L(f,P) \in \mathbb{Z}$, but $\nu(x) \in \mathbb{C} \setminus \mathbb{Z}$.

For further details and the proof of (ABL), we refer to [Atiyah-Bott 1965], [Atiyah-Bott, 1966], and [Atiyah-Bott 1967]. Incidentally, the proof is essentially simpler than the proof of the index formula (AS), and represents a "weak" version of the "heat equation proof" discussed above in III.3.D: One considers the zeta function $\zeta(z) := \mathrm{trace}(f^* \circ \Delta^{-z})$ whose value at $z = 0$ is easier to calculate, since it turns out that this zeta function is holomorphic not only for $\mathrm{Re}(z) > \dim X$, but also in all of $\mathbb{C}$ because of the

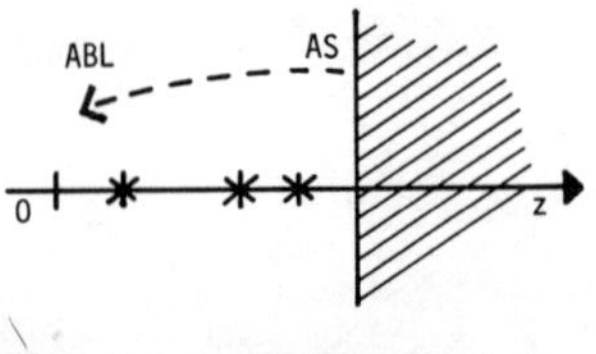

assumptions on the fixed points; see also the following section. □

L. Analysis on Symmetric Spaces

Let X be a closed C^∞ manifold and G a compact group of diffeomorphisms of X. One may think of $G = \{Id\}$ or $G = \{Id, g, g^2, \ldots, g^{v-1}\}$, where g is a diffeomorphism of order r (i.e., $g^r = Id$). G can also be a more general finite group, or a compact Lie group. Let P be a "G-invariant" elliptic operator over X, which commutes with the operations of the group elements (via the suitable bundle automorphisms), so that $\operatorname{Ker} P$ and $\operatorname{Coker} P$ are not only finite-dimensional complex vector spaces, but can be regarded as G-modules. The homomorphism $g \longmapsto g|\operatorname{Ker} P$, $g \in G$, is then a "finite-dimensional representation" of G, and $g \longmapsto \operatorname{trace}(g|\operatorname{Ker} P)$ is its "character". Corresponding considerations apply to $\operatorname{Coker} P$. If we define $L(g,P)$ as in the preceding section, then we have that the function

$$\operatorname{index}_G(P) : g \longmapsto L(g,P), \quad g \in G,$$

is a "virtual" character (i.e., an element of the "representation ring" $R(G)$). If $G = \{Id\}$, then $R(G) \cong K(\text{point}) \cong \mathbb{Z}$ and $\operatorname{index}_G(P) = \operatorname{index} P$. In the general case $G \supsetneq \{Id\}$, $R(G) \cong K(*) \cong *$ is a complicated object and correspondingly, $\operatorname{index}_G(P)$ is a "sharper" (also homotopy-) invariant than the integer $\operatorname{index}(P)$.

<u>Results</u>. In the framework of "equivariant" K-theory with groups $K_G(X)$ (esp. $K_G(TX)$) for the "symmetric" space X, one obtains an analogous index formula for $\operatorname{index}_G(P) \in R(G)$, and for each $g \in G$ a Lefschetz formula

$$L(g,P) = \sum_{i=1}^{\ell} \int_{X_i} \alpha_i$$

where the sum on the right is over the connected components $X_1, \ldots, X_\ell$ of $\operatorname{Fix}(g)$. Incidentally, $\operatorname{Fix}(g)$ is a mixed-dimensional submanifold of X, since one can introduce local coordinates in a suitable neighborhood of any fixed-point such that g operates linearly. Each individual term in the formula has the form of our index formula (see Section A or Chapter III.3 above).

For the precise definitions and the various formulations of these formulas, we refer to [Atiyah-Bott 1968] and [Atiyah-Segal 1968], as well as [Atiyah-Singer 1968 a/b] and [Atiyah 1974]. An abundance of applications mainly for the case of a complex manifold X with $P := \overline{D}$ and holomorphic g (and more generally, applications in the domain of elementary number theory) can be found in [Hirzebruch-Zagier 1974]. Here P is held fixed and attention is payed to the correct choice of G or g. Conversely, in [Bott 1965] = [Wallach 1973, 114-138] one finds an investigation of "homogeneous differential operators" on a fixed homogeneous space X = G/H, where H is a closed subgroup of G. Here the relevance lies in the fact that these formulas connect the fixed points of the action by G with global invariants of the symmetric space X. Thus "deep" formulas are obtained for the way in which (in the language of cohomology) individual characteristic classes of invariants can be synthesized to yield invariants of the entire cohomology ring of the manifold.

The proof follows essentially as in the argument carried out here in III.2/III.3. For the construction of the "topological" Index $K_G(TX) \to R(G)$, one needs an equivariant sharpening of the Bott Periodicity Theorem, which can be obtained by the same function analytic tools used in III.1 for the "classical" Bott Periodicity Theorem; incidentally, it is noteworthy that these tools are *necessary*, since (roughly) only this approach can be freed of the induction argument which cannot succeed in general for the equivariant case. See [Atiyah 1968a], and for a non-technical introduction [Atiyah 1967, 246 f.].

M. Further Applications

With the preceding overview, we have in no way encompassed all of the connections, alternative formulations, generalizations and special applications of the Atiyah-Singer Index Formula, but we have only touched those which have reached a certain definitive form in their development. For an insight into further perspectives and open problems, we refer to [Singer 1969], [Singer 1971], and [Atiyah 1976] and the literature given there. □

Part IV. The Index Formula and Gauge-Theoretical Physics

"The beauty and profundity of the geometry of fibre bundles were to a large extent brought forth by the (early) work of the man (S. S. Chern) who we are here to honor today. I must admit, however, that the appreciation of this beauty came to physicists only in recent years."

(C. N. Yang, in the Chern Symposium 1979 [Berkeley, June], Ed. W.-Y. Hsiang et al., Springer-Verlag (1980), p. 252.)

1. PHYSICAL MOTIVATION AND OVERVIEW

Both the goals and methods of physics are different from those of mathematics. Mathematicians have the rather nebulous goal of exploring and establishing that which is "logically possible" and "interesting", depending on the tastes of the individual. Physicists have the sharper goals of discovering, "explaining", and predicting actual phenomena in the physical world. The mode of reasoning in physics is rather fuzzy (yet publishable, if the end result fits the data or makes a promising prediction). If physicists were forced to be rigorous every step of the way, physics would not have advanced toward its goals nearly as much as it has. Generally speaking, although the creative process in mathematics is equally fuzzy in its initial stages, a "result" is not publishable until it has been proved with the framework of "commonly accepted logical standards." It is no wonder that mathematicians are ill at ease with heuristic arguments of physicists that have little chance of being cast into "strict" mathematical form. Physicists cannot be expected to have interest in mathematics that seems unrelated to physical phenomena. Of course, there are those (misnamed "mathematical physicists") who are interested primarily in strict (yet pretty) mathematics that seems to have physical relevance. In the following overview, it is hoped that the reader may gain some understanding of why the concepts covered in this book (e.g., elliptic operators, complex vector bundles, pseudo-differential operators, Hilbert spaces, distributions, etc.) have great physical relevance.

The reader is not expected to understand every detail in the following lengthy (yet, necessarily incomplete and historically vague) overview of "modern" physics. However, she or he may take whatever is digestible, realizing that this material is neither a prerequisite nor a substitute for the more precise (if drier) mathematics of the chapters that follow. Those who are untutored in relativity or quantum physics are likely to discover that the logical possibilities of the physical world can be every bit as beautiful and strange as those typically encountered in far-reaching mathematical diversions. There has been a lot of very nice mathematics that has been motivated by theoretical physics recently (Look through some issues of Communications in Mathematical Physics.); of course, there was little difference between the two disciplines before the early 1900's.

A. Classical Field Theory

In classical (as opposed to "quantum") physics, particles are viewed as pointlike objects that move along paths which are solution curves of systems of ordinary differential equations determined by a force field. For example, there is Newton's equation $m\vec{r}''(t) = \vec{F}(\vec{r}(t))$, where $\vec{F}: \mathbb{R}^3 \to \mathbb{R}^3$ is a given force field. The force may also depend on the velocity $\vec{r}'(t)$ as well as $\vec{r}(t)$. Indeed, an electromagnetic (E-M) field consists of a pair of vector fields $\vec{E}$ and $\vec{B}$ (which may be time-dependent); a particle of charge e moves according to the Lorentz force law

$$\frac{d}{dt}(m\vec{r}'(t)) = e[\vec{E}(\vec{r}(t),t) + \frac{1}{c}\vec{r}'(t) \times \vec{B}(\vec{r}(t),t)] .$$

The situation is complicated by the further empirical fact, that $\vec{E}$ and $\vec{B}$ must satisfy a system of partial differential equations, namely Maxwell's equations:

$$(1)\quad \vec{\nabla} \times \vec{E} + \frac{1}{c}\frac{\partial \vec{B}}{\partial t} = 0 \qquad\qquad (2)\quad \vec{\nabla} \cdot \vec{B} = 0$$

$$(3)\quad \vec{\nabla} \cdot \vec{E} = \rho \qquad\qquad (4)\quad \vec{\nabla} \times \vec{B} - \frac{1}{c}\frac{\partial \vec{E}}{\partial t} = \frac{1}{c}\vec{J},$$

where $\rho: \mathbb{R}^3 \to \mathbb{R}$ is proportional to the "charge density" of a "continuous medium" of charged particles, and $\vec{J}$ is essentially the "current density" of the medium (i.e., $\vec{J} = \rho\vec{v}$, where $\vec{v}$ is the velocity vector field of the medium). Thus, $\vec{E}$ and $\vec{B}$ are influenced by each other, as well as by the motions of the charged medium they are supposed to influence via the Lorentz force law.

Maxwell's equations can be simplified conceptually by considering the following 2-form, called the *E-M field strength*

$$F = cE_1 dx \wedge dt + cE_2 dy \wedge dt + cE_3 dz \wedge dt$$
$$+ B_1 dy \wedge dz + B_2 dz \wedge dx + B_3 dx \wedge dy.$$

One can easily check that Maxwell's equations (1) and (2) are equivalent to $dF = 0$. Defining the *source 1-form* j by

$$j = \rho dt - c^{-2}(J_1 dx + J_2 dy + J_3 dz),$$

the Maxwell equations (3) and (4) say that $\delta F = j$, where δ is the co-differential (the formal adjoint of d) relative to a *Lorentzian metric* (i.e., $c^2dt^2 - dx^2 - dy^2 - dz^2$). This means that Maxwell's equations can be immediately generalized to manifolds with Lorentzian metric tensors (i.e., *space-times*). Hence, Maxwell's equations fit quite naturally into general relativity. The Lorentz force law, which is <u>not</u> a consequence of Maxwell's equations, also can be written invariantly on a space-time M. Indeed, the "world-line" of a particle of mass m_0 and charge e is a curve $s \longmapsto \gamma(s) \in M$ which obeys the equation

$$m_0 \frac{D}{ds}[\gamma'(s)] = \frac{e}{c} F(\gamma'(s),\cdot)^{\#}, \tag{5}$$

where $\gamma'(s)$ is the tangent vector of γ at s, $\frac{D}{ds}$ denotes "covariant differentiation", and the sharp on the right side indicates that the co-vector $F(\gamma'(s),\cdot)$ has been converted to a vector by "raising indices" using the metric. In Minkowski space, $\frac{D}{ds}[\gamma'(s)]$ is simply $\gamma''(s)$, but s is not necessarily the time coordinate. Equation (5) implies that the length of $\gamma'(s)$ is constant; indeed

$$\frac{1}{2} m_0 \frac{d}{ds} |\gamma'(s)|^2 = m_0 \langle \frac{D\gamma'(s)}{ds}, \gamma'(s)\rangle = \frac{e}{c}\langle F(\gamma'(s),\cdot)^{\#},\gamma'(s)\rangle$$
$$= \frac{e}{c} F(\gamma'(s),\gamma'(s)) = 0,$$

since F is antisymmetric. However, note that equation (5) is not scale-invariant (i.e., if γ is replaced by $\bar{\gamma}$, where $\bar{\gamma}(s) = \gamma(as)$ $(a > 0,$ $a \neq 1)$, then (5) is no longer satisfied). In flat Minkowski space-time with metric $c^2dt^2 - dx^2 - dy^2 - dz^2$, equation (5) splits into a "spacial" part and a "time" component which are empirically correct only when $|\gamma'(s)|^2 = c^2$. Indeed, writing $\gamma(s) = (t(s),x(s),y(s),z(s)) = (t(s),\vec{r}(s))$, note that $|\gamma'(s)|^2 = c^2t'(s)^2 - |\vec{r}\,'(s)|^2 = c^2t'(s)^2[1 - |\vec{v}(t(s))|^2/c^2]$, where $\vec{v}(t(s)) = \frac{d}{dt}\vec{r}(t(s)) = \vec{r}\,'(s)\cdot t'(s)^{-1}$. Assuming $\beta := t'(s) > 0$, we then have $\beta = (1 - |\vec{v}|^2/c^2)^{-1/2}$ provided $|\gamma'(s)|^2 = c^2$. Thus, $\gamma'(s) = (1,\vec{v}(t(s)))\beta$ and (5) reduces to the pair of equations

$$\text{(a)}\quad \frac{d}{dt}(m_0\beta\vec{v}) = e(\vec{E} + \frac{\vec{v}}{c}\times\vec{B}) \qquad \text{(b)}\quad \frac{d}{dt}(m_0\beta c^2) = e\vec{E}\cdot\vec{v}.$$

Note that (a) is the Lorentz force law, where $m = m_0\beta$ is the so-called relativistic mass ($m \approx m_0$ = rest mass for $|\vec{v}| << c$; $m \to \infty$ as $|\vec{v}| \to c$). The right side of (b) is the rate at which the E-M field does work on the particle; note that $\vec{B}$ does no work since $(\vec{v}\times\vec{B})\cdot\vec{v} = 0$. Thus, the mc^2 ($= m_0\beta c^2$) on the left side of (b) must be the energy E of the particle (i.e., $E = mc^2$); note that $mc^2 - m_0c^2 = m_0c^2(\beta-1) = \frac{1}{2}m_0|\vec{v}|^2 + O(|\vec{v}|^4)$ (as $|\vec{v}| \to 0$) is the relativistic kinetic energy.

On abstract Minkowski space (i.e., a four-dimensional vector space M with metric determined by a scalar product $<\cdot,\cdot>$ of signature $(+,-,-,-)$), one can introduce many "inertial" coordinate systems (t,x,y,z) such that $<\cdot,\cdot> = c^2dt^2 - dx^2 - dy^2 - dz^2$. Any two such systems are related by an affine map $M \to M$, called a *Poincaré transformation*. The linear Poincaré transformations are known as *Lorentz transformations*. If $(\bar{t},\bar{x},\bar{y},\bar{z})$ is another inertial coordinate system with the same origin, say $0 \in M$, then $\bar{t}$ need not equal t (i.e., coordinate time has no absolute meaning). On the other hand, the condition $|\gamma'(s)|^2 = c^2$ does have a coordinate-free significance, for $\gamma\colon (a,b) \to M$. Indeed, the correct interpretation of the parameter s is that it measures the time on a clock *carried* by the particle with world line $\gamma(s)$; s is known as the *proper time* of the particle. By definition, $t'(s) = \beta = (1 - |\vec{v}|^2/c^2)^{-1/2}$, meaning that coordinate time "runs faster" for particles that are moving relative to the associated inertial system from which the coordinate time is measured. Assuming that the earth is approximately inertial (i.e., its world line does not deviate too much from the t-axis of some inertial coordinate system, assuming space-time is nearly Minkowskian), a high-velocity space traveler will find his twin brother older when he returns. Note that the space travel's world line is <u>not</u> close to being inertial, since it is far from any straight-line time axis. There is no contradiction, only a paradox; the result has in fact been verified using subatomic particles instead of humans.

Once more, things are complicated by the fact that the metric tensor $g_{\mu\nu}$ on space-time is not "flat" like that of Minkowski space. The deviation from flatness is due to the presence of E-M fields (e.g., radiation), hoards of particles (neutral and charged), etc. that live in space-time. One measure of curvature is the Ricci curvature tensor (defined in 2.E) "$R_{\mu\nu}$". The vanishing of the Ricci-tensor is necessary (but not sufficient, in dimension 4 or larger) in order that there exist a neighborhood of each point which is isometric to a subset of Minkowski space. The trace $S = g^{\mu\nu}R_{\mu\nu}$ is known as the scalar curvature. The Einstein field equation of general relativity is

$$R_{\mu\nu} - \frac{1}{2}Sg_{\mu\nu} = \frac{-8\pi K}{c^2}T_{\mu\nu},$$

where $T_{\mu\nu}$ is the "energy-momentum tensor" which is formed in a canonical fashion from the E-M field strength F on space-time and a continuous approximation of the 4-(energy/momentum) density of particle-like matter (cosmologists sometimes take these "particles" to be entire galaxies); finally, K is the ordinary gravitational constant in Newtonian gravity. In essence, the Einstein field equation (actually, ten component equations, since $R_{\mu\nu}$, $T_{\mu\nu}$, and $g_{\mu\nu}$ are symmetric in $\mu,\nu = 0,1,2,3$) tells us how the nongravitational energy-momentum of space-time $(T_{\mu\nu})$ influences the geometry of space-time. Neutral particles "move" along geodesics $\gamma(s)$ of space-time such that $|\gamma'(s)|^2 = c^2$. The apparent curvature of the geodesic, when projected onto what we perceive as "space", is due to gravity which is nothing more than the geometry of space-time.

The vanishing of the Ricci tensor (and hence the scalar curvature) does not imply that space-time is locally flat. Indeed, the full curvature tensor has 10 additional components that constitute the "Weyl conformal curvature tensor" (defined in 2.E). Thus, it is quite possible to have a curved space-time devoid of matter, etc. ($T_{\mu\nu} = 0$) which satisfies the so-called empty-space equation $R_{\mu\nu} - \frac{1}{2}Sg_{\mu\nu} = 0$. This equation can be formulated in terms of a variational principle. Indeed, let D be a compact domain in a space-time M with metric tensor g. Let L be a functional that assigns to each metric tensor g', the number $L(g') = \int_D S_{g'}\mu_{g'}$ where $S_{g'}$ is the scalar curvature of g' and $\mu_{g'}$ is the volume element. It is found (e.g., see [Bleecker, p. 125] for a coordinate-free proof) that g is a critical point of L within the space of g' agreeing with g on the boundary of D, if and only if the empty-space equations hold inside D. In order to obtain the full equation with non-vanishing $T_{\mu\nu}$, it is necessary to add additional terms to L for each type of particle or field that resides in space-time. These terms are known as "actions" or Lagrangians. The action for an E-M field F is proportional to $-\frac{1}{2}\int_D|F|_g^2\mu_g$, where

$$|F|_g^2 = \frac{1}{2}g^{\mu\nu}g^{\rho\sigma}F_{\mu\rho}F_{\nu\sigma}$$

(i.e., the standard "Lorentz-invariant" norm-square relative to the metric tensor g). In terms of an inertial coordinate system on Minkowski space, we can write F in terms of $\vec{E}$ and $\vec{B}$ as before; then $-\frac{1}{2}|F|^2 = \frac{1}{2}(|\vec{E}|^2 - |\vec{B}|^2)$. It is instructive to note that $|\vec{E}|$ and $|\vec{B}|$ can change under a change of inertial coordinates, but $|\vec{E}|^2 - |\vec{B}|^2$ is constant. The two-form F is a "covariant object", but (like coordinate time) $\vec{E}$ and $\vec{B}$ separately have no absolute significance. When $\frac{-k}{2}\int_D |F|^2_{g'}\mu_{g'}$ is added to $L(g') = \int_D S_{g'}\mu_{g'}$ and the first variation is set equal to 0, we obtain the equation

$$R_{\mu\nu} - \frac{1}{2} Sg_{\mu\nu} = -k(F^{\sigma}_{\mu}F_{\nu\sigma} - \frac{1}{4}|F|^2 g_{\mu\nu}).$$

For some k, this is actually the accepted field equation for a universe which is empty except for the E-M field F.

We can also get Maxwell's equation $\delta F = 0$ from a variational principle; note $j = 0$ in the absence of sources, which we assume here. Indeed, consider $\frac{1}{2}\int_D |F|^2_g \mu_g$ as a functional of F, instead of g. We assume that F and its variations satisfy the other Maxwell equation $dF = 0$. Letting D be simply-connected, we can write any variation F' of F as dA' for some 1-form A'. The associated variation of $\frac{1}{2}\int_D |F|^2 \mu$ is then

$$\int_D \langle F',F\rangle_\mu = \int_D \langle dA',F\rangle_\mu = \int_D \langle A',\delta F\rangle_\mu,$$

assuming that A' vanishes on the boundary of D. This variation is zero for all such A', exactly when $\delta F = 0$ in D. Succinctly, we get Einstein's field equation by setting to 0 the first variation of $L(g) - \frac{k}{2}\int_D |F|^2_g \mu_g$ with respect to g, and we get Maxwell's equation $\delta F = 0$ from the variation with respect to F.

The constraint $dF = 0$ implies that locally F can be written as $F = -dA$ (the minus is conventional) for some 1-form A. Such an A exists on any simply-connected domain and is called a *gauge potential* for F. However A is not unique, since for any function $\phi \in C^\infty(M)$, we have $d(A + d\phi) = dA = F$, whence $A + d\phi$ is also a gauge potential for F. The transformation $A \longmapsto A + d\phi$ is called a "gauge transformation". In terms of A, the equation $\delta F = 0$ becomes the more familiar wave equation $-\delta dA = 0$; in Minkowski space $-\delta d = c^{-2}\partial_t^2 - \partial_x^2 - \partial_y^2 - \partial_z^2$. This is very convenient for some purposes, since A is not restricted by any constraint, and one sees that the singularities of A propagate with speed c. However, A is not unique, and this defect is only partially overcome by the so-called *Lorentz condition* $\delta A = 0$ that can be imposed without loss of generality. There are plenty of functions ϕ for which $\delta(A + d\phi) = 0$ (i.e., those which satisfy the wave equation $\delta d\phi = 0$).

The E-M field strength F is not built into the metric $g_{\mu\nu}$, and Einstein spent many years trying (without success) to incorporate F into the geometry of space-time, thereby obtaining a "unified field theory". Actually, in [Kaluza 1921] and [O. Klein 1926], it was shown that E-M and gravity could be geometrized simultaneously (if not "unified" in the strict, and probably absurd, sense that all fields be included in the same irreducible representation) by attaching circles to the points of space-time, thereby enlarging it to 5-dimensions. In this truly remarkable Kaluza-Klein theory, the illusion that the universe has only four dimensions is not due to the smallness of the circles (although they may in fact be very small), but rather it is due to the perfect (in principle) homogeneity in the circular dimension. Off hand, it is difficult to believe that anything useful can come from adding an unobservable dimension, but indeed all "grand unified theories" (GUTs), popular these days, add at least 24 dimensions in order to unify all known non-gravitational forces.

In modern terminology (made precise in 2.A), the original Kaluza-Klein theory introduces a fiber bundle $\pi: P \to M$ over space-time, whose fibers $\pi^{-1}(x)$ are circles. The pull-back π^*g of the metric g on M is degenerate on the fibers. Thus, we need additional structures to complete π^*g to a Lorentzian metric on P. Of course there is a standard metric on a circle, but one needs to specify what vectors in P should be orthogonal to the fibers. This is conveniently accomplished by introducing a 1-form $\tilde{A}$ on P such that $\tilde{A}(\partial_\theta) = 1$, where ∂_θ is the standard vector field on P which is tangent to the circular fibers. A non-degenerate metric on P is then given by

$$\tilde{g}(X,Y) = g(\pi_*X, \pi_*Y) + \tilde{A}(X)\tilde{A}(Y)$$

or

$$\tilde{g} = \pi^*g + \tilde{A} \otimes \tilde{A}.$$

The subspaces which are orthogonal to the fibers are then precisely those annihilated by $\tilde{A}$. Moreover, $\tilde{A}$ should be invariant under a uniform rotation $R_\theta: P \to P$ in order to preserve the homogeneity of P with $\tilde{g}$ (i.e., to keep us from perceiving more than 4 dimensions).

We now come to the point of introducing this 5-dimensional "cylindrical" universe P and the 1-form $\tilde{A}$ with associated metric $\tilde{g}$. The most striking fact is that if γ is a geodesic in $(P,\tilde{g})$, then $\overline{\gamma} := \pi\circ\gamma$ turns out (cf. Bleecker, 1981) to be the path of an electrically charged particle moving according to the Lorentz force law, where the electromagnetic field strength F and charge/mass ratio are defined as follows. Let $\sigma\colon M \to P$ be a section ($\pi\circ\sigma = 1_M$). Then $A = \sigma^*\tilde{A}$ is a 1-form on M, and we set $F = -dA$. Even though A depends on the choice of σ (known as a choice of gauge), one can prove that if $\sigma'(x) := R_{\phi(x)}(\sigma(x))$ (i.e., σ' is obtained from σ via a pointwise dependent rotation through angle $\phi(x)$), then we have $A' := \sigma'^*\tilde{A} = A + d\phi$. Consequently, A just undergoes a gauge transformation, and $F = dA = dA + d^2\phi = dA'$ is invariant. Actually, nowadays $R_\phi\colon P \to P$ is more commonly known as a gauge transformation, and the change $A \longmapsto A + d\phi$ is induced from it. Having defined F, it remains to specify the ratio m/e for the particle with world line $\overline{\gamma}$; indeed, m/e turns out to be proportional to the length of the vertical component of γ' (i.e., $\tilde{g}(\gamma',\partial_\theta)$). Very neatly, it happens that $\tilde{g}(\gamma',\partial_\theta)$ is constant along γ, because of the invariance of $\tilde{g}$ under rotations R_θ. Hence, on P, force of electro-magnetism has been completely encoded into the metric $\tilde{g}$ via $\tilde{A}$; i.e., gravity and E-M have been "geometrized" on P in about the same way that gravity alone was geometrized on M. Even by itself, this fact concerning the geodesics on $(P,\tilde{g})$ would be sufficient to force one to take the 5-dimensional theory seriously, but yet another surprise emerges. The scalar curvature $S_{\tilde{g}}$ of $\tilde{g}$ on P is not just the same as that of g on M, but rather is the sum of the scalar curvature S_g of M and a multiple k (which can be adjusted by changing fiber lengths) of the self-action density of F:

$$S_{\tilde{g}} = S_g - \frac{k}{2}|F|^2.$$

Note that $S_{\tilde{g}}$ is constant on fibers, whence we can regard it as a function on M. Thus, the Einstein equation, with the electromagnetic energy momentum tensor, can be derived from the variational principle associated with the functional $g \longmapsto \int S_{\tilde{g}}\mu_g = \int S_g - \frac{k}{2}|F|^2\mu_g$. Hence, the Einstein field equation in a non-empty universe with an E-M field (but no matter) is induced from an "empty" bundle universe $(P,\tilde{g})$. All of this admits suitable generalization to the case where the fibers are not just circles, but rather, general Lie groups (typically the groups $SU(n)$ or $SO(n)$ in physical applications). The 1-forms (on these higher dimensional bundle spaces) are Lie-algebra-valued forms that differential geometers call "connections" and physicists call "gauge-potentials" (when they are pulled back to M via a section). The corresponding field strengths (known as "curvature" to geometers) are no longer $\mathbb{R}$-valued 2-forms on M, but rather have values in a certain vector bundle over M. There is the rather obvious hope that these field strengths have something to do with the other forces (e.g., "weak" forces which are responsible for weak interactions that cause, among other events, the decay of the free neutron; and the "strong" forces that are indirectly responsible for holding the nucleus together, and perhaps more directly involved in binding the quarks inside hadrons such as the proton, neutrons, pions, etc.). However, one must be wary about extrapolating classical field theory (which is all that we have discussed up to this point) to such small systems, which are governed by quantum theory.

B. Quantum Theory

Just as Newtonian mechanics breaks down for systems moving at high velocity (near that of light), classical field theory does not describe very small systems (say, of atomic dimensions or smaller) very well. The classical picture told us that there are "background" fields such as the field strength F of electromagnetism and the metric tensor of general relativity, and in sharp distinction to these, there were particles that move in trajectories determined by these fields. The background fields were thought to be rather diffuse and wave-like, while the particles had well-defined positions or trajectories. However, the reader has no doubt heard that under certain experimental conditions, it is found that electromagnetic radiation (light) produces results that are intuitively understood much better by assuming that light is made of streams of particles, known as photons. Indeed, when light falls on metalic surfaces in a vacuum, electrons are found to be emitted from the atoms, but the rate of emission is found to be much larger (e.g., by factors like 10^9, even under ordinary conditions) than what is theoretically calculated under the assumption that each atom absorbs *all* of the energy it receives from the portion of the (continuous) wave that contacts it. The most natural explanation of this photo-electric effect is that the wave is not continuous, but made up of chunks (quanta) that have sufficient energy to dislodge the electrons in the atoms that they happen to hit. In this way, not all atoms have to wait a long time to absorb enough energy to liberate an electron, and a much higher emission rate is not only possible, but actually happens immediately. Ironically, it was Einstein who was awarded a Nobel prize in

1922 in part for his explanation of the photo-electric effect in terms of the quantum theory of light, but he veered away from the dramatic development of quantum mechanics, preferring to work on the problem of unifying the classical field theories of electromagnetism and gravity (without adding the extra dimension of the Kaluza-Klein theory, which he deemed artificial); he did not succeed. Just as electromagnetic fields exhibit particle-like properties, it was also found that particles (e.g., electrons) exhibit wave-like properties. In the experiment where electrons are fired at a double slit, it is found that they collectively make a wave diffraction pattern on a screen behind the slit. Thus, the sharp "particle versus wave" dichotomy in classical physics must admit some fuzziness.

Quantum mechanics and quantum field theory grew out of the attempt to describe this state of affairs and to make predictions as accurately as possible. One of the most perplexing phenomena confronting the founders of quantum mechanics was that of atomic spectra. A spectrograph reveals that atoms emit and absorb light a certain fairly discrete wave lengths or energies. However, the classical planetary model, which has the electron circling the proton (say, for the hydrogen atom) under the inverse-square Coulomb law, predicts that the electron will radiate electromagnetic energy at a continuously varying wavelength and will actually spiral into the proton as it gives up its energy in a short time, rendering the atom unstable. Although it first seems rather far-fetched, one might hypothesize that the various energy levels of the atom are actually eigenvalues of some differential operator, just as the frequencies of a vibrating string with fixed ends are eigenvalues of $(\text{const.})\frac{d^2}{dx^2}$ operating on the space of functions vanishing at the ends. Of course, if the eigenvalues of the operator are to represent energies, the operator should have the physical dimension of energy. The Coulomb potential energy is $-Ze^2r^{-1}$, where e is the charge of an electron, Z is the number of protons in the nucleus, and r is the distance between the nucleus and the electron (only one, for simplicity). Moreover, the standard rotationally-invariant differential operator on $\mathbb{R}^3$ is the Laplacian. The most obvious operator formed from Δ and $-Ze^2r^{-1}$ and having dimensions of energy is $\alpha\Delta - Ze^2r^{-1}$, where α is some constant such that $\alpha\Delta$ has dimensions of energy. The point spectrum (assuming $\alpha < 0$) of this operator (densely defined on $L^2(\mathbb{R}^3)$) is $Ze^2/(4\alpha n^2)$, $n = 1,2,3,\ldots$. For each single-electron atom (ion, for $Z > 1$), there is a choice of α such that this spectrum is consistent with the observed energy spectrum (actually, differences of the above eigenvalues, as the electron jumps between levels). Indeed, it is found that α is a universal constant times μ^{-1}, where μ is the reduced mass (i.e., $\mu = m(1 + m/M)^{-1}$, m = mass of electron, M = mass of nucleus). Writing $\alpha = -k/\mu$, we find that the units of k must be $(\text{energy}\cdot\text{time})^2$. The constant $\hbar \approx 6.6256 \times 10^{-27}$ erg.-sec., introduced by Max Planck around 1900 in connection with black-body radiation, has dimensions energy·time, and the value $k = \frac{1}{2}\hbar^2$ is consistent with experiment. Hence, we have the striking coincidence that the point spectrum of the operator $-(\hbar^2/2\mu)\Delta - Ze^2/r$ coincides with the energy levels of the single-electron atom. The procedure, by which one replaces a classical observable (e.g., energy, momentum, position) by an operator whose eigenvalues are the possible values that are observed experimentally, is called "quantization". Thus, we have roughly succeeded in quantizing the energy of an electron in a Coulomb potential. Note that $-\hbar^2/2m(\partial_x^2 + \partial_y^2 + \partial_z^2)$ resembles the classical expression for kinetic energy $\frac{1}{2m}(p_x^2 + p_y^2 + p_z^2)$, $\vec{p}$ = momentum of particle of mass m, which suggests that the quantization of the classical momentum observable $\vec{p}$ should be $(\pm i\hbar\partial_x, \pm i\hbar\partial_y, \pm i\hbar\partial_z)$; actually, the minus sign is adopted. Thus, $-\frac{\hbar^2}{2m}\Delta$ should be the quantization of kinetic energy of a particle of mass m. Our atomic example then suggests that the quantization of a classical potential energy function $V(\vec{r})$ should be the multiplication operator $\hat{V}$ (on functions $\psi\colon \mathbb{R}^3 \to \mathbb{C}$) defined by $\hat{V}(\psi)(\vec{r}) = V(\vec{r})\psi(\vec{r})$. Since the quantization of a function of position $\vec{r} = (x^1, x^2, x^3)$ is multiplication by the function, the quantization of the x^i themselves (which are classical observables too) should be the multiplication operators $\hat{x}^i$ given by $\hat{x}^i(\psi)(\vec{r}) = x^i\psi(\vec{r})$, $\psi\colon \mathbb{R}^3 \to \mathbb{C}$. We have been very vague about the domains (spaces of functions) for these operators. In practice, physicists feel comfortable with this vagueness, as long as they can make physical sense of their results. For example, an eigenfunction of momentum operators $-i\hbar\partial_j$ $(j = 1,2,3)$ is of the form $e^{\frac{i}{\hbar}\vec{r}\cdot\vec{p}}$ (i.e., $-i\hbar\partial_j(e^{\frac{i}{\hbar}\vec{r}\cdot\vec{p}}) = p_j e^{\frac{i}{\hbar}\vec{r}\cdot\vec{p}}$), but this function of $\vec{r}$ is not in $L^2(\mathbb{R}^3)$. Also, for $\vec{r}_0 \in \mathbb{R}^3$, the delta distribution $\delta(\vec{r}-\vec{r}_0)$ is an "eigenfunction" for the position operators $\hat{x}^1$, $\hat{x}^2$, $\hat{x}^3$. Perhaps the most appropriate domain would be the space of tempered distributions (i.e., continuous linear functionals on the Schwartz space of rapidly decreasing functions); at least this encompasses the above examples. At any rate, the "functions" in the domain are called the various "states" of the particle. States which differ by a constant complex factor are identified (considered physically indistinguishable); thus, the space of states is an infinite-dimensional complex projective space. The following is a basic interpretive assumption of quantum mechanics; without it, one has trouble making physical sense of the theory.

Fundamental Assumption. Let $[\psi]$ be a state with representative ψ of $L^2(\mathbb{R}^3)$-norm $||\psi|| = 1$. Suppose a quantized observable (i.e., symmetric operator) A has an eigenvalue λ, then the probability that the observable is measured to be λ (when the particle is in state $[\psi]$) is the norm-square of the projection of ψ onto the eigenspace of λ. More generally, for observables that admit a (real) spectral resolution (i.e., projection-valued measure on $\mathbb{R}$), the probability that the observable is measured to be in a certain interval of $\mathbb{R}$ is the norm-square of the image of ψ under the spectral projection associated to the observable and interval. □

If the observable A has a complete set of eigenvectors u_n, $n = 1,2,\ldots$ (say $Au_n = \lambda_n u_n$), then the *expectation of measurements of* A (when the particle is in state ψ) is simply

$$\begin{aligned}\Sigma\, \lambda_n |\langle\psi,u_n\rangle|^2 &= \Sigma\, \lambda_n \langle\psi,u_n\rangle\overline{\langle\psi,u_n\rangle} \\ &= \Sigma\langle\psi,Au_n\rangle\overline{\langle\psi,u_n\rangle} = \Sigma\, \langle A\psi,u_n\rangle\overline{\langle\psi,u_n\rangle} \\ &= \langle A\psi,\psi\rangle .\end{aligned}$$

We get the same result for the more general case, when the observable has a spectral resolution.

The function $\vec{r} \longmapsto |\psi(\vec{r})|^2$ (where $||\psi|| = 1$) is the probability density for the position of the particle in state ψ; i.e., for a domain $D \subset \mathbb{R}^3$, $\int_D |\psi|^2$ is the expectation that the particle will be found in D, when the state is ψ. Indeed, let $\chi_D\colon \mathbb{R}^3 \to [0,1]$ be the characteristic function of D. Its quantization is $\hat{\chi}_D\colon L^2(\mathbb{R}^3) \to L^2(\mathbb{R}^3)$, given by $\hat{\chi}_D(\psi) = \chi_D\psi$. Classically, the observable χ_D is 1 if the particle is in D and 0 otherwise. Quantum mechanically, the expectation of the observable when the particle is in state ψ (i.e., the probability that the particle will be found in D) is $\langle\hat{\chi}_D\psi,\psi\rangle = \int_{\mathbb{R}^3} \hat{\chi}_D(\psi)\overline{\psi} = \int_D |\psi|^2$. For

$$\tilde{\psi}(\vec{p}) := (2\pi\hbar)^{-3/2} \int_{\mathbb{R}^3} \psi(\vec{r})e^{-i\vec{p}\cdot\vec{r}/\hbar}\, d^3r$$

(essentially the Fourier transform of ψ), $|\tilde{\psi}|^2$ is the momentum probability density for the particle in state ψ. Indeed, the Fourier inversion theorem yields $\psi(\vec{r}) = (2\pi\hbar)^{-3/2} \int \tilde{\psi}(\vec{p}) e^{i\vec{p}\cdot\vec{r}/\hbar} d^3p$, whence $(\hat{p}_j\psi)(\vec{r}) = -i\hbar\partial_j\psi(x) = (2\pi\hbar)^{-3/2} \int p_j\tilde{\psi}(\vec{p})e^{i\vec{p}\cdot\vec{r}} d^3p$; consequently, for a domain D in $\vec{p}$-space, the quantization of χ_D should be given by $(\hat{\chi}_D\psi)(\vec{r}) = (2\pi\hbar)^{-3/2} \int \chi_D(\vec{p})\tilde{\psi}(\vec{p})e^{i\vec{p}\cdot\vec{r}} d^3p$, and the expectation that the particle in state ψ have momentum in D is then (by Parseval's Theorem)

$$\langle\hat{\chi}_D\psi,\psi\rangle = \langle\chi_D\tilde{\psi},\tilde{\psi}\rangle = \int_D |\tilde{\psi}(p)|^2 d^3p,$$

as required. Note that pseudo-differential operators are essentially the quantizations of functions of momentum.

In classical physics, particles move along trajectories, and thus have well-defined positions and momenta at all times. In quantum mechanics, the position and momentum cannot both be simultaneously determined with arbitrary precision. This is a consequence of the (Heisenberg) uncertainty principle which (as we now show) follows from the above Fundamental Assumption. The uncertainty of the measurements of an observable A, when the particle is in state ψ $(||\psi|| = 1)$ is the standard deviation (i.e., the root of the expectation of $(A - \langle A\psi,\psi\rangle)^2$) of the measurements, namely

$$\Delta_\psi A := \langle(A - \langle A\psi,\psi\rangle)^2\psi,\psi\rangle^{1/2} = ||(A - \langle A\psi,\psi\rangle)(\psi)||.$$

The Schwarz inequality and a little algebra reveal that for any two observables A and B,

$$\Delta_\psi A \cdot \Delta_\psi B \geq \frac{1}{2} |\langle[A,B]\psi,\psi\rangle|.$$

This is the uncertainty principle. It says that non-commuting quantized observables cannot be simultaneous measured with arbitrary precision for every state. For example, $[\hat{x}^j,\hat{p}^k](\psi) = x^j(-i\hbar\partial_k\psi) + i\hbar\partial_k(x^j\psi) = i\hbar\delta^{jk}\psi$; i.e., $[\hat{x}^j,\hat{p}^k] = i\hbar\delta^{jk}$, whence $\Delta_\psi\hat{x}^j \cdot \Delta_\psi\hat{p}^j \geq \hbar/2$ for any state ψ $(||\psi|| = 1)$. The position and momentum components in the same direction cannot be determined simultaneously with arbitrary precision; e.g., the initial conditions for Newton's equation cannot be specified, and the philosophy of determinism wanes.

In order that quantum mechanics embrace special relativity, the quantization of the "position" 4-vector $x = (x^\mu) = (ct,\vec{r})$ (where $x^0 = ct$) should be $(c\hat{t},\hat{\vec{r}})$, where $\hat{t}(\tilde{\psi})(x) = t\tilde{\psi}(x)$; $\tilde{\psi}\colon \mathbb{R}^4 \to \mathbb{R}$. Moreover, the energy-momentum 4-vector $p = (p^\mu) = (E/c,\vec{p})$ ought to be quantized as $\hat{p}^{\mu} = i\hbar g^{\mu\nu}\partial_{\nu}$; note, $\hat{p}^{k} = -i\hbar\partial_{k}$, since $1 = g^{00} = -g^{kk}$, $k = 1,2,3$, and also $E/c = i\hbar g^{00}\partial_0 = i\hbar c^{-1}\partial_t$ or $\hat{E} = +i\hbar\partial_t$ (plus!). We already have seen that a reasonably good guess for the quantization of the nonrelativistic energy (kinetic + potential) is the operator $-\hbar^2/2m_0\Delta + \hat{V}$. The relativistic energy should be very close to the rest energy m_0c^2 plus the nonrelativistic energy, provided the potential V is adjusted to be zero when the particle is at rest, say by adding a constant to V

$$E \approx \frac{1}{2m_0}|\vec{p}|^2 + V(x) + m_0c^2.$$

Thus, we can expect that the time development of the state $\tilde{\psi}$ should be governed by

$$i\hbar\partial_t\tilde{\psi} = \frac{-\hbar^2}{2m_0}\Delta\tilde{\psi} + V(x)\tilde{\psi} + m_0c^2\tilde{\psi},$$

at least in a non-relativistic limit. If we define an "adjusted" wave function ψ by the equation $\tilde{\psi}(t,\vec{r}) = e^{-im_0c^2t/\hbar}\,\psi(t,\vec{r})$ (Note that ψ and $\tilde{\psi}$ define the same state, at each fixed time, as functions on $\mathbb{R}^3$.), then we obtain the well-known Schrödinger equation

$$i\hbar\partial_t\psi = \frac{-\hbar^2}{2m_0}\Delta\psi + V\psi. \tag{*}$$

An obvious defect of (*) is its non-invariance under Lorentz transformations. On the other hand, it fits in nicely with the Fundamental Assumption, in that $||\psi(t,\cdot)||^2 = \text{const.}$ because of (*). This can be verified either directly, by differentiating under the integral and using (*), or by noting that $t \longmapsto \psi(t,\cdot)$ is the trajectory of $\psi(\cdot,0) \in L^2(\mathbb{R}^3)$ under the 1-parameter group of unitary transformations generated by the skew-Hermitian operator $\frac{i}{\hbar}(\frac{\hbar^2}{2m_0}\Delta - \hat{V})$. It is also true that the time-dependent position expectation $\vec{r}(t) := \langle\hat{\vec{r}}\psi(t,\cdot),\psi(t,\cdot)\rangle$ satisfies "Newton's equation" $\vec{r}''(t) = -\langle\psi(t,\cdot)\vec{\nabla}V(\cdot),\psi(t,\cdot)\rangle$ as a consequence of (*). The primary reason why (*) and its many-particle generalizations are so successful in dealing with electrons in atoms (in spite of the non-relativistic nature of (*)) is that electrons "travel" at speeds only around (.01)c within atoms, according to a simple approximate classical calculation.

Perhaps it would have been better to insist on Lorentz invariance from the beginning, thereby replacing $(E/c)^2 - |\vec{p}|^2 = m_0^2c^2$ by its quantized analogue, the Klein-Gordon equation

$$-\hbar^2(c^{-2}\partial_t^2 - \Delta)\tilde{\psi} = m_0^2c^2\tilde{\psi}.$$

However, note that there is no vestige of a potential in this equation. If we are to introduce electromagnetism in some way, we must be careful to do it in a Lorentz-invariant manner. Writing the Klein-Gordon equation in the more covariant form,

$$-\hbar^2 g^{\mu\nu}\partial_\mu\partial_\nu\tilde{\psi} = m_0c^2\tilde{\psi}, \qquad (**)$$

the most obvious way of introducing E-M is simply by adding a multiple of the 4-vector potential (A_μ) to the 4-vector operator (∂_μ). However, in order that the equation be invariant under gauge transformations, we must subject $\tilde{\psi}$ to some kind of gauge transformation also. If $\phi\colon M \to \mathbb{R}$ is the function generating the gauge transformation, then we know $A_\mu \longmapsto A'_\mu = A_\mu + \partial_\mu\phi$. In order to keep $|\tilde{\psi}|^2$ constant, we might try $\tilde{\psi} \longmapsto \tilde{\psi}' = e^{i\phi}\tilde{\psi}$. Indeed, we have the identity

$$[\partial_\mu - i(A_\mu + \partial_\mu\phi)]e^{i\phi}\tilde{\psi} = e^{i\phi}[\partial_\mu - iA_\mu]\tilde{\psi},$$

or

$$[\partial_\mu - iA'_\mu]\tilde{\psi}' = ([\partial_\mu - iA_\mu]\tilde{\psi})'.$$

Thus, we obtain the desired invariance

$$-\hbar^2 g^{\mu\nu}(\partial_\mu - iA_\mu)(\partial_\nu - iA_\nu)\tilde{\psi} = m_0^2c^2\tilde{\psi} \qquad (**)'$$

$$\Longleftrightarrow$$

$$-\hbar^2 g^{\mu\nu}(\partial_\mu - iA'_\mu)(\partial_\nu - iA'_\nu)\tilde{\psi}' = m_0^2c^2\tilde{\psi}'.$$

In order for the units to work out, A must be replaced by (const.)$\cdot$A having the same dimension as ∂_μ, namely $(\text{length})^{-1}$. The choice of the constant that leads to the most correct physics is $e/\hbar c$. Thus, one incorporates E-M into the Klein-Gordon equation by replacing ∂_μ by $\partial_\mu - \frac{ie}{\hbar c}A_\mu$. Physicists call this "minimal replacement", while differential geometers recognize that this amounts to replacing ordinary derivatives by covariant derivatives. Note that since $\tilde{\psi}$ changes under a change of gauge, the "wave function" for a charged particle is more properly regarded as an equivariant $\mathbb{C}$-valued function on the Kaluza-Klein circle bundle P (or equivalently, a section of the associated complex vector bundle) rather than a $\mathbb{C}$-valued function on M; covariant differentiation must then be used. Observe that $|\tilde{\psi}|^2$ does not change under a change of gauge, but unfortunately it cannot represent a probability density. Indeed, choosing inertial coordinates $(t,\vec{r})$, it does <u>not</u> follow from the K-G equation that $\int_{\mathbb{R}^3}|\tilde{\psi}(t,\vec{r})|^2 d^3r$ is constant, as with the Schrödinger equation. Instead, one finds that $\int_{\mathbb{R}^3} i[\tilde{\psi}^*\partial_t\tilde{\psi} - \tilde{\psi}\partial_t\tilde{\psi}^*]$ is time-independent, but the integrand is no longer positive in general! If we write $\tilde{\psi}(t,\vec{r}) = e^{-im_0c^2t/\hbar}\psi(t,\vec{r})$, then $\hat{E}\tilde{\psi}(t,\vec{r}) = i\hbar\partial_t\tilde{\psi} = m_0c^2\tilde{\psi}(t,\vec{r}) + e^{-im_0c^2t/\hbar}\, i\hbar\partial_t\psi(t,\vec{r})$. For a slow-moving particle, we expect most of its energy to be rest energy m_0c^2; thus, the expectation of the second term ought to be small compared to the first. Assuming this, we have

$$\frac{i\hbar}{2m_0c^2}\int_{\mathbb{R}^3} \tilde{\psi}^*\partial_t\tilde{\psi} - \tilde{\psi}\partial_t\tilde{\psi}^* \approx \int_{\mathbb{R}^3} |\psi|^2.$$

Hence, in a "non-relativistic limit", the quantity $P(t,\vec{r}) = \frac{\hbar}{m_0c^2}\operatorname{Im}(\tilde{\psi}\partial_t\tilde{\psi}^*)$ becomes the Schrödinger probability density $|\psi|^2$. In general, $P(t,\vec{r})$ is not positive, and $eP(t,\vec{r})$ is given the interpretation of a charge probability density (what else?). Since $eP(t,\vec{r})$ is still not of fixed sign for a given e, we deduce that there ought to be particles of mass m and charge $-e$ that "contribute" to $eP(t,\vec{r})$. This is a general feature of relativistic equations; they predict that particles of mass m and charge e necessitate the existence of anti-particles of mass m and charge $-e$. Incidentally, the reader can verify that the K-G equation (**) for $\tilde{\psi}$ $(= e^{-im_0c^2t/\hbar}\psi)$ becomes the Schrödinger equation (*) (without potential V) for ψ, if one neglects the "nonrelativisticly small" third term in

$$c^{-2}\partial_t^2\tilde{\psi} = -\left(\frac{m_0^2c^2}{\hbar^2}\psi + \frac{2im_0}{\hbar}\partial_t\psi - c^{-2}\partial_t^2\psi\right)e^{-im_0c^2t/\hbar}.$$

In the case $(A_\mu) = (A_0,\vec{0})$ (static E-M potential), we recover Schrödinger's equation (*) with potential $-eA_0$, from (**)' (with A_μ replaced by $\frac{e}{\hbar c}A_\mu$), ignoring the "small" term in $\partial_t\tilde{\psi}$ (considered before) and terms involving A_0^2 and ∂_tA_0. Actually, if we set $A_0 = Ze/r$, $\vec{A} = \vec{0}$, and do not make any of the above approximations, it is possible to obtain the energy levels explicitly (see [L. I. Schiff 1968, p. 470]). To order 4 in the parameter $\gamma = \frac{Ze^2}{\hbar c}$, they are given by

$$E_{n,\ell} \approx m_0c^2\left[1 - \frac{\gamma^2}{2n^2} - \frac{\gamma^4}{2n^4}\left(\frac{n}{\ell+\frac{1}{2}} - \frac{3}{4}\right)\right].$$

Here $n = 1,2,3,\ldots$ and $\ell = 0,1,2,\ldots,n-1$ (total and azimuthal quantum numbers). The first term is the rest energy, the second term yields the nonrelativistic energy levels obtained by using the Schrödinger equation with potential $-Ze^2/r$ (note that these levels are independent of ℓ), and the third term is the first relativistic correction which breaks the degeneracy in ℓ for a given n. Experiments reveal that the degeneracy in ℓ is in fact broken, but the spread of the energy levels for a given n is significantly less than that predicted by the expression for $E_{n,\ell}$. The problem is that this most obvious relativistic K-G equation (with vector potential) is not the right one for electron wave functions. Indeed, there is a first-order relativistic equation that we missed -- the Dirac equation.

The search for such a first-order equation is partly motivated by the fact that the Schrödinger's equation only involves a first derivative with respect to t, and thus the evolution of the states can be described by a flow in the Hilbert space $L^2(\mathbb{R}^3)$, as mentioned above. Let us look for a first-order relativistically-invariant differential operator, say with constant coefficients γ^μ:

$$A = \gamma^0\partial_0 + \gamma^1\partial_1 + \gamma^2\partial_2 + \gamma^3\partial_3 \qquad (\partial_0 = \frac{1}{c}\,\partial_t).$$

Now, if A is relativistically invariant, then so is $A^2 = (\Sigma\, \gamma^\mu\partial_\mu)^2 = \frac{1}{2}\,\Sigma(\gamma^\mu\gamma^\nu + \gamma^\nu\gamma^\mu)\partial_\mu\partial_\nu$. Moreover, there are some obvious operators of this form which are relativistically invariant, namely multiples of the wave operator $\Sigma\, g^{\mu\nu}\partial_\mu\partial_\nu$, $g^{\mu\nu} = \mathrm{diag}(1,-1,-1,-1)$. Thus, suppose we impose the condition $\frac{1}{2}(\gamma^\mu\gamma^\nu + \gamma^\nu\gamma^\mu) = g^{\mu\nu}K$, for some constant K. One quickly finds that for $K \neq 0$, there are no real (or complex) constants γ^μ that satisfy this condition. With some perseverence, one can prove that if the γ^μ are allowed to be $n \times n$ matrices and $K = I_n$, then the first solutions are obtained when $n = 4$. Indeed, we could take

$$\gamma^0 = \begin{bmatrix} I & 0 \\ 0 & -I \end{bmatrix}, \quad \gamma^j = \begin{bmatrix} 0 & \sigma_j \\ -\sigma_j & 0 \end{bmatrix}, \quad j = 1,2,3, \text{ where}$$

$$\sigma_1 = \begin{bmatrix} 0 & 1 \\ 1 & 0 \end{bmatrix}, \quad \sigma_2 = \begin{bmatrix} 0 & -i \\ i & 0 \end{bmatrix}, \quad \sigma_3 = \begin{bmatrix} 1 & 0 \\ 0 & -1 \end{bmatrix}$$

A source of headaches is the fact that the γ^μ are not unique, since we can replace γ^μ by $B\gamma^\mu B^{-1}$ for any invertible B. At any rate, *Dirac's equation* for the 4-component wave function ψ is

$$\Sigma\ i\hbar\gamma^{\mu}\partial_{\mu}\psi = m_0\psi.$$

The electromagnetic 4-vector potential A_μ can be introduced in the most obvious way:

$$\Sigma\ i\hbar\gamma^{\mu}(\partial_{\mu} + \frac{ie}{c\hbar}A_{\mu})\psi = m_0\psi.$$

For $(A_\mu) = (\frac{Ze}{r}, \vec{0})$, this equation can be separated (cf. [L. I. Schiff 1968, p. 486]), and one finds that the energy levels $E_{n,|k|}$ (i.e., the values of $E > 0$ for which there is a solution of the form $e^{-iEt/\hbar}\,\tilde{\psi}(r)$ such that $\tilde{\psi}(\vec{r})$ decays suitably as $\vec{r} \to \infty$), are of the form (where $\gamma := \frac{Ze^2}{\hbar c}$)

$$E_{n,|k|} = m_0c^2\left[1 - \frac{\gamma^2}{2n^2} - \frac{\gamma^4}{2n^4}(\frac{n}{|k|} - \frac{3}{4}) + O(\gamma^6)\right],$$

where $|k|$ is an integer with $1 \leq |k| \leq n$. These values agree much better with observations than the K-G values. Although discrepencies still exist (e.g., the Lamb shift), they can be essentially accounted for within the more accurate context of quantum electrodynamics (QED), a quantum *field* theory. In this theory, the wave function ψ and the vector potential A are replaced by distributions which have values that are operators in some Hilbert space. Although the mathematics behind QED is shady, the formalities involved give rise to recipes (Feynman diagram rules) for computing physical quantities in terms of formal power series in the fine structure constant $\alpha = e^2/\hbar c \approx 1/137$. Such series are called renormalized perturbation series. The accuracy of QED is illustrated by the following comparison of the values of the "magnetic moment" of the electron:

Experiment:	$\frac{e\hbar}{2m_ec}(1.00115965241 \pm 20)$
QED:	$\frac{e\hbar}{2m_ec}(1.00115965221 \pm 60)$.

In spite of the mathematical problems with QED (e.g., the perturbation series may start diverging after a certain point, as an asymptotic series), such amazing agreement with experiment suggests that we must be doing something right.

The reader may have noticed that we did not indicate the sense in which the Dirac equation is relativistically (i.e., Lorentz)-invariant. For the Klein-Gordon equation Lorentz-invariance simply means that if $\tilde{\psi}$

is a solution and $L: \mathbb{R}^4 \to \mathbb{R}^4$ is Lorentz transformation, then $\tilde{\psi}\circ L$ will be a solution as well. For the Dirac equation, Lorentz-invariance holds in the sense that there is a "double-valued" representation of the Lorentz group by 4×4 complex matrices $r(L)$ such that if ψ solves Dirac's equation, then $r(L^{-1})\psi\circ L$ also solves it. Since this representation turns out to be the sum of two spin - $\frac{1}{2}$ representations, the Dirac equation takes into account (indeed predicts) the internal spin of the electron; it also forecasts the existence of the positron. In modern terminology, this means that the Dirac electron "wave function" (without E-M potential) is actually a section of a complex 4-dimensional vector bundle (the Dirac bispinor bundle) which is nontrivially associated to a double cover (spinor frames) of the bundle of Lorentz orthonormal frames over spacetime. If the E-M potential is included, then gauge invariance dictates that this vector bundle is associated with the fibered product of the spin frame bundle and the Kaluza-Klein $U(1)$-bundle. Indeed, all of the fundamental constituents of matter [namely, the quarks (up, down, strange, charm, top, and bottom - each in three different "colors") and the leptons (electron, muon and tau - each with its respective neutrino)] are believed to be generated by quantizations of sections of Dirac bispinor bundles associated to various fibered products of the spinor frame bundle with certain principal bundles having nonabelian gauge groups. (Actually, the neutrino wave functions can be reduced to 2 complex components, if neutrinos are massless.) If the speculations of grand unified field theorists are correct, then the 30 (or 32, if right-handed neutrinos exist) chiral halves of the bispinor fields of each of the three particle generations can be unified (at very high energies) into a single section of a 60 (or 64)-dimensional complex vector bundle obtained by tensoring each half of the Dirac bispinor bundle with a 15 (or 16) dimensional vector bundle associated to a principal bundle over space-time with a "simple" grand unification group G. For the popular theory with $G = SO(10)$ (or more precisely, $Spin(10)$), the representation giving rise to the associated vector bundle is a fundamental spinorial one of dimension $2^{10/2-1} = 16$. Only one of the closed 1-parameter subgroups of G corresponds to electric charge giving rise to the E-M force mediated by photons. Other subgroups correspond to other charges that respond to other forces. For example, $SU(3) \subset SU(5) \subset SO(10)$ is the gauge subgroup giving rise to the "color" force (mediated by "gluons") between the quarks inside strongly interacting particles (hadrons) such as the proton. The quantized form of this $SU(3)$-gauge theory is known as QCD (quantum-chromodynamics); as far as is known, this theory looks promising. The nonabelian aspect of SU(3) is thought to be responsible in part for the fact that quarks are confined in hadrons. Electrons and neutrinos may appear out in the open, because SU(3) acts trivially on the lepton sector of the representation, and hence these particles do not respond to the color force associated with SU(3). There is an $SU(2)_w \times U(1)_{em}$ (w = weak, em = electromagnetic) sitting inside SU(5) which is responsible for the electro-weak forces that cause (in particular) decay of isolated neutrons in about 15 minutes, on the average. It was primarily for their work in realizing that this electro-weak unification was feasible in the context of a quantized, spontaneously broken gauge theory, that Steven Weinberg and Abdus Salam were awarded a Nobel prize in 1980. As a point of historical interest, Oscar Klein was the first to introduce SU(2) gauge fields (commonly known as Yang-Mills fields) in [O. Klein 1938]; indeed, he even anticipated their use in modeling weak interactions. In partial confirmation of the work of many theorists, the weak gauge vector bosons W_+, W_-, and Z_0 (corresponding to the generators of $SU(2)_w$) were recently detected in experiments and have the predicted masses. Of course, the various simple grand unification groups G have some generators not in $SU(3)_c \times SU(2)_w \times U(1)_{em}$, and the associated gauge bosons are thought to be involved in processes that convert quarks to leptons, and hence lead to proton decay. The force of gravity is not represented by any subgroup of G, but rather is "gauged" (in a somewhat different sense) by the Lorentz group for the bundle of frames over space-time. In this way, "gravity" seems to resist attempts to unify it with other forces, and no universally convincing method of quantizing it has been forthcoming.

It is somewhat tricky to give a physical interpretation to the pre-quantized wave functions such as the 4-component bispinor electron field. One problem is that there are negative (as well as positive) energy solutions, a difficulty that is resolved after quantization when the negative energies are (in a sense) converted to positive energies for anti-electrons (i.e., positrons). It is true that in the positive energy sector, the first two of the four components become relatively large in a nonrelativistic limit and these may then be identified with the Schrödinger wave function. (Actually, in the case where an E-M potential is present, these make up the 2-component wave function of Pauli's equation that generalizes Schrödinger's equation for particles with spin.) Since $\psi^*\psi \geq 0$ and its integral over $\mathbb{R}^3$ is time-independent (as a consequence of Dirac's equation), one might be tempted to think of it as a probability density, but when ψ is quantized (i.e., turned into an operator-valued distribution), $\psi^*\psi$ reappears as a charge density of a collection of positrons and electrons, and so the positivity of $\psi^*\psi$ does not survive quantization. On the other hand, the prequantum energy density (which is indefinite) becomes positive after quantization! Space does not permit the inclusion of the details of quantizing interacting wave functions, even if the procedure were mathematically sound. It must be emphasized, however, that many classical fields (say, other than E-M fields) cannot be given a reasonable physical significance, until they are quantized. For example, although the gluon force of QCD is mediated by 0-mass gluons (corresponding to photons in QED) and is thus expected to have infinite range (decaying as r^{-2}, rather than exponentially), no one has directly observed long range effects of this force "out in the open" in the same way we sense E-M. Since it appears that the gluon force is confined to such small regions (i.e., about the size of a proton), it would appear very speculative to treat it as a classical, smooth (wave-like, diffuse) field; by the photoelectric effect, we know that even E-M does not necessarily behave as a wave, even at atomic dimensions. It is nevertheless believed that the classical solutions of the field equations for nonabelian gauge fields (particularly in 4-dimensional *Euclidean* space) will yield at least a first-order approximation to certain quantum effects for such fields, particularly with regard to "tunneling" phenomena. Although we cannot go into the details of how this works in quantum field theory, there is a similar situation in quantum mechanics, which can be understood through the following discussion.

Consider the Schrödinger operator $H = H_0 + V$, where $H_0 = -(\hbar^2/2m)\Delta$ and $V\colon \mathbb{R}^3 \to \mathbb{R}$ is a "nicely-behaved" potential. Then H is an (unbounded) self-adjoint operator defined on some dense domain in $L^2(\mathbb{R}^3,\mathbb{C})$, and one obtains a strongly continuous 1-parameter group of unitary transformations $\exp(-itH/\hbar)$. For f in the Schwartz space of rapidly decreasing functions on $\mathbb{R}^3$, the solution of the problem

$$i\hbar \frac{\partial}{\partial t}\psi = -(\hbar^2/2m)\Delta\psi + V\psi$$

$$\psi(x,0) = f(x)$$

is given by $\psi(x,t) = [\exp(-\frac{i}{\hbar}tH)f](x)$. Even though H_0 and V do not commute, a formula due to T. Kato and H. F. Trotter (cf., Pacific Math. J. 8(1958), p. 887) yields

$$\exp(-\frac{i}{\hbar} tH) = \lim_{k\to\infty} [\exp(-\frac{i}{\hbar}(\frac{t}{k})V)\exp(-\frac{i}{\hbar}(\frac{t}{k})H_0)]^k.$$

It is well-known (and not difficult to prove) that

$$\exp(-\frac{i}{\hbar} tH_0)[f](x) = (2\pi i\frac{\hbar t}{m})^{-3/2}\int_{\mathbb{R}^3}\exp(\frac{im|x-y|^2}{2\hbar t}) f(y)d^3y,$$

whence

$$\exp(-\frac{i}{\hbar}\frac{t}{k}V)\exp(-\frac{i}{\hbar}\frac{t}{k}H_0)[f](x)$$

$$= (2\pi i\frac{\hbar}{m}\frac{t}{k})^{-3/2}\int_{\mathbb{R}^3}\exp(\frac{i}{\hbar}[\frac{m}{2}|x-x_0|^2(t/k)^{-2} - V(x)]\frac{t}{k})f(x_0)d^3x_0.$$

For $x_0, x_1, \ldots, x_k \in \mathbb{R}^3$, let

$$A_t(x_0,x_1,\ldots,x_k) := \sum_{j=1}^{k}[\frac{m}{2}|x_j-x_{j-1}|^2(\frac{t}{k})^{-2} - V(x_j)]\frac{t}{k}.$$

Taking $x_k = x$, the Kato-Trotter formula then yields

$$\psi(x,t) = \lim_{k\to\infty}(2\pi i\hbar t/km)^{-3k/2}\int_{\mathbb{R}^{3k}} e^{\frac{i}{\hbar}A_t(x_0,\ldots,x_k)} f(x_0)d^3x_0\ldots d^3x_{k-1}.$$

Let $\gamma: [0,t] \to \mathbb{R}^3$ be a path with $\gamma(jt/k) = x_j$ and $\gamma(s) = (x_{j+1} - x_j)(\frac{k}{t}s-j) + x_j$ for $s \in [jt/k, (j+1)t/k]$. Then the "classical action" for the path γ is

$$\int_0^t \frac{m}{2}|\gamma'(s)|^2 - V(\gamma(s))ds$$

$$= \sum_{j=0}^{k-1}\int_{jt/k}^{(j+1)t/k}\frac{m}{2}|x_{j+1}-x_j|^2(\frac{t}{k})^{-2} - V(\gamma(s))ds$$

$$\approx \sum_{j=1}^{k}[\frac{m}{2}|x_j-x_{j-1}|^2(\frac{t}{k})^{-2} - V(x_j)](\frac{t}{k})$$

$$= A_t(x_0,\ldots,x_k).$$

Now, integrating with respect to $x_1,\ldots,x_{k-1}$ is like integrating over all polygonal paths having $k-1$ segments starting at x_0 at time 0 and ending at some arbitrary point x at time t. As $k \to \infty$ the variety of these paths is sufficiently arbitrary so that we (as R. P. Feynman) are tempted to rewrite the formula as

$$\psi(x,t) = \int_{\mathbb{R}^3} \left[\int_{P_t[x_0,x]} e^{\frac{i}{\hbar} A(\gamma)} d\gamma\right] f(x_0) d^3x_0$$

where $P_t[x_0,x]$ is the space of all paths $\gamma\colon [0,t] \to \mathbb{R}^3$ with $\gamma(0) = x_0$ and $\gamma(t) = x$, and $A(\gamma)$ is the action of γ, and $d\gamma$ is some kind of "measure" on $P_t[x_0,x]$. In other words, the weight function for the operator $e^{-\frac{i}{\hbar} tH}$ is (formally, of course)

$$K(x_0,x,t) = \int_{P_t[x_0,x]} e^{\frac{i}{\hbar} A(\gamma)} d\gamma.$$

The integrand oscillates the least about a path for which $A(\gamma)$ is stationary among paths in $P_t[x_0,x]$; i.e., paths which are solutions of Newton's equation $m\gamma'' = -\nabla V$. Hence, we expect $K(x_0,x,t)$ to be most greatly influenced by a classical path γ such that $\gamma(0) = x_0$ and $\gamma(t) = x$. As $\hbar \to 0$, this effect becomes more pronounced, and presumably we obtain classical mechanics in the limit, since $|K(x_0,x,t)|^2$ is the probability density that at time t the particle will be found at x, given that it was at x_0 at time 0. While the path integral is essentially a suggestive formalism for the rigorous Kato-Trotter limit, Mark Kac noticed that if one replaces the time variable t by a pure-imaginary parameter $-i\tau$, then

$$\frac{i}{\hbar} A_t(x_0,\ldots,x_k) = -\sum_{j=1}^{k} [\tfrac{m}{2} |x_j-x_{j-1}|^2(\tfrac{\tau}{k})^{-2} + V(x_j)]\tfrac{\tau}{k},$$

and we can write the path integral rigorously as an integral in terms of Wiener measure. The power in the exponent is now the negative of the "Euclidean" action $\int_0^\tau \frac{1}{2} m|\gamma'(u)|^2 + V(\gamma(u))du$ (i.e., the sign in front of V is now $+$). The greatest contributions to $K(x_0,x,-i\tau)$ are thus expected to come from the paths that minimize the Euclidean action; these will be among the solutions of $m\gamma''(u) = +\nabla V(\gamma(u))$ which satisfy $\gamma(0) = x_0$ and $\gamma(\tau) = x$, if such exist. We now interpret $K(x_0,x,-i\tau)$, at least formally. Suppose that $V(x)$ increases rapidly enough as $|x| \to \infty$ so that there is a complete orthonormal set of eigenfunctions $u_0,u_1,u_2,\ldots$ of $H = -\frac{\hbar^2}{2m}\Delta + V$ with eigenvalues (energies) arranged in increasing order, say $E_0 \le E_1 \le E_2 \le \ldots$ (degeneracy allowed). The weight "function" (i.e., distribution) of $e^{-\frac{i}{\hbar} tH}$ is given by

$$K(x_0,x,t) = \sum_{n=0}^{\infty} e^{-\frac{i}{\hbar}tE_n} u_n(x)\overline{u_n(x_0)}.$$

Indeed,

$$\psi(x,t) = \int_{\mathbb{R}^3}\left[\sum_{n=0}^{\infty} e^{-\frac{i}{\hbar}tE_n} u_n(x)\overline{u_n(x_0)}\right]f(x_0)dx_0$$

solves Schrödinger's equation (at least formally), and $\psi(x,0)$ is the eigenfunction expansion of $f(x)$. Replacing t by $-i\tau$, we obtain

$$K(x_0,x,-i\tau) = \sum_{n=0}^{\infty} e^{-\frac{\tau}{\hbar}E_n} u_n(x)\overline{u_n(x_0)}.$$

Letting $\tau \to \infty$, we formally obtain (for some $k > 0$)

$$K(x_0,x,-i\tau) \sim e^{-\frac{\tau}{\hbar}E_0}\left[\sum_{E_n=E_0} u_n(x)\overline{u_n(x_0)} + O(e^{-kt})\right],$$

where the sum is only over lowest energy states. If x_0 is a minimum for V, then the constant path $\gamma\colon u \longmapsto x_0$ clearly minimizes the Euclidean action $\int_0^\tau \frac{1}{2}m|\gamma'(u)|^2 + V(\gamma(u))du$ for any τ. Hence, such paths are likely to make $K(x_0,x_0,-i\tau)e^{\frac{\tau}{\hbar}E_0} \sim \sum_{E_n=E_0}|u_n(x_0)|^2$ larger at minima x_0 of V than at other x_0. Indeed, we expect the position probability densities $|u_n|^2$ to be larger at minima. Our classical intuition leads us to suspect that if there are N absolute minima for V, then there ought to be an N-fold degeneracy in the lowest energy level (i.e., $E_n = E_0$ for $n = 0,\ldots,N-1$); we expect that there should be N independent eigenfunctions, each peaked at a different minimum. A simple example in dimension 1 (i.e., $x \in \mathbb{R}$) suffices to prove us *wrong:*

Let $V(x) = \frac{1}{2}(x^2-x_0^2)^2$, as shown.

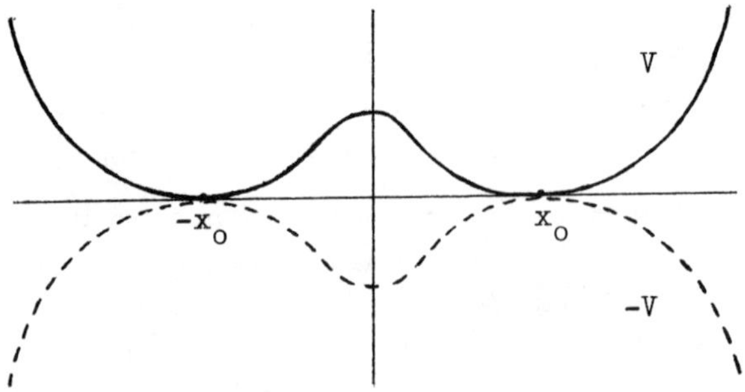

Note that the Schrödinger operator $H = \frac{-\hbar^2}{2m}\Delta + V$ commutes with the parity operator $(Pf)(x) = f(-x)$, and hence the eigenspaces of H split into even and odd functions ($+1$ and -1 eigenfunctions of P). Our intuition tells us that there ought to be two E_0-energy eigenfunctions, say ψ peaked at x_0 and $P\psi$ peaked at $-x_0$; $||\psi||^2 = ||P\psi||^2 = 1$. Let $\psi_e = (\psi + P\psi)/\sqrt{2}$ and $\psi_0 = (\psi - P\psi)/\sqrt{2}$ be the associated even and odd states. It then follows that for any $x \in \mathbb{R}$,

$$e^{\frac{\tau}{\hbar}E_0} K(-x,+x,-i\tau) \sim \psi_e(-x)\overline{\psi_e(x)} + \psi_0(-x)\overline{\psi_0(x)}$$

$$= |\psi_e(x)|^2 - |\psi_0(x)|^2.$$

Recall that $K(x_0,x,-i\tau)$ is the path integral of an exponential function with *real* exponent being $-\frac{1}{\hbar}(\int_0^\tau \frac{1}{2}m|\gamma'(u)|^2 + V(\gamma(u))du)$. Hence, undoubtedly $e^{\frac{\tau}{\hbar}E_0} K(-x,+x,-i\tau) \geq 0$, and we then have

$$0 \leq |\psi_e(x)|^2 - |\psi_0(x)|^2.$$

If the inequality is strict, even at a single point x, then we get the contradiction

$$0 < ||\psi_e||^2 - ||\psi_0||^2 = 1-1 = 0.$$

We argue that the inequality is strict near $x = x_0$, as follows. There is a sequence of classical paths for the Euclidean potential $-V$ (shown dashed in the above figure), say $x_k(u)$, $0 \leq u \leq \tau_k$ such that $x_k(0) \downarrow -x_0$, $x_k(\tau_k) \uparrow x_0$, and $\tau_k \to \infty$ as $k \to \infty$. The Euclidean actions are uniformly bounded as $k \to \infty$. Indeed they compare favorably with the action of the classical trajectory $x: (-\infty,\infty) \to (-x_0,x_0)$ obtained by letting a point slide down from atop the dashed hill at $-x_0$, eventually nearing $+x_0$ as $u \to \infty$. Physicists call such a trajectory an "instanton". The action of the instanton is

$$\int_{-\infty}^{\infty} \frac{1}{2}mx'(u)^2 + V(x(u))du = \int_{-\infty}^{\infty} 2V(x(u))du$$

$$= \int_{-x_0}^{x_0} 2V(x)\left(\frac{dx}{du}\right)^{-1}dx = \int_{-x_0}^{x_0} [2mV(x)]^{1/2}dx < \infty.$$

Since the actions of the paths x_k are uniformly bounded (essentially by that of the instanton) as $k \to \infty$, we suspect that the contribution of nearby paths to the path integral is significant enough so that as $\tau \to \infty$,

$$e^{\frac{\tau}{\hbar} E_0} K(-x_0,x_0,-i\tau) \sim \text{const.} > 0.$$

Thus, $|\psi_e(x_0)|^2 - |\psi_0(x_0)|^2 = \text{const.} > 0$ and we have a contradiction, as mentioned above. Hence one of ψ_e and ψ_0 is not a lowest energy E_0 eigenstate. In fact, ψ_0 is not, because otherwise we would have $e^{\frac{\tau}{\hbar} E_0} K(-x,x,-i\tau) \sim -|\psi_0(x)|^2$, in spite of the left side being non-negative. Indeed, explicit calculations establish that in the case at hand, the lowest energy level is nondegenerate with *even* eigenstate, as predicted by our argument. Physicists attribute the breaking of the degeneracy as a consequence of the existence of the instanton that connects the two minima $-x_0$ and x_0, leading to the positive contribution to $K(-x_0,x_0,-i\tau)$ (i.e., to the probability that the particle can "tunnel" through the potential barrier between $-x_0$ and x_0, having a positive probability of appearing about either point). The instanton has less effect as $\hbar \to 0$, and our classical intuition prevails in the limit.

Much of this discussion on the role of instantons in removing degeneracy and tunneling carries over, at least metaphorically, to quantum field theory. Since we will be primarily concerned with non-abelian gauge fields (e.g., where SU(2) is the gauge group), we confine ourselves to a very brief discussion of what instantons are in the context of the quantum theory of pure Yang-Mills fields and how they correspond to the ones we have discussed in relation to quantum mechanics. The configuration space in quantum mechanics is simply $\mathbb{R}^3$, but in the quantum theory of pure Yang-Mills fields, the configuration space is essentially the space of $\mathcal{SU}(2)$-valued vector potentials on $\mathbb{R}^3$, modulo gauge transformations. Of course, any SU(2)-bundle over $\mathbb{R}^3$ is just a product; hence a gauge transformation amounts to a function $\phi\colon \mathbb{R}^3 \to SU(2)$. We require that $\phi(x) \to \mathrm{Id} \in SU(2)$ as $x \to \infty$, and obtain a map $\phi'\colon S^3 \to SU(2) \cong S^3$ which is classified up up to homotopy by its degree. A gauge transformation ϕ is called homotopically trivial if ϕ' has degree zero. Now, more precisely, the configuration space consists of $\mathcal{SU}(2)$-valued vector potentials A on $\mathbb{R}^3$, modulo homotopically trivial gauge transformations; moreover, it is required that the field strengths F_A be square-integrable (i.e., as static Yang-Mills fields, they are required to have finite energy $\int_{\mathbb{R}^3} |F_A|^2 < \infty$). The functional $F \longmapsto \int_{\mathbb{R}^3} |F_A|^2$ plays the role of the potential energy function V in the quantum mechanical setting. Suppose $\phi_i : \mathbb{R}^3 \to SU(2)$ $(i = 1,2)$ are homotopically inequivalent (i.e., $\phi_1' \not\sim \phi_2'$). Then the vector potentials $A_i(x) := \phi_i^{-1}(x)d\phi_{ix}$ $(i = 1,2)$ are not equivalent in the configuration space. (Note $\phi_1' \cdot \phi_2'^{-1} \not\sim \mathrm{Id}.$), and yet both turn out to have vanishing field strength ($F_{A_1} = F_{A_2} = 0$). Thus, they represent different points that have zero "potential energy"; $\int_{\mathbb{R}^3} |F_{A_i}|^2 = 0$. As before, an instanton is then a curve (of points in the configuration space) which is parametrized by pure-imaginary time $i\tau$ $(-\infty < \tau < \infty)$, and absolutely minimizes the Euclidean Yang-Mills action (when we regard the curve as a single $SU(2)$-valued vector potential on Euclidean $\mathbb{R}^4$) among all curves connecting A_1 (as $\tau \to -\infty$) to A_2 (as $\tau \to +\infty$). Even up to gauge transformations (tending to Id at ∞) on $\mathbb{R}^4$, instantons are not unique for a given ϕ_1 and ϕ_2. In fact, the main thrust of the mathematics that follows is to prove (using the Atiyah-Singer Index Theorem) that the instantons connecting A_1 to A_2 form an 8k-3 dimensional family if $1 \leq k = |\deg \phi_1 - \deg \phi_2|$. It is important to know the dimension of the family so that full effect of the family (e.g., with regard to vacuum tunneling) has some fighting chance of being estimated. The interested reader may find some insight into this very tricky business in the contributions of Claude Bernard and E. Corrigan/P. Goddard to Volume 129 of Lecture Notes in Physics (Springer-Verlag) entitled "Geometrical and Topological Methods in Gauge Theories" (1980). Finally, it must be noted that the Lagrangians (actions) of other fields (e.g., fundamental fermionic Dirac type fields for quarks, electrons, etc.) must be added on to the pure self-actions of the gauge fields. The Atiyah-Singer Index Formula is also essential for treating the effects of these other fields (e.g., Euclidean fermionic lowest energy modes) on Green's functions of quantum field theory; e.g., see [A. S. Schwarz, 1979].

2. GEOMETRIC PRELIMINARIES

Here we collect some fundamental facts concerning the geometry and topology of fiber bundles that are essential to understanding gauge theories. In most recent journal articles on the subject, it is assumed that the reader knows this material or can readily dig it out from various sources. To cut down on the frustration, we develop the following topics, assuming a more modest background:

A. Principal G-Bundles

Let P be a manifold on which a Lie group G acts freely and smoothly to the right; without loss of generality, we may assume that G is a submanifold of the space of $N \times N$ complex (or real) matrices and the multiplication for G is matrix multiplication (e.g., $G = SU(N)$, $SO(N)$, etc.). Assume that the quotient space $M := P/G$ can be made into a manifold such that the projection $\pi: P \to M$ is smooth. Also suppose that P is locally a product in the following sense: For each $x \in M$, there is a neighborhood U of x in M and a diffeomorphism $T: \pi^{-1}(U) \to U \times G$ such that $T(pg) = (\pi(p), s(p))$, where $s: \pi^{-1}(U) \to G$ has the property $s(pg) = s(p)g$ for all $p \in \pi^{-1}(U)$ and $g \in G$.

If the conditions of the preceding paragraph hold, then we say that $\pi: P \to M$ is a *principal G-bundle*; P is called the *total space*; M is called the *base space*; and π is the *projection*. The diffeomorphism T is called a *local trivialization*. If there is a trivialization $T: \pi^{-1}(M) = P \to M \times G$ defined on all of P, then the G-bundle is called *trivial*. Let U be an open subset of M, and let $\sigma: U \to P$ be a map such that $\pi \circ \sigma = I_U$ (the identity on U). Then σ is called a *local section*. There is a one-to-one correspondence between local sections and trivializations. Indeed, given σ, define $T: \pi^{-1}(U) \to U \times G$ by $T(\sigma(x)g) = (x,g)$; and conversely given T, $\sigma: U \to P$ is uniquely defined by the same equation. It follows that a principal G-bundle is trivial precisely when it has a global section $\sigma: M \to P$. The corresponding statement for sphere bundles is false. For example, the unit tangent bundle, a Klein bottle, is nontrivial since the Klein bottle is nonorientable, but the Klein bottle does have a unit tangent vector field. This also shows that the unit tangent bundle of a Klein bottle cannot be made into a principal S^1 ($S^1 = U(1) = \{e^{i\theta}: \theta \in \mathbb{R}\}$)-bundle. Indeed, an orientation is precisely what is necessary in order to define a free S^1 action on the unit tangent bundle $S(M)$ of a Riemannian surface M; then $S(M) \to M$ will be trivial exactly when there is a unit tangent vector field on M.

B. Connections and Curvature

Connections on principal G-bundles can be defined in various ways. From a conceptual standpoint, the following definition is the best:

A *connection* on a principal G-bundle $\pi: P \to M$ smoothly assigns to each $p \in P$, a subspace H_p of the tangent space $(TP)_p$ such that $\pi_*: H_p \to (TM)_{\pi(p)}$ is an isomorphism. Moreover, if $R_g: P \to P$ is the right action of $g \in G$ on P (i.e., $R_g(p) = pg$), then it is required that $R_{g*}(H_p) = H_{pg}$ (i.e., the family H_p is R_g-invariant).

Now, one way of defining the various H_p would be to let them be the subspaces which are annihilated by a 1-form with values in a vector space V such that $\dim V = \dim P - \dim M = \dim G$. Fortunately, there is such a V that is quite natural, namely the Lie algebra $\mathcal{G}$ of G. Recall that the *Lie algebra* of G is the tangent space of G at $I \in GL(N,\mathbb{C})$; since $G \subset GL(N,\mathbb{C}) \subset M(N) \equiv$ vector space of $N \times N$ matrices, we can identify $\mathcal{G}$ with a linear subspace of $M(N)$. It turns out that $\mathcal{G}$ is closed under commutation (i.e., $A,B \in \mathcal{G}$ implies $[A,B] := AB - BA \in \mathcal{G}$); thus $\mathcal{G}$ is an algebra. For analytical purposes, the following definition of a connection, as a $\mathcal{G}$-valued 1-form, is very convenient:

A *connection (1-form)* for the G-bundle $\pi: P \to M$ is a $\mathcal{G}$-valued 1-form ω on P such that

1) For each $A \in \mathcal{G}$, we have $\omega(A^*) = A$, where A^* is the "fundamental vertical vector field" defined by $A^*_p = \frac{d}{dt} p\cdot exp(tA)\big|_{t=0}$.

2) For any $g \in G$, $p \in P$, and $X \in T_pP$, we have $\omega_{pg}(R_{g*}X) = g^{-1}\omega_p(X)g$; i.e., $R^*_g\omega = Ad_{g^{-1}}\omega$.

Note that the two definitions of connections are related in a very simple way. Namely, $H_p = \{X \in (TP)_p: \omega_p(X) = 0\}$. Condition 1) insures that $\pi_*: H_p \to (TX)_{\pi(p)}$ is an isomorphism (i.e., the kernel of ω_p does not intersect the fiber). Condition 2) guarantees that $R_{g*}(H_p) = H_{pg}$. Note that the "$Ad_{g^{-1}}$" is necessary in order that 2) be compatible with 1). Indeed, at $t = 0$, $R_{g*}(A^*_p) = \frac{d}{dt} p(exp\ tA)g = \frac{d}{dt} pgg^{-1}(exp\ tA)g = \frac{d}{dt} pg\ exp(tg^{-1}Ag) = (g^{-1}Ag)^*_{pg}$. Thus, $\omega_{pg}(R_{g*}A^*_p) = g^{-1}Ag = g^{-1}\omega_p(A^*_p)g$. Of course, when G is Abelian, "$Ad_{g^{-1}}$" can be dropped in 2), and ω is R_g invariant.

Given a connection ω for the G-bundle $\pi: P \to M$, we can decompose $X \in (TP)_p$ into horizontal and vertical parts, X^H and X^V respectively, where $\omega_p(X^H) = 0$, $\pi_*(X^V) = 0$, and $X = X^H + X^V$. If ϕ is a k-form on P with values in any vector space W, then we define a new k-form ϕ^H on P by

$$\phi^H(X_1,\ldots,X_k) = \phi(X_1^H,\ldots,X_k^H).$$

We define the *covariant derivative* $D^\omega\phi$ of ϕ relative to ω to be the W-valued (k+1)-form

$$D^\omega\phi := (d\phi)^H,$$

where d is the ordinary exterior derivative (which may be defined componentwise relative to a basis of W).

The *curvature* of the connection ω is simply

$$\Omega^\omega := D^\omega\omega.$$

Using conditions 1) and 2) for ω, one can prove (see [Bleecker 1981, 37-39] or [Kobayashi-Nomizu 1964, 77-78]) the "structural equation"

$$\Omega^\omega = d\omega + \frac{1}{2}[\omega,\omega],$$

where $[\omega,\omega](X,Y) := [\omega(X),\omega(Y)] - [\omega(Y),\omega(X)] = 2[\omega(X),\omega(Y)]$, the bracket being the commutator in G.

C. Equivariant Forms and Associated Bundles

Let $r: G \to GL(W)$ be a representation (i.e., homomorphism), where $GL(W)$ is the general linear group of a vector space W. For a G-bundle $\pi: P \to M$, there is a right action of G on $P \times W$ given by $(p,w)\cdot g = (pg, r(g^{-1})w)$. Let $[p,w]$ denote the orbit of (p,w), and let $P \times_G W$ be the set of all orbits. Let $\pi_W: P \times_G W \to M$ be given by $\pi_W([p,w]) = \pi(p)$. It is not difficult to verify that $\pi_W: P \times_G W \to M$ is a vector bundle. Note that any two points in the fiber above $\pi(p)$ have unique representatives of the form (p,w_1) and (p,w_2) and $[p,w_1] + [p,w_2] := [p,w_1+w_2]$. (It is a simple exercise to check that this addition is well defined.) Also, $P \times_G W$ is a manifold of dimension $\dim M + \dim W$, using the fact that the action of G on $P \times W$ is free. The fibers $\pi_W^{-1}(x)$ are isomorphic to W, but not canonically so, since the isomorphism $[p,w] \longmapsto w$ depends on the choice of $p \in \pi^{-1}(x)$. The vector bundle $\pi_W: P \times_G W \to M$ is called an *associated vector bundle of P* (relative to r).

Let $s: M \to P \times_G W$ be a section, and define $f: P \to W$ by the equation $s(\pi(p)) = [p,f(p)]$. Since $[p,f(p)] = s(\pi(p)) = s(\pi(pg)) =$

$[pg,f(pg)] = [p,r(g)f(pg)]$, we have $f(pg) = r(g)^{-1}f(p)$. Thus, we have an isomorphism $(s \longmapsto f)$

$$C^\infty(P \times_G W) \cong C(P,W) := \{f \in C^\infty(P,W) : f(pg) = r(g)^{-1}f(p)\}$$

$C(P,W)$ is called the space of *W-valued equivariant functions* on P. Whether one works with sections or equivariant functions, is largely a matter of taste or convenience.

We will need equivariant forms as well. Let $r: G \to GL(W)$ be a representation. We take $\Omega^k(P,W)$ to be the space of W-valued k-forms on P. The *space of equivariant W-valued k-forms on P*, denoted by $\overline{\Omega}^k(P,W)$, consists of all $\alpha \in \Omega^k(P,W)$ such that

1) $\alpha(X,\cdot,\ldots,\cdot) = 0$, if X is vertical (i.e., $\pi_*X = 0$).

2) $\alpha(R_{g*}X_1,\ldots,R_{g*}X_k) = r(g)^{-1}\alpha(X_1,\ldots,X_k)$ for all $g \in G$, and vector fields $X_1,\ldots,X_k$ (i.e., $R_g^*\alpha = r(g)^{-1}\alpha$).

Naturally, one takes $\overline{\Omega}^0(P,W)$ to be $C(P,W)$. One can verify that equivariance is preserved by covariant differentiation, namely $D^\omega: \overline{\Omega}^k(P,W) \to \overline{\Omega}^{k+1}(P,W)$. Moreover, there is a very convenient formula (see [Bleecker 1981, 44-45])

$$D^\omega\alpha = d\alpha + r'(\omega) \wedge \alpha$$

where $r': \mathcal{G} \to \mathcal{G}\ell(W)$ is the Lie algebra representation (i.e., derivative of r at I) and

$$(r'(\omega) \wedge \alpha)(X_1,\ldots,X_{k+1}) := \frac{1}{k!}\sum_\sigma (-1)^\sigma r'(\omega(X_{\sigma_1}))\alpha(X_{\sigma_2},\ldots,X_{\sigma_{k+1}}),$$

where σ runs over all permutations of $\{1,\ldots,k+1\}$.

The spaces of equivariant forms can be identified with spaces of forms on M with coefficients in the associated vector bundle. Indeed, suppose $Y_1,\ldots,Y_k$ are vector fields on M. The *horizontal lift* of Y_i is the unique vector field $\tilde{Y}_i$ on P such that $\omega(\tilde{Y}_i) = 0$ and $\pi_*\tilde{Y}_i = Y$. Since $R_{g*}\tilde{Y}_i = \tilde{Y}_i$, we have (for $\alpha \in \overline{\Omega}^k(P,W)$), $\alpha_{pg}(\tilde{Y}_1,\ldots,\tilde{Y}_k) = \alpha_{pg}(R_{g*}\tilde{Y}_1,\ldots,R_{g*}\tilde{Y}_k) = r(g)^{-1}\alpha_p(\tilde{Y}_1,\ldots,\tilde{Y}_k)$, whence the function $p \longmapsto \alpha_p(\tilde{Y}_1,\ldots,\tilde{Y}_k)$ is in $C(P,W) \cong C^\infty(P \times_G W)$. Thus, α yields a form $\overline{\alpha}$ on M with coefficients in $P \times_G W$ via the formula $\overline{\alpha}_{\pi(p)}(Y_1,\ldots,Y_k) = [p,\alpha_p(\tilde{Y}_1,\ldots,\tilde{Y}_k)] \in P \times_G W$. Indeed, we can make the identification $\overline{\Omega}^k(P,W) \cong \Omega^k(P \times_G W) :=$ space of k-forms on M with values in $P \times_G W$; the reader may check that this isomorphism is ω-independent. Moreover, we can regard D^ω as operating on $\Omega^k(P \times_G W)$, sending it to $\Omega^{k+1}(P \times_G W)$.

In order to define a formal adjoint to D^ω, we need to introduce some inner products. Let h be a Riemannian metric on M, and suppose k is an inner product on W for which $r: G \to GL(W)$ is orthogonal (i.e., $r(G) \subset O(W) := \{A \in GL(W): k(Aw_1, Aw_2) = k(w_1, w_2)$ for all $w_1, w_2 \in W\}$); such a k always exists if G is compact, by an averaging argument. Now, $C^\infty(P \times_G W)$ becomes a Riemannian vector bundle, since for $s, t \in C^\infty(P \times_G W)$ with associated $\sigma, \tau \in C(P, W)$, we have

$$\langle s,t\rangle_{\Pi(p)} := k(\sigma(p), \tau(p))$$

which is well defined, since σ and τ are equivariant and r is orthogonal. Hence, we have a pairing

$$\langle\ ,\ \rangle : C^\infty(P \times_G W) \times C^\infty(P \times_G W) \to C^\infty(M).$$

A pairing can also be defined for $\Omega^k(P \times_G W)$. Suppose $s \in C^\infty(P \times_G W)$ and $\beta \in \Omega^k(M, \mathbb{R})$. Then $s \otimes \beta \in \Omega^k(P \times_G W)$ is defined by $(s \otimes \beta)_x (Y_1, \ldots, Y_k) := \beta_x(Y_1, \ldots, Y_k) s(x)$. There is a natural pairing (induced by the Riemannian metric h) on $\Omega^k(M, \mathbb{R})$ given by the local formula

$$\langle\beta,\gamma\rangle = \frac{1}{k!} \beta^{i_1 \ldots i_k} \gamma_{i_1 \ldots i_k}$$

where $\gamma_{i_1 \ldots i_k} = \gamma(\partial_{i_1}, \ldots, \partial_{i_k})$ for local coordinate vector fields $\partial_{i_1}, \ldots, \partial_{i_k}$, and $\beta^{i_1 \ldots i_k} := h^{i_1 j_1} \ldots h^{i_k j_k} \beta_{j_1 \ldots j_k}$ where h^{ij} are the entries of the inverse of the matrix $(h_{ij}) = (h(\partial_i, \partial_j))$; repeated indices are summed from 1 to $\dim M$. Now, we define the pairing

$$\langle\ ,\ \rangle : \Omega^k(P \times_G W) \times \Omega^k(P \times_G W) \to C^\infty(M)$$

to be the unique one such that

$$\langle s \otimes \beta, t \otimes \gamma\rangle = \langle s,t\rangle\langle\beta,\gamma\rangle.$$

Rather than introducing spaces of sections with compact support, let us assume that M is compact. Then, we have the inner product $(\ ,\ )$ on $\Omega^k(P \times_G W)$ given by

$$(\alpha_1, \alpha_2) = \int_M \langle\alpha_1, \alpha_2\rangle |\nu| \in \mathbb{R},$$

where $|\nu|$ is the density on M relative to h (i.e., locally, $|\nu| = |\det(h_{ij})|^{1/2} dx_1 \ldots dx_n$, $n = \dim M$). We write

$$||\alpha||^2 := (\alpha,\alpha) \quad \text{and} \quad |\alpha|^2 := \langle\alpha,\alpha\rangle \in C^\infty(M).$$

Suitable modifications can easily be made to handle the case where W is complex with Hermitian scalar product k.

To explicitly construct a formal adjoint of $D^\omega\colon \Omega^k(P \times_G W) \to \Omega^{k+1}(P \times_G W)$, we need the (Hodge) *star operator* $*\colon \Omega^m(M, \mathbb{R}) \to \Omega^{n-m}(M, \mathbb{R})$ which is defined to be the unique linear map such that for $\alpha,\beta \in \Omega^m(M, \mathbb{R})$ $(m = 1,\ldots,n = \dim M)$ and M oriented with volume element ν relative to h, we have

$$\alpha \wedge *\beta = \langle\alpha,\beta\rangle\nu.$$

Of course, $*$ can be defined pointwise. Moreover, there is a local formula (proven in [Bleecker 1981, p. 5])

$$(*\beta)_{j_{m+1}\cdots j_n} = \frac{1}{m!}\,|\det(h_{ij})|^{1/2}\,\beta^{j_1\cdots j_m}\,\varepsilon_{j_1\cdots j_m j_{m+1}\cdots j_n}$$

where ε is antisymmetric in its indices, with $\varepsilon_{12\ldots n} = 1$ and $\nu(\partial_1,\ldots,\partial_n) > 0$. It is also proved that

$$*^2\beta := *(*\beta) = \operatorname{sign}(\det(h_{ij}))(-1)^{m(n-m)}\beta.$$

For a Riemannian (positive definite) h, $\operatorname{sign}(\cdot\cdot) = 1$, but for a Lorentzian h, we have $\operatorname{sign}(\cdot\cdot) = -1$. The distinction is important for $n = \dim M = 4$, where $*^2 = \mathrm{Id}$ on $\Omega^2(M)$ for h Riemannian, but $*^2 = -\mathrm{Id}$ on $\Omega^2(M)$ for h Lorentzian (i.e., of signature (1,3)). Consequently, one has a decomposition for Riemannian 4-manifolds:

$$\Omega^2(M, \mathbb{R}) = \Omega^2_+(M, \mathbb{R}) \oplus \Omega^2_-(M, \mathbb{R})$$

where $\Omega^2_\pm(M, \mathbb{R}) := \{\beta \in \Omega^2(M, \mathbb{R})\colon *\beta = \pm\beta\}$; $\Omega^2_+(M, \mathbb{R})$ (resp. $\Omega^2_-(M, \mathbb{R})$) is called the space of *self-dual* (resp. *anti-self-dual*) 2-forms. For Lorentzian M, there is a similar notion, but only after complexification (i.e., consider $\Omega^2(M,\mathbb{C})$ and $*\beta = \pm i\beta$). Unless otherwise indicated, we will stick to Riemannian metrics.

Of course, we can extend $*$ to the spaces $\Omega^m(P \times_G W) \cong C^\infty(P \times_G W) \otimes \Omega^m(M, \mathbb{R})$ via

$$*(s \otimes \beta) = s \otimes (*\beta).$$

Moreover, for $\dim M = 4$, we can decompose $\Omega^2(P \times_G W)$ into self-dual and anti-self-dual parts. One should note that Ω_+ and Ω_- are reversed if the orientation of M is changed (i.e., $\nu \rightsquigarrow -\nu$). Moreover, $*$ is "conformally invariant" on $\Omega^2(P \times_G W)$ for $\dim M = 4$, since the local formula for $*$ given above is unchanged in this case, if h is replaced by λh, $\lambda \in C^\infty(M, \mathbb{R}^+)$; note that index raising involves multiplying by h^{ij} which is changed to $\lambda^{-1}h^{ij}$ for λh.

Proposition 1. The formal adjoint of $D^\omega: \Omega^m(P \times_G W) \to \Omega^{m+1}(P \times_G W)$ on compact, oriented, Riemannian n-manifold M is the *covariant codifferential* $\delta^\omega := -(-1)^{n(m+1)} *D^\omega*$. In other words, for $\alpha \in \Omega^m(P \times_G W)$ and $\alpha' \in \Omega^{m+1}(P \times_G W)$, we have $(D^\omega\alpha, \alpha') = (\alpha, \delta^\omega\alpha')$.

Proof: By linearity, it suffices to prove the result in the case $\alpha = s \otimes \beta$, $\alpha' = s' \otimes \beta'$, where $s, s' \in C^\infty(P \times_G W)$, $\beta \in \Omega^m(M, \mathbb{R})$, $\beta' \in \Omega^{m+1}(M, \mathbb{R})$. Note that $\gamma := \langle s,s'\rangle(\beta \wedge *\beta') \in \Omega^{n-1}(M, \mathbb{R})$, whence $\int_M d\gamma = 0$ by Stokes' Theorem; a short (10-line) proof of Stokes' Theorem for closed manifolds (without boundary) is found in [Bleecker 1981, 12-13]. We are done, if we prove $d\gamma = (\langle D^\omega\alpha,\alpha'\rangle - \langle\alpha,\delta^\omega\alpha'\rangle)\nu$. For this, we "simply" compute:

$$d\gamma = d(\langle s,s'\rangle)\beta \wedge *\beta' + \langle s,s'\rangle(d\beta \wedge *\beta' + (-1)^m\beta \wedge d*\beta')$$

$$= (\langle D^\omega s(\cdot),s'\rangle \wedge \beta + \langle s,s'\rangle d\beta) \wedge *\beta'$$

$$+ (\langle s,D^\omega s'(\cdot)\rangle \wedge \beta \wedge *\beta' + (-1)^m\langle s,s'\rangle \wedge d*\beta').$$

The first term is $\langle D^\omega s(\cdot) \wedge \beta + s \otimes d\beta,\ s' \otimes \beta'\rangle\nu = \langle D^\omega(s \otimes \beta),\ s' \otimes \beta'\rangle\nu$, while the second is

$$(-1)^m\beta \wedge \langle s,D^\omega s'(\cdot)\rangle \wedge *\beta' + (-1)^{mn}\langle s,s'\rangle\ \beta \wedge *^2 d*\beta'$$

$$= (-1)^{mn}(\beta \wedge **[\langle s,D^\omega s'(\cdot)\rangle \wedge *\beta'] + \langle s,s'\rangle\beta \wedge **d*\beta')$$

$$= (-1)^{mn}\langle\beta \otimes s,\ *[D^\omega s' \wedge *\beta' + s' \otimes d*\beta']\rangle\nu$$

$$= (-1)^{mn}\langle\beta \otimes s,\ *D^\omega*(s' \otimes \beta')\rangle\nu$$

$$= -\langle\beta \otimes s,\ \delta^\omega(s' \otimes \beta')\rangle\nu, \text{ as needed.} \quad \square$$

Remark. In order to obtain formulas for D^ω and δ^ω in local coordinates, let $\sigma: U \to P$ be a local section, and suppose $\alpha \in \Omega^m(P \times_G W)$. For each $x \in U$, we have an isomorphism $(P \times_G W)_x \to W$ given by $[\sigma(x),w] \longmapsto w$. This yields an isomorphism $\Omega^m(P \times_G W|U) \cong \Omega^m(U,W)$ which we designate by $\alpha \longmapsto \tilde{\alpha}$. One can easily check that if $\overline{\alpha} \in \overline{\Omega}^m(\pi^{-1}(U),W)$ is the equivariant form corresponding to α, then $\tilde{\alpha} = \sigma^*(\overline{\alpha})$. Using the above mentioned formula $D^\omega\overline{\alpha} = d\overline{\alpha} + r'(\omega) \wedge \overline{\alpha}$, and the fact that $(*\alpha)^\sim = *\tilde{\alpha}$, we obtain

$$(D^\omega\alpha)^\sim = \sigma^*(D^\omega\bar{\alpha}) = \sigma^*(d\bar{\alpha} + r'(\omega)\wedge\bar{\alpha})$$
$$= d\sigma^*\bar{\alpha} + r'(\sigma^*\omega)\wedge\sigma^*\bar{\alpha}$$
$$= d\tilde{\alpha} + r'(\sigma^*\omega)\wedge\tilde{\alpha}.$$

Writing $\tilde{\alpha} = \frac{1}{m!}\Sigma\,\tilde{\alpha}_{i_1\ldots i_m}\,dx^{i_1}\wedge\ldots\wedge dx^{i_m}$ in local coordinates, where antisymmetry is assumed in $\tilde{\alpha}_{i_1\ldots i_m}$, we obtain

$$(D^\omega\alpha)^\sim_{j_1\ldots j_{m+1}} = \sum_k (-1)^{k+1}[\partial_{j_k}(\alpha_{j_1\ldots\hat{j}_k\ldots j_{m+1}}) + r(\sigma^*\omega)_{j_k}\alpha_{j_1\ldots\hat{j}_k\ldots j_{m+1}}].$$

For $\delta^\omega\alpha$, we have

$$-(-1)^{n(m+1)}(\delta^\omega\alpha)^\sim = (*D^\omega*\alpha)^\sim = *(D^\omega*\alpha)^\sim$$
$$= *(d*\tilde{\alpha} + r(\sigma^*\omega)\wedge*\tilde{\alpha}).$$

Now, it is straightforward to check that if b is a 1-form and c is a m-form, then

$$*(b\wedge*c)_{i_1\ldots i_{m-1}} = (-1)^{n(m-1)}b^i c_{i\,i_1\ldots i_{m-1}}$$

Thus, we have (since applying "d" amounts to wedging with the "1-form" $\Sigma\,\partial_i(\cdot)dx^i$, except that ∂_i does not commute with the function $|h|^{1/2} \equiv |\det(h_{ij})|^{1/2}$)

$$(\delta^\omega\alpha)^\sim_{i_1\ldots i_{m-1}} = -|h|^{-1/2}\,\partial_i(|h|^{1/2}\,\tilde{\alpha}^i_{i_1\ldots i_{m-1}}) - r(\sigma^*\omega_i)\tilde{\alpha}^{i\,i_1\ldots i_{m-1}}.$$

D. Gauge Transformations

A *gauge transformation* of a principal G-bundle $\pi: P \to M$ is a diffeomorphism $F: P \to P$ such that

(i) $F(pg) = F(p)g$

(ii) $\pi(F(p)) = \pi(p)$.

Note that condition (ii) says that fibers are mapped into themselves. If (ii) is dropped, (i) still implies that there is a map $f: M \to M$ such that $\pi(F(p)) = f(\pi(p))$; F is called an *automorphism* of P in this more general case. We denote the group of automorphisms of P by $\mathrm{Aut}(P)$ and the subgroup of gauge transformations by $GA(P)$.

$\mathrm{Aut}(P)$ acts on the space $C(P)$ of connections on P and the spaces $\bar{\Omega}^k(P,W)$ of horizontal equivariant W-valued forms on P (relative to a representation $r: G \to GL(W))$ via pull-back:

$$F\cdot\alpha := (F^{-1})^*\alpha, \quad F \in \mathrm{Aut}(P), \quad \alpha \in \begin{cases} C(P) \\ \overline{\Omega}^k(P,W) \end{cases}$$

To prove $F\cdot\alpha \in C(P)$ for $\alpha \in C(P)$, note that F^{-1} preserves the vertical fields A^* for $A \in \mathcal{G}$ (since $F^{-1}(p \exp tA) = F^{-1}(p) \exp tA)$ whence $(F\cdot\alpha)(A^*) = \alpha(F_*^{-1}A^*) = \alpha(A^*) = A$; moreover, since $F^{-1}\circ R_g = R_g\circ F^{-1}$, we have

$$R_g^*F^{-1*}\alpha = F^{-1*}R_g^*\alpha = F^{-1*}(g^{-1}\alpha g) = g^{-1}F^{-1*}\alpha g.$$

It is also easy to prove that $F^{-1}\cdot\alpha \in \overline{\Omega}^k(P,W)$ for $\alpha\in\overline{\Omega}^k(P,W)$.

Recall (Section C) that the spaces $\overline{\Omega}^k(P,W)$ have a pairing $\langle,\rangle$, if the rep. $r: G \to GL(W)$ is orthogonal relative to an inner product on W and a metric tensor h is given on M. This pairing is invariant under $GA(P)$. Indeed, for $F \in GA(P)$, $p \in P$, and $s \in \overline{\Omega}^0(P,W)$, we have $|s(F^{-1}(p))| = |s(pg)| = |r(g^{-1})s(p)| = |s(p)|$; since each $\alpha \in \overline{\Omega}^k(P,W)$ is a sum of terms of the form $s \otimes \pi^*\beta$, $\beta \in \Omega^k(M)$, the result holds for $k > 0$ too, because $|F^{-1*}(s \otimes \pi^*\beta)|^2 = |s\circ F^{-1} \otimes F^*\pi^*\beta|^2 = |s\circ F^{-1} \otimes \pi^*\beta|^2 = |s\circ F^{-1}|^2\,|\beta|^2 = |s|^2|\beta|^2 = |s \otimes \pi^*\beta|^2$ (recall $\pi\circ F = \pi \Rightarrow F^*\circ\pi^* = \pi^*$). In particular, $GA(P)$ acts by isometries on the pre-Hilbert spaces $\overline{\Omega}^k(P,W) \cong \Omega^k(P \times_G W)$.

Since we have taken G to be a Lie group of $N \times N$ matrices, we may consider the adjoint action of G on itself (i.e., $g\cdot g_0 = gg_0g^{-1})$ to be a representation of the form $\mathrm{Ad} : G \to GL(\mathrm{End}(\mathbb{C}^N))$; this is in contrast to the adjoint representation of G on $\mathcal{G}$ $(g\cdot A = g^{-1}Ag)$ $\mathcal{A}d$: $G \to GL(\mathcal{G}) \subset GL(\mathrm{End}(\mathbb{C}^N))$. We let $C(P,G)$ be the set of all $f: P \to G$ such that $f(pg) = g^{-1}f(p)g$; i.e., $C(P,G)$ is the subset of $\Omega^0(P,\mathrm{End}(\mathbb{C}^N))$ (relative to Ad) consisting of Ad-equivariant functions with values in $G \subset \mathrm{End}(\mathbb{C}^N)$. Note that $C(P,G)$ is a group via $(f_1\cdot f_2)(p) = f_1(p)f_2(p)$.

<u>Proposition 1</u>. For $f \in C(P,G)$, define $\Phi(f): P \to P$ by $\Phi(f)(p) = pf(p)$. Then $\Phi: C(P,G) \to GA(P)$ is an isomorphism of groups.

<u>Proof</u>: Note $\Phi(f)(pg) = pgf(pg) = pgg^{-1}f(p)g = \Phi(f)(p)g$, whence $\Phi(C(P,G)) \subset GA(P)$; Φ^{-1} clearly exists too. Also $\Phi(f_1\cdot f_2)(p) = p[f_1(p)f_2(p)] = \Phi(f_1)(p)\cdot f_2(p) = \Phi(f_1)(p\cdot f_2(p)) = \Phi(f_1)(\Phi(f_2)(p)) = \Phi(f_1)\circ\Phi(f_2)(p)$. □

There is a map $\mathrm{Exp}: C(P,\mathcal{G}) \to C(P,G)$, given simply by $\mathrm{Exp}(s)(p) = \exp(s(p)) = \Sigma_{n=0}^{\infty} \frac{1}{n!} s(p)^n$; note $s(p) \in \mathcal{G} \subset \mathrm{End}(\mathbb{C}^N)$. Consequently, we also have a map $\Phi\circ\mathrm{Exp}: C(P,\mathcal{G}) \to GA(P)$. Just as $\exp: \mathcal{G} \to G$ is a local diffeomorphism from a neighborhood of $0 \in \mathcal{G}$ to one of $I \in G$, $\Phi \circ \mathrm{Exp}$ yields a natural bijection of a neighborhood of $0 \in C(P,\mathcal{G})$ to a neighborhood of $\mathrm{Id} \in GA(P)$, say relative to C^0-topologies. Note that for $s \in C(P,\mathcal{G})$, we have $\Phi(\mathrm{Exp}(ts))(p) = p \exp(ts(p))$ for $t \in \mathbb{R}$. Since $GA(P)$ acts on the linear space $\overline{\Omega}^k(P,W)$, it makes sense to take the derivative (where $\alpha \in \overline{\Omega}^k(P,W)$)

$$s\cdot\alpha := \frac{d}{dt}\left[\Phi(\mathrm{Exp}\ ts)\cdot\alpha\right]_{t=0} \in \overline{\Omega}^k(P,W).$$

Indeed, $C(P,\mathcal{G})$ may be regarded as the "Lie algebra" of the infinite-dimensional "Lie group" $GA(P)$, and $s \longmapsto (\alpha \longmapsto s\cdot\alpha) \in \mathrm{End}(\overline{\Omega}^k(P,W))$ is the "Lie algebra representation" for the "Lie group representation" $GA(P) \to GL(\overline{\Omega}^k(P,W))$.

Proposition 2. For $\alpha \in \overline{\Omega}^k(P,W)$, relative to a representation $r\colon G \to GL(W)$, and for $s \in C(P,\mathcal{G})$, we have

$$(s\cdot\alpha)_p = r'(s(p))\circ\alpha_p,$$

where $r'\colon \mathcal{G} \to \mathrm{End}(W)$ is the Lie algebra representation of r.

Proof: Without loss of generality, let $\alpha = u \otimes \pi^*\beta$ for $u \in C(P,W)$ and $\beta \in \Omega^k(M)$. At $t = 0$,

$$\begin{aligned}
(s\cdot\alpha)_p &= \frac{d}{dt}[\Phi(\mathrm{Exp}\ ts)\cdot\alpha]_p \\
&= \frac{d}{dt}[\Phi(\mathrm{Exp}\ ts)^{-1*}\ u \otimes \pi^*\beta]_p \\
&= \frac{d}{dt}[u(p\ \exp(-ts(p))) \otimes \pi^*\beta_p] \\
&= \frac{d}{dt}[r(\exp(ts(p)))u(p) \otimes \pi^*\beta_p] \\
&= r'(s(p))\circ\alpha_p. \quad \square
\end{aligned}$$

Proposition 3. For $\omega \in C(P)$ and $f \in C(P,G) \subset C(P, \mathrm{End}(\mathbb{C}^N))$, we have

$$f^{-1}\cdot\omega := \Phi(f)^*\omega = f^{-1}df + f^{-1}\omega f$$

or more precisely, for $X \in T_pP$,

$$(f^{-1}\cdot\omega)(X) = f(p)^{-1}df_p(X) + f(p)^{-1}\omega(X)f(p).$$

Proof: Let $\gamma\colon \mathbb{R} \to P$ be a curve with $\gamma'(0) = X$. At $t = 0$, we have

$$\begin{aligned}\Phi(f)_*(X) &= \frac{d}{dt}\Phi(f)(\gamma(t)) = \frac{d}{dt}\gamma(t)f(\gamma(t)) \\ &= \frac{d}{dt}pf(\gamma(t)) + \frac{d}{dt}\gamma(t)f(p) \\ &= \frac{d}{dt}pf(p)f(p)^{-1}f(\gamma(t)) + \frac{d}{dt}R_{f(p)}(\gamma(t)) \\ &= [f^{-1}(p)df_p(X)]^*_{pf(p)} + R_{f(p)*}(X).\end{aligned}$$

Note that $f^{-1}(p)f(\gamma(t))$ is a curve in G through the identity, whence $f^{-1}(p)df_p(X) \in \mathcal{G}$. Thus

$$\begin{aligned}(f^{-1}\cdot\omega)(X) &= \omega(\Phi(f)_*X) \\ &= f^{-1}(p)df_p(X) + \omega(R_{f(p)*}X) \\ &= f^{-1}(p)df_p(X) + f(p)^{-1}\omega(X)f(p). \qquad \square\end{aligned}$$

Corollary. For $s \in C(P,\mathcal{G})$ and $\omega \in \mathcal{C}(P)$, we have at $t = 0$

$$s\cdot\omega := \frac{d}{dt}[(\operatorname{Exp} ts)\cdot\omega] = ds + [\omega,s] = D^{\omega}s.$$

Proof: Let $f = \operatorname{Exp} ts$ in Prop. 3, and differentiate with respect to t at $t = 0$. Note that $\frac{d}{dt} f^{-1}df = \frac{d}{dt}\operatorname{Exp}(-ts)d\operatorname{Exp}(ts) = \frac{d}{dt}(I-ts)(t\,ds) = ds$ at $t = 0$. $\square$

E. Curvature in Riemannian Geometry

Let M be a C^∞ n-manifold. A *linear frame* of M at $x \in M$ is an isomorphism $u\colon \mathbb{R}^n \to (TM)_x$; note that $u(e_1),\ldots,u(e_n)$ is a basis of $(TM)_x$. Let $L(M)_x$ be the set of all linear frames at x, and set $L(M) = \bigcup_{x\in M}\{L(M)_x\}$. For $A \in GL(\mathbb{R}^n)$, we have $u\circ A \in L(M)_x$ for any $u \in L(M)_x$. Indeed $L(M)$ admits a natural differentiable structure such that $\pi\colon L(M) \to M$ $(\pi^{-1}(x) = L(M)_x)$ is a principal $GL(\mathbb{R}^n)$-bundle; it is called the *bundle of linear frames*. The reader is most likely familiar with tensor representations of $GL(\mathbb{R}^n)$. For example, let $S^2(\mathbb{R}^n)$ be the vector space of symmetric bilinear forms on $\mathbb{R}^n$. We have the tensor representation $r\colon GL(\mathbb{R}^n) \to GL(S^2(\mathbb{R}^n))$ given by $r(A)(s)(v_1,v_2) = s(A^{-1}v_1,A^{-1}v_2)$ for $v_1,v_2 \in \mathbb{R}^n$, $s \in S^2(\mathbb{R}^n)$. The reader can easily check that the associated vector bundle $L(M) \times_{G_n} S^2(\mathbb{R}^n)$ $(G_n \equiv GL(\mathbb{R}^n))$ is the bundle of symmetric bilinear forms on the tangent spaces of M (e.g., $[u,s](X,Y) = s(u^{-1}(X),u^{-1}(Y))$ for $u \in (LM)_x$, $s \in S^2(\mathbb{R}^n)$, and $X,Y \in (TM)_x$). In the case of the "defining" representation $GL(\mathbb{R}^n) \to GL(\mathbb{R}^n)$

(where $A \mapsto A$), we have $L(M) \times_{G_n} \mathbb{R}^n \cong TM$ via $[u,v] \mapsto u(v)$, $u \in L(M)$, $v \in \mathbb{R}^n$. The *fundamental 1-form* on $L(M)$ is the element $\phi \in \overline{\Omega}^1(L(M), \mathbb{R}^n)$ (see Section C for notation) defined by

$$\phi_u(X) = u^{-1}(\pi_* X), \quad u \in LM, \quad X \in (TL(M))_u.$$

We know that $\overline{\Omega}^1(L(M), \mathbb{R}^n)$ is isomorphic to $\Omega^1(L(M) \times_{G_n} \mathbb{R}^n) \cong \Omega^1(TM) =$ the space of 1-forms on M with coefficients in the tangent bundle; in other words $\overline{\Omega}^1(L(M), \mathbb{R}^n)$ is isomorphic to the space of endomorphisms of TM. It turns out that ϕ corresponds to the identity endomorphism; indeed, for $u \in L(M)$, $Y \in (TL(M))_u$, we have $\overline{\phi}_x(\pi_* Y) := [u, \phi_u(Y)] = [u, u^{-1}(\pi_* Y)] \mapsto u(u^{-1}\pi_* Y) = \pi_* Y$ for $Y \in (TL(M))_u$.

A connection ω for the $GL(\mathbb{R}^n)$-bundle $\pi: L(M) \to M$ is called a *linear connection*. The *torsion* of ω is $D^\omega \phi \in \Omega^2(L(M), \mathbb{R}^n) \cong \Omega^2(TM) =$ space of 2-forms on M with coefficients in TM (i.e., tangent-vector-valued two-forms). Now, suppose h is a Riemannian metric on M. Then $u \in L(M)$ is called an *orthonormal frame* if $u: \mathbb{R}^n \to (TX)_x$ is an isometry (i.e., $h(u(v),u(w)) = v \cdot w = v_1 w_1 + \ldots + v_n w_n$). The set $F(M)$ of all orthonormal frames is a principal $O(n)$-bundle called the *orthonormal frame bundle* of M relative to h. If ω is a linear connection on $L(M)$ whose horizontal subspaces at points of $F(M)$ are tangent to $F(M)$, then ω is called a *metric connection* relative to h. It is easily seen that ω is metric precisely when the restriction $\omega|F(M)$ is a connection on $F(M)$; in this case, $\omega|F(M)$ is $\mathcal{O}(n)$-valued.

Fundamental Lemma. For each Riemannian metric h on M, there is a unique linear connection on $L(M)$ which is metric relative to h and has vanishing torsion.

Remark. The proof can be found in almost any book on differential geometry. The one in [Bleecker 1981, 79-80] is closest to the notation here. The connection of the Lemma is called the *Levi-Civita connection* of h, and we denote its restriction to $F(M)$ by θ; there is no harm in calling θ the Levi-Civita connection too, since θ extends uniquely to the corresponding Levi-Civita connection on $L(M)$.

The curvature of the connection θ on the $O(n)$-bundle $F(M) \to M$ is denoted by $\Omega^\theta := D^\theta \theta = d\theta + \frac{1}{2}[\theta,\theta] \in \overline{\Omega}^2(F(M), \mathcal{O}(n))$. Here $\mathcal{O}(n) := \{A \in \mathrm{End}(\mathbb{R}^n) : A^T = -A\}$ is the Lie algebra of $O(n)$. Now $\mathcal{O}(n)$ is naturally isomorphic to $\Lambda^2(\mathbb{R}^n) :=$ the vector space of skew-symmetric bilinear forms on $\mathbb{R}^n$ via $A \mapsto A'$ where $A'(v,w) = v \cdot Aw$. This means that $\Omega^\theta \in \overline{\Omega}^2(F(M),\mathcal{O}(n))$ can be identified with a two-form, say $\tilde{\Omega}^\theta(\cdot,\cdot)$, with values in the bundle of skew-symmetric (relative to h) endomorphisms of TM, or with values in the bundle of exterior 2-forms. Using the latter interpretation, the covariant 4-tensor R on M is defined by

$$R(X,Y,Z,W) := \tilde{\Omega}^\theta(X,Y)(Z,W)$$

for X,Y,Z,W vector fields on M. R is called the *Riemann-Christoffel curvature tensor* of h. Since $\tilde{\Omega}^\theta$ is a (2-form)-valued 2-form, we know that R(X,Y,Z,W) is skew-symmetric in X and Y, and in Z and W. Because of the vanishing torsion of the Levi-Civita connection, one can prove the so-called "first Bianchi identity"

$$R(X,Y,Z,W) + R(X,W,Y,Z) + R(X,Z,W,Y) = 0.$$

which (for our notation) is proved in [Bleecker 1981, 115]. As a consequence of the above identities, we have yet another [loc. cit.]

$$R(X,Y,Z,W) = R(Z,W,X,Y).$$

The *Ricci curvature* of h is the contraction of R in the first and third slots; i.e., $\mathrm{Ricc}_x(Y,W) = \sum_i R(E_i,Y,E_i,W)$, where $E_1,\ldots,E_n$ is an orthonormal basis of $(TM)_x$. In terms of components relative to a coordinate system, the Ricci tensor is written as $R_{ij} := R^k_{ikj}$. Using $R(X,Y,Z,W) = R(Z,W,X,Y)$, we see that Ricc is a symmetric 2-tensor. The trace of Ricc is the *scalar curvature*, denoted by $S := R^i_i$ in local coordinates.

We will need to study the space $\mathcal{R}$ of covariant 4-tensors on $\mathbb{R}^n$ that satisfy the identities

1) $R_{hijk} = -R_{ihjk} = R_{ihkj}$

2) $R_{hijk} + R_{hkij} + R_{hjki} = 0.$

It is not hard to check that this vector space $\mathcal{R}$ of "curvature tensors" has dimension $\frac{1}{2}\frac{n(n-1)}{2}(\frac{n(n-1)}{2}+1) - \binom{n}{4} = \frac{1}{12}n^2(n^2-1)$; recall that (1) and (2) imply $R_{hijk} = R_{jkhi}$ and note that (2) only adds an additional constraint when h,i,j,k are distinct.

Let $r: \mathcal{R} \to \mathcal{S}^2$ (:= space of symmetric tensors) be the "Ricci map" $r(R)_{ik} = R^j_{ijk}$ and let $s: \mathcal{R} \to \mathbb{R}$ be the "scalar map" $s(R) = r(R)^i_i = R^{ji}_{ji}$. There is a bilinear symmetric map $v: \mathcal{S}^2 \times \mathcal{S}^2 \to \mathcal{R}$ given by

$$4(P \vee Q)_{hijk} = P_{hj}Q_{ik} - P_{ij}Q_{hk} + P_{ik}Q_{hj} - P_{hk}Q_{ij}.$$

The diligent reader will check that the Bianchi identity (2) holds for PvQ.

Every $R \in \mathcal{R}$ can be written in the form $R = R_1 + R_2 + R_3$ where (if $I = (\delta_{ij})$ and $n \geq 3$)

$$R_1 = \frac{2s(R)}{n(n-1)} (IvI)$$

$$R_2 = \frac{4}{n-2} (r(R)vI) - \frac{4s(R)}{n(n-2)} (IvI)$$

$$R_3 = R - R_1 - R_2 = R - \frac{4}{n-2}(r(R)vI) + \frac{2s(R)}{(n-2)(n-1)}(IvI).$$

The significance of this decomposition is seen as follows. One can easily check that the adjoint of the Ricci map $r: \mathcal{R} \to S^2$ is the composition $r^*: S^2 \to S^2 \times \{I\} \xrightarrow{v} \mathcal{R}$ (i.e., $\langle r(R),P\rangle = \langle R,PvI\rangle$). Now $\mathcal{R}$ has the orthogonal decomposition $\mathcal{R} = \mathrm{Im}\, r^* \oplus \mathrm{Ker}\, r$. Clearly, $R_1 + R_2 \in \mathrm{Im}\, r^*$ and R_1 lies in the $O(n)$-invariant subspace of $\mathcal{R}$ spanned by IvI. Using the easily derived formula $\langle PvI,P'vI\rangle = \frac{1}{4}\, \mathit{tr}(P)\mathit{tr}(P') + \frac{1}{4}(n-2)\langle P,P'\rangle$, it may be verified that $\langle R_1,R_2\rangle = 0$. Thus, R_2 lies in the $O(n)$-invariant subspace of $\mathrm{Im}(r^*)$ orthogonal to IvI. Moreover, using the formula $r(PvI) = \frac{1}{4}[\mathit{tr}(P)I + (n-2)P]$, it is easy to check that $r(R_3) = 0$. Thus, R_1, R_2 and R_3 are the components of R that lie in certain orthogonal invariant subspaces of $\mathcal{R}$. Noting that $s(PvI) = \mathit{tr}(PvI) = \frac{1}{2}(n-1)\mathit{tr}(P)$, we see $s(R_2) = 0$. Since $s(R_3) = 0$ as well, the one-dimensional space spanned by IvI is $(\mathrm{Ker}\, s)^{\perp}$ and our decomposition of $\mathcal{R}$ is $\mathcal{R} = \mathcal{R}_1 \oplus \mathcal{R}_2 \oplus \mathcal{R}_3 := (\mathrm{Ker}\, s)^{\perp} \oplus (\mathrm{Ker}\, s \cap \mathrm{Ker}(r)^{\perp}) \oplus \mathrm{Ker}(r)$. Although, we will not need to know that $\mathcal{R}_1$, $\mathcal{R}_2$, and $\mathcal{R}_3$ are irreducible $O(n)$-invariant subspaces, this follows from the fact that $\mathcal{R}$ is irreducible ~~as~~ as a $GL(\mathbb{R},n)$-module (corresponding to the Young diagram $\begin{array}{|c|c|}\hline 1 & 3 \\ \hline 2 & 4 \\ \hline\end{array}$) and r and s are the only independent contractions of $\mathcal{R}$ (cf. [H. Weyl p. 153 ff.]).

Since our decomposition is $O(n)$-invariant, the curvature tensor of a Riemannian manifold admits a corresponding decomposition given by the same formulas. The various parts of $R = R_1 \oplus R_2 \oplus R_3$ have names:

R_1 = constant-curvature part

R_2 = traceless-Ricci part

R_3 = W = Weyl conformal curvature tensor.

R_1 gets its name as follows: The Gaussian curvature of the surface in M obtained by taking the union of all geodesics which are tangent to a 2-dimensional subspace V of $(TM)_x$ is $R(X,Y,X,Y)$, where X and Y are orthogonal unit vectors in V. In the case where $R_2 = R_3 = 0$, we have $R(X,Y,X,Y) = \frac{s(R)}{n(n-1)}$ which is independent of the choice of V in $(TM)_x$ (i.e., constant). R_2 gets its name from the fact that it is completely determined by the Ricci tensor of R, and its own trace $s(R_2)$ vanishes. $R_3 = W$ has the property that the associated (1,3)-tensor $W^\# = (W^h{}_{ijk})$ is invariant under a conformal change of metric $h \longmapsto e^{2\sigma}h$, $\sigma \in C^\infty(M)$. Indeed, if the $\overline{R}$ is the curvature of $e^{2\sigma}h$, then a somewhat tedious computation (see [Eisenhart 1966, §28]) yields

$$R^\# - \overline{R}^\# = 4(\tilde{\sigma} v h)^\# + 2|\nabla\sigma|^2(hvh)^\#$$

where $\tilde{\sigma} := \nabla^2\sigma - d\sigma \otimes d\sigma$ is a symmetric tensor ($\nabla^2\sigma :=$ double covariant derivative of σ = Hessian of σ with respect to h). Since the analogue of $I = (\delta_{ij})$ on $\mathbb{R}^n$ is h on the manifold, we see that $R^\# - \overline{R}^\#$ lies in $R_1^\# \oplus R_2^\#$ and so $W^\# = \overline{W}^\#$. Also in [Eisenhart 1966, §28], it is shown that (for $\dim M \geq 4$) h is conformally flat (i.e., locally, there is a σ such that $e^{2\sigma}h$ is flat) if and only if $W \equiv 0$. It should be noted that $W \equiv 0$ if $\dim M = 3$, since in $\mathbb{R}^3$ we have $\dim R_3 = \dim R - \dim(R_1 \oplus R_2) = \frac{1}{12}3^2(3^2-1) - 6 = 0$; however, this does not imply that h is conformally flat. If $\dim M = 2$, then the existence of isothermal coordinates means that any h is conformally flat; also $R_2 = R_3 = 0$ for $\dim M = 2$.

By means of the metric h, we can regard the curvature R as an automorphism of $\Lambda^2(M, \mathbb{R})$, namely R sends the 2-form α to the form $R(\alpha)_{hi} := R_{hi}{}^{jk}\alpha_{jk}$. Since $\langle R(\alpha),\beta\rangle = \frac{1}{2}R_{hijk}\alpha^{jk}\beta^{hi} = \frac{1}{2}R_{jkhi}\beta^{hi}\alpha^{jk} = \langle\alpha,R(\beta)\rangle$, we have that R is a symmetric endomorphism. If M is an oriented 4-manifold, then we have another automorphism of $\Omega^2(M, \mathbb{R})$, namely $*: \Omega^2(M, \mathbb{R}) \to \Omega^2(M, \mathbb{R})$. While the decomposition $R = R_1 \oplus R_2 \oplus R_3$ of the space of curvature tensors on $\mathbb{R}^n$ consists of $O(n)$-irreducible subspaces, in the case $n = 4$ we can prove that R_3 is not $SO(4)$-irreducible because of the star operator on $\Lambda^2 :=$ the space of anti-symmetric bilinear forms on $\mathbb{R}^4$. To see this, first recall that $\Lambda^2 = \Lambda^2_+ \oplus \Lambda^2_-$ where Λ^2_+ and Λ^2_- are the ± 1 eigenspaces of $*$. Since $*$ is given by $(*\alpha)_{ji} = \frac{1}{2}\varepsilon^{hi}{}_{jk}\alpha_{hi}$, we have $tr* = \frac{1}{4}\varepsilon^{hi}{}_{hi} = 0$, since ε is totally anti-symmetric. Since $tr* = 0$, we see that $\dim\Lambda^2_- = \dim\Lambda^2_+ = \frac{1}{2}\dim\Lambda^2 = 3$. Concretely, a basis for $\Lambda^2_\pm$ is given by

$$\{e^1 \wedge e^2 \pm e^3 \wedge e^4,\ e^2 \wedge e^3 \pm e^1 \wedge e^4,\ e^1 \wedge e^3 \pm e^4 \wedge e^2\}.$$

Considering elements of R as endomorphisms of Λ^2, we can write any

$R \in \mathcal{R}$ in block form relative to the decomposition $\Lambda^2_+ \oplus \Lambda^2_-$:

$$R = \begin{bmatrix} A & B \\ B^T & C \end{bmatrix}$$

where $A = A^T$ and $C = C^T$, since we know R is symmetric. The automorphism $*$ is given by

$$* = \begin{bmatrix} I & 0 \\ 0 & -I \end{bmatrix}.$$

We know that $\dim \mathcal{R} = \frac{1}{12}(4^2)(4^2-1) = 20$, but the dimension of the space of all 6×6 symmetric matrices is $\frac{1}{2}6(6+1) = 21$. Indeed $*$ is orthogonal to $\mathcal{R}$, because by the Bianchi identity

$$0 = \varepsilon^{hijk}(R_{hijk} + R_{hkij} + R_{hjki}) = 3\varepsilon^{hijk}R_{hijk} = \langle R,*\rangle.$$

Thus, $*$ spans the orthogonal complement of $\mathcal{R}$ in $S^2(\Lambda^2)$:= the space of symmetric endomorphisms of Λ^2. Also, note that $0 = \langle R,*\rangle = \mathit{tr}\, A - \mathit{tr}\, C$, while $\mathit{tr}\, A + \mathit{tr}\, C = \mathit{tr}\, R = \frac{1}{2} R^{ij}{}_{ij} = \frac{1}{2}s(R)$. Thus, $\mathit{tr}\, A = \mathit{tr}\, C = \frac{1}{4}s(R)$, and writing $\tilde{A} = A - \frac{1}{3}(\mathit{tr}\, A)I$ and $\tilde{C} = C - \frac{1}{3}(\mathit{tr}\, C)I$, we have

$$R = \begin{bmatrix} \frac{1}{12}s(R) & 0 \\ 0 & \frac{1}{12}s(R) \end{bmatrix} + \begin{bmatrix} 0 & B \\ B^T & 0 \end{bmatrix} + \begin{bmatrix} \tilde{A} & 0 \\ 0 & 0 \end{bmatrix} + \begin{bmatrix} 0 & 0 \\ 0 & \tilde{C} \end{bmatrix}.$$

As $s(R)$, B, and (traceless, symmetric) $\tilde{A}$ and $\tilde{C}$ vary independently, each of the above matrices in the sum runs over a subspace of $\mathcal{R}$; from left to right, let these subspaces be C_1, C_2, C_3, and C_4. From the fact that $SO(4)$ acting on Λ^2 commutes with $*$, we have that the decomposition $\Lambda^2 = \Lambda^2_+ \oplus \Lambda^2_-$ is $SO(4)$-invariant and each of the C_i $(i = 1,\dots,4)$ is $SO(4)$-invariant.

<u>Theorem</u>. We have the following

$C_1 = \mathcal{R}_1$:= {constant curvature parts}

$C_2 = \mathcal{R}_2$:= {traceless Ricci parts} $= \{R \in S^2(\Lambda^2)\colon R\circ * = -*\circ R\}$

$C_3 + C_4 = \mathcal{R}_3$:= {Weyl parts} $= \{R \in S^2(\Lambda^2)\colon \langle R,*\rangle = 0,$
$\mathit{tr}(R) = 0,\ R\circ * = *\circ R\}$.

<u>Proof</u>: We begin with an elementary observation. Suppose $r\colon O(n) \to O(V)$ is an irreducible representation. Then either $r|SO(n)$ is irreducible, or V is the direct sum of two $SO(n)$-invariant irreducible subspaces of equal dimension. Indeed, let $\{0\} \neq V' \subset V$ be an irreducible $SO(n)$-

invariant subspace, and let $A \in O(n)$ with $\det A = -1$. Since $SO(n) = A\,SO(n)A^{-1}$, it follows that $r(A)(V')$ is $SO(n)$-invariant. Since V' is irreducible, either $r(A)(V') \cap V' = V'$ or $r(A)(V') \cap V' = \{0\}$. In the first case, V' is $O(n)$-invariant, and so $V = V'$ is $SO(n)$-irreducible. In the second case, $r(A)(V') + V'$ is a direct sum and $O(n)$-invariant, whence $V = r(A)(V') \oplus V'$. If W is a proper $SO(n)$-invariant subspace of $r(A)(V')$, then for any $B \in SO(n)$, $r(B)r(A)^{-1}W = r(BA^{-1})W = r(A^{-1}B')W = r(A^{-1})r(B')W = r(A^{-1})W$, whence $r(A)^{-1}W$ is a proper $SO(n)$-invariant subspace of V', a contradiction. Thus, $r(A)(V')$ is $SO(n)$-irreducible, as desired.

Now, we know that R_1, R_2, and R_3 are $O(4)$-irreducible. Moreover (for any $n > 2$), we have $\dim(R_1 \oplus R_2) = \frac{1}{2}n(n+1)$. Indeed, $R_1 \oplus R_2 = \mathrm{Ker}(r)^{\perp} = \mathrm{Im}(r^*)$, and $r^*\colon S^2 \to R$ is injective since $PvI = 0 \Rightarrow 0 = r(PvI) = \frac{1}{4}[tr(P)I + (n-2)P] \Rightarrow (n-2)P = -tr(P)I$, and taking the traces, we get $(n-2)tr\,P = -n\,tr(P)$, whence for $n > 1$ we have $tr\,P = 0$ and $(n-2)P = 0$; and so $P = 0$ for $n > 2$. Now, for $n = 4$, $\dim R_1 = 1$, $\dim R_2 = \frac{1}{2}4(4+1)-1 = 9$, $\dim R_3 = \dim R - 10 = 10$. Thus, R_1 and R_2 are $SO(4)$-irreducible (being of odd dimension), while R_3 is either $SO(4)$-irreducible, or it splits into two 5-dimensional $SO(4)$-irreducible subspaces. (Actually, the latter happens, since otherwise the dimension of one of the C_i would be at least 10.) The C_i of largest dimension is C_2 ($\dim C_2 = 9$), and so $C_2 = R_2$. The two 5's of R_3 must then be C_3 and C_4. Finally, $R_1 = C_1$ is clear, as well as the remaining equalities. □

Using the more suggestive notation, $W^+ := C_3$ and $W^- := C_4$, we have the decomposition $R_3 = W^+ \oplus W^-$ for $\mathbb{R}^4$ with the usual orientation. On any oriented, Riemannian 4-manifold (M,h), we then have the corresponding decomposition of the curvature tensor:

$$R = R_1 + R_2 + W^+ + W^-.$$

Now (M,h) is called *self-dual* if $W^- = 0$, and *anti-self-dual* if $W^+ = 0$. By changing the orientation, W^+ and W^- are interchanged (since $*$ changes to $-*$), and these notions are reversed.

F. Bochner-Weitzenböck Formulas

Let $\pi\colon P \to M$ be a principal G-bundle and let ω be a connection 1-form on P. Suppose $r\colon G \to GL(W)$ is a representation, and let $\overline{\Omega}^k(P,W)$ be the space of equivariant W-valued k-forms on P (or, equivalently, the space of k-forms on M with coefficients in the associated bundle $P \times_G W$).

Unlike ordinary exterior derivatives, $(D^{\omega})^2 \neq 0$ on $\overline{\Omega}^k(P,W)$, in general.

Proposition 1. For $\alpha \in \overline{\Omega}^k(P,W)$, we have $D^{\omega}(D^{\omega}\alpha) = r'(\Omega^{\omega}) \wedge \alpha$.

Proof:

$$\begin{aligned} D^{\omega}(D^{\omega}\alpha) &= d(D^{\omega}\alpha) + r'(\omega) \wedge D^{\omega}\alpha \\ &= d(d\alpha + r'(\omega) \wedge \alpha) + r'(\omega) \wedge (d\alpha + r'(\omega) \wedge \alpha) \\ &= r'(d\omega) \wedge \alpha - r'(\omega) \wedge d\alpha + r'(\omega) \wedge d\alpha + r'(\omega) \wedge r'(\omega) \wedge \alpha \\ &= [r'(d\omega) + r'(\omega) \wedge r'(\omega)] \wedge \alpha = r'(d\omega + \tfrac{1}{2}[\omega,\omega]) \wedge \alpha \\ &= r'(\Omega^{\omega}) \wedge \alpha, \end{aligned}$$

using the formulas of Section C. □

Let h be a Riemannian metric on the compact C^{∞} n-manifold M, and let θ be the Levi-Civita connection for h on the orthonormal frame bundle $\pi_F : F(M) \to M$. Let $P \circ F(M) := \{(p,u) \in P \times F(M): \pi(p) = \pi_F(u)\}$. The group $G \times O(n)$ acts freely on $P \circ F(M)$ in the obvious way, making $\tilde{\pi}: P \circ F(M) \to M$ a principal $G \times SO(n)$-bundle; it is known as the *fibered-product* of the two principal bundles. The forms ω and θ lift to forms on $P \circ F(M)$ via pull-back using the projections $P \circ F(M) \to P$ and $P \circ F(M) \to F(M)$. The sum of the two pull-back forms is a $\mathcal{G} \oplus \mathcal{O}(n)$-valued form that we denote by $\omega \oplus \theta$; one easily checks that $\omega \oplus \theta$ is a connection for $\tilde{\pi}: P \circ F(M) \to M$. Let $t^{r,s}$ be a tensor representation $t^{r,s}: O(n) \to GL(T^{r,s})$, where $T^{r,s} := \mathbb{R}^n \otimes \overset{r}{\ldots} \otimes \mathbb{R}^n \otimes \hat{\mathbb{R}}^n \otimes \overset{s}{\ldots} \otimes \hat{\mathbb{R}}^n$ ($\hat{\mathbb{R}}^n$:= dual space of $\mathbb{R}^n$) and the $O(n)$ action is the usual one:

$$t^{r,s}(A)(v_1 \otimes \ldots \otimes v_r \otimes \hat{w}_1 \otimes \ldots \otimes \hat{w}_s) = Av_1 \otimes \ldots \otimes \hat{w}_s \circ A^{-1}.$$

Then $r \otimes t^{r,s}: G \times O(n) \to W \otimes T^{r,s}$ is a representation, and we may consider the spaces $C(P \circ F(M), W \otimes T^{r,s})$, $\overline{\Omega}^k(P \circ F(M), W \otimes T^{r,s})$ of equivariant $W \otimes T^{r,s}$-valued functions or forms. Actually, the space $\overline{\Omega}^k(P \circ F(M), W \otimes T^{r,s})$ can be identified with the subspace of $C(P \circ F(M), W \otimes T^{r,s+k})$ consisting of equivariant functions with values in the subspace of $W \otimes T^{r,s+k}$ consisting of tensors which are anti-symmetric in the last k covariant slots; i.e., a tensor field on M with coefficients in a tensor bundle can be identified with a higher-order tensor. For this reason, we need only work within the spaces $C(P \circ F(M), W \otimes T^{r,s})$, or equivalently, $C^{\infty}(P \circ F(M) \times_{G \times O(n)} W \otimes T^{r,s})$ which is the space of tensor fields of type (r,s) on M with coefficients in $P \times_G W$. For brevity we denote this space by $\mathcal{T}^{r,s}(W)$. The covariant derivative $D^{\omega\oplus\theta}$ on $C(P\circ F(M), W\otimes T^{r,s})$ then gives rise to a first order linear differential operator

$$\nabla^{\omega\oplus\theta} : \mathcal{T}^{r,s}(W) \to \mathcal{T}^{r,s+1}(W).$$

Let $\Omega^s(W)$ be the subspace of $\mathcal{T}^{0,s}(W)$ consisting of the anti-symmetric tensors. The covariant derivative $D^{\omega\oplus\theta}$ on $\overline{\Omega}^k(P\circ F(M), W)$ corresponds to an operator

$$D^{\omega\oplus\theta} : \Omega^k(W) \to \Omega^{k+1}(W).$$

One should be careful to note that $D^{\omega\oplus\theta}$ is <u>not</u> $\nabla^{\omega\oplus\theta}|\Omega^k(W)$ (except if $k = 0$); hence we have used different symbols. However, $D^{\omega\oplus\theta}$ is $\nabla^{\omega\oplus\theta}$ followed by antisymmetrization. More precisely, we have:

<u>Proposition 2</u>. For $\alpha \in \Omega^k(W) \cong \Omega^k(P\times_G W)$, we have $D^\omega\alpha = D^{\omega\oplus\theta}\alpha$. Moreover,

$$(D^\omega\alpha)(X_1,\ldots,X_{k+1}) = \sum_{p=1}^{k+1}(-1)^{p+1}(\nabla^{\omega\oplus\theta}_{X_p}\alpha)(X_1,\ldots,\hat{X}_p,\ldots,X_{k+1}).$$

<u>Proof:</u> It suffices to check this in the case $\alpha = s\otimes\beta$, $s \in C^\infty(P\times_G W)$, $\beta \in \Omega^k(M)$. We know $D^\omega\alpha = D^\omega s\wedge\beta + s\otimes d\beta$. However, one can prove that

$$d\beta(X_1,\ldots,X_{k+1}) = \sum_{p=1}^{k+1}(-1)^{p+1}(\nabla^\theta_{X_p}\beta)(X_1,\ldots,\hat{X}_p,\ldots,X_{k+1});$$

indeed, choosing "normal" coordinates about $x \in M$ and taking $X_1,\ldots,X_{k+1}$ to be among the coordinate vector fields, the equation is the coordinate definition of $d\beta$ at x. Since $\nabla^{\omega\oplus\theta}_X\alpha = D^\omega_X s\otimes\beta + s\otimes\nabla^\theta_X\beta$, the right side of the desired equation is

$$(D^\omega s\wedge\beta + s\otimes d\beta)(X_1,\ldots,X_{k+1}) = (D^\omega\alpha)(X_1,\ldots,X_{k+1}). \qquad \square$$

Contractions of tensor fields are most efficiently displayed in terms of their components, say relative to a local coordinate system. Since the B-W formula and its proof involve many contractions, we introduce components, indices, and certain conventions. For a local coordinate system $x^1,\ldots,x^n$ on $U\subset M$, with coordinate vector fields $\partial_1,\ldots,\partial_n$ and 1-forms $dx^1,\ldots,dx^n$, the components of $t \in \mathcal{T}^{r,s}(W)$ are defined by

$$t^{j_1\ldots j_s}_{i_1\ldots i_r} := t(dx^{j_1},\ldots,dx^{j_s}, \partial_{i_1},\ldots,\partial_{i_r}).$$

Note that these components are sections of $(P \times_G W)|U$. The components of $\nabla^{\omega\oplus\theta}(t) \in T^{r,s+1}(W)$ will be denoted by

$$t^{j_1\cdots j_s}_{i_1\cdots i_r|k} := (\nabla^{\omega\oplus\theta}_{\partial_k} t)(dx^{j_1},\ldots,\partial_{i_r})$$

which is not the same as $D^\omega(t^{(j)}_{(i)})(\partial_k)$ unless the $x^1,\ldots,x^n$ are "normal" about a point y and everything is evaluated at y; we do not make this assumption in our derivations. The components of $\nabla^{\omega\oplus\theta}(\nabla^{\omega\oplus\theta} t) \in T^{r,s+2}(W)$ are denoted by $t^{(i)}_{(j)|k_1k_2}$, etc., where the new index corresponding to the latest covariant derivative occurs farthest to the right. Indices may be raised or lowered in the standard way using the metric tensor; e.g.,

$$t^{ij}{}_{hk} = h^{im} t_m{}^j{}_{hk}.$$

Note that it does not matter whether an index is raised before or after covariant differentiation by $\nabla^{\omega\oplus\theta}$, since $\nabla^\theta h = 0$ for the Levi-Civita connection θ of h; and so for $t = s \otimes \beta$, $s \in C^\infty(P \times_G W)$, $\beta \in T^{r,s}(M; \mathbb{R})$, we have $\nabla^{\omega\oplus\theta}(t \otimes h) = D^\omega s \otimes \beta \otimes h + s \otimes \nabla^\theta(\beta) \otimes h = (\nabla^{\omega\oplus\theta} t) \otimes h$. Finally, repeated indices are always assumed to be summed from 1 to n, and "$\wedge$" means that the index under it is omitted.

Proposition 3. For $\alpha \in \Omega^k(W) \cong \Omega^k(P \times_G W)$,

$$(D^\omega\alpha)_{i_0\cdots i_k} = \sum_{p=0}^{k} (-1)^p \alpha_{i_0\cdots\hat{i}_p\cdots i_k|i_p}$$

$$(\delta^\omega\alpha)_{i_1\cdots i_{k-1}} = -\alpha^i{}_{i_1\cdots i_{k-1}|i}$$

Moreover, $D^{\omega\oplus\theta}\alpha = D^\omega\alpha$ and $\delta^{\omega\oplus\theta}\alpha = \delta^\omega\alpha$, whence these are given by the same formulas.

Proof: The first equation is just a restatement of Prop. 2, and $\delta^{\omega\oplus\theta}\alpha = -(-1)^{n(k+1)} * D^{\omega\oplus\theta} * \alpha = \delta^\omega\alpha$, since $D^{\omega\oplus\theta}\alpha = D^\omega\alpha$ also by Prop. 2. For the second equation, note that $(\delta^\omega\alpha)_{i_1\cdots i_{k-1}} = -(-1)^{n(k+1)} * (D^\omega(*\alpha))$, but $D^\omega(\cdot) = D^\omega_{\partial_i} \otimes dx^i \wedge (\cdot)$; thus the same manipulations used at the end of Section C apply here to give the desired result. □

Now, recall that we also have the operator $\nabla^{\omega\oplus\theta}$ on $\Omega^k(W) \subset T^{0,k}(W)$ which is really $D^{\omega\oplus\theta}$ if $\Omega^k(W)$ is regarded as a space of tensor-valued *zero*-forms. The adjoint of $\nabla^{\omega\oplus\theta}$ is then $\delta^{\omega\oplus\theta}$ on tensor-valued one-forms; however, to avoid confusion, let us denote the adjoint of $\nabla^{\omega\oplus\theta}$

by $(\nabla^{\omega\oplus\theta})^*$. These remarks, together with Prop. 3, then yield the formula:

Corollary. For $\alpha \in \Omega^k(W)$, we have

$$(\nabla^{\omega\oplus\theta})^*(\nabla^{\omega\oplus\theta})(\alpha)_{i_1\ldots i_k} = -\alpha_{i_1\ldots i_k|}{}^{i}{}_{i} . \quad \square$$

The following formula allows us to compare the two positive operators $\Delta^\omega := \delta^\omega D^\omega + D^\omega\delta^\omega$ and $(\nabla^{\omega\oplus\theta})^*\nabla^{\omega\oplus\theta}$ on $\Omega^k(W \times_G P) \cong \Omega^k(W) \subset T^{0,k}(W)$.

Bochner-Weitzenböck Formula: For $\alpha \in \Omega^k(P \times_G W)$, we have

$$\begin{aligned}(\Delta^\omega\alpha)_{i_1\ldots i_k} &= [(\nabla^{\omega\oplus\theta})^*\nabla^{\omega\oplus\theta}\alpha]_{i_1\ldots i_k} \\ &- \sum_{p=1}^{k} (-1)^p R_{i_p h}\alpha^h{}_{i_1\ldots\hat{i}_p\ldots i_k} \\ &- \sum_{p\neq q=1}^{k} (-1)^p R^h{}_{i_q}{}^q{}_{i_p i}\,\alpha^i{}_{i_1\ldots h_q\ldots\hat{i}_p\ldots i_k} \\ &+ \sum_{p=1}^{k} (-1)^p r'(\Omega^\omega{}_{i_p i})\cdot\alpha^i{}_{i_1\ldots\hat{i}_p\ldots i_k} .\end{aligned}$$

Proof: Using Prop. 3, we have

$$\begin{aligned}(\delta^\omega D^\omega\alpha)_{i_1\ldots i_k} &= -(D^\omega\alpha)^i{}_{i_1\ldots i_k|i} \\ &= -\alpha_{i_1\ldots i_k|}{}^{i}{}_{i} - \sum_{p=1}^{k} (-1)^p\alpha^i{}_{i_1\ldots\hat{i}_p\ldots i_k|i_p i}\end{aligned}$$

$$\begin{aligned}(D^\omega\delta^\omega\alpha)_{i_1\ldots i_k} &= \sum_{p=1}^{k} (-1)^{p+1}(\delta^\omega\alpha)_{i_1\ldots\hat{i}_p\ldots i_k|i_p} \\ &= -\sum_{p=1}^{k} (-1)^{p+1}\alpha^i{}_{i_1\ldots\hat{i}_p\ldots i_k|ii_p} \ ; \text{ thus}\end{aligned}$$

$$(\Delta^\omega\alpha)_{i_1\ldots i_k} = -\alpha_{i_1\ldots i_k|}{}^{i}{}_{i} + \sum_{p=1}^{k} (-1)^p(\alpha^i{}_{i_1\ldots\hat{i}_p\ldots i_k|ii_p} - \alpha^i{}_{i_1\ldots\hat{i}_p\ldots i_k|i_p i}).$$

The first term is $[(\nabla^{\omega\oplus\theta})^*\nabla^{\omega\oplus\theta}\alpha]_{i_1\ldots i_k}$. Regarding α as a tensor-valued *zero*-form, we have (using Prop. 3 with W replaced by $W \otimes \Lambda^k$ and $\omega \oplus \theta$ replacing ω, $P\circ F(M)$ replacing P, etc.)

$$(\alpha^{i}{}_{i_1\ldots\hat{i}_p\ldots i_k|ii_p} - \alpha^{i}{}_{i_1\ldots\hat{i}_p\ldots i_k|i_pi})$$
$$= D^{\omega\oplus\theta}(D^{\omega\oplus\theta}\alpha)^{i}{}_{i_1\ldots\hat{i}_p\ldots i_k i_p i}.$$

However, by Prop. 1, we have

$$D^{\omega\oplus\theta}(D^{\omega\oplus\theta}\alpha) = \rho'(\Omega^{\omega\oplus\theta})\cdot\alpha$$

where α is still regarded as a tensor-valued zero-form, $P\circ F(M)$ replaces P, etc., and

$$\rho\colon G \times O(n) \to GL(W \otimes \Lambda^k)$$

is given by

$$\rho(g,A)(w \otimes \beta)_{i_1\ldots i_k} = (A^{-1})^{h_1}{}_{i_1}\ldots(A^{-1})^{h_k}{}_{i_k}\beta_{h_1\ldots h_k}r(g)(w),$$

whence (where $g \in G$, $A \in O(n)$)

$$\rho'(g,A)(w \otimes \beta)_{i_1\ldots i_k} = \beta_{i_1\ldots i_k}r'(g)(w) - \sum_{q=1}^{k} A^{h_q}{}_{i_q}\beta_{i_1\ldots h_q\ldots i_k}.$$

Thus,

$$[\rho'(\Omega^{\omega\oplus\theta})\cdot\alpha]_{i_0 i_1\ldots\hat{i}_p\ldots i_k i_p i}$$
$$= r'(\Omega^{\omega}{}_{i_p i})\cdot\alpha_{i_0\ldots\hat{i}_p\ldots i_k} - \sum_{\substack{q=0\\ q\neq p}}^{k} R^{h_q}{}_{i_q i_p i}\alpha_{i_0\ldots h_q\ldots\hat{i}_p\ldots i_k},$$

where we do not mean to imply that $q < p$, but we do assume $q \neq p$ in the sum. Now, the sum splits into two parts, namely the term where $q = 0$ (recall $p \geq 1$) and the part $\Sigma^{k}_{q\geq 1,q\neq p}$, and the sum is then

$$R^{h_0}{}_{i_0 i_p i}\,\alpha_{h_0 i_1\ldots\hat{i}_p\ldots i_k} + \sum_{\substack{q=1\\ q\neq p}}^{k} R^{h_q}{}_{i_q i_p i}\,\alpha_{i_0\ldots h_q\ldots\hat{i}_p\ldots i_k}.$$

Raising i_0 and contracting it with i, we obtain the desired result upon back-substitution into previous equations. □

We consider some special cases.

1. <u>(k = 1):</u> Here, $(\Delta^{\omega}\alpha)_j = ((\nabla^{\omega\oplus\theta})^*\nabla^{\omega\oplus\theta}\alpha)_j + R_{ji}\alpha^{i} - r'(\Omega^{\omega}_{ji})\cdot\alpha^{i}$,

since here the second sum in the formula is non-existent. Writing

$R_{ji} = (\mathrm{Ricc})_{ji}$ and considering both Ricc and $r'(\Omega^{\omega})$ as elements of $\mathrm{End}(\Omega^1(W))$, we have

$$(\Delta^{\omega}\alpha,\alpha) = ||\nabla^{\omega\oplus\theta}\alpha||^2 + (\mathrm{Ricc}\ \alpha,\alpha) - (r'(\Omega^{\omega})\alpha,\alpha).$$

It follows that if $\mathrm{Ricc} - r'(\Omega^{\omega}) \in \mathrm{End}(\Omega^1(W))$ is positive (e.g., if it is non-negative at each point and positive at some point), then $\Delta^{\omega}\alpha = 0 \Rightarrow \alpha = 0$. In particular, if $G = \{I\}$ is trivial, we obtain the well-known result that a compact Riemannian manifold with positive Ricci curvature admits no nonzero harmonic 1-form (originally due to S. Bochner).

2. <u>(k = 2)</u>: Here, we obtain

$$\begin{aligned}(\Delta^{\omega}\alpha)_{i_1i_2} &= [\nabla^{\omega\oplus\theta}{*}\nabla^{\omega\oplus\theta}\alpha]_{i_1i_2} \\ &+ R_{i_1i}\alpha^{i}{}_{i_2} - R_{i_2i}\alpha^{i}{}_{i_1} \\ &+ R^{h}{}_{i_2i_1j}\alpha^{j}{}_{h} - R^{h}{}_{i_1i_2j}\alpha^{j}{}_{h} \\ &- (r'(\Omega^{\omega}{}_{i_1i})\cdot\alpha^{i}{}_{i_2} - r'(\Omega^{\omega}{}_{i_2i})\cdot\alpha^{i}{}_{i_1}).\end{aligned}$$

In terms, of the operation $v\colon S^2 \times S^2 \to R$ (see Section D), the second line is $2[(\mathrm{Ricc}\ vh)(\alpha)]_{i_1i_2}$. The third line is $(R^{h}{}_{i_2i_1j} + R^{h}{}_{i_1ji_2})\alpha^{j}{}_{h} = R^{h}{}_{ji_1i_2}\alpha^{j}{}_{h} = -R(\alpha)_{i_1i_2}$, where R is regarded as in $\mathrm{End}(\Omega^2(W))$. The final line is appropriately denoted as $-[r'(\Omega^{\omega}),\alpha]_A$ where the "A" is a reminder that the brackets do not have anything to do with those for G, but rather denote the anti-symmetrization of a "composition" of endomorphisms of TM with coefficients in other bundles (i.e., $P \times_G \mathrm{End}(W)$ and $P \times_G W$). Thus,

$$\begin{aligned}\Delta^{\omega}\alpha &= (\nabla^{\omega\oplus\theta})^*\nabla^{\omega\oplus\theta}\alpha + 2(\mathrm{Ricc}\ vh)\alpha \\ &\quad - R(\alpha) - [r'(\Omega^{\omega}),\alpha]_A.\end{aligned} \tag{*}$$

3. <u>(k = 2, dim M = 4)</u>: In this case, recall from Section D that $2(\mathrm{Ricc}\ vh) - R = \frac{1}{3}S(hvh) - W$, where $W = R_3$ is the Weyl tensor and S is the scalar curvature; this is simply the definition of W when $n = 4$. Thus, here (*) becomes

$$\Delta^{\omega}\alpha = (\nabla^{\omega\oplus\theta})^*\nabla^{\omega\oplus\theta}\alpha + \frac{1}{3}S\alpha - W(\alpha) - [r'(\Omega^{\omega}),\alpha]_A.$$

We need to examine the consequences of this result in the case where α is anti-self-dual, Ω^{ω} is self-dual, and h is self-dual (i.e., recall that this means $W_- = 0$). Then $W(\alpha) = W_+(\alpha) = 0$, and we can prove $[r'(\Omega^{\omega}),\alpha]_A = 0$, as follows. The index lowering operation $\mathcal{O}(4) \to \Lambda^2(4)$ given by $B^i{}_j \longmapsto B_{ij}$ is an equivalence of the adjoint representation $SO(4) \to GL(\mathcal{O}(4))$ and the tensor representation $SO(4) \to GL(\Lambda^2(4))$, meaning that it preserves the $SO(4)$ actions. We know that $\Lambda^2(4) = \Lambda^2_+ \oplus \Lambda^2_-$ is a decomposition into $SO(4)$-invariant subspaces, which then must translate into a corresponding decomposition $\mathcal{O}(4) = \mathcal{O}_+ \oplus \mathcal{O}_-$ which is invariant under the adjoint representation. On the Lie algebra level, this means $[\mathcal{O}(4),\mathcal{O}_\pm] \subset \mathcal{O}_\pm$, whence $[\mathcal{O}_+,\mathcal{O}_-] = 0$ and $[\mathcal{O}_\pm,\mathcal{O}_\pm] \subset \mathcal{O}_\pm$. Hence $[r'(\Omega^{\omega}),\alpha]_A = [r'(\Omega^{\omega}_+),\alpha_-]_A = 0$; note that the vector bundle coefficients play no role in the above argument. Consequently, in the case where $\alpha = \alpha_-$, $W = W_+$, $\Omega^{\omega} = \Omega^{\omega}_+$, we have

$$\Delta^{\omega}\alpha = (\nabla^{\omega\oplus\theta})^*(\nabla^{\omega\oplus\theta})\alpha + \frac{1}{3}S\alpha.$$

Taking the L^2-inner product with α, gives the following important result.

Corollary. For any orthogonal (or unitary) representation $r: G \to O(W)$, self-dual metric h on the compact Riemannian 4- manifold M, connection ω on the principal G-bundle $\pi: P \to M$ with self-dual curvature, and anti-self-dual 2-form $\alpha \in \Omega^2_-(P \times_G W)$, we have

$$(\Delta^{\omega}\alpha,\alpha) = ||\nabla^{\omega\oplus\theta}\alpha||^2 + \frac{1}{3}(S\alpha,\alpha),$$

where S is the scalar curvature of h. Consequently, if $S \geq 0$, then $\Delta^{\omega}\alpha = 0$ implies $||\nabla^{\omega\oplus\theta}\alpha||^2 = 0$ (whence $\nabla^{\omega\oplus\theta}\alpha = 0$, and so $|\alpha|^2$ is constant). If in addition $S \not\equiv 0$, then $0 = \frac{1}{3}|\alpha|^2\int_M S\nu$, whence $\alpha = 0$.

G. Chern Classes as Curvature Forms

Let $E \to M$ be a C^{∞} complex, Hermitian vector bundle. A frame of E_x $(x \in M)$ is an isometry $u: \mathbb{C}^m \to E_x$ (i.e., an isomorphism of complex vector spaces such that $\langle u(z),u(w)\rangle = z_1\bar{w}_1 + \dots + z_m\bar{w}_m$). If u is a frame and $A \in U(m)$, then $u\circ A$ is a frame. Indeed, the set $F(E)$ consisting of all frames at all points can be made into a C^{∞} manifold such that $\pi: F(E) \to M$ is a principal $U(m)$-bundle, namely the *bundle of unitary frames of* E. Without difficulty, one can prove that $E \cong P \times_{U(m)} \mathbb{C}^m$ where $U(m) \to GL(\mathbb{C}^m)$ is the representation defined by inclusion.

Let $\tilde{\omega}$ be a connection on $F(E)$; $\tilde{\omega}$ is a $\mathcal{U}(m)$-valued 1-form on $F(E)$ with the requisite properties (see Section B) and recall $\mathcal{U}(m) :=$

Lie algebra of $U(m) = \{A \in \operatorname{End}(\mathbb{C}^m): A^* = -A\}$. We will give a prescription for the Chern classes of $E \to M$ in terms of the curvature $\tilde{\Omega}$ of $\tilde{\omega}$.

We begin by defining functions $s_k: GL(m) \to \mathbb{C}$ by means of the equation

$$\det(A + tI) = \sum_{k=0}^{m} s_k(A)t^{m-k}.$$

Each $s_k(A)$ is a homogeneous polynomial of degree k in the entries of the variable A. Indeed,

$$s_k(A) = \frac{1}{k!}\sum_{(i),(j)} \delta^{j_1\ldots j_k}_{i_1\ldots i_k} a^{i_1}{}_{j_1}\cdots a^{i_k}{}_{j_k}$$

where $A = (a^i{}_j)$; $(i) = (i_1,\ldots,i_k)$, $i_1,\ldots,i_k$ distinct elements of $\{1,\ldots,m\}$ and similarly for (j); and $\delta^{(i)}_{(j)}$ is $+1$ (resp. -1) if (i) is an even (resp. odd) permutation of (j), and $\delta^{(i)}_{(j)} = 0$ if $\{i_1,\ldots,i_k\} \neq \{j_1,\ldots,j_k\}$. The s_k are $\operatorname{Ad}(GL(m,\mathbb{C}))$-invariant in the sense that $s_k(BAB^{-1}) = s_k(A)$, since $\det(BAB^{-1} + tI) = \det(B(A+tI)B^{-1}) = \det(A + tI)$ for any B $GL(m,\mathbb{C})$. Now the curvature $\tilde{\Omega}$ can be regarded as a matrix of $\mathbb{C}$-valued 2-forms, say $\tilde{\Omega} = (\Omega^i{}_j)$ $(1 \le i,j \le m)$ such then $\Omega^i{}_j = -\overline{\Omega}^j{}_i$, and we define

$$s_k(\tilde{\Omega}) = \frac{1}{k!}\sum_{(i),(j)} \delta^{j_1\ldots j_k}_{i_1\ldots i_k} \Omega^{i_1}{}_{j_1} \wedge\cdots\wedge \Omega^{i_k}{}_{j_k}$$

which is a $\mathbb{C}$-valued $2k$-form on $F(E)$. It follows (from the *Ad* $U(m)$-invariance of s_k and the fact that $\tilde{\Omega}(A^*,\cdot) = 0$ for any $A \in \mathcal{U}(m)$) that $s_k(\tilde{\Omega}) = \pi^*(\underline{s}_k(\tilde{\Omega}))$ for a unique $\mathbb{C}$-valued $2k$-form $\underline{s}_k(\tilde{\Omega})$ on M; see [Bleecker 1981, 161-165] = [Kobayashi-Nomizu 1969, §12] for the details in a more general context. Indeed, the $\underline{s}_k(\tilde{\Omega})$ are closed and determine de Rham cohomology classes $[\underline{s}_k(\tilde{\Omega})] \in H^{2k}(M,\mathbb{C})$ which are independent of $\tilde{\omega}$ (loc. cit.). The k-th Chern class is given by

$$c_k(E) = \left(\frac{i}{2\pi}\right)^k[\underline{s}_k(\tilde{\Omega})].$$

The factor of i^k ensures that $c_k(E)$ is actually in $H^{2k}(M,\mathbb{R})$; indeed, for A skew-Hermitian (i.e., in $\mathcal{U}(m)$), iA is Hermitian and $i^k s_k(A) = s_k(iA)$ is essentially the k-th elementary symmetric function of the real eigenvalues of iA. The factor of $(2\pi)^{-k}$ is a normalization which ensures that $c_k(E) \in H^{2k}(M,\mathbb{Z})$; see [Kobayashi-Nomizu 1969, §12].

Of course, once the Chern classes are determined, the Chern character $ch(E) \in H^*(M, \mathbb{Q})$ may be defined using them (cf. III.4.A). Alternatively, we can get $ch(E)$ directly as follows. For $A \in \mathcal{U}(m)$, write

$$\mathcal{tr}(\exp \frac{i}{2\pi} At) = \sum_{k=0}^{\infty} r_k(A)t^k,$$

where $r_k(A) = [k!\,(2\pi)^k]^{-1}\mathcal{tr}([iA]^k)$ explicitly. Then the r_k are $Ad\ U(m)$-invariant, as before with the s_k, and $ch(E)$ is given by

$$ch(E) = \bigoplus_{k=0}^{\infty} [\underline{r_k}(\tilde{\Omega})].$$

Remarks. a) In the case where $F(E)$ is reducible to an $SU(m)$-bundle, say $F(E)_0$, we can choose $\tilde{\omega}$ on $F(E)$ such that $\tilde{\omega}|F(E)_0$ is a ($\mathcal{SU}(m)$-valued) connection 1-form on $F(E)_0$, and $\tilde{\Omega}$ will then be $\mathcal{SU}(m)$-valued. Since $s_1(A) = \mathcal{tr}\ A = 0$ for $A \in \mathcal{SU}(m)$, it follows that $c_1(E) = 0$ in this case.

b) Note that $8\pi^2 r_2(A) = \mathcal{tr}(-A^2)$, whence $ch(E)_2 = \frac{-1}{8\pi^2}[\mathcal{tr}(\tilde{\Omega} \wedge \tilde{\Omega})]$, where we have regarded $\tilde{\Omega}$ as a two-form on M with coefficients in End(E), and a composition of endomorphisms is implicit in the wedge.

c) Other characteristic classes can be represented in terms of forms. The Euler class of an oriented Riemannian 2n-manifold M is represented by the *Gauss-Bonnet form*

$$\frac{(-1)^n}{2^{2n}\pi^n n!} \sum_{(i)} \varepsilon_{i_1 \ldots i_{2n}} \Omega^{\theta}_{i_1 i_2} \wedge \ldots \wedge \Omega^{\theta}_{i_{2n-1} i_{2n}}$$

where Ω^{θ} is the curvature form of any connection θ on the bundle $F(M)$ of oriented orthonormal frames; here we regard Ω^{θ} as an End(TM)-valued 2-form on M and Ω^{θ}_{ij} are the $\mathbb{R}$-valued entries of Ω^{θ} relative to a local orthonormal frame field on M. The fact that the form is independent of the choice of local frame field stems from the fact that the "Pfaffian"

$$A \longmapsto \frac{1}{2^n n!} \sum_{(i)} \varepsilon_{i_1 \ldots i_{2n}} A_{i_1 i_2} \ldots A_{i_{2n-1} i_{2n}}$$

is an Ad-invariant polynomial on $\mathcal{O}(2n)$. The Gauss-Bonnet Theorem asserts that the integral of the Gauss-Bonnet form is $\chi(M)$, when M is compact. For the record, the *Pontryagin classes* of M are represented by the 4k-forms

$$\frac{1}{(2\pi)^{2k}(2k)!}\sum_{(i)(j)} \delta^{i_1\cdots i_{2k}}_{j_1\cdots j_{2k}}\, \Omega^{\theta}{}_{i_1j_1}\wedge\cdots\wedge\Omega^{\theta}{}_{i_{2k}j_{2k}},$$

where it is no longer necessary to restrict the dimension of M to be even. For more details on this approach to characteristic classes, see [Milnor-Stasheff 1974, 289-314] and [Kobayashi-Nomizu 1969, §12].

H. Holonomy

Let ω be a connection on the principal G-bundle $\pi: P \to M$. Fix a point $p_0 \in P$, and let P_0 be the set of all points $p \in P$ which can be joined to p_0 by a smooth horizontal curve $\gamma: [a,b] \to P$ (i.e., $\gamma(a) = p_0$, $\gamma(b) = p$, and $\omega(\gamma'(t)) = 0$ for $t \in [a,b]$). Set $G_0 :=$ $\{g \in G: p_0g \in P_0\}$. It can be proved (cf. [Kobayashi-Nomizu 1963, 83-85]) that P_0 is an immersed submanifold of P, and $P_0 \to M$ is a principal G_0-bundle; P_0 is called the *holonomy bundle* of ω through p_0 and G_0 is the *holonomy group* of ω at p_0. If G_0 is a proper subgroup of G, then ω is said to be *reducible* to G_0; if $G_0 = G$, then ω is *irreducible*.

Let $I_\omega := \{F \in GA(P): F^*\omega = \omega\}$, the isotropy subgroup at ω for the action of $GA(P)$ on the space $C(P)$ of connections on P. Under the isomorphism $\Phi: C(P,G) \to GA(P)$ of Section D (i.e., $\Phi(f)(p) = pf(p)$), we can also regard I_ω as a subgroup of $C(P,G)$. I_ω is closely related to the holonomy group G_0.

<u>Proposition 1</u>. The homomorphism $I_\omega \to G$ given by $\Phi(f) \longmapsto f(p_0)$ maps I_ω isomorphically onto the centralizer $C(G_0) := \{g \in G: gg_0 = g_0g$ for all $g_0 \in G_0\}$ of G_0.

<u>Proof</u>: Let $\Phi(f) \in I_\omega$ and $g_0 \in G_0$. To prove $f(p_0) \in C(G_0)$, we need to show that $f(p_0)g_0 = g_0f(p_0)$. Let γ be a horizontal curve joining p_0 to p_0g_0. Then $\gamma\cdot f(p_0)$ is a horizontal curve joining $p_0f(p_0)$ to $p_0g_0f(p_0)$. Now, $\Phi(f)\circ\gamma$ is a horizontal (since $\Phi(f)^*\omega = \omega$) curve joining $p_0f(p_0)$ to $\Phi(f)(p_0g_0) = \Phi(f)(p_0)g_0 = p_0f(p_0)g_0$. Since $\pi\circ(\gamma\cdot f(p_0)) = \pi\circ(\Phi(f)\circ\gamma)$, we have that $\gamma\cdot f(p_0)$ and $\Phi(f)\circ\gamma$ project to the same curve on M and have the same initial point $p_0f(p_0)$. Hence, by uniqueness of "horizontal lifts" (cf. [Kobayashi - Nomizu 1963, 69]), the endpoints must agree (i.e., $p_0g_0f(p_0) = p_0f(p_0)g_0$), and so $f(p_0) \in$ $C(G_0)$. To see that $I_\omega \longmapsto C(G_0)$ is injective, note that $\Phi(f) \in I_\omega$ implies that $D^\omega f = 0$ (regarding $f \in C(P,\mathrm{End}(\mathbb{C}_N))$) by Prop. 3 of Section D. Such an f is equal to the constant $f(p_0)$ on P_0. Since P_0 meets each fiber of P and f is equivariant, we know that f is completely determined by its value $f(p_0)$ on P_0, and so $I_\omega \to C(G_0)$ is

injective. To prove that $I_\omega \to C(G_0)$ is onto, let $g' \in C(G_0)$ and define $f: P_0 \to G$ to be the constant map with value g'. The reader can easily check that f extends to a well defined element of $C(P,G)$ via the formula $f(q_0g) = g^{-1}f(q_0)g$, for $q_0 \in P_0$, $g \in G$. Moreover, since f is constant on P_0, we have $df|P_0 = 0$; and so $D^\omega f = 0$ on P_0, since the tangent spaces of P_0 contain the horizontal subspaces of ω. Since $D^\omega f$ is equivariant, we know $D^\omega f = 0$ on P as well; thus, $\Phi(f) \in I_\omega$ and $I_\omega \to C(G_0)$ is onto. □

Proposition 2. The curvature Ω^ω restricted to the holonomy bundle P_0 has values in the Lie algebra $\mathcal{G}_0$ of the holonomy group G_0.

Proof: Note that $\omega|P_0$ clearly has values in $\mathcal{G}_0$. Thus, $(d\omega)|P_0 = d(\omega|P_0)$ has values in $\mathcal{G}_0$. But since the horizontal subspaces of ω at points in P_0 are tangent to P_0, we have (for $X,Y \in T_qP_0$), $\Omega^\omega(X,Y) = d\omega(X^H,Y^H) \in \mathcal{G}_0$. //

3. GAUGE-THEORETIC INSTANTONS

In this chapter, we formally investigate the space of gauge-equivalence classes (moduli) of Yang-Mills instantons whose possible physical significance was informally discussed near the end of IV.1. The literature specifically on this subject (e.g., [Atiyah-Hitchin-Singer 1978], [Bernard-Christ-Guth-Weinberg 1977], [A. S. Schwarz 1979], etc.) is primarily intended for experts who can supply the details which must perforce be omitted in any research article of reasonable proportions. The surveys on the general subject of geometry in classical field theories (e.g., [Eguchi-Gilkey-Hanson 1980], [Madore 1981], etc.) contain a wealth of useful information and are great for finding one's way through the literature. The article [T. Parker 1982] is particularly good at exposing the relevant techniques of non-linear functional analysis in gauge theories, as well as treating the coupled version of the removable singularity theorem for Yang-Mills fields (cf. [K. Uhlenbeck 1982]). These works inspired much of the material in Section D, without which there is no real proof that the moduli space is a manifold or that the dimension (computed in Section C using the index theorem) has any real significance. The computation of the index of the relevant operator in C avoids the use of spinors, and hence is good for those not familiar with spinors, Dirac operators, and the $\hat{A}$ genus; it applies to 4-manifolds with or without spin structures.

We close the chapter with Section E. There we sketch the main ideas of the proof of a very remarkable and recent theorem of [S. K. Donaldson 1983] which uses the moduli space to investigate the topological structure of smooth, compact, simply-connected 4-manifolds without non-zero anti-self-dual harmonic 2-forms.

A. The Yang-Mills Functional

In order to define the Yang-Mills functional on the space of connections on a principal G-bundle, we need an inner product on the Lie algebra $\mathcal{G}$ of G which is invariant under the adjoint representation Ad: $G \to GL(\mathcal{G})$ $(Ad_g(A) = gAg^{-1})$. For $A \in \mathcal{G}$, let $ad(A)\colon \mathcal{G} \to \mathcal{G}$ be defined by $ad(A)(B) = [A,B]$. Let $k\colon \mathcal{G} \times \mathcal{G} \to \mathbb{R}$ be the symmetric bilinear form given by

$$k(A,C) = -tr(ad(A) \circ ad(C)), \quad A,C \in \mathcal{G}.$$

Note that $k(Ad_g A, Ad_g C) = k(A,C)$ from the easily proved fact $ad(Ad_g A) = Ad_g \circ ad(A) \circ Ad_g{-1}$; $-k$ is known as the *Killing form* of $\mathcal{G}$. G (or $\mathcal{G}$) is called *semi-simple* if k is nondegenerate. We assume in this chapter that G is semi-simple and compact. In this case, we can prove that k is actually positive-definite. Indeed, let k_0 be any positive definite inner product. Define $\overline{k}_0 = \int_G Ad_g^* k_0 dg$, where dg is a biinvariant measure on G and $(Ad_g^* k_0)(A,C) := k_0(Ad_g A, Ad_g C)$; the integral exists since G is compact. Then relative to a $\overline{k}_0$ orthonormal basis of $\mathcal{G}$, the matrix of Ad_g will be orthogonal, since $\overline{k}_0$ is preserved by Ad_g. The infinitesimal generators $ad(A)$ will then be represented by skew-symmetric matrices, whence $k(A,A) = -tr(ad(A) \circ ad(A)) = \Sigma\, ad(A)_{ij}^2 \geq 0$, and k is definite as claimed.

Let M be a compact, oriented, manifold with Riemannian metric h. We define the *Yang-Mills* functional $YM\colon C(P) \to \mathbb{R}$ on the space $C(P)$ of connections on the principal G-bundle P over M, by

$$YM(\omega) = \frac{1}{2} ||\Omega^\omega||^2 = \int_M \frac{1}{2} |\Omega^\omega|^2 ,$$

where $\Omega^\omega \in \Omega^2(P \times_G \mathcal{G})$ is the curvature of ω (cf. Section 2B) and $|\Omega^\omega|^2 = \langle\Omega^\omega, \Omega^\omega\rangle$ is the pairing introduced in Section 2C relative to k on $\mathcal{G}$ and h on M.

Note that $\omega \in C(P)$ is not in $\overline{\Omega}^1(P,\mathcal{G})$ since $\omega(A^*) = A \neq 0$ in general. However, if $\omega' \in C(P)$, then $\omega - \omega' \in \overline{\Omega}^1(P,\mathcal{G})$; indeed for any $\tau \in \overline{\Omega}^1(P,\mathcal{G})$, we have $\omega + \tau \in C(P)$ and all connections are obtained as τ varies. In other words, $C(P)$ is an affine space modelled on $\overline{\Omega}^1(P,\mathcal{G})$, or equivalently $\Omega^1(P \times_G \mathcal{G})$. For $t \in \mathbb{R}$, $\omega \in C(P)$, and $\tau \in \bar{\Omega}^1(P,\mathcal{G})$, let $\omega_t = \omega + t\tau \in C(P)$; i.e., $t \mapsto \omega_t$ is a curve of connections. At $t = 0$,

$$\frac{d}{dt}\Omega^{\omega_t} = \frac{d}{dt}(d\omega_t + \frac{1}{2}[\omega_t,\omega_t])$$

$$= d\tau + [\omega,\tau] = D^{\omega}\tau.$$

Thus at $t = 0$, we have (see 2C for δ^{ω})

$$\frac{d}{dt}\mathrm{YM}(\omega_t) = \frac{d}{dt}\frac{1}{2}(\Omega^{\omega_t},\Omega^{\omega_t})$$

$$= (D^{\omega}\tau,\Omega^{\omega}) = (\tau,\delta^{\omega}\Omega^{\omega}).$$

Since, all inner products are assumed positive definite, it follows that all these directional derivatives will be 0 (for arbitrary τ), precisely when

$$\delta^{\omega}\Omega^{\omega} = 0.$$

This is called the (source-free) *Yang-Mills equation*; we have shown that it characterizes critical points of the Yang-Mills functional.

If $\dim M = 4$, then we may speak of self-dual curvatures Ω^{ω} (i.e., those for which $*\Omega^{\omega} = \Omega^{\omega}$). For these, the Yang-Mills equation always holds, since $\delta^{\omega}\Omega^{\omega} = -*D^{\omega}*\Omega^{\omega} = -*D^{\omega}\Omega^{\omega} = 0$ by the Bianchi identity ($D^{\omega}\Omega^{\omega} = 0$). Indeed, by the following arguments, we can show that if Ω^{ω} is self-dual, then ω is an absolute minimum of $\mathrm{YM}: C(P) \to \mathbb{R}$.

If we complexify, forming $\mathcal{G}_c := \mathcal{G} \otimes \mathbb{C}$, and extend k on $\mathcal{G}$ to a Hermitian inner product k_c on $\mathcal{G}_c$ in the obvious way (i.e., $k_c(zA,wB) = z\bar{w}\,k(A,B)$), then $Ad: G \to O(\mathcal{G})$ extends to a unitary representation $Ad_c: G \to U(\mathcal{G}_c)$. Thus, $E := P \times_G \mathcal{G}_c \to M$ is a Hermitian vector bundle and we may speak of its Chern classes $c_i(E) \in H^{2i}(M, \mathbb{Z})$ and Chern character $ch(E) \in H^*(M, \mathbb{Q})$. For our purposes, we use the differential-geometric definition of these notions given in Section 2G. There, in the closing remarks, we noted that $ch(E)_2$ is represented by $\frac{-1}{8\pi^2}\,tr(\tilde{\Omega}^{\omega} \wedge \tilde{\Omega}^{\omega})$ where $\tilde{\Omega}^{\omega}$ is regarded as in $\Omega^2(M,\mathrm{End}(E))$, and the wedge implicitly involves composition of endomorphisms. Thus, by the definition of k on $\mathcal{G}$ as $k(A,A) = -tr(ad(A)^2)$, it follows that $ch(E)_2$ is represented by

$$\frac{1}{8\pi^2}\langle\Omega^{\omega},*\Omega^{\omega}\rangle\nu \qquad (\nu = \text{Vol. elt. of } (M,h))$$

for $\dim M = 4$. Also note that $c_1(E) = 0$, in the case at hand, since G is represented by real matrices in $U(\mathcal{G}_c)$ relative to a basis of $\mathcal{G}$

(which is also a basis of G_c), whence G is represented by matrices in $SU(G_c)$; this means that the unitary frame bundle $F(P \times_G G_c)$ is reducible to an SU(n)-bundle and $c_1(E) = 0$, by the closing remark a) in Section 2G. Thus, $ch(E)_2 = \frac{1}{2}[c_1(E)^2 - 2c_2(E)] = -c_2(E)$, in the case at hand. Now, we have

$$\begin{aligned}||\Omega^\omega - *\Omega^\omega||^2 &= ||\Omega^\omega||^2 + ||*\Omega^\omega||^2 - 2(\Omega^\omega,*\Omega^\omega) \\ &= 2||\Omega^\omega||^2 - 2(\Omega^\omega,*\Omega^\omega).\end{aligned}$$

Consequently,

$$2YM(\omega) = (\Omega^\omega,*\Omega^\omega) + \frac{1}{2}||\Omega^\omega - *\Omega^\omega||^2.$$

We have already shown that $(\Omega^\omega,*\Omega^\omega)$ is a topological invariant $(8\pi^2\, ch(E)_2[M])$. Thus, YM is absolutely minimized by any ω with Ω^ω self-dual (if such exists). Such connections ω are called *self-dual connections* even though Ω^ω is self-dual, not ω. Self-dual connections have also been given the name "instantons", particularly when the base is S^4. This deserves some explanation. We saw toward the end of 1B that generally speaking an instanton is a minimum-action solution of a Euclidean action principal that "joins" (via the pure-imaginary time parameter $i\tau$) two potential minima in the configuration space. The term "instanton" is by no means limited to connections. However, when it is applied to connections (or gauge potentials), it means a minimum of the YM functional for connections on bundles over Euclidean space with certain asymptotic conditions at ∞. Now, any bundle over $\mathbb{R}^4$ is trivial, whence it would seem that a trivial flat connection would always yield the minimum, but the asymptotic conditions (described at the end of 1B) generally force the connection to be nonflat. Now, the YM functional is invariant under conformal changes in the metric on the base (of dimension 4); indeed we can rewrite $YM(\omega) = \int_M \frac{1}{2}\, tr(\tilde{\Omega}^\omega \wedge *\tilde{\Omega}^\omega)$ and we already know that $*$ is conformally invariant in dimension 4 (cf. 2C before Prop. 1). Thus, self-dual connections over $\mathbb{R}^4$ can be regarded as self-dual connections on $S^4 - \{\infty\}$ via (conformal) stereographic projection. It turns out that the physically reasonable asymptotic conditions at ∞ are precisely those that enable one (via a theorem in [K. Uhlenbeck 1982]) to extend (possibly non-trivially) the bundle and the connection across the north pole (∞) on S^4 to get a self-dual connection (instanton) on S^4. Of course, there is no mathematical reason for restricting the base to be S^4, rather than a general compact orientable Riemannian 4-manifold.

B. Instantons on Euclidean 4-Space

In the sections that follow, we will construct manifolds of self-dual connections and compute their dimensions. However, these results require the existence of such connections, or else they might be just sophisticated statements about elements of the empty set. Here, our goal will be to produce self-dual connections for SU(2)-bundles $P \to S^4$ with arbitrary nonpositive (integral) Chern number $c_2(P \times_{SU(2)} \mathbb{C}^2)[S^4]$. Rather than simply list some families of solutions, we will motivate our construction from the standpoint of Riemannian geometry, but first we give a brief history of various constructions.

The first such instanton that came from the physics community was the BPST instanton of [A. A. Belavin, A. M. Polyakov, A. S. Schwarz, Y. S. Tyupkin 1975]. Actually, the BPST instanton was given on $\mathbb{R}^4$, but it is easy to extend it to a connection for a nontrivial SU(2)-bundle over S^4. When this is done, the BPST instanton becomes the well-known universal connection on the quaternionic Hopf bundle $S^7 \to S^4$. More precisely, the BPST solution depends on 5-parameters, and we get this 5-parameter family by moving the universal connection around via lifts (to S^7) of conformal transformations of S^4; the isometries leave the universal connection fixed, but the 5-parameter family of "boosts" is effective. The Chern number of the bundle is -1, and so the BPST family realizes this case. We will consider it more explicitly later. The first solutions for arbitrary negative Chern number -k seem to have been produced by E. Witten, after which G. 't Hooft (unpublished) generalized the solutions to 5k parameter families. Using the fact that these families could be augmented to conformally invariant families, in [R. Jackiw, C. Nohl, and C. Rebbi 1976] the families were enlarged to 5k + 4 parameters. (For k = 1 and k = 2, not all of these parameters are effective.) In [Atiyah-Hitchin-Singer 1977] and [A. S. Schwarz 1977], the Index Theorem was applied to the problem, and it was found that the maximal number of effective parameters is actually 8k-3 ($k \geq 1$). Not long after, the analysis was applied to arbitrary G-bundles over S^4 (G simple, compact); see [Atiyah-Hitchin-Singer 1978] and [Bernard-Christ-Guth-Weinberg 1977] for this story. The problem of actually constructing the most general (8k-3)-parameter family of solutions was solved using techniques originating with the twistor theory of Roger Penrose and some algebraic geometry (cf. [Atiyah-Ward

1977]). A construction involving "only" linear algebra was finally developed in [Atiyah-Hitchin-Drinfeld-Manin 1978]; the motivation for this construction lies rather deeply within algebraic geometry. The interested reader may substantially augment this very brief history by consulting the excellent surveys [Eguchi-Gilkey-Hanson 1980] and [J. Madore 1981].

In what follows, we will stumble upon the Jackiw-Nohl-Rebbi $5k+4$ family in very natural ways. The starting point is the observation just before the final Corollary in 2F that the Lie algebra $\mathcal{O}(4)$ splits into subalgebras $\mathcal{O}_+$ and $\mathcal{O}_-$ corresponding to the self-dual and anti-self-dual summands in $\Lambda^2(4) = \Lambda^2_+ \oplus \Lambda^2_-$ under the natural isomorphism $\mathcal{O}(4) \cong \Lambda(4)$. Each of the summands $\mathcal{O}_+$ and $\mathcal{O}_-$ is isomorphic to $\mathcal{SU}(2)$ (the Lie algebra of $SU(2)$). We can make these isomorphisms explicit as follows. The anti-self-dual space Λ^2_- has the orthogonal basis $\{(e_2 \wedge e_3 - e_1 \wedge e_4),\ (e_3 \wedge e_1 - e_2 \wedge e_4),\ (e_1 \wedge e_2 - e_3 \wedge e_4)\}$. These basis vectors correspond to the following elements of $\mathcal{O}(4)_-$ ('t Hooft matrices)

$$\begin{bmatrix} 0 & 0 & 0 & -1 \\ 0 & 0 & 1 & 0 \\ 0 & -1 & 0 & 0 \\ 1 & 0 & 0 & 0 \end{bmatrix}, \quad \begin{bmatrix} 0 & 0 & -1 & 0 \\ 0 & 0 & 0 & -1 \\ 1 & 0 & 0 & 0 \\ 0 & 1 & 0 & 0 \end{bmatrix}, \quad \begin{bmatrix} 0 & 1 & 0 & 0 \\ -1 & 0 & 0 & 0 \\ 0 & 0 & 0 & -1 \\ 0 & 0 & 1 & 0 \end{bmatrix}.$$

In conformance with the literature, these are denoted by $\bar{\eta}_1, \bar{\eta}_2, \bar{\eta}_3$ from left to right. One easily checks that $[\bar{\eta}_a, \bar{\eta}_b] = -2\varepsilon_{abc}\bar{\eta}_c$. Moreover, for the Pauli matrices

$$\sigma_1 = \begin{bmatrix} 0 & 1 \\ 1 & 0 \end{bmatrix}, \quad \sigma_2 = \begin{bmatrix} 0 & -i \\ i & 0 \end{bmatrix}, \quad \sigma_3 = \begin{bmatrix} 1 & 0 \\ 0 & -1 \end{bmatrix},$$

we have $i\sigma_a \in \mathcal{SU}(2)$ with $[i\sigma_a, i\sigma_b] = -2\varepsilon_{abc} i\sigma_c$. Thus, $\bar{\eta}_a \to i\sigma_a$ yields an explicit isomorphism from $\mathcal{O}(4)_-$ to $\mathcal{SU}(2)$. By changing the signs of the elements of the fourth row and fourth column of each 't Hooft matrix, we obtain the 't Hooft matrices η_1, η_2, η_3, which give us an isomorphism $\mathcal{O}(4)_+ \cong \mathcal{SU}(2)$. Recall from Section 2E that the Levi-Civita connection for a Riemannian 4-manifold M is an $\mathcal{O}(4)$-valued 1-form θ on the bundle $F(M)$ of orthonormal frames. Now, relative to the decomposition and isomorphism $\mathcal{O}(4) = \mathcal{O}_+ \oplus \mathcal{O}_- \cong \mathcal{SU}(2) \oplus \mathcal{SU}(2)$, we can regard θ as a sum of $\mathcal{SU}(2)$-valued forms θ_+ and θ_-. Suppose that M is simply $\mathbb{R}^4$ with some metric tensor h. Then we have a global section $\tau : \mathbb{R}^4 \to F(\mathbb{R}^4)$ of the (trivial) frame bundle for h, and pull-backs $A^\pm := \tau^*\theta^\pm$ which are $\mathcal{SU}(2)$-valued 1-forms on $\mathbb{R}^4$. Letting $F^\pm = dA^\pm + \frac{1}{2}[A^\pm, A^\pm]$ be the field strengths, we may ask for conditions on h such that $*F^+ = F^+$ or $*F^- = F^-$. Fortunately, these are not difficult to determine.

Indeed, $F^+ \oplus F^-$ is essentially the decomposition $R = \frac{1}{2}(1 + *)R + \frac{1}{2}(1 - *)R$ of the curvature tensor (for θ) regarded as a $\Lambda^2(4) \cong \mathcal{O}(4) \cong \mathcal{O}(4)_+ \oplus \mathcal{O}(4)_-$-valued 2-form on $\mathbb{R}^4$. Note that here "$*$" is acting on the *values* of R in $\Lambda^2(4)$ (i.e., not on R as a 2-form). In order to denote $*$ of R regarded as a 2-form, we write $*$ on the right. In other words, relative to an orthonormal (relative to h) frame field on $\mathbb{R}^4$, we have for components

$$(*R)_{hijk} = \frac{1}{2}\,\varepsilon_{hipq}R^{pq}{}_{jk}$$

$$(R*)_{hijk} = \frac{1}{2}\,R_{hi}{}^{pq}\,\varepsilon_{pqjk}.$$

Consequently, $F^{\pm} \leftrightarrow \frac{1}{2}(1 \pm *)R$ will be self-dual, when $(1\pm*)R* = (1\pm*)R$. Writing $* = \begin{bmatrix} I & 0 \\ 0 & -I \end{bmatrix}$ and $R = \begin{bmatrix} A & B \\ B^t & C \end{bmatrix}$ as in 2E, this becomes

$$\begin{bmatrix} A\pm A & -(B\pm B) \\ B^t\mp B^t & -(C\mp C) \end{bmatrix} = \begin{bmatrix} A\pm A & B\pm B \\ B^t\mp B^t & C\mp C \end{bmatrix}$$

Thus, F^+ is self-dual $\Leftrightarrow B = 0$, while F^- is self-dual $\Leftrightarrow C = 0$. It turns out to be most productive to construct metrics h with $C = 0$. Now, $C = 0 \Leftrightarrow W^- = 0$ and S = scalar curvature = $4\,\mathrm{tr}C = 0$. In particular, any conformally-flat metric with 0 scalar curvature will do. (In that case, $W^+ = 0$, too, but this is not needed to obtain a self-dual F^-.) We compute the Levi-Civita connection for the conformally flat metric

$$h = f^2((dx^1)^2 + \dots + (dx^n)^2) = f^2 ds^2 \qquad (0 < f \in C^\infty(\mathbb{R}^n)) \tag{1}$$

on $\mathbb{R}^n$, eventually specializing to the case $n = 4$. Note that $\frac{1}{f}\partial_1,\dots,\frac{1}{f}\partial_n$ is a global section σ of the frame bundle $\pi\colon F(\mathbb{R}^n) \to \mathbb{R}^n$ for h. Let $\tilde{\phi}$ be the pull-back $\sigma^*\phi$ of the $\mathbb{R}^n$-valued canonical 1-form ϕ on $F(\mathbb{R}^n)$ given by $\phi_u(X) = u^{-1}(\pi_*X)$, $X \in T_uF(\mathbb{R}^n)$, $u \in F(\mathbb{R}^n)$. Now $(\sigma^*\phi)_x(\frac{1}{f}\partial_i) = \phi_{\sigma(x)}(\sigma_*(\frac{1}{f}\partial_i)) = \sigma(x)^{-1}(\frac{1}{f}\partial_i) = e_i$ = i-th standard unit vector in $\mathbb{R}^n$. Thus, $\sigma^*\phi := (\phi^1,\dots,\phi^n)$ is the dual coframe of $\frac{1}{f}\partial_1,\dots,\frac{1}{f}\partial_n$; i.e., $\phi^i = f\,dx^i$. The Levi-Civita connection θ is the unique $\mathcal{O}(n)$-valued connection 1-form on $F(M)$ with vanishing torsion (i.e., $d\phi + \theta \wedge \phi = 0$). Writing $\sigma^*\theta = (\theta^h{}_i)$, we then obtain

$$d\phi^h = -\sum_{j=1}^{n} \theta^h{}_i \wedge \phi^i;\quad \theta^h{}_i = -\theta^h{}_i.$$

On the other hand,

$$\begin{aligned} d\phi^h &= df \wedge dx^h = \sum_i \partial_i f\, dx^i \wedge dx^h \\ &= -\sum_i f^{-1}\partial_i f\, dx^h \wedge \phi^i \\ &= -\sum_i (f^{-1}\partial_i f\, dx^h - f^{-1}\partial_h f\, dx^i) \wedge \phi^i. \end{aligned}$$

By the uniqueness of the Levi-Civita connection, we have

$$\theta^h{}_i = f^{-1}(\partial_i f\, dx^h - \partial_h f\, dx^i). \tag{2}$$

Since $\Omega^\theta = d\theta + \frac{1}{2}[\theta,\theta] = d\theta + \theta \wedge \theta$, we obtain

$$(\sigma^*\Omega^\theta)^h{}_i = d\theta^h{}_i + \sum_p \theta^h{}_p \wedge \theta^p{}_i.$$

Now,

$$\begin{aligned} d\theta^i{}_j &= -f^{-2}(\partial_q f\, dx^q) \wedge (\partial_i f\, dx^h - \partial_h f\, dx^i) \\ &\qquad + f^{-1}(\partial_s\partial_i f\, dx^s \wedge dx^h - \partial_s\partial_h f\, dx^s \wedge dx^i) \end{aligned}$$

and

$$\theta^h_p \wedge \theta^p_i = f^{-2}(\partial_p f\, dx^h - \partial_h f\, dx^p) \wedge (\partial_i f\, dx^p - \partial_p f\, dx^i).$$

Thus,

$$\begin{aligned} R^h{}_{ijk} &= -f^{-2}[\partial_q f\, \partial_i f(\delta^q_j\delta^h_k - \delta^q_k\delta^h_j) - \partial_q f\partial_h f(\delta^q_j\delta^i_k - \delta^q_k\delta^i_j) \\ &\quad + f^{-1}[\partial_s\partial_i f(\delta^s_j\delta^h_k - \delta^s_k\delta^h_j) - \partial_s\partial_h f(\delta^s_j\delta^i_k - \delta^s_k\delta^i_j)] \\ &\quad + f^{-2}(\partial_p f\partial f_i(\delta^h_j\delta^p_k - \delta^h_k\delta^p_j) + \partial_h f\partial_p f(\delta^p_j\delta^i_k - \delta^p_k\delta^i_j) \\ &\qquad - \Sigma(\partial_p f)^2(\delta^h_j\delta^i_k - \delta^h_k\delta^i_j)]. \end{aligned}$$

Lowering the index h means multiplying by f^2, and then using the notation of 2E, we have

$$R = (R_{hijk}) = 8[(df \otimes df) \vee I] - 4f[\nabla^2 f \vee I] - 2|\nabla f|^2[I \vee I]. \tag{3}$$

Consequently, the scalar curvature is

$$\begin{aligned} S &= 8f^{-4}\,\tfrac{1}{2}(n-1)|\nabla f|^2 - f^{-3}4\,\tfrac{1}{2}(n-1)\Delta f - 2f^{-4}|\nabla f|^2\,\tfrac{1}{2}(n-1)n \\ &= (4-n)(n-1)f^{-4}|\nabla f|^2 - 2(n-1)f^{-3}\Delta f. \end{aligned}$$

For $n = 4$, we then have $S = 0 \iff \Delta f = 0$ (i.e., f is harmonic).

By construction, $F^- := \frac{1}{2}(1 - *)R$ is self-dual for $\Delta f = 0$; moreover, F^- is the field strength of $A^- := \tau^*\theta^- \leftrightarrow \frac{1}{2}(1 - *)(\theta^i{}_j)$. We can express A^- directly in terms of f, as follows. The projection of the $\mathcal{O}(4)$-valued 1-form $\tau^*\theta$ onto the subspace of $\mathcal{O}_-$-valued 1-forms is

$$\frac{1}{4}\sum_{a=1}^{3} [(\tau^*\theta)^{ij}\,\bar{\eta}_{aij}]\bar{\eta}_a,$$

where the $\frac{1}{4}$ comes from the fact that $\bar{\eta}_a{}^{ij}\bar{\eta}_{aij} = 4$ ("a" fixed, sum over i and j). Under the isomorphism $\mathcal{O}(4)^- \to SU(2)$, this projection is sent to

$$A^- = \frac{i}{4}\sum_{a=1}^{3} (\bar{\eta}_a{}^{ij}\theta_{ij})\sigma_a = \frac{i}{2}\sum_{a=1}^{3} \bar{\eta}_{ai}{}^j\partial_j(\log f)\sigma_a dx^i, \tag{4}$$

since by equation (2), $\theta_{ij} = \partial_j(\log f)dx^i - \partial_i(\log f)dx^j$. According to our above expression (3) for R, we get

$$F^-_{jk} = \frac{i}{4}\sum_{a=1}^{3} \sigma_a\bar{\eta}_{ah}{}^i(P \vee I)^h{}_{ijk}, \tag{5}$$

where

$$P := (8\, df \otimes df - 4f\nabla^2 f - 2|\nabla f|^2 I). \tag{6}$$

We now consider some obvious choices for the dilation factor f with $\Delta f = 0$. Of course $f = 1$ yields the standard metric (see (1)) with $R = 0$ and $A^- = 0$. Consider $f = \lambda^2 r^{-2}$ $(r^2 = x_1^2 + \dots + x_4^2,\ 0 < \lambda \in \mathbb{R})$. Then one computes $R = 0$ also, either by direct substitution, or by noting that $\lambda^4 r^{-4} ds^2$ is the pull-back of the flat metric ds^2 under inversion $x \longmapsto \lambda^2|x|^{-2}x$ of $\mathbb{R}^4$ in the sphere $|x| - \lambda$. If we add, taking $f = 1 + \lambda^2 r^{-2}$, then $\Delta f = 0$, but $f^2 ds^2$ is no longer flat! Indeed, we obtain

$$P = -4\,\lambda^2\nabla^2(r^{-2}) \tag{7a}$$

$$\nabla^2(r^{-2})_{ij} = \begin{cases} 8x_i x_j r^{-6} & i \neq j \\ (8x_i^2 - 2r^2)r^{-6} & i = j \end{cases} \tag{7b}$$

To avoid confusion in what follows, we will denote norms relative to $h = f^2 ds^2$ by $|\ |_h$ and norms relative to ds^2 by $|\ |_e$. We need to compute the integral $\int |F^-|_e^2 \nu_e$ which is really the same as $\int |F^-|_h^2 \nu_h$ by conformal invariance of YM (cf. Section A). In $|F^-|_e^2$ or $|F^-|_h^2$, we have also used the metric $k(A,B) = -tr(ad(A)\circ ad(B))$ on $SU(2)$;

one can check that $k(i\sigma_a, i\sigma_b) = 8\delta_{ab}$. Since $\bar{\eta}_a \longmapsto i\sigma_a$ induces an isomorphism of Lie algebras, we know $k(\bar{\eta}_a, \bar{\eta}_b) = 8\delta_{ab}$ for the Killing form $-k$ on $\mathcal{O}(4)$; this is twice the usual contraction inner product $\bar{\eta}_a^{ij}\bar{\eta}_{bij} = 4\delta_{ab}$. Moreover, in the case at hand, F^- and F^+ are of equal magnitude since $\frac{1}{2}(1+*)\begin{bmatrix} 0 & B \\ B^t & 0 \end{bmatrix} = \begin{bmatrix} 0 & B \\ 0 & 0 \end{bmatrix}$ and $\frac{1}{2}(1-*)\begin{bmatrix} 0 & B \\ B^t & 0 \end{bmatrix} = \begin{bmatrix} 0 & 0 \\ B^t & 0 \end{bmatrix}$.

Combining these facts, we have

$$\begin{aligned} |F^-|_e^2 &= \sum_{i<j} (F_{ij}^-)^2 = \tfrac{1}{2}(|F^-|_e^2 + |F^+|_e^2) \\ &= \frac{1}{2}\cdot 2 \sum_{i<j;a,b} (R^a{}_{bij})^2 = \frac{1}{2} \sum_{i,j,a,b} (R^a{}_{bij})^2 \\ &= \tfrac{1}{2}|R^\#|_e^2 = \tfrac{1}{2}|(PvI)^\#|_e^2 = \tfrac{1}{2} f^{-4}|PvI|_e^2 \\ &= \tfrac{1}{4} f^{-4}|P|_e^2, \end{aligned} \tag{8}$$

where we used $|PvI|_e^2 = \frac{1}{4}\,\mathrm{tr}(P)^2 + \frac{1}{4}(n-2)|P|_e^2$ (cf. 2E) and $\mathrm{tr}(P) = 0$.

From (7b), we get $|\nabla^2(r^{-2})|_e^2 = 48r^{-8}$; thus (7a) and (8) yield

$$|F^-|_e^2 = \tfrac{1}{4}(1 + \lambda^2 r^{-2})^{-4}\cdot 16\cdot\lambda^4\cdot 48r^{-8} = \frac{192\,\lambda^4}{(r^2+\lambda^2)^4}\,. \tag{9}$$

We compute

$$\begin{aligned} \int_{\mathbb{R}^4} (r^2+\lambda^2)^{-4} &= \mathrm{Vol.}(S^3)\int_0^\infty r^3(r^2+\lambda^2)^{-4}dr \\ &= \ldots(\text{set } y = \lambda^2+r^2)\ldots = 2\pi^2 \int_{\lambda^2}^\infty (y-\lambda^2)y^{-4}\,\tfrac{1}{2}\,dy \\ &= -\pi^2(\tfrac{1}{2}\,y^{-2} - \tfrac{1}{3}\,\lambda^2 y^{-3})\Big|_{\lambda^2}^\infty = \tfrac{1}{6}\,\pi^2\lambda^{-4}. \end{aligned}$$

If we replace ∞ by $2\lambda^2$, we get $\frac{1}{12}\pi^2\lambda^{-4}$; i.e., the integral over the ball of radius $r = \lambda$ is half of the integral over $\mathbb{R}^4$. Thus, we have

$$\int_{\mathbb{R}^4}|F^-|_e^2 = 32\pi^2 = 2\int_{|r|\leq\lambda}|F^-|_e^2. \tag{10}$$

We have found a finite-action self-dual gauge potential (instanton) A^- defined on $\mathbb{R}^4-\{0\}$ by (4) with $f = 1 + \lambda^2 r^{-2}$. Since half of the action is in the ball $r \leq \lambda$, the *size* of the instanton is λ. Although the results in [K. Uhlenbeck 1982a] tell us that the trivial bundle $P_0 := (\mathbb{R}^4 - \{0\}) \times SU(2)$ and the connection (say ω, such that $\sigma^*\omega = A^-$ for some section σ) can be smoothly extended across 0 and ∞ to form a (nontrivial) bundle P over S^4 with smooth connection, it is instructive to work explicitly in the case at hand. In particular, the singularity in $A^- = \sigma^*\omega$ at 0 can be removed by a change in the choice of gauge σ. More precisely, suppose $g\colon \mathbb{R}^4 - \{0\} \to SU(2)$ is C^∞, and let $\tilde{\sigma}(x) = \sigma(x)g(x)$. Let $H \in GA(P_0)$ be a gauge transformation such that $H \circ \sigma = \tilde{\sigma} = \sigma g$, and let $h\colon P_0 \to SU(2)$ be associated with H via $H(p) = ph(p)$. Then Prop. 3 of 2D yields $H^*\omega = h^{-1}dh + h^{-1}\omega h$, and applying σ^* to both sides, yields

$$\tilde{A}^- := \tilde{\sigma}^*\omega = g^{-1}dg + g^{-1}A^-g. \tag{11}$$

Then it is claimed that there is a g such that $\tilde{A}^-$ is smoothly extendable across $0 \in \mathbb{R}^4$. The $g\colon \mathbb{R}^4 - \{0\} \to SU(2)$ that does the trick is defined by

$$g(x) = \frac{1}{|x|}\begin{bmatrix} x_4 - ix_3 & -x_2 - ix_1 \\ x_2 - ix_1 & x_4 + ix_3 \end{bmatrix} = \frac{1}{|x|}(x_4 I - i\vec{\sigma}\cdot\vec{x}).$$

A rather lengthy computation (left as an exercise for the patient), reveals that for this g,

$$\tilde{A}^- = \left(\frac{r^2}{\lambda^2+r^2}\right)g^{-1}dg = (\lambda^2+r^2)^{-1}\sum_a i\sigma_a \eta^a{}_{ij}x^i dx^j \tag{12}$$

which is smooth even at $x = 0$. We can show that A^- itself extends across ∞ as follows. We have $A^- = g\tilde{A}^-g^{-1} - (dg)g^{-1} = (-\lambda^2/\lambda^2+r^2)(dg)g^{-1}$ by (12). Note that g is invariant under inversion $I(x) = |x|^{-2}x$, whence $(dg)g^{-1}$ is invariant, and so A^- pulls back to $\left(\frac{-\lambda^2}{\lambda^2+r^{-2}}\right)(dg)g^{-1} = \left(\frac{-\lambda^2}{\lambda^2r^2+1}\right)r^2(dg)g^{-1}$, but $r^2(dg)g^{-1}$ is smooth by inspection; thus, $I^*(A^-)$ extends across 0, and A^- extends across ∞. Observe that g serves as a clutching function for the trivial bundles over $S^4 - \{0\}$ and $S^4 - \{\infty\}$ on which A^- and $\tilde{A}^-$ are defined. Since, g is essentially the identity on S^3 (regarding SU(2) as the unit quaternions), we suspect that the Chern number of the associated vector bundle $E' = P \times_{SU(2)} \mathbb{C}^2$ for the resulting SU(2)-bundle P over S^4 is -1. Actually, we can prove this as follows. By definition E' is obtained from the inclusion representation $r\colon SU(2) \to GL(\mathbb{C}^2)$. Now, $r \otimes r\colon SU(2) \to GL(\mathbb{C}^2 \otimes \mathbb{C}^2) \cong GL(\mathrm{Hom}\,\mathbb{C})$ decomposes into two representations, namely $Ad\colon SU(2) \to GL(\mathcal{SU}(2)_c)$ and the trivial representation; indeed any $C \in \mathrm{Hom}(\mathbb{C}^2)$ can be written uniquely in the form $C = zB + wI$ where $z,w \in ₵$, $B \in \mathcal{SU}(2)$. Defining $E = P \times_{SU(2)} \mathcal{SU}(2)_c$, we then have $E' \otimes E' \cong E \oplus 1$, and so $ch(E')^2 = ch(E) + ch(1)$. As previously noted in 2.G, $c_1(E') = c_1(E) = 0$ (E' and E are reduced to $SU(2)$). Writing $ch(E') = 2 + kx$ (where x is the positive generator of $H^4(S^4)$, and $k \in \mathbb{Z}$), we obtain $(2+kx)^2 = ch(E')^2 = ch(E) + 1$, and so $ch(E) = 3 + 4kx$, in which case

$$4k = ch(E)[S^4] = \frac{1}{8\pi^2}\int_{S^4} \langle\Omega^\omega, *\Omega^\omega\rangle \nu \tag{13}$$

as shown in Section A. The general formula $ch = \dim + c_1 + \frac{1}{2}(c_1^2 - 2c_2) + \ldots$ yields $c_2(E')[S^4] = -ch(E')[S^4] = -k$. Hence, for a self-dual connection ω on an $SU(2)$-bundle over S^4 (actually, we may replace S^4 by any compact, orientable, Riemannian 4-manifold), equation (13) yields

$$k = -c_2(E')[S^4] = \frac{1}{32\ \pi^2}\int_{S^4} |\Omega^\omega|^2 \nu. \tag{14}$$

Thus, (10) and (14) tell us that the instanton over S^4 constructed from A^- and $\tilde{A}^-$ is defined on an $SU(2)$-bundle P with $c_2(E')[S^4] = -1$. Of course, (14) shows that only those $SU(2)$-bundles with $c_2(E)[S^4] \le 0$ are capable of supporting a self-dual connection. So far we have only realized the case where $k = 1$; $k = 0$ is trivial.

It should come as no great surprise that to realize Chern number $-k$, we should consider the dilation $f = 1 + \Sigma_{i=1}^k \lambda_i^2|x-x_i|^{-2}$, where $x_1,\ldots,x_k \in \mathbb{R}^4$ are distinct. Recall from (8) that $|F^-|_e^2 = \frac{1}{4} f^{-4}|P|_e^2$; writing $f_i := \lambda_i^2|x-x_i|^{-2}$, we have

$$\begin{aligned} P &= (8\ df \otimes df - 4f\nabla^2 f - 2|\nabla f|_e^2 I) \\ &= -4\sum_i \nabla^2 f_i + \sum_{i,j} 8\ df_i \otimes df_j - 4f_i\nabla^2 f_j - 2\langle df_i, df_j\rangle_e I. \end{aligned} \tag{15}$$

The terms of the sum for which $i = j$ vanish, since $f_i^2 ds^2$ is flat (cf. discussion after (6)). Setting $r_i(x) := |x-x_i|$, we have

$$f_i = \lambda_i^2 r_i^{-2}, \quad |df_i|_e^2 = 4\lambda_i^4 r_i^{-6}, \quad |\nabla^2 f_i|_e^2 = 48\ \lambda_i^4 r_i^{-8}. \tag{16}$$

Since the "i = j" terms can be ignored, we have

$$|P|_e^2 = 16|\nabla^2 f_i|_e^2 + O(r_i^{-6}) = 16\cdot 48\ \lambda_i^4 r_i^{-8} + O(r_i^{-6}) \tag{17}$$

near x_i as $r_i \to 0$. Thus, we have

$$|F^-|_e^2 = \frac{1}{4} f^{-4}|P|_e^2 = (1 + \lambda_i^2 r_i^{-2} + \sum_{j\neq i} \lambda_j^2 r_j^{-2})^{-4}[4\cdot 48\lambda_i^4 r_i^{-8} + O(r_i^{-6})]$$
$$= [r_i^2 + \lambda_i^2 + \sum_{j\neq i} \lambda_j^2 (r_i/r_j)^2]^{-4}\ 192\ \lambda_i^4 + O(r_i^2). \tag{18}$$

Hence $|F^-|_e^2$ is bounded (indeed, smooth) about the x_i. Also, as $r = |x| \to \infty$, we have

$$|F^-|_e^2 \leq 4f^{-4} \sum_i |\nabla^2 f_i|_e^2 + O(r^{-10}) \leq Cr^{-8}, \tag{19}$$

and so $\int |F^-|_e^2 < \infty$. Moreover, from (16) we know that C in (19) tends to 0 as $\lambda_i \to 0$, $i = 1,\ldots,k$. By using K. Uhlenbeck's theorem (or by directly using some clutching functions g_i about x_i, as before), we know that A^- (defined on $S^4 - \{x_1,\ldots,x_k,\infty\}$ is actually the pull-back of a smooth self-dual connection ω on a SU(2)-bundle P over S^4. Thus, we know that $\int |F^-|_e^2$ is a constant function of the variables $\lambda_1,\ldots,\lambda_k > 0$ and distinct $x_1,\ldots,x_k$; indeed we know from (14) that the integral is $32\ \pi^2 N$ for some $N = 0,1,2,\ldots$. To prove that $N = k$ (i.e., ω realizes the case of Chern number $-k$), we proceed as follows. Select the x_i to be far enough apart so that if B_i is the ball of radius 1 about x_i, then $r_i/r_j \leq \varepsilon$ in B_i for $j \neq i = 1,\ldots,k$. Using formulas (15)-(18), we see that

$$\int_{B_i} |F^-|_e^2 = 192\ \lambda_i^4 \int_{B_i} [r_i^2 + \lambda_i^2]^{-4} + O(\varepsilon)$$
$$= 32\ \pi^2 + O(\lambda_i) + O(\varepsilon) \quad \text{as} \quad \varepsilon,\lambda_i \to 0, \tag{20}$$

where we have used the analysis leading to (10) for the second equality. By (19) and the fact that C in (19) tends to 0 as $\lambda_i \to 0$ for $x \in B' := \mathbb{R}^4 - \bigcup_i B_i$, we obtain $\int_{B'} |F^-|_e^2 \to 0$ as the $\lambda_i \to 0$. Combining this with (20) we get that (for ε and the λ_i sufficiently small) $\int |F^-|_e^2$ can be made arbitrarily close to $32\ \pi^2 k$. Since the only number of the form $32\ \pi^2 N$ that is arbitrarily close to $32\ \pi^2 k$ is $32\ \pi^2 k$, we conclude $\int |F^-|_e^2 = 32\ \pi^2 k$, as desired.

We have constructed a family of self-dual connections on an SU(2)-bundle P over S^4 such that $c_2(P \times_{SU(2)} \mathbb{C}^2) = -k$. The family depends on $5k$ parameters $\lambda_1,\ldots,\lambda_k$, $x_1,\ldots,x_k$ within the function $f = (1 + \sum_i \lambda_i^2 |x-x_i|^{-2})$. Recall that $f^2 ds^2$ on $\mathbb{R}^4 - \{x_1,\ldots,x_k\} = S^4 - \{\infty,x_1,\ldots,x_k\}$ is a conformally flat 0-scalar curvature metric. A conformal transformation (say, relative to the standard metric on S^4)

$C: S^4 \to S^4$ pulls the metric f^2ds^2 back to a metric $C^*(f^2ds^2)$ on $S^4 - \{C^{-1}(\infty), C^{-1}(x_1), \ldots, C^{-1}(x_k)\}$. Since f^2ds^2 and $C^*(f^2ds^2)$ are isometric via C, we know that $C^*(f^2ds^2)$ is also conformally flat with 0 scalar curvature. However, $C^*(f^2ds^2)$ will not be of the same form as f^2ds^2 unless $\infty \in \{C^{-1}(\infty), C^{-1}(x_1), \ldots, C^{-1}(x_k)\}$, in spite of the fact that $C^*(f^2ds^2)$ will still yield a self-dual connection for some $SU(2)$-bundle over S^4 by our original analysis. The problem is that the family $(1 + \Sigma\lambda_i^2|x-x_i|^{-2})^2ds^2$ is not invariant under the inversion $I: x \longmapsto x/|x|^2$. Indeed

$$I^*[(1 + \Sigma\lambda_i^2|x-x_i|^{-2})^2ds^2) = (1 + \Sigma\lambda_i^2|I(x)-x_i|^{-2})^2|x|^{-4}ds^2$$

$$= (|x|^{-2} + \Sigma\lambda_i^2|x|^{-2}||x|^{-2}x-x_i|^{-2})^2ds^2.$$

From the identity $|x|^2\,||x|^{-2}x-x_i|^2 = |x_i|^2|x - |x_i|^{-2}x_i|^2$, the above (assuming $x_i \neq 0$) is

$$= (|x|^{-2} + \Sigma\lambda_i^2|x_i|^{-2}|x - |x_i|^{-2}x_i|^{-2})^2ds^2.$$

If some $x_i = 0$, then $\lambda_i^2|x|^{-2}||x|^{-2}x-x_i|^{-2} = \lambda_i^2$. In any event, $I^*(f^2ds^2)$ is in one of the forms

$$[\Sigma_0^k\ \alpha_i^2|x-y_i|^{-2}]^2ds^2, \qquad [\beta_0^2 + \Sigma_1^k\ \alpha_i^2|x-y_i|^{-2}]^2ds^2.$$

The second form is a limiting case of the first as $y_0 \to \infty$ and $\alpha_0 = \beta_0|y_0|$. The above forms together make a family which is invariant under translations, dilations, and inversions (i.e., all conformal transformations). Indeed, this family yields a $5(k+1)$ parameter family of self-dual connections, but not all of these parameters are effective. Multiplying f by a constant does not change the connection, which means that at most $5k+4$ parameters are effective. Actually, there are $5k+4$ effective parameters only for $k \geq 3$; for $k = 1$, there are only 5 corresponding to the position and size of the BPST instanton; and for $k = 2$, there are 13 effective parameters (cf. [Jackiw-Nohl-Rebbi 1976]). In the next section, we prove (using the Index Theorem) that these families for $k \geq 3$ do not capture all the gauge equivalence classes (moduli) of self-dual connections. The correct number of parameters is $8k-3$, for all $k \geq 1$. There was really no guarantee that the procedure that we used (i.e., generating connections from conformally flat, 0-scalar curvature metrics) would give us all of the instantons. Nevertheless, it could conceivably happen that the only instantons of physical consequence are those that arise in this fashion. The metrics f^2ds^2 are interesting in their own right. Unlike the connection A^-, these metrics are genuinely singular at the singularities of f. Indeed, it is not hard to see that the metric f^2ds^2 makes a punctured neighborhood of one of these points x_i asymptotically isometric to the *exterior* of a ball in Euclidean space, with x_i corresponding to ∞. Since the Weyl tensor W is a measure of the purely gravitational energy density (the other fields determining the Ricci tensor), we can regard f^2ds^2 as a "gravitational instanton" connecting the asymptotically Euclidean regions about the x_i.

C. Linearization of the "Manifold" of Moduli of Self-Dual Connections

Recall from 2D that the action of $GA(P)$ on the pre-Hilbert spaces $\overline{\Omega}^k(P,W)$ is by isometries. Moreover, for $F \in GA(P)$, the curvature of $F^*\omega$ is $d(F^*\omega) + \frac{1}{2}[F^*\omega,F^*\omega] = F^*(d\omega + \frac{1}{2}[\omega,\omega]) = F^*\Omega^\omega$, for $\omega \in C(P)$ = space of connections. Thus, we have $YM(F^*\omega) = \frac{1}{2}||F^*\Omega^\omega||^2 = \frac{1}{2}||\Omega^\omega||^2 = YM(\omega)$, which means that $YM: C(P) \to \mathbb{R}$ is invariant under the action of $GA(P)$ on $C(P)$. In particular, the set of critical points of YM is preserved by this action, as well as the (possibly empty) set of self-dual connections (absolute minima of YM, if they exist); this can also be verified using the fact that $*$ commutes with the action of $GA(P)$ on $\Omega^2(M,P \times_G G) \cong \overline{\Omega}^2(P,G)$. The quotient space $M := C(P)/GA(P)$ is a natural object for study. In particular, one would like to know the extent to which we may regard M as a manifold, say modeled on some infinite dimensional Fréchet space. Also, if $C(P)^+$ is the space of self-dual connections (i.e., with self-dual curvature), it would be interesting to compute the dimension of "the space of moduli of self-dual connections" $M^+ := C(P)^+/GA(P)$, at least where it is a submanifold of M. In the heuristic discussion which follows, we will talk rather loosely about infinite dimensional manifolds, and their tangent spaces and normal spaces. However, this discussion will motivate a precise theorem with a precise proof. In Section D, we will provide the framework within which it makes sense to speak of M^+ as being a submanifold of M.

Heuristic discussion: The "tangent space" at $\omega \in C(P)$ of the orbit $GA(P)\cdot\omega := \{F\cdot\omega: F \in GA(P)\} = \{\Phi(f)\cdot\omega: f \in C(P,G)\}$ (cf. 2D for notation) is $T_\omega[GA(P)\cdot\omega] := \{\frac{d}{dt}\Phi(\text{Exp } ts)\cdot\omega|_{t=0} : s \in C(P,G)\} = \{D^\omega s: s \in C(P,G)\}$, according to the Corollary of Prop. 3 of 2D. Note that $T_\omega[GA(P)\cdot\omega] \subset \overline{\Omega}^1(P,G)$ which is the vector space on which the affine space $C(P)$ is modeled. For $\tau \in \overline{\Omega}^1(P,G)$ and $s \in C(P,G)$, we have $(\tau,D^\omega s) = (\delta^\omega\tau,s)$, whence the "normal space" to $GA(P)\cdot\omega$ at ω is $N_\omega[GA(P)\cdot\omega] := \{\tau \in \overline{\Omega}^1(P,G): \delta^\omega\tau = 0\}$. Let $S_\omega := \{\omega+\tau : \tau \in N_\omega[GA(P)]\}$ be the "slice of the action". We suspect that every connection ω' in a "suitably small neighborhood" of ω will be gauge-equivalent to a connection in S_ω; i.e., there is some $F \in GA(P)$ such that $F\cdot\omega' \in S_\omega$. The subgroup $I_\omega := \{F \in GA(P): F\cdot\omega = \omega\}$ leaves S_ω setwise fixed ($F \in I_\omega \Rightarrow F\cdot(\omega+\tau) = \omega+F\cdot\tau$ and $\delta^\omega(F\cdot\tau) = F\cdot\delta^\omega\tau$). Thus, unless I_ω acts trivially on S_ω, there will be gauge-equivalent connections in S_ω, and so S_ω will not parametrize $M = C(P)/GA(P)$ in a 1-1 fashion even locally about ω. Note that if $g \in A :=$ center of G, then $R_g: P \to P$ is in $GA(P)$ and for any $\omega' \in C(P)$ we have $R_g^*\omega' = \omega'$ by Prop. 3 of 2D. A condition which implies that I_ω consists of only these "central" gauge transformations is that ω be irreducible; this is immediate from Prop. 1 of 2H. The center of a semi-simple, compact G must be finite. Thus, for irreducible ω, there can be no one-parameter subgroup in I_ω; if this latter condition holds, we call ω *weakly-irreducible*. Now, suppose $\omega \in C(P)^+$ is a weakly-irreducible self-dual connection. Then $\omega' \in C(P)^+ \cap S_\omega \Leftrightarrow *\Omega^{\omega'} = \Omega^{\omega'}$ and $\delta^\omega(\omega'-\omega) = 0$. Writing $\tau = \omega'-\omega$, we have $\Omega^{\omega'} = d\omega' + \frac{1}{2}[\omega',\omega'] = d\omega + \frac{1}{2}[\omega,\omega] + d\tau + [\omega,\tau] + \frac{1}{2}[\tau,\tau] = \Omega^\omega + D^\omega\tau + \frac{1}{2}[\tau,\tau]$. Thus,

$$\begin{aligned} *\Omega^{\omega'} = \Omega^{\omega'} &\Leftrightarrow (1-*)(D^\omega\tau + \tfrac{1}{2}[\tau,\tau]) = 0 \\ \omega' \in S_\omega &\Leftrightarrow \delta^\omega\tau = 0. \end{aligned} \tag{*}$$

Let $t \longmapsto \omega + \tau(t)$ be a smooth curve in $C(P)^+ \cap S^\omega$ with $\tau(0) = 0$; then, substituting into (*) and differentiating with respect to t at $t = 0$, we get $(1-*)(D^\omega[\tau'(0)]) = 0$ and $\delta^\omega[\tau'(0)] = 0$. Hence, if $C(P)^+ \cap S_\omega$ is a submanifold of $C(P)$, then its tangent space at ω is $\{\tau \in \overline{\Omega}^1(P,G): (1-*)D^\omega\tau = 0 \text{ and } \delta^\omega\tau = 0\}$.

The preceding discussion motivates the following key result which specifies the dimension of the hypothetical manifold $C(P)^+ \cap S_\omega$ near a weakly-irreducible self-dual connection ω defined on a principal G-bundle P (G compact and semi-simple) over a suitable base. The original proof in [Atiyah-Hitchin-Singer 1978] makes use of the Dirac operator on spinor fields and the $\hat{A}$ genus, along with the Index Theorem. The proof here is spinor-free, and is essentially Hodge-theoretic; the Index Theorem is used mainly to handle the "twist" in the bundle $P \times_G G$.

Theorem 1. Let $\omega \in C(P)$ be a weakly-irreducible self-dual connection on a principal G-bundle P over a self-dual, compact, oriented Riemannian 4-manifold with non-negative scalar curvature $S \not\equiv 0$. Then the vector space of all $\tau \in \bar{\Omega}^1(P,G)$ satisfying $(1-*)D^\omega\tau = 0$ and $\delta^\omega\tau = 0$ has dimension

$$2\mathrm{ch}(P \times_G G_c)[M] - \frac{1}{2}\dim(G)(\chi(M) - \mathrm{sgn}(M)),$$

where $\chi(M)$ is the Euler characteristic of M and $\mathrm{sgn}(M)$ is the signature of M.

Proof: Let E be the associated complex vector bundle $P \times_G G_c$, $G_c :=$ $G \otimes \mathbb{C}$. As in Section 2C, the space $\Omega^k(E)$ of E-valued k-forms on M can be identified with $\bar{\Omega}^k(P,G_c)$, a space of equivariant k-forms on P. Let $\Omega^2_-(E)$ denote the space of anti-self-dual E-valued 2-forms on M. Define $T\colon \Omega^1(E) \to \Omega^0(E) \oplus \Omega^2_-(E)$ by

$$T(\tau) = \delta^\omega\tau \oplus (1-*)D^\omega\tau, \quad \tau \in \Omega^1(E) \cong \bar{\Omega}^1(P,G_c).$$

Of course, $\Omega^0(E) \oplus \Omega^2_-(E)$ can be regarded as the space of sections of the single bundle $E \otimes (\Lambda^0(M) \oplus \Lambda^2_-(M))$, and so T is a differential operator (of order 1). Since $\dim(\Lambda^1(\mathbb{R}^4)) = 4 = \dim \Lambda^0(\mathbb{R}^4) + \dim \Lambda^2_-(\mathbb{R}^4)$, we know that T will be elliptic if its symbol $\sigma(T)$ is injective. From the local formulas for D^ω and δ^ω in 2C, it follows that $\sigma(T) = \mathrm{Id}_E \otimes \sigma$, where

$$\sigma\colon \pi_0^*\Lambda^1(M) \to \pi_0^*(\Lambda^0(M) \oplus \Lambda^2_-(M))$$

($\pi_0\colon S(M) \to M$, $S(M) :=$ unit cosphere bundle) is given (for $\xi \in S(M)_x$, $\eta \in \Lambda^1(M)_x$) by

$$\sigma_\xi(\eta) = -\langle\xi,\eta\rangle \oplus (1-*)(\xi \wedge \eta).$$

Now, $\sigma_\xi(\eta) = 0 \Longleftrightarrow \langle\xi,\eta\rangle = 0$ and $*(\xi \wedge \eta) = \xi \wedge \eta$. If $*(\xi \wedge \eta) = (\xi \wedge \eta)$, then by definition of $*$ (cf., 2C) $|\xi \wedge \eta|^2\nu = (\xi \wedge \eta) \wedge *(\xi \wedge \eta) =$ $(\xi \wedge \eta) \wedge (\xi \wedge \eta) = -\xi \wedge (\eta \wedge \eta) \wedge \xi = 0$, whence $\xi \wedge \eta = 0$, and so $\eta = c\xi$ for some $c \in \mathbb{R}$. However, if $\langle\xi,\eta\rangle = 0$, then $c = 0$ and $\eta = 0$, whence σ and $\sigma(T) = \mathrm{Id}_E \otimes \sigma$ are injective; and T is elliptic.

Our goal is to compute $\dim(\mathrm{Ker}\, T)$, since $\mathrm{Ker}\, T$ is the vector space in the statement of the theorem. Of course, $\dim \mathrm{Ker}\, T = \mathrm{index}(T)$ if $\mathrm{Ker}\, T^* = 0$, where $T^* =$ adjoint of T. To prove $\mathrm{Ker}\, T^* = 0$, we first derive a formula for T^*: For $\alpha \in \Omega^0(E)$, $\beta \in \Omega^2_-(E)$, $\tau \in \Omega^1(E)$,

$$\begin{aligned}(T^*(\alpha \oplus \beta),\tau) &= (\alpha \oplus \beta, T(\tau)) \\ &= (\alpha \oplus \beta,\ \delta^\omega\tau \oplus (1-*)D^\omega\tau) = (\alpha,\delta^\omega\tau) + (\beta,(1-*)D^\omega\tau)\end{aligned}$$

$$= (\alpha,\delta^{\omega}\tau) + (\beta,D^{\omega}\tau) = (D^{\omega}\alpha + \delta^{\omega}\beta,\tau),$$

and so $T^*(\alpha \oplus \beta) = D^{\omega}\alpha + \delta^{\omega}\beta$. Since $(D^{\omega}\alpha,\delta^{\omega}\beta) = (D^{\omega}D^{\omega}\alpha,\beta) = ([\Omega^{\omega},\alpha],\beta) = 0$ (because $\Lambda^2_+ \perp \Lambda^2_-$), we have $T^*(\alpha \oplus \beta) = 0 \Rightarrow D^{\omega}\alpha = 0$ and $0\ \delta^{\omega}\beta = -{*}D^{\omega}{*}\beta = -{*}D^{\omega}\beta$ (i.e., $D^{\omega}\alpha = 0$ and $D^{\omega}\beta = 0$). If $D^{\omega}\alpha = 0$, then (in the notation of 2D) $\{\Phi(\mathrm{Exp}(t\alpha)) : t \in \mathbb{R}\}$ is a 1-parameter group of gauge transformations fixing ω. Indeed, by Prop. 3 of 2D, we have

$$\frac{d}{dt}\,\Phi(\mathrm{Exp}\ t\alpha)\cdot\omega = \frac{d}{dt}[\mathrm{Exp}(-t\alpha)d\ \mathrm{Exp}(t\alpha) + \mathrm{Exp}(-t\alpha)\omega\ \mathrm{Exp}(t\alpha)]$$
$$= \mathrm{Exp}(-t\alpha)(d\alpha + [\omega,\alpha])\mathrm{Exp}(t\alpha);$$

thus, $D^{\omega}\alpha = 0$ implies $\Phi(\mathrm{Exp}\ t\alpha)\cdot\omega$ is constant, and hence equals ω for all t. The semi-simplicity of G and the weak-irreducibility of ω then imply that $\alpha = 0$. Indeed, suppose $\alpha(p) \neq 0$. Because of the weak-irreducibility of ω, we know that $\mathrm{Exp}(t\alpha(p))$ is in the center of G for all t; and so $[\alpha(p),A] = 0$ for all $A \in \mathcal{G}$. Then $0 = tr[ad(\alpha(p))\circ ad(A)] = -k(\alpha(p),A)$, and so k is nondegenerate in violation of semi-simplicity. Hence, $D^{\omega}\alpha = 0 \Rightarrow \alpha = 0$, under the assumptions. To complete the proof that $\mathrm{Ker}(T^*) = 0$, we need to prove "$\delta^{\omega}\beta = 0 \Rightarrow \beta = 0$" for an anti-self-dual 2-form β. We have already seen that $\delta^{\omega}\beta = 0 \Rightarrow D^{\omega}\beta = 0$, whence $\Delta^{\omega}\beta = (\delta^{\omega}D^{\omega} + D^{\omega}\delta^{\omega})\beta = 0$. Thus, $\beta = 0$ by the final Corollary in 2F. Hence, $\mathrm{Ker}(T^*) = 0$, and $\dim(\mathrm{Ker}\ T) = \mathrm{index}(T)$.

To compute index T, we note that the symbol $\sigma(T)$ is $\mathrm{Id}_E \otimes \sigma(T_0)$ where $T_0\colon \Omega^1(M) \to \Omega^0(M) \oplus \Omega^2_-(M)$ is the "untwisted" (i.e., coefficients no longer in E) operator given by $T_0(\eta) = \delta\eta \oplus (1-*)d\eta$ for a real-valued 1-form η on M. Using the Index Theorem (cf. III.4.A), we have

$$\mathrm{index}\ T = \{(\phi^{-1}\ \mathrm{ch}[\sigma(T)]) \cup \tau(TM \otimes \mathbb{C})\}[M]$$
$$= \{(\phi^{-1}\mathrm{ch}[\mathrm{Id}_E \otimes \sigma(T_0)]) \cup \tau(TM \otimes \mathbb{C})\}[M]$$
$$= \{\mathrm{ch}(E) \cup (\phi^{-1}\mathrm{ch}[\sigma(T_0)]) \cup \tau(TM \otimes \mathbb{C})\}[M].$$

Now, $\mathrm{ch}(E) = \dim \mathcal{G} + \mathrm{ch}(E)_2$, since $\mathrm{ch}(E)_1 = c_1(E) = 0$; indeed, $E = P \times_G \mathcal{G}_c$ is reducible to $SU(\dim \mathcal{G})$, because $Ad\colon G \to GL(\mathcal{G})$ is orthogonal relative to k and Remark a) at the end of 2G applies. It is not particularly easy to show that the 0-degree part of $\phi^{-1}(\mathrm{ch}[\sigma(T_0)]) \cup \tau(TM \otimes \mathbb{C})$ is 2; we will prove this last, but assuming it, we then have

$$\mathrm{index}\ T = 2\ \mathrm{ch}(E_2)[M] - \dim G\ \mathrm{index}\ (T_0).$$

To prove $\mathrm{index}\ T_0 = -\frac{1}{2}(\chi(M) - \mathrm{sgn}(M))$, let $b_k := \dim H^k(M,\mathbb{R})$. By Hodge

Theory (cf. III.4.D), we may take $H^k(M, \mathbb{R})$ to be the space of harmonic k-forms on M (i.e., $\beta \in H^k(M, \mathbb{R}) \Leftrightarrow d\beta = 0$ and $\delta\beta = 0$). Now $H^2(M, \mathbb{R})$ splits into the (± 1) eigenspaces of $*$, say $H^2_{\pm}(M, \mathbb{R})$ of dimensions $b_2^{\pm}$; $b_2 = b_2^+ + b_2^-$. For $\alpha \in H^2_+(M, \mathbb{R})$ and $\beta \in H^2_-(M, \mathbb{R})$, we have $\alpha \wedge \alpha = \alpha \wedge *\alpha = |\alpha|^2 \nu$ and $\beta \wedge \beta = -\beta \wedge *\beta = -|\beta|^2\nu$, while $\alpha \wedge \beta = -\langle\alpha,\beta\rangle\nu = -\langle\beta,\alpha\rangle\nu = -\beta \wedge *\alpha = -\beta \wedge \alpha$ implies $\alpha \wedge \beta = 0$; thus, $\mathrm{sgn}(M) = b_2^+ - b_2^-$. Since $\chi(M) = b_0 - b_1 + b_2 - b_3 + b_4 = 2-2b_1 + b_2$, we have

$$-\frac{1}{2}(\chi(M) - \mathrm{sgn}(M)) = b_1 - (1 + b_2^-).$$

Thus, it suffices to prove $b_1 = \dim \operatorname{Ker} T_0$ and $1 + b_2^- = \dim \operatorname{Ker} T_0^*$. If $\alpha \in \operatorname{Ker} T_0$, then $\delta\alpha = 0$ and $(1-*)d\alpha = 0$; thus, $0 = \delta(1-*)d\alpha = \delta d\alpha - \delta * d\alpha = \delta d\alpha + *d^2\alpha = \delta d\alpha$. Now $||\delta\alpha||^2 = (d\delta\alpha,\alpha) = 0$ and $||d\alpha||^2 = (\delta d\alpha,\alpha) = 0$, whence $\alpha \in H^1(M, \mathbb{R})$. Since $H^1(M, \mathbb{R}) \subset \operatorname{Ker} T_0$ is clear, we have $b_1 = \dim \operatorname{Ker} T_0$. Suppose $\beta_0 \oplus \beta_2 \in \operatorname{Ker} T_0^*$; i.e., $d\beta_0 + \delta\beta_2 = 0$. Then $0 = \delta(d\beta_0 + \delta\beta_2) = \delta d\beta_0 = \Delta\beta_0$, and β_0 is constant. Thus, $0 = d\beta_0 + \delta\beta_2 = \delta\beta_2 = -*d*\beta_2 = *d\beta_2$, using the fact that β_2 is anti-self-dual. Hence, $\beta_2 \in H^2_-(M, \mathbb{R})$. Hence, $\operatorname{Ker} T_0^* \subset H^0(M, \mathbb{R}) \oplus H^2_-(M, \mathbb{R})$, and the other inclusion is clear.

We now prove that the 0-degree part of $\phi^{-1}(\mathrm{ch}[\sigma(T_0)]) \cup \tau(TM \otimes \mathbb{C})$ is 2. The 0-degree part of $\tau(TM \otimes \mathbb{C})$ is 1 (directly from its definition in III.4.A), and so we need to show $\phi^{-1}(\mathrm{ch}[\sigma(T_0)])$ has 0-degree part = 2. We use the fact that the Thom class $U \in H^4(B(M),S(M))$ and Euler class $\chi(TM) \in H^4(M)$ are related by $\pi^*\chi(TM) = i^*U$, where $\pi: B(M) \to M$ and $i: (B(M),\phi) \to (B(M),S(M))$. For $E = \Lambda^1(M)_c$ and $F = \Lambda^0(M)_c \oplus \Lambda^2_-(M)_c$, we prove

$$\chi(TM) \cup \phi^{-1}(\mathrm{ch}[\sigma(T_0)])_0 = \mathrm{ch}(E)_2 - \mathrm{ch}(F)_2.$$

This identity may seem strange, because the left side seems to depend on $\sigma(T_0)$, while the right side does not. To dispel any doubts, we exhibit the commutative diagram and the computation. (The coefficients are rational.)

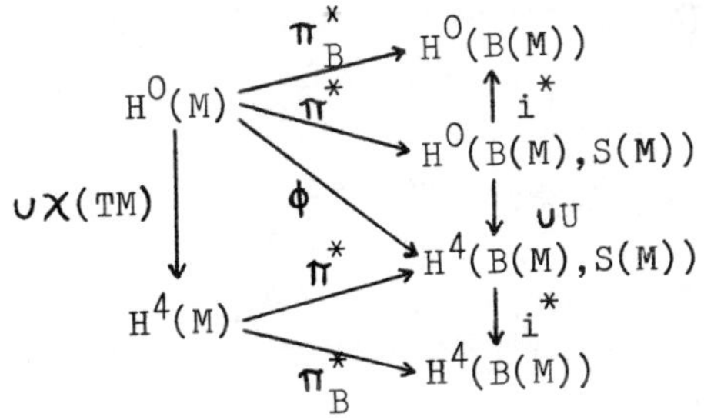

Since ϕ^{-1} lowers degree by 4, $\phi^{-1}(\mathrm{ch}[\sigma(T_0)])_0 = \phi^{-1}(\mathrm{ch}[\sigma(T_0)]_2)$. Thus, we have

$$\pi_B^*(\chi(TM) \cup \phi^{-1}(\mathrm{ch}[\sigma(T_0)])_0) = \pi_B^*\chi(TM) \cup \pi_B^*\phi^{-1}(\mathrm{ch}[\sigma(T_0)]_2)$$
$$= i^*U \cup i^*\pi^*\phi^{-1}(\mathrm{ch}[\sigma(T_0)]_2) = i^*(U \cup \pi^*\phi^{-1}(\mathrm{ch}[\sigma(T_0)]_2))$$
$$= i^*(\mathrm{ch}([\sigma(T_0)]_2)) = \mathrm{ch}(\pi_B^*E)_2 - \mathrm{ch}(\pi_B^*F)_2 = \pi_B^*(\mathrm{ch}(E)_2 - \mathrm{ch}(F)_2).$$

Since M is homotopic to $B(M)$, we know π_B^* is an isomorphism, and the identity is proven. To show $\phi^{-1}(\mathrm{ch}[\sigma(T_0)])_0 = 2$ we need to prove that $\mathrm{ch}(E)_2 - \mathrm{ch}(F)_2 = 2\chi(TM)$. This identity is conveniently proved in terms of forms; see the final remarks in 2G. Note that the Hermitian vector bundles, $E = \Lambda^1(M) \otimes \mathbb{C}$, $F = (\Lambda^0(M) \oplus \Lambda^2_-(M)) \otimes \mathbb{C}$, are complexifications of Riemannian bundles. Hence the principal unitary frame bundles of E and F reduce to principal orthogonal frame bundles, say $O(E)$ and $O(F)$. When the curvature forms of the unitary frame bundles are restricted to $O(E)$ and $O(F)$, they have values in spaces of skew-symmetric real matrices (i.e., Lie algebras of orthogonal groups). Hence, in verifying $\mathrm{ch}(E)_2 - \mathrm{ch}(F)_2 = 2\chi(TM)$, it suffices to work with real skew-symmetric matrices when checking the corresponding identity on the level of invariant polynomials. The representation of $SO(4)$ associated with TM is just the identity $SO(4) \to SO(4)$. The dual rep. is $B \longmapsto (B^{-1})^T$, which is also the identity for $B \in SO(4)$; hence the rep. of $SO(4)$ associated with $\Lambda^1(M)$ is also $\mathrm{Id}\colon SO(4) \to SO(4)$. The rep. for $\Lambda^0(M) \oplus \Lambda^2_-(M)$ is the direct sum of the trivial rep. and the rep. $SO(4) \to GL(\Lambda^2_-(4)) \cong GL(\mathcal{O}_-(4))$. On the Lie algebra level, this later rep. can be expressed in terms of the 't Hooft matrices $\eta_a \in \mathcal{O}_-(4)$ given early in 3B. Indeed, the Lie algebra rep. is $A \to S(A)$, where $S(A)$ is the 3×3 matrix given by $[A,\bar{\eta}_b] = \sum_a S(A)_{ab}\bar{\eta}_a$. We compute:

$$[A,\bar{\eta}_b] = [\tfrac{1}{4}\sum_c (A\cdot\bar{\eta}_c)\bar{\eta}_c,\bar{\eta}_b] = \sum_{a,c} \tfrac{1}{4}(A\cdot\bar{\eta}_c)(-2\varepsilon_{cba}\bar{\eta}_a) = \tfrac{1}{2}\sum_{a,c}(A\cdot\bar{\eta}_c)\varepsilon_{abc}\bar{\eta}_a.$$

Thus, $S(A)_{ab} = \sum_c \frac{1}{2}(A\cdot\bar{\eta}_c)\varepsilon_{abc}$; in particular, $S(A)_{12} = \frac{1}{2}(A\cdot\bar{\eta}_3) = A_{12} - A_{34}$, $S(A)_{23} = A_{23} - A_{14}$ and $S(A)_{31} = A_{31} - A_{24}$. Hence for $A \in \mathcal{O}(4)$,

$$\tfrac{1}{2}\,tr(-A^2) - \tfrac{1}{2}\,tr(-S(A)^2) = (\sum_{i<j} A_{ij}^2) - (A_{12} - A_{34})^2$$
$$- (A_{23} - A_{14})^2 - (A_{31} - A_{24})^2 = 2(A_{12}A_{34} + A_{23}A_{14} + A_{31}A_{24})$$

$$= \frac{1}{4} \sum_{(i)} \varepsilon_{i_1 i_2 i_3 i_4} A_{i_1 i_2} A_{i_3 i_4}.$$

Dividing by $4\pi^2$,

$$\frac{1}{8\pi^2} \, tr(-A^2) - \frac{1}{8\pi^2} \, tr(-S(A)^2) = 2 \cdot \frac{1}{32\,\pi^2} \sum_{(i)} \varepsilon_{i_1 i_2 i_3 i_4} A_{i_1 i_2} A_{i_3 i_4}$$

which gives the desired result $ch(E)_2 - ch(F)_2 = 2\chi(TM)$ upon substitution of the curvature form (say of the Levi-Civita connection of M) for A. □

Remark. When $M = S^4$, we have $\chi(M) = 2$, $sgn(M) = 0$, and for P an $SU(2)$-bundle over S^4 with $c_2(P \times_{SU(2)} \mathbb{C}^2) = -k$, we saw in 3B that $ch(P \times_{SU(2)} \mathcal{SU}(2)_c) = 4k$. Thus, if $\omega \in C(P)$ is a weakly-irreducible self-dual connection on P, we then have that the vector space $\{\tau \in \bar{\Omega}^1(P,\mathcal{G})\colon \delta^\omega\tau = 0,\ (1-*)D^\omega\tau = 0\}$ has dimension $8k - 3$. Any self-dual connection ω constructed in 3B on P with $k \geq 1$ is necessarily irreducible. Indeed, suppose the holonomy group of such an ω is a proper subgroup $G_0 \subsetneq SU(2)$. The Lie algebra $\mathcal{G}_0$ of G_0 can have dimension at most 1; indeed if $v_1, v_2 \in \mathcal{G}_0$ are independent, then $v_1, v_2, [v_1,v_2]$ are independent in $\mathcal{G}_0$ and $\mathcal{G}_0 = \mathcal{SU}(2)$ is mapped via exp onto $SU(2)$ which would then be contained in G_0, a contradiction. By Prop. 2 of 2H, the restriction $\Omega^\omega|P_0$ has values in $\mathcal{G}_0$, and we know that $\Omega^\omega|P_0$ is invariant under G_0 since $Ad\colon G_0 \to GL(\mathcal{G}_0)$ is trivial. Thus, $\Omega^\omega|P_0 = (\pi^*\alpha)|P_0$ for a unique $\mathcal{G}_0$-valued 2-form α on M. For a local section $\sigma\colon M \to P_0$, we have (since $D^\omega\Omega^\omega = 0$ by the Bianchi identity) $0 = \sigma^*(D^\omega\Omega^\omega) = \sigma^*(d\Omega^\omega + [\omega,\Omega^\omega]) = d\sigma^*(\Omega^\omega) + [\sigma^*\omega,\sigma^*\Omega^\omega] = d\alpha + 0$, since $\sigma^*\omega$ and $\sigma^*\Omega^\omega$ have values in the 1-dimensional $\mathcal{G}_0$. Since $*\Omega^\omega = \Omega^\omega$, we have $*\alpha = \alpha$ and $\delta\alpha = -*d*\alpha = -*d\alpha = 0$. Thus, $\Delta\alpha = (d\delta + \delta d)\alpha = 0$ and $\alpha = 0$ (contradicting $\Omega^\omega \neq 0$ for $k \geq 1$), because there are no harmonic 2-forms on S^4 by Hodge theory, since $H^2(S^4) = 0$; alternatively, one can use the B-W formula [see the special case 3 ($k = 2$, $\dim M = 4$) following the B-W formula in 2F, where ω there is absent or flat, $S > 0$, and $W = 0$]. The results of Section D imply that the $(8k-3)$-dimensional vector space can be identified with the tangent space at $[\omega]$ of an actual *manifold* of moduli of self-dual connections.

D. Manifold Structure for Moduli of Self-Dual Connections

Let $C(P)^+$ be the set of weakly-irreducible self-dual connections on a principal G-bundle P over a compact, oriented, Riemannian 4-manifold M. Assuming $C(P)^+ \neq \emptyset$ and other hypotheses of Theorem 1 of Section C (i.e., M is self-dual with positive scalar curvature $S \not\equiv 0$, or more generally that there be no nonzero Δ^ω-harmonic forms in $\Omega^2_-(P \times_G G)$), we will prove that $C(P)^+ \cap S^\omega$ is a submanifold of $C(P)$ with dimension $2ch(P \times_G G)[M] - \frac{1}{2}\dim G(\chi(M) - \mathrm{sgn}(M))$ in a neighborhood of ω; recall $S_\omega = \{\tau+\omega \in C(P): \delta^\omega\tau = 0\}$, and we continue to assume that G is semi-simple. We go on to show that $C(P)^+_m/GA(P)$ (where $C(P)^+_m \subset C(P)^+$ is defined later) can be made into a Hausdorff topological space such that each class $[\omega] \in C(P)^+_m/GA(P)$ possesses a neighborhood U such that for each $[\omega'] \in U$, there is a unique connection $\phi_\omega([\omega']) \in C(P)^+_m \cap S^\omega$ with $[\phi_\omega(\omega')] = [\omega']$. Moreover, we prove that the collection $\{\phi_\omega: \omega \in C(P)^+\}$ is an atlas, making $C(P)^+_m/GA(P)$ a C^∞ manifold. In doing all of this, we need to state some basic results about L^p-Sobolev Banach spaces of sections of vector bundles. Proofs or references to proofs can be found in the excellent survey article [M. Cantor 1981]. In the following review of these results, M is a *compact* Riemannian n-manifold.

Let $E \to M$ be a C^∞ Hermitian vector bundle. A covariant derivative operator $D: C^\infty(E) \to \Omega^1(E)$ (say coming from a connection on the unitary frame bundle for E) and the covariant derivative $\nabla^\theta: T^{p,q}(M) \to T^{p,q+1}(M)$ (determined by the Levi-Civita connection θ) induce a covariant derivative $\nabla = D \otimes 1 + 1 \otimes \nabla^\theta: C^\infty(E \otimes T^{p,q}(M)) \to C^\infty(E \otimes T^{p,q+1}(M))$. For $k \geq 0$, $p \geq 1$, and $f \in C^\infty(E)$, we set $|f|_{p,k} = \Sigma_{j=0}^k(\int_M|\nabla^j f|^p\nu)^{1/p}$, where $\nabla^j f = \nabla \circ \overset{j}{\ldots} \circ \nabla f \in C^\infty(E \otimes T^{0,j}(M))$ and $|\nabla^j f|$ is computed using the Hermitian structure on E and the Riemannian metric on M. The Sobolev space $W^{p,k}(E)$ is the completion of $C^\infty(E)$ with the norm $|\ |_{p,k}$. The Banach space $W^{p,k}(E)$ coincides with the subspace of $L^p(E)$ $(= W^{p,0}(E))$ consisting of sections that have "weak derivatives" of orders $\leq k$ in L^p. More precisely, $f \in W^{p,k}(E) \Longleftrightarrow$ for each $j \leq k$, there is a $v_j \in L^p(E \otimes T^{0,j}(M))$ such that for all $w \in C^\infty(E \otimes T^{0,j}(M))$, we have $\int_M\langle v_j,w\rangle\nu = \int_M\langle f,(\nabla^*)^j w\rangle\nu$ where $(\nabla^*)^j$ is the formal adjoint of ∇^j. We say "$\nabla^j f = v_j$ in the weak (or distributional) sense". We have the following standard results (cf. [M. Cantor 1981] and [R. S. Palais 1968]).

Proposition 1. Let $D: C^\infty(E) \to C^\infty(F)$ be a linear differential operator of order m (E,F Hermitian vector bundles over M). Then D has a unique continuous extension $D_{p,k+m}: W^{p,k+m}(E) \to W^{p,k}(F)$ for each $k \geq 0$. Moreover, if D is elliptic, then there is a constant $C > 0$ such that if $u \in L^p(E)$ and $Du \in W^{p,k}(F)$ exists weakly, then $u \in W^{p,k+m}(E)$ and $|u|_{k+m,p} \leq C(|Du|_{p,k} + |u|_{p,k})$; C only depends on D, p, and k.

Proposition 2. We have the following compact (completely continuous) inclusions:

A. Let $n = \dim M$, and suppose $0 \leq m < k - \frac{n}{p}$. Then $W_{p,k}(E) \subset C^m(E)$, where $C^m(E)$ is the Banach space of m-times (strongly) differentiable sections, the norm being $||f||_m := \sum_{j=0}^{m} \sup |\nabla^j f|$.

B. For $k > m$ and $k - \frac{n}{p} > m - \frac{n}{q}$, we have $W^{p,k}(E) \subset W^{q,m}(E)$.

Proposition 3. Let $Q: E_1 \times E_2 \to E_3$ be a smooth bilinear map of Hermitian vector bundles over M (still compact, $\dim M = n$). Then Q induces a bounded bilinear map $W^{p,k}(E_1) \times W^{p,j}(E_2) \to W^{p,j}(E_3)$ for $k > n/p$, $0 \leq j \leq k$. (Note that by Prop. 2(A), $W^{p,k} \subset C^0$, whence this induced map may be defined pointwise.)

Proposition 4. Let $D: C^\infty(E) \to C^\infty(F)$ be an elliptic differential operator of order m and let $(D^*)_{p,k+m}: W^{p,k+m}(F) \to W^{p,k}(E)$ be the Sobolev extension of the formal adjoint D^* of D. Then $W^{p,k}(E) = \operatorname{Ker} D \oplus (D^*)_{p,k+m}(W^{p,k+m}(F))$, the direct sum of *closed* subspaces.

Given the estimate in Prop. 1, this last result is not difficult to prove. Moreover, note that Prop. 1 implies that if $u \in L^p(E)$ and $Du = 0$ weakly, then $u \in W^{p,k}(E)$ for all $k \geq 0$, and hence $u \in C^\infty(E)$ by Prop. 2(A) (i.e., weak solutions of $Du = 0$ are actually C^∞ "strong" solutions). Prop. 4 is indispensable in verifying the hypotheses of the following implicit function theorem for Banach spaces in applications where the differential of the map is an elliptic operator.

Implicit Function Theorem 1. Let $F: B_1 \to B_2$ be C^k $(1 \leq k \leq \infty)$ between Banach spaces, and assume

(1) $F_{*x}: (TB_1)_x \to (TB_2)_{F(x)}$ is a surjection.

(2) $(TB_1)_x = \operatorname{Ker}(F_{*x}) \oplus H$, where H is closed.

Then $F^{-1}(F(x))$ is a C^k submanifold of B_1 in a neighborhood of x, and its tangent space at x is $\operatorname{Ker}(F_{*x})$.

We will also need the following version, whose proof is also in [Lang 1972].

Implicit Function Theorem 2. Let B_1, B_2, B_3 be Banach spaces and let $F: B_1 \times B_2 \to B_3$ be a C^k map $(1 \leq k \leq \infty)$ with $F(x_1,x_2) = x_3$. Suppose that the partial derivative of F in the B_2-direction at (x_1,x_2) (i.e., $D_2F(x_1,x_2): B_2 \to B_3$) is a bicontinuous isomorphism. Then there are neighborhoods U_1 of x_1 and U_2 of x_2 such that there is a unique C^k map $G: U_1 \to U_2$ whose graph $\{(x_1,G(x_1)): x_1 \in U_1\}$ is $F^{-1}(x_3) \cap (U_1 \times U_2)$.

In the rest of this section, the assumptions of the opening paragraph are in force (e.g., $\dim M = 4$ now). We set $E := P \times_G \mathcal{G}$ and make the identification of the space $\overline{\Omega}^k(P,\mathcal{G})$ (of equivariant forms on P) with $\Omega^k(E)$.

Theorem 1. Let $\omega \in \mathcal{C}(P)^+$ and recall that $S_\omega := \{\tau \in \overline{\Omega}^1(P,\mathcal{G}): \delta^\omega\tau = 0\}$. For $1 \leq p < \infty$, $0 \leq k < \infty$ and $p(k+1) > 4$, $\mathcal{C}(P)^+ \cap S_\omega$ is a C^∞ submanifold of $\mathcal{C}(P)^{p,\overline{k+1}} := \omega + W^{p,k+1}(E \otimes \Lambda^1(M))$ in a neighborhood U of ω, with $\dim(U \cap \mathcal{C}(P)^+ \cap S_\omega)$ specified in Theorem 1 of Section C.

Proof: Recall that $\omega+\tau \in \mathcal{C}(P)^+ \cap S_\omega \Leftrightarrow 0 = R(\tau) := (\delta^\omega\tau, (1-*)(D^\omega\tau + \frac{1}{2}[\tau,\tau])$. For $\tau \in W^{p,k+1}(E \otimes \Lambda^1(M))$, we have $[\tau,\tau] \in W^{p,k+1}(E \otimes \Lambda^2(M))$ by Prop. 3 (since $p(k+1) > 4$) and $\tau \longmapsto [\tau,\tau] \in W^{p,k+1}(E \otimes \Lambda^2(M)) \subset W^{p,k}(E \otimes \Lambda^2(M))$ defines a bounded bilinear map. Combining this with Prop. 1, we have a continuous extension $R_{p,k+1}: W^{p,k+1}(E \otimes \Lambda^1(M)) \to W^{p,k}(E \otimes [\Lambda^0(M) \oplus \Lambda^2_-(M)])$ of R. Note that $R_{p,k+1}$ is actually C^∞, since it is the sum of bounded linear maps $\delta^\omega_{p,k+1}$ and $(1-*)D^\omega_{p,k+1}$ and a bounded quadratic map $\tau \longmapsto (1-*)[\tau,\tau]$. The differential of $R_{p,k+1}$ at $\tau = 0$ is just the linear part of $R_{p,k+1}$, namely the Sobolev extension $T_{p,k+1}$ of the elliptic operator $T = \delta^\omega \oplus (1-*)D^\omega$ introduced in the proof of Theorem 1 of Section C. Under the assumptions that ω is weakly-irreducible and the space of Δ^ω-harmonic forms in $\Omega^2_-(E)$ is trivial, we know that $\operatorname{Ker} T^* = 0$. Then, applying Prop. 4 with $D = T^*$, we see that $T_{p,k+1}$ is onto. Using Prop. 4 with $D = T$ gives us the splitting needed to apply Implicit Function Theorem 1 to the map $R_{p,k+1}$ at $\tau = 0$. Then $R^{-1}_{p,k+1}(0)$ is a C^∞ submanifold of $W^{p,k+1}(E \otimes \Lambda^1(M))$ in a neighborhood U of 0, and $\dim(R^{-1}_{p,k+1}(0) \cap U) = \dim \operatorname{Ker} T$. It remains to prove that $R^{-1}_{p,k+1}(0)$ consists of C^∞ elements of $W^{p,k+1}(E \otimes \Lambda^1(M))$, and that for suitable U, we have $R^{-1}_{p,k+1}(0) \cap U \subset \mathcal{C}(P)^+$. Since T is elliptic, Prop. 1 and Prop. 3 yield (for $\tau \in W^{p,k+1}(E \otimes \Lambda^1(M))$ with $R_{p,k+1}(\tau) = 0$) $|\tau|_{p,k+2} \leq C(|T_{p,k+1}\tau|_{p,k+1} + |\tau|_{p,k+1}) = C(|[\tau,\tau]|_{p,k+1} + |\tau|_{p,k+1}) \leq C(C'|\tau|^2_{p,k+1} + |\tau|_{p,k+1})$, and replacing k by $k+1, k+2, \ldots$, we have that τ is in $W^{p,j}(E \otimes \Lambda^1(M))$ for all j, whence τ is C^∞

by Prop. 2(A). It remains to show that U can be chosen so that the C^∞ elements of U are weakly-irreducible. For this, note that $\tau \longmapsto (D^{\omega+\tau}: \Omega^0(E) \to \Omega^1(E))$ induces a continuous map from $W^{p,k+1}(E \otimes \Lambda^1(M))$ to the space $B(W^{p,k+1}(E), W^{p,k}(E \otimes \Lambda^1(M)))$ of bounded linear maps. Indeed, $D^{\omega+\tau}(\alpha) = d\alpha + [\omega+\tau,\alpha] = D^\omega\alpha + [\tau,\alpha]$; and so the continuous map is actually affine, and its image is in the space of bounded linear maps by Props. 1, 2, and 3. Since $D^\omega: W^{p,k+1}(E) \to W^{p,k}(E \otimes \Lambda^1(E))$ is injective (recall ω is weakly-irreducible), it then follows that $D^{\omega+\tau}$ will be in the open subset of injective bounded linear maps for $|\tau|_{p,k+1} < \varepsilon$ sufficiently small. Then, choosing $U = \{\omega+\tau: |\tau|_{p,k+1} < \varepsilon\}$, we ensure that the C^∞ elements of U are weakly-irreducible. □

Theorem 2. Let $\omega \in C(P)$ be weakly-irreducible. For $1 \leq p < \infty$, $0 \leq k < \infty$, and $pk > 4$, there are constants $C_1, C_2 > 0$ such that for every $\tau \in W^{p,k+1}(E \otimes \Lambda^1(M))$ with $|\tau|_{p,k+1} < C_1$, there is a unique $s \in W^{p,k+2}(E)$ with $|s|_{p,k+2} < C_2$ such that $\delta^\omega_{p,k+1}(\mathrm{Exp}(s)\cdot(\omega+\tau)-\omega) = 0$ (cf. Prop. 3 of 2D for notation); s is a C^∞ function of τ.

Proof: Let $C^\infty(E) \times C^\infty(E \otimes \Lambda^1(M)) \to C^\infty(E)$ be the map $(s,\tau) \longmapsto \delta^\omega(\mathrm{Exp}(s)\cdot(\omega+\tau)-\omega)$. Writing $f = \mathrm{Exp}(s) \in C(P,G) \subset C(P,\mathrm{End}(\mathbb{C}^N))$, as in Prop. 3 of 2D, we have $f\cdot\omega = fdf^{-1} + f\omega f^{-1}$. Expanding this equation into a power series in s, $D^\omega s$ $(= ds + [\omega,s])$, and τ, we obtain after a computation (where $ad_s(\cdot) = [s,\cdot]$, $ad_s^n = (ad_s)^n$)

$$\mathrm{Exp}(s)\cdot(\omega+\tau)-\omega = -D^\omega s + \tau - \sum_{n=1}^{\infty} \frac{1}{(n+1)!}\, ad_s^n(D^\omega s) + \sum_{n=1}^{\infty} \frac{1}{n!}\, ad_s^n(\tau).$$

Recall the fact that a map $x \longmapsto h(x)$ of Banach spaces is k-times differentiable at x_0 if $h(x)$ is a k-th degree polynomial plus a term $O(|x-x_0|^{k+1})$. Using this and repeated application of Prop. 3, we see that the map $(s,\tau) \longmapsto \mathrm{Exp}(s)\cdot(\omega+\tau)-\omega$ extends to a C^∞ (indeed, analytic) map $W^{p,k+2}(E) \times W^{p,k+1}(E \otimes \Lambda^1(M)) \to W^{p,k+1}(E \otimes \Lambda^1(M))$ for $p(k+2) > 4$. Composing this map with the linear map $\delta^\omega_{p,k+1}$, we get a C^∞ map $F: W^{p,k+2}(E) \times W^{p,k+1}(E \otimes \Lambda^1(M)) \to W^{p,k}(E \otimes \Lambda^1(M))$ that extends the map $(s,\tau) \longmapsto -\delta^\omega(\mathrm{Exp}(s)\cdot(\omega+\tau)-\omega)$. The derivative F_* at $(s,\tau) = (0,0)$ is given by the first-order terms; i.e., $F_*(s,\tau) = -\delta^\omega D^\omega s + \delta^\omega\tau$. Hence the partial derivative of F (at $(0,0)$) in the $W^{p,k+2}(E)$ direction is the Sobolev extension of $-\Delta^\omega = -\delta^\omega D^\omega$, namely $-\Delta^\omega_{p,k+2}: W^{p,k+2}(E) \to W^{p,k}(E)$. Now, $\Delta^\omega: C^\infty(E) \to C^\infty(E)$ is elliptic, formally self-adjoint, and has trivial kernel, since ω is weakly-irreducible. Hence, by Prop. 4, $W^{p,k}(E) = \operatorname{Ker} \Delta^\omega \oplus \operatorname{Im} \Delta^\omega_{p,k+2} = \operatorname{Im} \Delta^\omega_{p,k+2}$; and so $\Delta^\omega_{p,k+2}$ is onto. By the open mapping theorem, $\Delta^\omega_{p,k+2}$ has a continuous inverse. Thus, Implicit Function Theorem 2 applies to give us the constants C_1 and C_2, and the C^∞ function G sending τ to s. □

Note that Theorem 2 provides us with a local slice for the action of the group $GA(P)^{p,k+2} \subset W^{p,k+2}(P \times_G End(\mathbb{C}^N))$ on the space $C(P)^{p,k+1}$; i.e., every connection $\omega' \in C(P)^{p,k+1}$ in a suitably small neighborhood of $\omega \in C(P)^+$ is gauge-equivalent to a connection in $S^\omega_{p,k+1} := \{\omega+\tau: \delta^\omega_{p,k+1}\tau = 0, \tau \in W^{p,k+1}(E \otimes \Lambda^1(M))$ via an element $\operatorname{Exp} s$ in $GA(P)^{p,k+2}$ close to the identity (i.e., $|s|_{p,k+2}$ small). The proof that (for $pk > \dim M$) $GA(P)^{p,k}$ is actually a Lie group, modelled on the Banach space $W^{p,k}(E)$ in such a way that $\operatorname{Exp}_{p,k}: W^{p,k}(E) \to GA(P)^{p,k}$ is a local diffeomorphism, can be found in [P. K. Mitter and C. M. Viallet 1981] for the case $p = 2$, but their proof works for $pk > \dim M$ as well. Although s is a C^∞ function of τ, we still need to prove that s is C^∞ if τ is C^∞.

Theorem 3. In Theorem 2, s is C^∞ if τ is C^∞, provided C_1 is chosen small enough.

Proof: From the proof of Theorem 2, we know that s obeys the equation $\Delta^\omega s = \delta^\omega D^\omega s = \delta^\omega(\tau + R(s,D^\omega s,\tau))$, where $R(s,D^\omega s,\tau)$ is a power series in $s, D^\omega s,\tau$ which is explicitly given by

$$R(s,D^\omega s,\tau) = -\sum_{n=1}^{\infty} \frac{1}{(n+1)!}\, ad_s^n(D^\omega s) + \sum_{n=1}^{\infty} \frac{1}{n!}\, ad_s^n(\tau).$$

Now,

$$\delta^\omega R(s,D^\omega s,\tau) = -\sum_{n=1}^{\infty} \frac{1}{(n+1)!}\, ad_s^n(\Delta^\omega s) + Q(s,D^\omega s,\delta^\omega\tau,\tau),$$

where Q is a series involving the indicated arguments. Thus, s satisfies the equation

$$\left[1 + \sum_{n=1}^{\infty} \frac{1}{(n+1)!}\, ad_s^n\right]\Delta^\omega s = \delta^\omega\tau + Q(s,D^\omega s,\delta^\omega\tau,\tau),$$

where we note that s is C^2 by Prop. 2 (since $s \in W^{p,k+2}(E)$ and $2 < k + 2 - \frac{4}{p}$ holds for $pk > 4$) and hence is a "strong" solution of this equation. Assume that $\max|s|$ is sufficiently small so that $\left[1 + \sum_{n=1}^{\infty} \frac{1}{(n+1)!}\, ad_s^n\right]_x : E_x \to E_x$ can be inverted at all $x \in M$. Then the equation implies that (for $k' \geq k$) $|\Delta^\omega s|_{p,k'} < \infty$ (i.e., $\Delta^\omega s \in W^{p,k'}(E)$) provided $|s|_{p,k'+1} < \infty$ and $|\tau|_{p,k'+1} < \infty$; this requires some work and thought which is left as an exercise. By Prop. 1, we then have $|s|_{p,k'+2} \leq C(|\Delta^\omega s|_{p,k'} + |s|_{p,k'}) < \infty$, provided $|s|_{p,k'+1} < \infty$ and $|\tau|_{p,k'+1} < \infty$ (and $\max|s|$ sufficiently small as above). If τ is C^∞, then $|\tau|_{p,k'+1} < \infty$ for all k', and by induction, $|s|_{p,k'} < \infty$ for all k', whence s is C^∞ provided $\max|s|$ is small enough. Now, the infimum of the choices of the constant C_2 in Theorem 2 (for a given small enough C_1) goes to 0 as $C_1 \to 0$; this can be established from a look at the proof of the implicit function theorem. Thus, for small enough C_1, we can choose C_2 small enough so that $|s|_{p,k+2} < C_2$ ensures that $\max|s|$ is sufficiently small; recall $\max|s| \leq \text{const.}\cdot|s|_{p,k+2}$, since the inclusion $C^0(E) \subset W^{p,k+2}(E)$ is continuous by Prop. 2. □

We have proved that every C^∞ connection in a sufficiently small neighborhood (in $C(P)^{p,k+1}$, $pk > 4$) of a weakly-irreducible $\omega \in C(P)$ is gauge equivalent, via a unique small (i.e., close to I in $GA(P)^{p,k+2}$) gauge transformation which is necessarily smooth, to a connection in S_ω. We have yet to introduce hypotheses which will enable us to prove that no two smooth connections near ω in S_ω are gauge equivalent. For this we need a stronger condition on ω than weak-irreducibility. If we insist that ω be irreducible, then at some later point, we would have to face the problem of proving that every connection near an irreducible connection is irreducible; while this is quite believable, there seems to be no simple proof. To get around the difficulty, we introduce a milder irreducibility condition which is still stronger than weak-irreducibility. To this end, note that the adjoint representation $Ad\colon G \to GL(\mathcal{G})$ induces a representation $r = Ad \otimes Ad^*\colon G \to GL(\text{End } \mathcal{G})$ given by $r(g)(h) - Ad_g \circ h \circ Ad_g{-1}$. Now End $\mathcal{G}$ splits into two invariant subspaces $H' := \{h \in \text{End } \mathcal{G}\colon r(g)(h) = h \text{ for all } g \in G\}$ and $H'' = H'^\perp$ = the subspace of End $\mathcal{G}$ orthogonal to H' with respect to the inner product on End $\mathcal{G}$ induced from k on $\mathcal{G}$. We call $\omega \in C(P)$ *mildly-irreducible* if there is no nonzero element of H'' which is fixed by all transformations $r(g_0)$ as g_0 ranges over the holonomy group G_0 of ω. Clearly, irred. ⇒ mildly-irred. Moreover, mildly-irred. ⇒ weakly-irred.. Indeed, suppose $g \in C(G_0)$; then $Ad_g \in \text{End } \mathcal{G}$ is invariant under all elements of $r(G_0)$, and the same is true of the projection of Ad_g onto H''. Hence, mild-irreducibility implies $Ad_g \in H'$. Thus, for all $\bar{g} \in G$, $Ad_{\bar{g}} Ad_g Ad_{\bar{g}^{-1}} = Ad_g$ or $Ad_{\bar{g}g\bar{g}^{-1}g^{-1}} = \text{Id}$. Since G is connected, we then know that $\bar{g}g\bar{g}^{-1}g^{-1} \in Z$, using the fact that $\exp\circ Ad_g = Ad_g\circ\exp$ for any $g \in G$. However, since G is semi-simple, Z is discrete; and it follows that $\bar{g}g\bar{g}^{-1}g^{-1}$ = identity of G, because $\bar{g}$ can be connected to the identity. Thus, $g \in Z$, and we have proven $C(G_0) = Z$. Hence, not only do we have "mildly-irred. ⇒ weakly-irred.", but also "mildly-irred. ⇒ $C(G_0) = Z$ (i.e., I_ω consists of only R_g, $g \in Z$, and ω has minimal gauge symmetry)." Since I_ω acts trivially on S_ω (indeed, on all of $C(P)$) for ω mildly-irreducible, we suspect that no two connections sufficiently close to ω in S_ω will be gauge equivalent.

<u>Theorem 4</u>. Let $\omega \in C(P)$ be mildly-irreducible. Then there is a constant $C > 0$ such that if $\omega_1,\omega_2 \in C(P) \cap S_\omega$ with $|\omega-\omega_i|_{p,k+1} \leq C$ ($i = 1,2$; $pk > 4$) and there is $F \in GA(P)^{p,k+2} \subset W^{p,k+2}(P \times_G \mathrm{End}(\mathbb{C}^N))$ such that $F^*\omega_1 = \omega_2$, then $F \in I_\omega$ (i.e., $F = R_g$ for $g \in Z$, and $\omega_1 = \omega_2$). Also, C can be chosen small enough to ensure that $\bar{\omega} \in C(P)$ is mildly-irred., if $|\omega-\bar{\omega}|_{p,k+1} \leq C$.

<u>Proof</u>: Let $\tau = \omega_2-\omega_1 \in \Omega^1(E)$; $E = P \times_G G$. Let $f \in C(P,G)^{p,k+2}$ correspond to F (i.e., $F(p) = pf(p)$). By Prop. 3 of 2D, $F^*\omega_1 = \omega_2 \Leftrightarrow f^{-1}df + f^{-1}\omega_1 f = \omega_2 = \omega_1 + \tau \Leftrightarrow f^{-1}D^{\omega_1}f = \tau$. The idea is to prove that if τ is chosen small enough, then F will be close to some R_g ($g \in Z$) and $g^{-1}\cdot f$ will be close enough to the constant map (to the identity of G) so that we can apply Theorem 2 to conclude that $R_g^{-1}\circ F = \mathrm{Id}$. The devices we use here to carry this out were inspired by [Atiyah-Hitchin - Singer 1978]. From the splitting $\mathrm{End}\, G = H' \oplus H''$ introduced above, we have a decomposition $\mathrm{End}\, E = \mathrm{End}'\, E \oplus \mathrm{End}''\, E$ (orthogonal subbundles). We establish that $\bar{\omega} \in C(P)$ is mildly-irred. $\Leftrightarrow \mathrm{Ker}[D^{\bar{\omega}}: C^\infty(\mathrm{End}''\, E) \to \Omega^1(\mathrm{End}''\, E)] = 0$. Indeed, suppose $\bar{\omega} \in C(P)$ and $D^{\bar{\omega}}\alpha = 0$. Then α (regarded as in $C(P,H'')$) is constant on the holonomy bundle P_0 of ω through some $p_0 \in P$ (cf. Section 2H). Since α is equivariant as well, we have (for $g_0 \in G_0$:= holonomy group of $\bar{\omega}$) $r(g_0^{-1})\alpha(p_0) = \alpha(p_0g_0) = \alpha(p_0)$, whence $\alpha(p_0) \in H''$ is fixed by all members of $r(G_0)$, and $\alpha(p_0) = 0$ since $\bar{\omega}$ is mildly-irreducible; thus, $\alpha = 0$. The converse is equally clear; given $h \in H''$ invariant under $r(G_0)$, define α = const = h on P_0, and note that <u>α</u> extends to an element of $C(P,H'')$ in $\mathrm{Ker}\, D^{\omega}$. Observe that $\bar{\omega} \longmapsto D^{\bar{\omega}}$ gives rise to a C^∞ map $C(P)^{p,k+1} \to B(W^{p,k+1}(\mathrm{End}''\, E), W^{p,k}(\mathrm{End}''\, E \otimes \Lambda^1(M))$ (bounded linear maps). Since the subset of injective bounded linear maps is open, we see that the mildly-irreducible connections form an open subset of $C(P)$ in the $W^{p,k+1}$ topology; this takes care of the final assertion of the Theorem, but the rest remains.

Since $\operatorname{Ker} D^{\omega_1} = 0$, we know that the smallest eigenvalue λ of the positive self-adjoint elliptic operator $\Delta^{\omega_1} = \delta^{\omega_1} D^{\omega_1}: C^\infty(\mathrm{End}''E) \to C^\infty(\mathrm{End}''E)$ is positive. Thus, $||D^{\omega_1} f''||^2 = (\Delta^{\omega_1} f'', f'') \geq \lambda ||f''||^2$ for $f'' \in C^\infty(\mathrm{End}''E)$. Recall that $f^{-1} D^{\omega_1} f = \tau$. Let $\tilde{f} \in C^\infty(\mathrm{End}\, E)$ be the image of $f \in C(P,G)$ induced by the homomorphism (with kernel Z) $Ad\colon G \to GL(\mathrm{End}(\mathcal{G}))$. We can decompose $\tilde{f} = f' + f''$ corresponding to $\mathrm{End}\, E = \mathrm{End}'E \oplus \mathrm{End}''E$, and we can map the equation $f^{-1} D^{\omega_1} f = \tau$ to $\tilde{f}^{-1} D^{\omega_1} \tilde{f} = \tilde{\tau}$ in $\Omega^1(\mathrm{End}\, E \otimes \Lambda^1(M))$. Since $Ad(G) \subset O(\mathrm{End}(\mathcal{G}))$, we have that $\tilde{f}$ (or $\tilde{f}^{-1}$) yields a pointwise norm-preserving automorphism of the bundle $\mathrm{End}\, E$. Hence, we have (pointwise),

$$|\tilde{\tau}|^2 = |\tilde{f}^{-1} D^{\omega_1} \tilde{f}|^2 = |D^{\omega_1} \tilde{f}|^2 = |D^{\omega_1} f'|^2 + |D^{\omega_1} f''|^2, \tag{*}$$

and upon integrating this, we obtain

$$||\tilde{\tau}||^2 \geq ||D^{\omega_1} f''||^2 \geq \lambda ||f''||^2. \tag{**}$$

From Prop. 2, we know that there is a constant K so that $\max|\tilde{\tau}| \leq K|\tilde{\tau}|_{p,k+1}$. Since $d|f''|^2 = 2\langle D^{\omega_1} f'', f''\rangle$, we have

$$|d|f''|^2| < 2|D^{\omega_1} f''|\ |f''| \leq 2|\tilde{\tau}|\ |f''| \leq 2K|\tilde{\tau}|_{p,k+1}|f''|$$

using (*). Hence, for $|\tilde{\tau}|_{p,k+1}$ (or $|\tau|_{p,k+1}$) sufficiently small, we can assure that $|d|f''|^2| < \varepsilon$ on M; note that $|f''| < |\tilde{f}| = \text{const.}$, indep. of $f \in C(P,G)$. This implies that $|f''|^2$ can be made arbitrarily close to a constant function as $|\tau|_{p,k+1} \to 0$. In fact, by (**) and $\max|\tilde{\tau}| \leq K|\tilde{\tau}|_{p,k+1}$, we know this constant function must be zero. Consequently, $\tilde{f} = f' + f''$ can be made arbitrarily uniformly close to the subbundle $\mathrm{End}'E$. However, we saw in the discussion before this theorem that the only elements in H' of the form Ad_g were ones where $g \in Z$. (Thus, $Ad(G) \cap H' = \{I\}$.) Hence, $f \in GA(P)^{p,k+2}$ is forced to be near to a constant map $P \to \{g\}$, $g \in Z$, if $|\tau|_{p,k+1}$ is small enough. Thus, $f^{-1} \circ R_g$ is of the form $\mathrm{Exp}(s)$ for some $s \in W^{p,k+2}(E)$, and we obtain

$$\mathrm{Exp}(s) \cdot \omega_1 = f^{-1} \cdot (R_g \cdot \omega_1) = f^{-1} \cdot \omega_1 = F^* \omega_1 = \omega_2.$$

By the same computation as in the proof of Theorem 3, the equation $\mathrm{Exp}(s) \cdot \omega_1 - \omega_1 = \omega_2 - \omega_1 = \tau$ can be expanded out to yield

$$\left[1 + \sum_{n=1}^{\infty} \frac{1}{(n+1)!}\, ad_s^n\right](D^{\omega_1} s) = -\tau.$$

We have already seen that $\max|s| \to 0$ as $|\tau|_{p,k+1} \to 0$; thus, for sufficiently small $|\tau|_{p,k+1}$, $|ad_s|$ will be small enough so that the series in ad_s can be inverted, say

$$\left[1 + \sum_{n=1}^{\infty} \frac{1}{(n+1)!}\, ad_s^n\right]^{-1} = 1 + \sum_{n=1}^{\infty} c_n ad_s^n$$

for constants c_n. Then, we have $D^{\omega_1}s = \left[1 + \sum_{n=1}^{\infty} c_n ad_s^n\right]\tau$. For $j = 0,\ldots,k+1$, Prop. 3 yields the result

$$|D^{\omega_1}s|_{p,j} \le \left(1 + \sum_{n=0}^{\infty} c_n K^n |s|_{p,j}^n\right)|\tau|_{p,k+1}$$

for some constant $K > 0$, independent of s and τ. Hence, if "$|\tau|_{p,k+1} \to 0 \Rightarrow |s|_{p,j} \to 0$", then "$|\tau|_{p,k+1} \to 0 \Rightarrow |D^{\omega_1}s|_{p,j} \to 0$." Now, by definition of the $W^{p,k}$ spaces, we have $|s|_{p,j+1} \le C(|D^{\omega_1}s|_{p,j} + |s|_{p,0})$, and we already know $|s|_{p,0} \to 0$ as $|\tau|_{p,k+1} \to 0$ (since $\max|s| \to 0$ and M is compact); thus, if "$|\tau|_{p,k+1} \to 0 \Rightarrow |s|_{p,j} \to 0$", then "$|\tau|_{p,k+1} \to 0 \Rightarrow |s|_{p,j+1} \to 0$". Since we know "$|\tau|_{p,k+1} \to 0 \Rightarrow |s|_{p,0} \to 0$", by induction "$|\tau|_{p,k+1} \to 0 \Rightarrow |s|_{p,k+2} \to 0$". Hence, for $|\tau|_{p,k+1}$ sufficiently small, we have s in the "uniqueness neighborhood" of Theorem 2, in which case $f^{-1}\circ R_g = \mathrm{Exp}\, s = \mathrm{Id}$ (since ω_1 and ω_2 are both in S_ω). □

Let $C(P)_m^+$ be the set (assumed nonvoid) of mildly-irreducible, self-dual, C^∞ connections on P. We now show how to naturally introduce a C^∞ atlas on $M^+ := C(P)_m^+/GA(P)$ (using the local slices S_ω, $\omega \in C(P)_m^+$) in such a way that the topology induced on M^+ by the atlas is Hausdorff; M^+ will have the dimension $2\,\mathrm{ch}(E)[M] = \frac{1}{2}\dim G(\chi(M) - \mathrm{sgn}(M))$ found in Theorem 1 of Section C.

For each $\omega \in C(P)_m^+$, we choose a neighborhood (in the $W^{p,k+1}$-norm, $kp > 4$) U_ω of ω in $C(P)^+ \cap S_\omega$ such that $U_\omega \to M^+$ (given by $\omega' \longmapsto [\omega] := \{F\cdot\omega\colon F \in GA(P)\}$) is injective; U_ω exists by Theorem 4. Let $[U_\omega]$ denote the image of U_ω in M^+, and let $\phi_\omega\colon [U_\omega] \to U_\omega$ be the inverse. We know from Theorem 1 that we can (and do) choose U_ω small enough so that U_ω is a C^∞ submanifold of $C(P)^{p,k+1}$. To show that the ϕ_ω constitute an atlas, we need to prove that $\phi_\omega\circ\phi_{\omega'}^{-1}\colon \phi_{\omega'}([U_\omega] \cap [U_{\omega'}]) \to \phi_\omega([U_\omega] \cap [U_{\omega'}])$ is C^∞ if $[U_\omega] \cap [U_{\omega'}] \ne \emptyset$. Select $\omega'' \in U_{\omega'}$ such that $[\omega''] \in [U_\omega]$. Then, there is $F \in GA(P)$ such that $F\cdot\omega'' \in U_\omega$. Since the map $\tilde\omega \longmapsto F\cdot\tilde\omega$ is a C^∞ diffeomorphism of $C(P)^{p,k+1}$ (check this, using Prop. 3 of 2D and Prop. 3 of this section), we know that $F\cdot U_{\omega'}$

is a C^∞ submanifold of $C(P)^{p,k+1}$ containing $F\cdot\omega'' \in U_{\omega}$. A neighborhood of $F\cdot\omega''$ in $F\cdot U_{\omega'}$ will be contained in the ball of radius C_1 about ω in Theorem 2, provided we choose U_ω small enough (we assume this has been done). Then Theorem 2 provides us with a C^∞ map $\omega+\tau \longmapsto \mathrm{Exp}(s)\cdot(\omega+\tau)$ which will carry this neighborhood of $F\cdot\omega''$ smoothly into U_{ω}, proving that $\phi_\omega\circ\phi_{\omega'}^{-1}$ is C^∞ at the arbitrary $\omega'' \in \phi_{\omega'}([U_\omega] \cap [U_{\omega'}])$. Note that Theorem 3 is needed to ensure that $\omega+\tau$ is equivalent to $\mathrm{Exp}(s)\cdot(\omega+\tau)$ by an element of $GA(P) \subset C^\infty(P,P)$, provided $\omega+\tau$ is C^∞; we know that the elements of $F\cdot U_{\omega'}$ are C^∞. The topology on M^+ is the smallest topology that makes the maps ϕ_ω continuous. To show that M^+ is Hausdorff, select ω and ω' with $[\omega] \neq [\omega']$, and set $\tau = \omega'-\omega$. We argue as in [Atiyah-Hitchin-Singer 1978] by first noting that $[\omega] \neq [\omega']$ means that $0 \neq f^{-1}\cdot\omega-\omega' = f^{-1}d\omega + f^{-1}\omega f + \omega + \omega-\omega' = f^{-1}D^\omega f - \tau$ for all $f \in GA(P)$. As in the proof of Theorem 4, we can consider the $\tilde{f} \in C^\infty(\mathrm{End}\ E)$ $(E := P \times_G G)$ associated with f; and $\tilde{\tau} \in \Omega^1(\mathrm{End}\ E)$ associated with τ. If we can show that $||\tilde{f}^{-1}D^\omega\tilde{f} - \tilde{\tau}||$ is bounded away from zero as f varies, then $f^{-1}\cdot\omega$ will be bounded away from ω' in the $|\ |_{2,0}$ norm which is weaker than $|\ |_{p,k+1}$ for $pk > 4$ by Prop. 2; thus the orbits of ω and ω' will be bounded away from each other in $C(P)^{p,k+1}$, and we will have that M^+ is not only Hausdorff, but actually a metric space. Now, since $\tilde{f}$ yields an isometry of End E, we have $||\tilde{f}^{-1}D^\omega\tilde{f} - \tilde{\tau}||^2 = ||D^\omega\tilde{f} - \tilde{f}\tilde{\tau}||^2 \geq \lambda||f^\perp||^2$ where $f^\perp$ is the projection of $\tilde{f}$ onto the orthogonal complement of $\mathrm{Ker}[(D^\omega-\tilde{\tau})^*(D^\omega-\tilde{\tau})]$ and λ is the smallest positive eigenvalue of $(D^\omega-\tilde{\tau})^*(D^\omega-\tilde{\tau})$. We need to show $||f^\perp||$ is bounded away from 0. Suppose there is a sequence $f_n \in GA(P)$ such that $||f_n^\perp|| \to 0$. Since $||\tilde{f}_n-f_n^\perp||^2 + ||f_n^\perp||^2 = ||\tilde{f}_n||^2 = \text{const.}$, we see that $\tilde{f}_n - f_n^\perp$ is a bounded sequence in $K := \mathrm{Ker}[(D^\omega-\tilde{\tau})^*(D^\omega-\tilde{\tau})]$ that has a convergent subsequence (which we may assume is itself), since $\dim K < \infty$. Calling the limit $f_\infty \in K$, we have $||f_\infty-\tilde{f}_n||^2 = ||f_\infty-(\tilde{f}_n-f_n^\perp)||^2 + ||f_n^\perp||^2 \to 0$, whence $\tilde{f}_n \to f_\infty$ in L^2. Suppose that the smooth f_∞ did not have values in $P \times_G \mathrm{Ad}(G) \subset \mathrm{End}\ E$, say $f_\infty(x) \notin P \times_G \mathrm{Ad}(G)$. Then f_∞ would be bounded away from $P \times_G \mathrm{Ad}(G)$ in a neighborhood V of $x \in M$, and "$|\tilde{f}_n-f_\infty| \geq c > 0$ in V" would contradict $||\tilde{f}_n-f_\infty|| \to 0$. Thus, $f_\infty \in C^\infty(P \times_G \mathrm{Ad}(G))$ and $D^\omega f_\infty - \tilde{\tau}f_\infty = 0$, contradicting $f^{-1}D^\omega f - \tau \neq 0$ for all $f \in GA(P)$. In summary, we have proved the following.

Theorem 5. Let $\pi\colon P \to M$ be a principal G-bundle (G compact, semi-simple) over a compact, oriented, self-dual, Riemannian 4-manifold with non-negative scalar curvature $S \not\equiv 0$. Then the space $C(P)_m^+/GA(P)$ of moduli of mildly-irreducible self-dual connections has (if non-empty) the structure of a Hausdorff C^∞ manifold of dimension

$$2\,\mathrm{ch}(E)[M] - \frac{1}{2}\dim G(\chi(M) - \mathrm{sgn}(M))$$

where $E = P \times_G G$.

Remark. If $\phi \neq C_0(P)_m^+ :=$ the subset of $\omega \in C(P)_m^+$ such that $\mathrm{Ker}[\Delta^\omega$ on $\Omega_-^2(E)] = 0$, then the same conclusion holds for $C_0(P)_m^+/GA(P)$, and we may drop the self-duality and positive scalar curvature assumptions on M. It is likely that $C_0(P)_m^+ = C(P)_m^+$ for a generic class of metrics on M; cf. [Freed-Freedman-Uhlenbeck 1983, §3].

E. Gauge-Theoretic Topology in Dimension Four

As a result of the encouragement from the physics community to expand our knowledge of classical gauge theory, mathematicians have also profoundly increased their knowledge of the structure of smooth 4-manifolds. A thorough exposition of these developments would require a book. Indeed, several extensive monographs have been under preparation (e.g., [H. B. Lawson 1983] and [Freed-Freedman-Uhlenbeck 1983]); these should be readily available by the time the present work is in print. Hence our discussion here is merely intended to provide a good intuitive grasp of the crucial ideas and a hunger for more of the details.

Let M be a compact, simply-connected, oriented C^∞ 4-manifold. From Hodge theory, we know that the cohomology space $H^2(M, \mathbb{R})$ can be identified with the space of harmonic 2-forms on M. As such, we have a symmetric bilinear form s_M on $H^2(M, \mathbb{R})$ given by $s_M(\alpha,\beta) = \int_M \alpha \wedge \beta$. For $\alpha,\beta \in H^2(M, \mathbb{Z})$, it happens that $s_M(\alpha,\beta)$ is an integer. This is a consequence of the fact that the dual version of s_M on $H_2(M, \mathbb{Z})$ is "intersection number" of oriented compact surfaces in M; alternatively, s_M is essentially the $\mathbb{Z} \cong H^4(M, \mathbb{Z})$-valued cup product in the more general topological setting. Poincare duality implies that s_M is non-degenerate on the *lattice* (free abelian group) $H^2(M, \mathbb{Z})$, meaning that the map $s_M^\#$: $H^2(M, \mathbb{Z}) \to \mathrm{Hom}(H^2(M, \mathbb{Z}), \mathbb{Z})$, given by $s_M^\#(\alpha)(\cdot) = s_M(\alpha,\cdot)$, is an isomorphism. One easily verifies that non-degeneracy means that the matrix $[s_M(e_i,e_j)]$ (where $e_1,\ldots,e_r$ are independent generators) has an inverse with integer entries (i.e., $\det[s_M(e_i,e_j)] = \pm 1$). The study of such forms is important, because two oriented, compact, simply-connected, topological 4-manifolds M, N are homotopy-equivalent precisely when $(H^2(M, \mathbb{Z}), s_M)$ and $(H^2(N, \mathbb{Z}), s_N)$ are isometric (i.e., $s_M = F^* s_N$ for $F\colon H^2(M, \mathbb{Z}) \cong H^2(N, \mathbb{Z})$); cf. [Milnor-Husemoller 1973, 100-105]. More recently, M. H.

Freedman has shown that every element of B (:= set of isometry classes of nondegenerate, symmetric, $\mathbb{Z}$-valued bilinear forms on lattices) is realized by some oriented, simply-connected, topological (i.e., not necessarily having a differentiable structure) 4-manifold; up to homeomorphism, there are at most two such manifolds representing a given form (cf. [M. H. Freedman 1983]). The problem of classifying the elements of B is taken up in [Milnor-Husemoller 1973, 15-55], where proofs of assertions made in the following paragraph may be found.

The set B is divided into *type I* forms b (or classes [b]) for which there is an element $x \in X_b$:= lattice on which b is defined, such that $b(x,x)$ is odd; and *type II* forms, for which $b(x,x)$ is even for all $x \in X_b$. Moreover, $[b] \in B$ is called *indefinite* if it takes on both positive and negative values; otherwise [b] (or b) is called *definite*. The indefinite forms are easy to list. The indefinite type I forms have diagonal matrices relative to some basis of X_b:

$$[b(e_i,e_j)] = \overset{p}{<1> \oplus \cdots \oplus <1>} \oplus \overset{n}{<-1> \oplus \cdots \oplus <-1>}$$

$$= \operatorname{diag}(1,\ldots,1,-1,\ldots,-1), \quad pn \geq 1.$$

For indefinite type II forms, we can arrange that $[b(e_i,e_j)] = H \oplus \overset{a}{\cdots} \oplus H \oplus E_8 \oplus \overset{b}{\cdots} \oplus E_8$, where $H := \begin{bmatrix} 0 & 1 \\ 1 & 0 \end{bmatrix}$ and

$$E_8 := \begin{bmatrix} 2 & 1 & 0 & 0 & 0 & 0 & 0 & 0 \\ 1 & 2 & 1 & 0 & 0 & 0 & 0 & 0 \\ 0 & 1 & 2 & 1 & 0 & 0 & 0 & 0 \\ 0 & 0 & 1 & 2 & 1 & 0 & 0 & 0 \\ 0 & 0 & 0 & 1 & 2 & 1 & 1 & 0 \\ 0 & 0 & 0 & 0 & 1 & 2 & 0 & 0 \\ 0 & 0 & 0 & 0 & 1 & 0 & 2 & 1 \\ 0 & 0 & 0 & 0 & 0 & 0 & 1 & 2 \end{bmatrix}.$$

Note that E_8 is positive-definite, and so we must require $a > 0$; also E_8 is related to the exceptional Lie group having the same name. Oddly enough, the definite forms are quite difficult to list; the number of isometry classes grows rapidly with the rank (e.g., for rank 40, there are over 10^{51} different classes).

Any indefinite type I class can be realized by a smooth manifold. Indeed, $H^2(P_2(\mathbb{C}))$ is generated by the first Chern class c_1, and $\int_{P_2(\mathbb{C})} c_1 \wedge c_1 = \pm 1$ depending on whether $P_2(\mathbb{C})$ is given its standard orientation or the opposite; we write $\pm P_2(\mathbb{C})$. Recall that the connected sum $M_1 \# M_2$ of two n-manifolds is the manifold obtained by removing a ball from each and gluing the exposed spherical boundaries together. It is not too hard to show that

$$+P_2(\mathbb{C}) \# \overset{p}{\ldots} \# +P_2(\mathbb{C}) \# -P_2(\mathbb{C}) \# \overset{n}{\ldots} \# -P_2(\mathbb{C})$$

has form $\mathrm{diag}(1,\overset{p}{\ldots},1, -1,\overset{n}{\ldots},-1)$, whence all type I indefinite forms are realized by such connected sums. The form for $S^2 \times S^2$ is $H = \begin{bmatrix} 0 & 1 \\ 1 & 0 \end{bmatrix}$. Indeed, the generators of $H^2(S^2 \times S^2, \mathbb{Z})$ are the pull-backs of the Chern forms (normalized volume elements) on $S^2 \cong P_1(\mathbb{C})$ under the projections $S^2 \times S^2 \to S^2$ onto either factor; examine the wedges, and note $H \sim -H$. A theorem of V. A. Rohlin (cf. [Freedman-Kirby 1978]) asserts that a compact, simply-connected C^∞ 4-manifold with s_M of type II must have *signature* (i.e., the signature of s_M extended to $H^2(M, \mathbb{R}) \supset H^2(M, \mathbb{Z})$) which is divisible by 16. Consequently, there is no smooth M representing E_8; the values of "a" and "b" such that $aH \oplus bE_8$ is realized by a smooth M are not entirely known yet. Of course, aH is realized by $S^2 \times S^2 \# \overset{a}{\ldots} \# S^2 \times S^2$, and the quartic K-3 surface ($\dim_{\mathbb{C}} = 2$) in $P_3(\mathbb{C})$ given by $z_0^4 + \ldots + z_3^4$ happens to realize $3H \oplus 2E_8$. What about the unwieldy variety of positive-definite forms? Thanks to the following theorem of [S. K. Donaldson, 1983], only one positive-definite form of a given rank is realized by a smooth manifold.

<u>Theorem (Donaldson)</u>. Let M be a compact, smooth, simply-connected, oriented 4-manifold with positive-definite form s_M. Then s_M is the standard diagonal form of type I (i.e., $s_M \sim \langle 1\rangle \oplus \ldots \oplus \langle 1\rangle$). Consequently, M is a connected sum of $+P_2(\mathbb{C})$s.

As mentioned in the opening paragraph, we will only attempt a sketch of the proof.

<u>Lemma</u>. Let b be a non-degenerate positive-definite symmetric bilinear form on a lattice X of dimension $r(b)$ (rank of b), and let $n(b)$ be the number of pairs of solutions $\pm a \in X$ of $b(x,x) = 1$. Then $n(b) \le r(b)$ with equality only when b is equivalent to the standard form $\langle 1\rangle \oplus \ldots \oplus \langle 1\rangle$ ($r(b)$-times).

<u>Proof</u>: Suppose $b(a,a) = 1$ and let $c \in X$. Then $c = b(c,a)a \oplus c'$, where $c' = c - b(c,a)a \in a^\perp := \{x \in X\colon b(a,x) = 0\}$. Since $b(c,c) = b(c,a)^2 + b(c',c')$, we see that $b(c,c) = 1$ only if $c = \pm a$ or $c \in a^\perp$. Hence, by induction, $n(b) = n(b|a^\perp) + 1 \le r(b|a^\perp) + 1 = r(b)$, the case $r(b) = 1$ being trivial. We obtain equality only in the case where the process may be continued $r(b)$-times, in which case b is standard. □

Lemma 1 tells us that Theorem 1 follows if the dimension of $H^2(M, \mathbb{Z})$ (i.e., $r(s_M)$) equals $n(s_M)$. Since s_M is assumed positive definite, we know that $r(s_M) = \sigma(s_M) :=$ signature of s_M extended to $H^2(M, \mathbb{Z}) \otimes \mathbb{R}$. There is something that can be said of signatures: An *oriented cobordism* of two compact, oriented 4-manifolds N and N' is a compact, oriented 5-manifold N'' having boundary consisting of N and $-N'$ (i.e., N' receives the opposite orientation from N''). Then we have $\sigma(s_N) = \sigma(s_{N'})$; cf. [R. Stong 1958, 220-222]. Using this, Theorem 1 will follow if we can prove that there is an oriented cobordism between M and $n(s_M)$ copies of $\pm P_2(\mathbb{C})$ (say p copies of $+P_2(\mathbb{C})$ and q copies of $-P_2(\mathbb{C})$). Indeed, we would have $r(s_M) = \sigma(s_M) = q-p \leq q+p = n(s_M)$, whence Lemma 1 gives $r(s_M) = n(s_M)$, and s_M is standard.

Oddly enough, the desired cobordism emerges from gauge theoretic analysis. Indeed, let $\pi: P \to M$ be a principal $SU(2)$-bundle with $c_2(E)[M] = -1$ where $E := P \times_{SU(2)} \mathbb{C}^2$; we will construct such a bundle later. For a dense set of metrics on M, we will find that the space M^+ of moduli of self-dual connections (instantons) has the structure of an oriented 5-manifold with a finite number (actually $n(s_M)$) of singular points which have neighborhoods homeomorphic to cones on $P_2(\mathbb{C})$ (i.e., $P_2(\mathbb{C}) \times [0,1)/(P_2(\mathbb{C}) \times \{0\})$). Of course, the nonsingular points have neighborhoods which are 5-balls (i.e., cones on S^4). The singular points actually correspond to the moduli of reducible connections (necessarily with circular holonomy groups), as we will see. While M^+ is noncompact, it turns out that it has a natural compactification with ideal boundary diffeomorphic to M. This boundary consists of limits of sequences of instantons whose strengths are increasingly concentrated near the various points of M. We have already seen in 3B that such arbitrarily concentrated instantons exist for the bundle $P \to S^4$ with $k = 1$; indeed, for this special case, it happens that M^+ is the 5-ball with ideal boundary S^4 (Modulo connectedness, this follows from the results of 3B). By truncating M^+ at the singular points, we then obtain a cobordism between M and $n(s_M)$ copies of $P_2(\mathbb{C})$. The fact that this cobordism may be oriented is proved using the simple-connectivity of M (cf. [Donaldson 1983]). We will see that $H^1(M, \mathbb{R}) = 0$ is used to obtain the cones over $P_2(\mathbb{C})$, instead of over some other space.

Without getting mired in technicalities, we will consider the reasoning behind some of the above claims. The construction of even a single irreducible instanton for $\pi: P \to M$ (as above) is no easy matter. Such was first carried out in [C. H. Taubes 1982]. Roughly, one grafts a highly concentrated instanton (with $k = 1$) from S^4 onto a neighborhood of a point of M. While the result may no longer be self-dual, it is close enough to self-duality that one can deduce that there is an instanton nearby; one uses an implicit function theorem and the positivity of s_M as shown below. More precisely, for $\varepsilon < \pi/2$, let B_ε be the open geodesic ball of radius ε about the south pole 0 of $S^4 = \mathbb{R}^4 \cup \{\infty\}$ ($\infty \leftrightarrow$ north pole). For $x \in M$ and ε sufficiently small, we can map the geodesic ball $B_\varepsilon(x)$ of radius ε about x smoothly onto B_ε, using geodesic polar coordinates. We can smoothly extend this map so that $B_{2\varepsilon}(x)$ is mapped diffeomorphically onto $S^4 - \{\infty\}$ and $M \setminus B_{2\varepsilon}(x)$ is mapped to the point ∞. The resulting map $f\colon M \to S^4$ has degree 1, and hence the pull-back of the $k = 1$ principal $SU(2)$-bundle over S^4 yields a $k = 1$ principal $SU(2)$-bundle $\pi\colon P \to M$. From the results of 3B, for each $\lambda \in (0,\pi)$ there is a $k = 1$ instanton over S^4 which has exactly half of its total action concentrated in B_λ about 0. Pulling this instanton back to M using the map f (or its bundle-covering), we obtain a connection ω_λ on P over M which however may not be self-dual. Writing an arbitrary $\omega \in C(P)$ as $\omega = \omega_\lambda + \tau$, $\tau \in \overline{\Omega}^1(P,\mathcal{SU}(2))$, we have that ω will be self-dual (i.e., $(1-*)\Omega^\omega = 0$) if $(1-*)(D^{\omega_\lambda}\tau + \frac{1}{2}[\tau,\tau]) = -(1-*)\Omega^{\omega_\lambda}$. To obtain an elliptic equation, we constrain τ to be of the form $\delta^{\omega_\lambda}\alpha$ for $\alpha \in \overline{\Omega}^2_-(P,\mathcal{SU}(2)) \cong \Omega^2_-(P \times_G G)$, $G = SU(2)$. We then obtain a nonlinear elliptic equation in α with linear highest order part being $(1-*)D^{\omega_\lambda}\delta^{\omega_\lambda}\alpha$. Intuitively, we expect that if λ is chosen small enough (actually, it turns out best to take $\lambda = \varepsilon^2$ and choose ε sufficiently small), then $(1-*)\Omega^{\omega_\lambda}$ will be small enough in some $W_{p,k}$ Sobolev space so that one can find a small solution α, provided the kernel of $(1-*)D^{\omega_\lambda}\delta^{\omega_\lambda}$ is trivial. (Then, at least the linearization of the equation is solvable.) Now, since ω_λ is flat outside a small neighborhood $B_{2\varepsilon}(x)$ of x, we expect that the first eigenvalue of $(1-*)D^{\omega_\lambda}\delta^{\omega_\lambda}$ will approach that of the untwisted operator $(1-*)d\delta$ on $\Omega^2_-(M,\mathbb{R})$. The hypothesis that s_M is positive means that there are no non-zero harmonic anti-self-dual forms on M, whence the first eigenvalue of $(1-*)d\delta$ is positive, as desired. Of course, this argument is no substitute for the hard technicalities of [Taubes 1982]. As we will see, there are only a finite number of moduli of reducible self-dual connections; hence for all ε sufficiently small, the instantons constructed are irreducible.

Using transversality theory (cf. [Freed-Freedman-Uhlenbeck 1983, §3]), one obtains a generic set of metrics on M such that the set M_0^+ of moduli of irreducible self-dual connections is a smooth 5-manifold. Essentially, one shows that, for a generic set of metrics, Δ^ω on $\Omega_-^2(E)$ has trivial kernel for all $[\omega] \in M^+$. Then we know that the conclusion of Theorem 5 of 3D will hold; for $G = SU(2)$, it is easy to check that mild (and weak)-irreducibility are equivalent to irreducibility. Henceforth, we assume that the metric on M is chosen from this generic set. Let V_ε be the subset of M_0^+ consisting of those instantons with more than half of their total action in a ball of radius ε about some point in M. For ε sufficiently small, it happens that for each $\omega \in V_\varepsilon$ there is a unique ball of smallest radius $r(\omega)$ (about a point $x(\omega) \in M$) that contains exactly half of the total action of ω. With some work, one proves that the map $V_\varepsilon \to M \times (0,\varepsilon)$ given by $\omega \longmapsto (x(\omega), r(\omega))$ is a diffeomorphism. Using several key analytical results on instantons (cf. [K. Uhlenbeck 1982a, 1982b]), one can prove that $M^+ \setminus V_\varepsilon$ is compact (we need that for the cobordism). The following rough sketch of the proof of the compactness is instructive. Generally the space of solutions of a homogeneous linear elliptic equation over a compact manifold is finite dimensional, and hence the subset of solutions of L^2-norm $\leq$ constant is compact. In the case of the self-duality equation for instantons, we lack linearity, and we achieve ellipticity only locally by imposing a gauge condition (i.e., $\delta^{\omega_0}(\omega-\omega_0) = 0$). Nevertheless, the results of [K. Uhlenbeck 1982b] indicate that for any small ball $B \subset M$, there is a constant C such that if the actions $\int_B ||F_i||^2$ of a sequence of self-dual connections ω_i are uniformly bounded by C, then there is a sequence $\tilde{\omega}_n$ of connections, which are gauge-equivalent to ω_{i_n} in a subsequence of ω_i, such that $\tilde{\omega}_n$ converges to a self-dual connection in some suitable Sobolev $W_{p,k}$ sense, at least on a proper sub-ball. The constant C is not particularly sensitive to the size of B because of the conformal invariance of the YM (action) functional. Let us call a point $x \in M$ "singular" if there is no ball B about x such that $\liminf_{i\to\infty} \int_B ||F_i||^2 \leq C$. Since $k = 1$, we know that $\int_M ||F_i||^2 = 32\,\pi^2$, whence there can be at most $C/32\,\pi^2$ such "singular" points. By a Cantor diagonal argument, one obtains $\tilde{\omega}_n$ converging on compact subsets of the complement $M \setminus \{x_1,\ldots,x_N\}$ of the set of "singular" points to a self-dual connection ω_∞ on $M \setminus \{x_1,\ldots,x_N\}$. However, the action of ω_∞ is bounded, and hence the removable singularity theorem of [K. Uhlenbeck 1982a] tells us that we can topologically change (if necessary) the bundle $P \to M$ by means of clutching functions about the "singular" points so that ω_∞ extends smoothly to a connection, say ω_∞ on the new bundle $P_\infty \to M$. If there are no "singular" points, then the original sequence $[\omega_i] \in M^+$ will have a subsequence converging to $[\omega_\infty] \in M^+$, since $P_\infty = P$ in this case. If there are singular points, we will now argue that there can be only one, and in fact the subsequence eventually must stay in V_ε; i.e., we get $M^+ \setminus V_\varepsilon$ compact, as desired. Thus, suppose that there are singular points. If B is any small ball about one of them, then we have $C + \int_{M\setminus B} ||F_i||^2 \le \int_M ||F_i||^2 = 32\,\pi^2$, and taking the limit for the subsequence, we have $\int_{M\setminus B} ||F_\infty||^2 \le 32\,\pi^2 - C$ independent of B; thus, $\int ||F_\infty||^2 < 32\,\pi^2$, and $F_\infty \equiv 0$ with P_∞ trivial, since $\int_M ||f_\infty||^2 \in 32\,\pi^2\,\mathbb{Z}$. It should come as no great surprise that the degree of the clutching map about a "singular" point (which is needed to construct P_∞) is given by a formula $(32\,\pi^2)^{-1} \int_B ||F_i||^2 - \int_{\partial B} T(\omega_i)$ where $T(\omega_i)$ is a 3-form (essentially a Chern-Simons invariant - cf. [Chern-Simons 1974]) which is constructed from $\omega_i - \omega_\infty$ in a canonical way. The boundary term vanishes in the limit of the subsequence, and so $\int_B ||F_i||^2 \to 32\,\pi^2 n$ for some integer $n > 0$. Consequently, there can be at most one singular point since $\int_M ||F_i||^2 = 32\,\pi^2$, and if there is a singular point, the converging subsequence of $[\omega_i]$ must eventually lie in V_ε, as desired.

So far, we have argued that $M^+ \setminus V_\varepsilon$ is compact, and upon removing the set of moduli of reducible connections, we obtain a smooth 5-manifold $M_0^+ \setminus V_\varepsilon$ with boundary diffeomorphic to M. The sketch will be completed by showing that there are exactly $n(s_M)$ moduli of reducible self-dual connections, and they have neighborhoods in M^+ that are cones on $P_2(\mathbb{C})$. Suppose ω is a reducible connection on $\pi\colon P \to M$ $(k = 1)$. Then the holonomy group G_0 of ω at $p_0 \in P$ is a proper subgroup of $SU(2)$ and cannot be discrete, since otherwise Ω^ω would vanish by Prop. 2 of 2H and $0 = \int_M |\Omega^\omega|^2 = 32\,\pi^2$. The only proper non-discrete subgroups of $SU(2)$ are the subgroups isomorphic to the circle group, and any two such subgroups are conjugate, as the reader can verify. Since the holonomy group at $p_0 g$ is $g^{-1}G_0 g$, we may assume that G_0 is actually the particular subgroup

$$\left\{ \begin{bmatrix} e^{i\theta} & 0 \\ 0 & e^{-i\theta} \end{bmatrix} : \ \theta \in \mathbb{R} \right\}$$

which we identify with $U(1) := \{e^{i\theta}\colon \theta \in \mathbb{R}\}$. The curvature Ω^ω, when restricted to the holonomy bundle P_0, is then a form with values in the

Lie algebra $i\mathbb{R}$ of $U(1)$, whence there is a unique 2-form iF on M such that $\pi^*(iF) = \Omega^\omega|P_0$; this uses the fact that $U(1)$ is abelian. We know that $dF = 0$ (Bianchi identity), and $[-(2\pi)^{-1}F] \in H^2(M, \mathbb{Z})$ is the Chern class of the $U(1)$-bundle $P_0 \to M$ (cf. 2G). If ω is self-dual (i.e., $*F = F$), then $\delta F = -*d*F = -*dF = 0$, whence F is harmonic; conversely, if F is harmonic, then ω is self-dual by the assumption that s_M is positive-definite. Also, for ω self-dual, we have $\int_M F \wedge F = 4\pi^2$ (i.e., $[(2\pi)^{-1}F] \in H^2(M, \mathbb{Z})$ is a solution of $s_M(x,x) = 1$). Indeed, the inner product $k :=$ -Killing form on $SU(2)$ gives the element $\begin{bmatrix} i & 0 \\ 0 & -i \end{bmatrix}$ a norm-square of 8, whence $|\Omega^\omega|^2 = 8|F|^2$, and

$$32\,\pi^2 = \int_M |\Omega^\omega|^2\nu = \int_M 8|F|^2\nu = \int_M 8F \wedge *F = 8\int_M F \wedge F.$$

Suppose $H\colon P \to P$ is a gauge transformation. Then it is easy to check that $\omega' := H^{-1*}\omega$ has holonomy bundle $H(P_0)$ through $H(p_0)$, and the holonomy group of ω' at $H(p_0)$ is the same G_0 $(= U(1)) \subset SU(2)$. Since $\pi \circ H = \pi$, we have

$$(\pi^*iF)|H(P_0) = H^{-1}*(\pi^*iF|P_0) = H^{-1}*(\Omega^\omega|P_0) = \Omega^{\omega'}|H(P_0),$$

and so "the" form F is the same for gauge-equivalent reducible connections. For a given reducible ω, "the" form F is not quite uniquely determined because there are actually two holonomy bundles with group $U(1) = \left\{\begin{bmatrix} e^{i\theta} & 0 \\ 0 & e^{-i\theta} \end{bmatrix}\right\}$. Indeed, note that conjugating by

$$g_\phi := \begin{bmatrix} 0 & e^{-i\phi} \\ -e^{i\phi} & 0 \end{bmatrix}$$

gives us $U(1)$ back again, but each element of $U(1)$ is sent to its inverse. Hence, $P_0 g_\phi$ will be a holonomy bundle for ω through $p_0 g_\phi$ and will have holonomy group $U(1)$. $P_0 g_\phi$ is easily checked to be independent of ϕ, and it is the only other holonomy bundle with group $U(1)$ for ω. The mapping $P_0 \to P_0 g_\phi$ given by $p \longmapsto pg_\phi$ is not an isomorphism of $U(1)$-bundles, but rather reverses the sense of rotation. If we had chosen $P_0 g_\phi$ instead of P_0, then the associated 2-form on M would be $-F$ instead of F, as can be verified. In summary, we have constructed a function from the set R of moduli of reducible self-dual connections to the set S of $(\pm)$-pairs of solutions of $s_M(x,x) = 1$, $x \in H^2(M, \mathbb{Z})$, namely $\omega \longmapsto \pm[(2\pi)^{-1}F]$. In fact, this map $R \to S$ is seen to be bijective as follows. From [Milnor-Husemoller 1973, 103-105], we know that M is homotopic to a bouquet of 2-spheres $S^2 \vee \dots \vee S^2$ ($r(s_M)$ times) with a 4-cell attached. Consequently (cf. [Spanier 1966, 399-405]), the homotopy class of a map from M to the classifying space $P_\infty(\mathbb{C})$ for $U(1)$-bundles will be determined by the induced homology map $H_2(M, \mathbb{Z}) \to H_2(P_\infty(\mathbb{C}), \mathbb{Z}) \cong \mathbb{Z}$. In other words, the Chern class (determining the dual map $H^2(P_\infty(\mathbb{C}), \mathbb{Z}) \to H^2(M, \mathbb{Z})$) of a line bundle will characterize that bundle up to isomorphism; moreover, we see that any element of $H^2(M, \mathbb{Z})$ is realizable as the Chern class of some line-bundle. Thus, if ω and ω' are two reducible connections on P with the same associated forms ($F = F'$), then the two holonomy bundles P_0 and P_0' must be isomorphic principal $U(1)$-bundles. Any isomorphism between them extends uniquely to a gauge transformation $H \in GA(P)$ which carries P_0 to P_0'. Then $H^{-1}*\omega$ and ω' will be two connections with the same holonomy bundle P_0' and the same associated form F on M. The difference $(H^{-1}*\omega - \omega')|P_0'$ is the pull-back of a unique 1-form α on M, and $d\alpha = iF - iF' = 0$. Since $H^1(M, \mathbb{R}) = 0$, we then have $\alpha = i du$ for some $u \in C^\infty(M)$. The gauge transformation on P_0' provided by e^{iu} will then carry $H^{-1}*\omega|P_0'$ to $\omega'|P_0'$, and it uniquely extends to a gauge transformation of P which carries $H^{-1}*\omega$ to ω' (i.e., ω and ω' are gauge-equivalent). Thus, two reducible self-dual connections that have the same associated form F on M are necessarily gauge equivalent; and we have shown that $R \to S$ is injective. To prove that $R \to S$ is onto, let $x \in H^2(M, \mathbb{Z})$ with $s_M(x,x) = 1$. We know that there is a line bundle L such that $c_1(L) = x$ and a line bundle L^* with transition functions conjugate to those of L (L^* = dual bundle of L). Thus, $L \oplus L^*$ is a 2-dimensional complex vector bundle, and letting some Hermitian structure on L induce one on L^*, we have a Hermitian structure on $L \oplus L^*$. Then we may consider the principal $U(2)$-bundle $\tilde{P} \to M$ of unitary frames. Now because of the special structure of $L \oplus L^*$, $\tilde{P}$ reduces to a principal $SU(2)$-bundle $P' := \{(e_1 \oplus f_1^*, e_2 \oplus f_2^*) \in \tilde{P}: f_2^*(e_1) - f_1^*(e_2) = 1\}$ over M. Since $c_2(L \oplus L^*) = c_1(L) \cup c_1(L^*) = x \cup (-x) = -x \cup x$ and $s_M(x,x) = 1$, we have $c_2(L \oplus L^*)[M] = -1$ and $P' \to M$ is isomorphic to $P \to M$. We can then find a subbundle P_0 of P corresponding to the special frames of the form $(e \oplus 0, 0 \oplus e^*)$. Since P_0 can be identified with the principal $U(1)$-bundle of unitary frames of L, a connection ω_0 on P_0 will have associated form F with $[-(2\pi)^{-1}F] = x$. By Hodge theory, we know that an exact two-form $d\alpha$ can be found such that $F + d\alpha$ is harmonic (and hence self-dual, by assumption). Replacing ω_0 by $\omega_0 + i\pi^*\alpha$ on P_0, we may then assume F is self-dual. It is easy to see that any connection on a principal subbundle extends uniquely to a connection on the ambient bundle. Thus, ω_0 extends to a reducible self-dual connection on P with associated form $-(2\pi)^{-1}F$ representing the solution x, and $R \to S$ is now proved to be bijective.

We complete our sketch by arguing that small neighborhoods of the moduli $[\omega_1],\ldots,[\omega_N]$ $N = n(s_M)$, of reducible self-dual connections on M^+ can be found which are cones on $P_2(\mathbb{C})$. Let $[\omega]$ be one of these moduli. Although ω is not weakly-irreducible, the proof of Theorem 1 of 2D can be generalized to give us the fact that the self-dual connections ω' satisfying $\delta^\omega(\omega'-\omega) = 0$ form a manifold in a neighborhood of ω. The proof uses the fact that, while the elliptic operator $T = \delta^\omega \oplus (1-*)D^\omega$ is not onto, it is still Fredholm so that the zeros of the map R in Theorem 1 of 3D will still form a submanifold with tangent space Ker T. The index of T is still $8k-3 = 5$, because T varies continuously with ω in the connected space $C(P)$ and we know the index is 5 at an irreducible self-dual connection. (Alternatively, ω was arbitrary in our computation of the index of T in the proof of Theorem 1 of 3C.) Recall that the metric g on M was chosen to be such that Δ^ω on $\overline{\Omega}^2_-(P,SU(2)) \cong \Omega^2_-(E)$ has trivial kernel for all self-dual ω. Thus, from the proof of Theorem 1 of 3C, we have that Ker T* is the subspace of $\Omega^0(E) \oplus \Omega^2_-(E)$ consisting of the infinitesimal gauge transformations $\alpha \in \Omega^0(E)$ which preserve ω (i.e., $D^\omega\alpha = 0$). Since the centralizer of the holonomy group $U(1) = \left\{\begin{bmatrix} e^{i\theta} & 0 \\ 0 & e^{-i\theta}\end{bmatrix}\right\}$ is U(1) again, we know(cf. Prop. 1 of 2H) that the group I_ω of gauge transformations fixing ω is isomorphic to U(1); hence I_ω is the circle group consisting of all gauge transformations of the form $p \longmapsto p\exp(t\alpha(p))$ for some nonzero $\alpha \in \mathrm{Ker}[D^\omega$ on $\Omega^0(P,SU(2))]$ and $t \in \mathbb{R}$ (cf. proof of Theorem 1 of 3C). Thus, $\dim \mathrm{Ker}\, T = \mathrm{index}\, T + \dim \mathrm{Ker}\, T^* = 6 = \dim \tilde{S}_\omega$, where $\tilde{S}_\omega = \{\omega' \in C(P)\colon \delta^\omega(\omega'-\omega) = 0,\ \omega'$ self-dual, and ω' close to $\omega\}$. Note that I_ω acts on $\tilde{S}_\omega$ and Ker T via pull-back. It can be proved that the map $\omega' \longmapsto [\omega']$ induces a homeomorphism of S_ω/I_ω to a neighborhood of $[\omega]$ in M^+ (indeed a diffeomorphism except at $\{\omega\}$). Since the projection of a neighborhood of 0 in Ker T onto S_ω via the orbits of GA(P) on $C(P) \cong \omega + \overline{\Omega}^1(P,SU(2))$ turns out to be an I_ω-equivariant diffeomorphism, it suffices to prove that the action of I_ω on the 6-dimensional space Ker T is a complex circle group action for some complex structure on Ker T; then $S_\omega/I_\omega \cong \mathbb{C}^3/U(1)$ will be a cone on $P_2(\mathbb{C})$. Now Ker T consists of those $\beta \in \overline{\Omega}^1(P,SU(2))$ for which $\delta\beta = 0$ and $(1-*)D^\omega\beta = 0$.

The elements of $\overline{\Omega}^1(P_0,\mathcal{SU}(2))$ extend uniquely to elements of $\overline{\Omega}^1(P,\mathcal{SU}(2))$ (i.e., the two spaces are isomorphic). Under the adjoint rep. of

$$U(1) = \left\{\begin{bmatrix} e^{i\theta} & 0 \\ 0 & e^{-i\theta}\end{bmatrix}\right\} \quad \text{on } \mathcal{SU}(2), \text{ we have the decomposition } \mathcal{SU}(2) = V \oplus W,$$

$$V = \mathbb{R}\begin{bmatrix} i & 0 \\ 0 & -i\end{bmatrix} \quad \text{and} \quad W = \left\{\begin{bmatrix} 0 & z \\ -\bar{z} & 0\end{bmatrix} : z \in \mathbb{C}\right\}.$$

Note that $U(1)$ acts trivially on the real space V, while $U(1)$ acts via multiplication by $e^{2\theta i}$ on the z variable for the complex vector space W. The space $\overline{\Omega}^1(P_0,\mathcal{SU}(2))$ then splits into $\overline{\Omega}^1(P_0,V) \oplus \overline{\Omega}^1(P_0,W)$. Since the $U(1)$ action on V is trivial, we can identify $\overline{\Omega}^1(P_0,V)$ with $\Omega^1(M,\mathbb{R})$, and D^ω, δ^ω become d, δ respectively. Since $H^1(M,\mathbb{R}) = 0$ by assumption, we know that there are no harmonic forms on M, and so $\operatorname{Ker} T$ must be contained entirely in the complex vector space $\overline{\Omega}^1(P_0,W)$ (i.e., $\operatorname{Ker} T$ is a complex vector space with $\dim_{\mathbb{C}} \operatorname{Ker} T = 3$). Since a generator $\alpha \in \Omega^0(P,\mathcal{SU}(2))$ of I_ω satisfies $D^\omega\alpha = 0$, we know that α is constant on P_0. Indeed we may take $\alpha \equiv \begin{bmatrix} i & 0 \\ 0 & -i\end{bmatrix}$ on P_0, since we know $e^{t\alpha}$ preserves $\omega|P_0$ and the unique extension of $e^{t\alpha}$ to P preserves ω. The action of $e^{t\alpha}$ on $\overline{\Omega}^1(P_0,W)$ is via multiplication by e^{2ti}, since for any $p \in P_0$ and $\beta \in \overline{\Omega}^1(P_0,W)$, we have $(e^{t\alpha}\cdot\beta)_p = r(e^{t\alpha(p)})\beta_p = e^{2ti}\beta_p$, where r is the rep. of $U(1)$ on W (cf. proof of Prop. 2 of 2D). Thus, the orbit space of the action of I_ω on $\operatorname{Ker} T$ is a cone on $P_2(\mathbb{C})$, as needed. As mentioned, the space which remains, when these open cones and the collar V_ε of concentrated instantons are removed from M^+, supplies the desired oriented cobordism between M and $n(s_M)$ copies of $\pm P_2(\mathbb{C})$. This completes our sketch.

We will close with the interesting observation that, as a consequence of Donaldson's Theorem, $\mathbb{R}^4$ admits an exotic differentiable structure which renders it not diffeomorphic to $\mathbb{R}^4$ with its usual differentiable structure. Although more details may be found in [Lawson 1983], we will briefly indicate how this can happen. By Freedman's result, there is a topological, compact, simply-connected manifold M with $s_M \sim E_8 \oplus \langle 1\rangle$. Donaldson's Theorem tells us that M admits no differentiable structure. Nevertheless, R. Gompf has shown (cf. [Lawson 1983]) that this M has the property that it is smoothable if some point $p \in M$ is removed. Indeed, it turns out that there is a neighborhood U of p such that $U\setminus\{p\}$ is diffeomorphic to $\tilde{U}\setminus h(S^2)$, where $\tilde{U}$ is a neighborhood of the image

$h(S^2)$ of $S^2 = P_1(\mathbb{C})$ under a homeomorphism $h: P_2(\mathbb{C}) \to P_2(\mathbb{C})$. Note that $P_1(\mathbb{C}) := \{[z_1,z_2,0] \in P_2(\mathbb{C})\} \subset P_2(\mathbb{C})$ with $P_2(\mathbb{C}) \setminus P_1(\mathbb{C}) \cong \mathbb{C}^2$, via $[z_1,z_2,z_3] \longmapsto (z_1/z_3, z_2/z_3)$. Thus, $P_2(\mathbb{C}) \setminus h(S^2) = h(P_2(\mathbb{C}) \setminus P_1(\mathbb{C}))$ is homeomorphic to $\mathbb{R}^4$. We argue that $P_2(\mathbb{C}) \setminus h(S^2)$ cannot be diffeomorphic to the usual $\mathbb{R}^4$. The usual $\mathbb{R}^4$ has the property that every compact set can be enclosed by a smoothly embedded 3-sphere. The complement of $\tilde{U} \setminus h(S^2)$ in $P_2(\mathbb{C}) \setminus h(S^2)$ is compact, but there can be no smooth S^3 enclosing it. Otherwise, we could remove the exterior of this S^3 and glue on a 4-disk; transferring the result back to M, we obtain an impossible smoothing of M.

Appendix. What are Vector Bundles?

Let X be a topological space. A *family of vector spaces* over X is a topological space E together with

(i) a continuous surjective map $p: E \to X$ and

(ii) a vector space structure of finite dimension in each $E_x = p^{-1}(x)$, $x \in X$, which carries the topology induced by E.

By vector spaces, we mean complex vector spaces, unless explicitly indicated otherwise.

The mapping p is called the *projection*, E is called the *total space* of the family, X is the *parameter space* or *base space* of the family, and for $x \in X$, E_x is the *fiber* over x. A *section* of a family $p: E \to X$ is a continuous map $s: X \to E$ such that $(p \circ s)(x) = x$ for all $x \in X$.

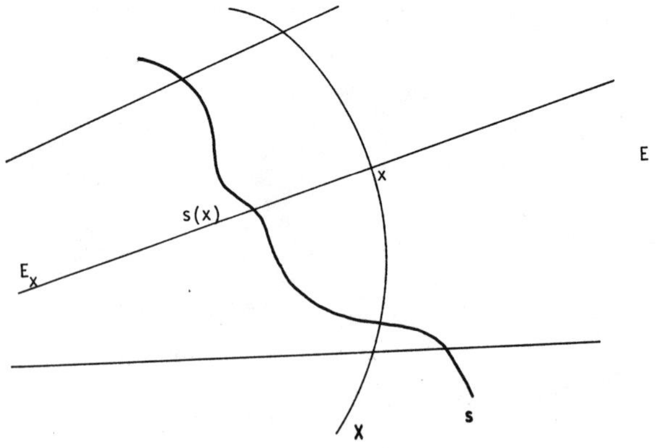

A *homomorphism* from one family $p: E \to X$ to another $q: F \to X$ is a continuous map $\phi: E \to F$ such that

(i) $q \circ \phi = p$,

(ii) $\phi_x: E_x \to F_x$ is a linear map for each $x \in X$. We write $\phi \in \mathrm{Hom}(E,F)$.

We say that such a ϕ is an *isomorphism* when ϕ is bijective and ϕ^{-1} is continuous. E and F are called *isomorphic* when there is an isomorphism between them. We write $\phi \in \mathrm{Iso}(E,F)$ and $E \cong F$. □

Exercise 1. a) Let V be a finite-dimensional vector space; e.g, $V = \mathbb{C}^N$. Show that a family of vector spaces over X is obtained by taking $E := X \times V$ with $p: E \to X$ being the projection on the first factor. This is the *product family* V_X with fiber V.

b) If F is a family which is isomorphic to a product family, then one calls F *trivial*. Show that a trivial family of finite dimensional (real) vector spaces is obtained, if $E := \{(x,y,-\lambda y,\lambda x): x,y,\lambda \in \mathbb{R}$ and $x^2 + y^2 = 1\}$ and $p(x,y,*,*) := (x,y)$. In general, prove that a bundle F is isomorphic to a product family V_X, if and only if one can find N sections $s_i: X \to F$ such that $s_1(x),\dots,s_N(x)$ forms a basis for F_x for each $x \in X$; here $N = \dim V$.

c) Let Y be a subset of X and E a family of vector spaces over X with projection p. Show that $p^{-1}(Y) \to Y$ is a family over Y. We call this the *restriction* of E to Y, and write $E|Y$ for this family.

d) More generally: Let Y be an arbitrary topological space and $f: Y \to X$ a continuous map. As follows, define the *induced family* $f^*(p): f^*(E) \to Y$: Take $f^*(E)$ to be the subspace of $Y \times E$ consisting of points (y,e) with $f(y) = p(e)$; the projection and the vector space structure on the fibers are self-evident. Show: For each further map $g: Z \to Y$, there is a natural isomorphism $g^*f^*(E) \overset{\leftarrow}{\cong} (fg)^*(E)$ which one obtains by mapping each point of the form (z,e) with $z \in Z$ and $e \in E$ to the point $(z,g(z),e)$. If $f: Y \to X$ is the inclusion, then there is an isomorphism $E|Y \cong f^*(F)$ given by mapping $e \in E|Y$ to the point $(p(e),e)$.

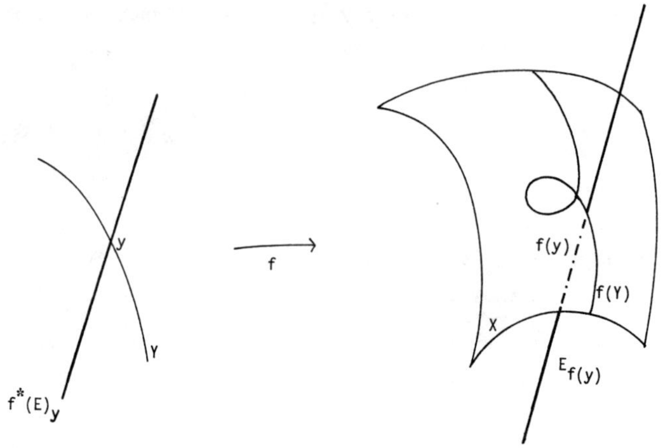

□

A family of vector spaces is called *locally trivial,* if each $x \in X$ possesses a neighborhood U such that $E|U$ is trivial. A locally trivial family is called a *vector bundle*; trivial families are called *trivial bundles*. If $f: Y \to X$ and E is a vector bundle over X, then clearly $f^*(E)$ is a vector bundle over Y which we call the *induced bundle*.

Note: If E is a vector bundle over X, then $\dim(E_x)$ is a locally constant function on X, and hence is constant on each connected component of X. If $\dim(E_x)$ is constant on all of X, then one says that E has *dimension* equal to the common fiber dimension $\dim(E_x)$. If X is a manifold, then the real dimension of E (regarded as a topological space) equals $\dim(X) + 2\dim(E_x)$. Vector bundles of fiber dimension 1 are also called *line bundles*.

Since a vector bundle is locally trivial, each section can be written locally as a vector-valued function on the base space. For a vector bundle E, we denote the space of sections of E by $C^0(E)$; $C^0(E)$ forms a vector space in a natural way via pointwise addition, etc. □

Exercise 2. a) Let V be a (complex) vector space and $\mathbb{P}V$ its associated "projective space" of all one-dimensional linear subspaces of V. We can write $\mathbb{P}V = (V-\{0\})/\sim$, where $\sim$ is the equivalence relation $v \sim w \Leftrightarrow \lambda v = w$ for some $\lambda \in \mathbb{C}$. We define $H_V \subset \mathbb{P}V \times V$ as the set of all (x,v) such that $x \in \mathbb{P}V$, $v \in V$, and v belongs to the complex line x. Show that H_V is a vector bundle in a natural way. (The construction goes back to Heinz Hopf.)

b) Go through the corresponding construction in the (more intuitive) category of real vector bundles when V is real (e.g., $\mathbb{R}^n$), and show that H_V is a real subbundle of $\mathbb{P}V \times V$ of fiber dimension 1, and that H_V is nontrivial if $\dim V \geq 2$.

c) For the moment, we remain in the real category and consider the following family (parametrized by $\theta \in [0,2\pi]$) of integro-differential equations for C^∞ functions on the unit interval which satisfy the boundary condition $f(0) = f(1)$:

$$(\cos\theta)f(x) + (\sin\theta)\frac{df}{dx} = (\cos\theta)\int_0^1 f(x)dx, \quad 0 \leq \theta \leq \pi,$$

$$(\cos\theta)f(x) + (\sin\theta)(Lf(x)) = (\cos\theta)\int_0^1 f(x)dx + (\sin\theta)\int_0^1 Lf(x)dx,$$
$$\pi \leq \theta \leq 2\pi.$$

Here, L is an operator with $L^2 = -\mathrm{Id}$. Show that the solutions of the family of equations forms a real vector bundle over the circle $S^1 := \mathbb{R}/2\pi\mathbb{Z}$ that is nontrivial and isomorphic to the bundle $H_{\mathbb{R}^2}$. (Actually, *every* real line bundle over S^1 is either trivial or isomorphic to $H_{\mathbb{R}^2}$; see also [Bröcker-Jänich 1973/1982, 3.23.9].)

<u>Hint for b)</u>: In contrast to the complex numbers, $\{-1\}$ cannot be deformed into $\{1\}$ without going through $\{0\}$. Thus, a real bundle is nontrivial, if it remains connected after the zero section is removed.

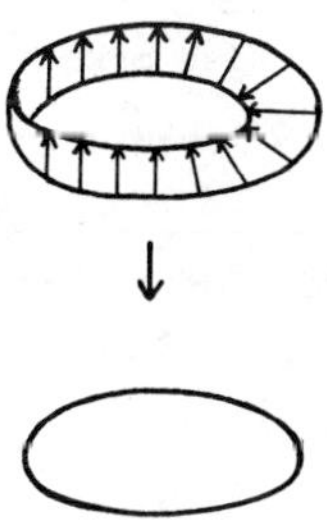

<u>For c)</u>: First show that for $0 \leq \theta \leq \pi$, the solutions are the constant functions $c\mathbf{1}$, and for $\pi \leq \theta \leq 2\pi$ the solutions are the functions $c(\cos\theta)\mathbf{1} - (\sin\theta)L(\mathbf{1})$, $c \in \mathbb{R}$. With the topology of $S^1 \times C^0(I)$ or of $S^1 \times \mathbb{R}^2$ (since every solution can be written in the form $c_1\mathbf{1} + c_2L(\mathbf{1})$), construct a family of 1-dimensional (real) vector spaces over S^1 and show the local triviality. For this use the initial value map of $f \longmapsto f(0)$. Note that this function can vanish for $f \neq 0$, namely if $\cos\theta_0 = (\sin\theta_0)L(\mathbf{1})(0)$. How can one procede in a neighborhood of θ_0? Distinguish the cases $L(\mathbf{1})(0) \gtreqless 0$. Incidentally, how does one obtain an L with $L^2 = -1$? Start with the Fourier series

$f(x) = a_0 + a_\nu \sin \nu x + b_\mu \cos \mu x$, and replace a_ν by $b_{\nu+1}$ and b_μ by $-a_{\mu-1}$; see also [Singer 1970]. □

Remark. a) and b) describe the origin of the bundle concept in analytic and projective geometry. Part c) is characteristic for many function analytic situations with "jumps" where the passage from one side to another (from one solution curve to another of the same equation) cannot be understood within the given space but requires an extension of the system (e.g., by parametrization). A basic model for such a process is present in the geometry of number fields (see Hint for b)). Many classical results of analysis - especially concerning the dependence of the solutions of a functional equation on the variation of its coefficients, and on the zeros and poles of its solutions - can be aptly formulated in the language of vector bundles. Conversely, the theorem of Grothendieck, for example, that every holomorphic vector bundle E on the Riemannian number sphere $S^2 = \mathbb{P}(\mathbb{C}^2)$ can be represented as a Whitney sum $E_1 \oplus \dots \oplus E_n$ of line bundles (Am. J. Math. 79(1957), 121-138) was known to analysts at the beginning of the century: See G. Birkhoff, Math. Ann. 54(1913), 122-139, where Grothendieck's theorem appears as a theorem about matrices of analytic functions. G. B. was lead to this theorem through his investigation of the singular points of ordinary differential equations; further see D. Hilbert, Gött. Nachr. (1905), 307-338, who gave a proof of Grothendieck's theorem for $N = 2$ in his "Fundamentals of a General Theory of Integral Equations" in connection with the "Riemannian Problem". (Contributed by M. Schneider). □

Exercise 3. Show that the usual operations for vector spaces in linear algebra also make sense for vector bundles. In particular, investigate the *direct sum* $E \oplus F$, the *tensor product* $E \otimes F$, the *homomorphism bundle* $L(E,F)$, and the dual bundle $E^* := L(E,\mathbb{C}_X)$, for vector bundles E and F over the same base. Also, carry over the concepts of subspace and quotient space from linear algebra to the corresponding concepts of a *subbundle* F of E and a *quotient bundle* E/F.

Hint: Make use of the fact that the corresponding operations in the "structure group" $GL(N,\mathbb{C})$ are continuous! Example: set $E^* := \bigcup E_x^*$ and introduce a topology on $E^*|U$ which makes $U \times \mathbb{C}^N \xrightarrow{\phi^*} E|U$ a homeomorphism, where $\phi: E|U \to U \times \mathbb{C}^N$ is a local trivialization for E over the open subset $U \subset X$. Let $\psi: E|V \to V \times \mathbb{C}^N$ be another trivialization. Do ϕ and ψ define the same topology on $E^*|U \cap V$? Does the continuity of $(\psi\circ\phi^{-1})^*: (U \cap V) \times \mathbb{C}^N \to (U \cap V) \times \mathbb{C}^N$ follow from that of $\psi\circ\phi^{-1}$:

$(U \cap V) \times \mathbb{C}^N \to (U \cap V) \times \mathbb{C}^N$? For this, write the two chart changes in the form $U \cap V \to GL(N,\mathbb{C})$ and prove (!) that $GL(N,\mathbb{C}) \overset{*}{\to} GL(N,\mathbb{C})$ is continuous. □

We denote the set of isomorphism classes of vector bundles over X by $\mathrm{Vect}(X)$, and let $\mathrm{Vect}_N(X)$ be the subset of $\mathrm{Vect}(X)$ consisting of the classes of bundles of dimension N. $\mathrm{Vect}(X)$ is an abelian semigroup under the operation $\oplus$. In $\mathrm{Vect}_N(X)$, there is a naturally distinguished element, namely the class of the trivial bundle of dimension N. A vector bundle over a point is a vector space, and hence $\mathrm{Vect}(X)$ can be identified with the semigroup $\mathbb{Z}_+$ of non-negative integers, in this case. However, in the general case, when there are nontrivial bundles (see Exercise 2 above), the isomorphism classes of vector bundles are not determined by their dimensions. □

Theorem 1. (i) If $f: X \to Y$ is a homotopy equivalence, then the transformation $f^*: \mathrm{Vect}(Y) \to \mathrm{Vect}(X)$ is bijective. (Assume X and Y compact.) (ii) If X is contractible, then every bundle over X is trivial, and $\mathrm{Vect}(X)$ is isomorphic to the non-negative integers.

$f: X \to Y$ is a *homotopy-equivalence,* if there is a continuous map $g: Y \to X$ such that $g \circ f \sim \mathrm{Id}_X$ and $f \circ g \sim \mathrm{Id}_Y$ ("~" means "homotopic"). X and Y are then called *homotopy-equivalent.*

Two mappings $f,h: X \to Y$ are *homotopic,* if there is a continuous map $F: X \times I \to Y$ $(I = [0,1])$ such that $F_0 := F|X \times \{0\} = f$ and $F_1 := F|X \times \{1\} = g$. The set of *homotopy classes* of maps $X \to Y$ is denoted by $[X,Y]$.

X is called *contractible,* if X is homotopy-equivalent to a point.

Proof: (ii) follows easily from (i), since $\mathrm{Vect}_N(P)$ consists only of the isomorphism class of the trivial bundle of dimension N, in the case where P is a point.

(i) follows from the fact that $F_0^*E \cong F_1^*E$, if $F: X \times I \to Y$ is a homotopy and E is a vector bundle over Y. We give a proof of this in three steps:

Step 1: Let H be a vector bundle over $X \times I$ and $s \in C^0(H|X \times \{\tau\})$, $\tau \in I$. We show that s can be extended to a section $S \in C^0(H)$ with $S|X \times \{\tau\} = s$. Since a section of a vector bundle can be regarded locally as a graph of a continuous vector-valued function, one can locally apply the Tietze Extension Theorem [Dugundji 1966]: We can find a neighborhood U about each (x,τ), and a section $t \in C^0(H|U)$ such that t and s coincide on $(X \times \{\tau\}) \cap U$. By the compactness of X, we can

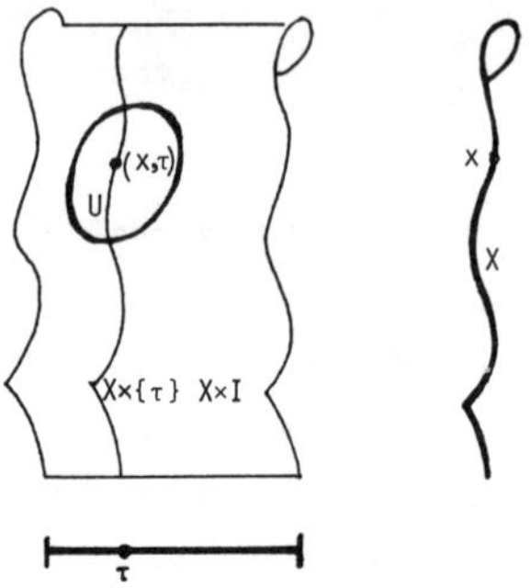

obtain a finite system $\{U_j\}$, $\{t_j\}$ with $X \times \{\tau\} \subset \cup U_j$. If $\{\phi_j\}$ is a C^0 partition of unity subordinate to $\{U_j\}$ (see also Theorem II.2.1), then we set

$$s_j := \begin{cases} \phi_j t_j & \text{on} \quad U_j \\ 0 & \quad (X\times I)\setminus U_j . \end{cases}$$

By construction $s_j \in C^0(H)$; hence, $\Sigma s_j \in C^0(H)$ is a well-defined extension of S.

Step 2: From Step 1, we conclude that two vector bundles G and G' over $X \times I$ which are isomorphic over $X \times \{\tau\}$, are also isomorphic in a neighborhood of $X \times \{\tau\}$: Each $s \in \mathrm{Iso}(G|X \times \{\tau\}, G'|X \times \{\tau\})$ can be regarded as a section of $H|X \times \{\tau\}$, where H is the bundle $L(G,G')$ of linear maps from fibers of G to co-responding fibers of G'. Let $S \in C^0(H)$ be an extension of s. Then the set $U := \{z \in X \times I$: $S_z: G_z \to G'_z$ is bijective$\}$ is open in $X \times I$ (by the classical zero-determinant argument) and contains all of $X \times \{\tau\}$ by construction. Since the inverse map of $GL(N,\mathbb{C})$ is continuous, it follows that the mapping $z \longmapsto (S_z)^{-1}$ is continuous, and hence a bundle isomorphism is defined on U.

Step 3. We now set $G := F^*E$ and $G' := p^*(F_\tau)^*E$, where $F_\tau(x) := F(x,\tau)$ and $p: X \times I \to X$ is the projection. By Exercise 1d), G and G' are isomorphic over $X \times \{\tau\}$, and by Step 2, they are also isomorphic in a whole neighborhood, which we can take to be a strip $X \times \delta(\tau)$, by the compactness of X. For all $\rho \in \delta(\tau)$, we

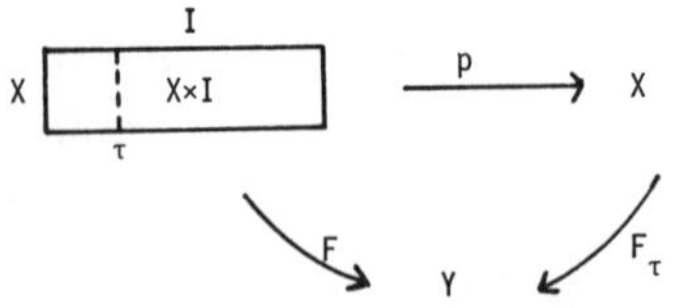

then have $(F_\rho)^*E \cong (F_\tau)^*E$. Since the unit interval I is compact and connected, we obtain that the isomorphism classes of $(F_\tau)^*E$ do not depend on τ. □

Remark. The statements proved in the first two steps of the proof, also apply to more general situations, and are occasionally formulated as independent theorems; e.g., see [Atiyah-Bott 1964b, 233f] = [Atiyah 1967a, 16]. One can also express the result of our proof somewhat more generally (we write $X \times Z$ instead of $X \times I$): Each vector bundle E over the topological space $X \times Z$ (X and Z compact) can be regarded as a "continuous" family of vector bundles E_z over X where the parameter z is in Z, and the isomorphism classes of E_z in $\mathrm{Vect}(X)$ are locally constant. □

Vector bundles are often given via a *clutching construction*: Let $X = X_1 \cup X_2$ and $A = X_1 \cap X_2$, where all spaces are compact. Let E_i be a vector bundle over X_i and $\phi: E_1|A \to E_2|A$ an isomorphism. Then we define the vector bundle $E_1 \cup_\phi E_2$ over X as follows. As a topological space $E_1 \cup_\phi E_2$ is the quotient space of the disjoint union $E_1 + E_2$ under the equivalence relation which identifies $e_1 \in E_1|A$ with $\phi(e_1) \in E_2|A$. We take X to be the corresponding quotient space of $X_1 + X_2$; then we obtain a natural projection $p: E_1 \cup_\phi E_2 \to X$ and $p^{-1}(x)$ has a natural vector space structure. □

Exercise 4. a) Show that $E_1 \cup_\phi E_2$ is a vector bundle.

b) Show that the isomorphism classes of $E_1 \cup_\phi E_2$ depend solely on the homotopy class of the isomorphism $\phi: E_1|A \to E_2|A$.

Hint for a): It remains only to show that $E_1 \cup_\phi E_2$ is locally trivial. Outside of A, this is clear. In order to extend a trivialization of E_1 in a neighborhood $V_1 \subset X_1$ of a point $a \in A$ to a trivialization of E_2 over V_2 with $a \in V_2 \subset X_2$, argue as in Step 2 of Theorem 1. See also [Atiyah-Bott 1964b, 235] = [Atiyah 1967a, 21]. For b): Reduce to Theorem 1. See also the given sources. □

We now give $\mathrm{Vect}_N(X)$ a homotopy-theoretic interpretation, when X can be represented as the suspension $S(Y)$ of another space Y.* Here the *suspension* $S(Y)$ is the union of two cones over Y. Thus, we write $S(Y) := C^+(Y) \cup C^-(Y)$, where $C^+(Y) := Y \times [0,\frac{1}{2}]/Y \times \{0\}$ and $C^-(Y) := Y \times [\frac{1}{2},1]/Y \times \{1\}$. Then $Y = C^+(Y) \cap C^-(Y)$. We note that the suspension $S(S^n)$ of the n-sphere is homeomorphic to the (n+1)sphere S^{n+1}. □

*With the concept of Grassmann manifolds, one can give a homotopy-theoretic definition of $\mathrm{Vect}_N(X)$ for *arbitrary* X; see [Atiyah 1967a, 24-30].

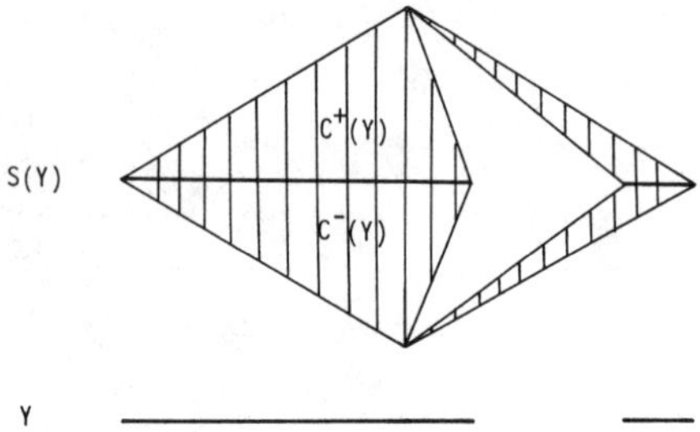

Theorem 2. The clutching of trivial bundles over $C^+(Y)$ and $C^-(Y)$ defines a natural isomorphism $[Y, GL(N,\mathbb{C})] \xrightarrow{\cong} \mathrm{Vect}_N(S(Y))$.

Proof: (i) Each $\ell: Y \to GL(N,\mathbb{C})$ yields a bundle over SY via the clutching of the N-dimensional trivial bundles over the two cones, and homotopic maps ℓ_0 and ℓ_1 yield isomorphic bundles; see the proof of Theorem 1(i). (ii) Conversely, we have the composition

$$\mathrm{Vect}_N(SY) \to \mathrm{Vect}_N(C^-Y) \oplus \mathrm{Vect}_N(C^+Y) \to [X,GL(N,\mathbb{C})],$$

the left arrow is given by the restrictions of the bundle, where one obtains trivial bundles since $C^{\pm}Y$ are contractible (see Theorem I(ii).). If $\alpha^{\pm}$ are such trivializations, then the right arrow is defined by taking the homotopy class of $(\alpha^+Y)(\alpha^-Y)^{-1}: Y \to GL(N,\mathbb{C})$, which actually only depends on the homotopy classes of $\alpha^{\pm}$ and hence only on the isomorphism class in $\mathrm{Vect}_N(SY)$. (iii) By construction the functions given in (i) and (ii) are inverse to each other. □

Exercise 5. Show $H_{\mathbb{C}^2} \cong \mathbb{C}_{B_0} \cup_a \mathbb{C}_{B_\infty}$, where $H_{\mathbb{C}^2}$ is the complex line bundle over $\mathbb{P}\mathbb{C}^2 = \mathbb{C} \cup \{\infty\} = S^2 = B_0 \cup B_\infty$ defined in Exercise 2a); here (z_0,z_1) are homogeneous coordinates for $\mathbb{P}\mathbb{C}^2$ with $(0,1) = \infty$, and $z = z_1/z_0$ is the coordinate for $\mathbb{C}$, $B_0 := \{z \in \mathbb{C}: |z| \leq 1\}$, and $B_\infty := \{z \in \mathbb{C}: |z| \geq 1\} \cup \{\infty\}$ (the two canonical hemispheres of S^2). Finally, $a: z \mapsto z$, $|z| = 1$, is the standard map $S^1 \to \mathbb{C}^* = GL(1,\mathbb{C})$. □

Exercise 6. Show that for each bundle E, there is a bundle F such that $E \oplus F$ is trivial.

Hint: Show first with the help of a finite open cover of the compact parameter space X and a suitable partition of unity, that $C^0(E)$ contains an "ample" subspace; i.e., a subspace $V \subset C^0(E)$ such that each point of E is in the image of a section $s \in V$. If $\dim V = N$, then we have an epimorphism $\phi: X \times \mathbb{C}^N \to E$, and consequently there is an isomorphism $E \oplus F \cong X \times \mathbb{C}^N$ where F is the kernel bundle of ϕ; see also [Atiyah 1967a, 26 f.]. □

Exercise 7. Let X be a topological space which possesses in addition the structure of a C^∞ manifold of dimension n (see II.2).

a) Show that the tangent bundle TX is a (real) vector bundle over X; do the same for the "normal bundle" NX, when X is a submanifold of a Riemannian manifold Y.

b) When may one call a continuous vector bundle over X, whose total space is a C^∞ manifold, a C^∞ vector bundle? Show that each (continuous) vector bundle over X is isomorphic to a C^∞ vector bundle.

Hint for a): First investigate the case $X = S^1$ and show that TS^1 is isomorphic to the real line bundle defined in Exercise 1b. In the general case, this "direct" method is also possible, since one can (see Theorem II.2.2c) embed X into a high dimensional Euclidean space, and thus realize TX as a (real) subbundle of a higher dimensional trivial bundle. It is easier (especially since TX is in general not trivial; e.g., for $X = S^2$, see Exercise III.1.11) to carry out a local analysis, where each chart u from the C^∞ atlas yields a local trivialization of TX via the differential forms $(du_1,\ldots,du_n)$; see Chapter II.2. See also the discussion below.

For b): For the definition of C^∞ vector bundles, see also [Bröcker-Jänich 1982, Ch. 3], and for the topological equivalence of the categories of C^0 and C^∞ vector bundles, see the Whitney Approximation Theorem in [Bröcker-Jänich 1982, 66]. Details of the argument are in [Hirsch 1976, 101], where it is shown that one can make E itself into a C^∞ vector bundle. □

Remark. From Exercise 7b, it follows (without loss of generality) that we only need to investigate C^∞ vector bundles, if the base is C^∞. Exercise 6 does not say that we always encounter trivial bundles (compare with the analogous - but deeper - embedding theorem for manifolds, Theorem II.2.2c): Roughly, the carrying along of additional "irrelevant" parameters is not only redundant, but can also produce so much "noise" that this noise destroys the structure of the problem or renders it unrecognizable. This is the case with the index problem for elliptic operators, whose solution consists exactly in distinguishing certain vector bundles generated by the symbol of the operator; see Part III. □

It is possible to construct, from the tangent bundles, *bundles of "exterior differential forms"* by means of multilinear algebra. These are an important tool in describing physical laws mainly in the areas of electromagnetism and special relativity. This is the case when empirical relationships are to be expressed in terms of an integral in such a way that the physicist or engineer can pursue qualitative and quantitative changes resulting from modifications of the integrand or the domain of integration. For this reason, exterior bundles are of interest also in differential topology; see Chapter III.4.

We briefly summarize (details are found in [UNESCO 1970, 111-161] and the literature given there): For a real n-dimensional vector space V, we form the vector space $\Lambda^p(V)$ of p-fold "*skew-symmetric tensors*" ("forms"); these are the multilinear maps

$$\overbrace{V^* \times \ldots \times V^*}^{p \text{ times}} \to \mathbb{R}, \quad p \in N, \quad V^* := L(V, \mathbb{R}),$$

which change, under a permutation of the arguments, by a factor equal to the sign of the permutation. One sets $\Lambda^0(V) := \mathbb{R}$ and obtains $\Lambda^1(V) = V$, $\Lambda^{n-1}(V) \cong V$, $\Lambda^n(V) \cong \mathbb{R}$ and $\Lambda^p(V) = 0$ for $p > n$. For $v \in \Lambda^p(V)$ and $w \in \Lambda^q(V)$, we define

$$(v \wedge w)(a_1, \ldots, a_{p+q}) := \frac{1}{p!q!} \sum_\sigma \operatorname{sgn}(\sigma) v \otimes w(a_{\sigma(1)}, \ldots, a_{\sigma(p+q)})$$

(sum over all permutations), which gives the *exterior multiplication* $\Lambda^p(V) \times \Lambda^q(V) \to \Lambda^{p+q}(V)$. This multiplication makes $\Lambda^*(V) := \Sigma_{p=0}^n \Lambda^p(V)$ a graded algebra, the *exterior algebra* of V.

If $e_1, \ldots, e_n$ is a basis of V, then the $\binom{n}{p}$ forms $e_{i_1} \wedge \ldots \wedge e_{i_p}$ with $1 \le i_1 < \ldots < i_p \le n$ yield a basis for $\Lambda^p(V)$. With this property $\Lambda^p(V)$ is occasionally defined (in order to avoid the suggestive but tedious definition via maps) as the space of "p-vectors": The space of formal linear combinations of the p-tuples of basis vectors $e_{i_1} \wedge \ldots \wedge e_{i_p}$ with only the relation $e_{\sigma(i_1)} \wedge \ldots \wedge e_{\sigma(i_p)} = \operatorname{sgn}(\sigma) e_{i_1} \wedge \ldots \wedge e_{i_p}$.

A scalar product (= inner product) for V induces a scalar product $<\ ,\ >$ for $\Lambda^p(V)$, and declaring an orthonormal basis $e_1, \ldots, e_n$ of V to be positively oriented yields an explicit isomorphism $\Lambda^n(V) \cong \mathbb{R}$ via $e_1 \wedge \ldots \wedge e_n \longrightarrow 1$, which only depends on the chosen orientation. Via the condition $u \wedge (*v) = <u,v>$ for all $u \in \Lambda^p(V)$, an isomorphism (the "*star operator*") $* : \Lambda^p(V) \to \Lambda^{n-p}(V)$ is defined. □

Exercise 8. Let X be a closed C^∞ manifold of dimension n with cotangent bundle T^*X.

a) Show that the family of vector spaces $\Lambda^p(T_x^*)$, $x \in X$, yields a real vector bundle of fiber dimension $\binom{n}{p}$ over X in a natural way. We denote this bundle by $\Lambda^p(T^*X)$.

b) Customarily, one writes $\Omega^p(X) := C^\infty(\Lambda^p(T^*X))$. For a C^∞ function f, consider the differential df (see also Section II.2.B) and show that the operator $d: \Omega^0(X) \to \Omega^1(X)$ "extends" to a linear differential operator $d: \Omega^p(X) \to \Omega^{p+1}(X)$ of first order (see Chapter II.2) for each p.

c) Show that for an oriented Riemannian manifold X, we have the "Hodge duality" $*: \Omega^p(X) \overset{\cong}{\to} \Omega^{n-p}(X)$.

d) Prove that for a compact, oriented, n-dimensional, Riemannian manifold with boundary, we have Stokes' Theorem $\int_X d\omega = \int_{\partial X} \omega|\partial X$; $\omega \in \Omega^{n-1}(X)$.

Hint for a): In principle, use the same mechanism as in Exercise 3. Note the simple transformation rules, e.g., for 1-forms

$$v = \sum_{i=1}^n a_i du^i|_x = \sum_{i=1}^n b_i dw^i|_x, \qquad x \in X,$$

where

$$b_i := \sum_{j=1}^n a_j \frac{\partial(u \circ w^{-1})^j}{\partial x^i}(w(x)),$$

and u and w are charts for X in a neighborhood of x.

For b): d is characterized by the Leibniz rule $d(v \wedge w) = dv \wedge w + (-1)^p v \wedge dw$; $v \in \Omega^p$, $w \in \Omega^q$. How is d written in local coordinates?

For d): [Guillemin-Pollack 1974, 182-187]. Incidentally, here one really needs the orientation. □

The idea of a vector bundle originates in the analysis of non-Euclidean manifolds. Otherwise, according to Theorem 1(ii), there are only trivial bundles, since Euclidean space is contractible. As an example, let $c: I \to X$ be a differentiable path in a manifold X. Classical mechanics considers the velocity vector $\dot{c}(t)$ for $t \in I$. For the physicist it was always clear (by reasons of physics) how $\dot{c}(t)$ is multiplied by a scalar, how $\dot{c}_1(t_1)$ and $\dot{c}_2(t_2)$ are added or how the equality of $\dot{c}_1(t_1)$ and $\dot{c}_2(t_2)$ is checked when $c_1, c_2: I \to X$ are two paths in X with $c_1(t_1) = c_2(t_2)$. From the point of view of physics no confusion between a velocity vector and a position vector was conceivable, but the earliest mathematical abstractions could not express the difference:

If the position vectors $c(t)$ are represented as triples of real numbers $(c_1(t), c_2(t), c_3(t))$, then the velocity vector is $(\dot{c}_1(t), \dot{c}_2(t), \dot{c}_3(t))$ where $\dot{c}_i(t)$ is the derivative of c_i at t. Thus $c(t)$ and

$\dot{c}(t)$ are both elements of the one vector space $\mathbb{R}^3$. This low level of abstraction was fully sufficient as long as X was Euclidean space (actually an affine space - but choose a base point and call it 0). In this case there is indeed a natural interpretation of the velocity vector $\dot{c}(t)$ at the space point $c(t)$ as a velocity vector $\dot{\tilde{c}}(t)$ at the point $\tilde{c}(t) = 0$. In fact, consider the translated path $\tilde{c}: I \to X$ with $\tilde{c}(\tau) := c(\tau) - c(t)$, $\tau \in I$. This means that the tangent spaces at the various points of X can be identified canonically (via the retraction $r: X \to \{0\}$) with the tangent space at the point 0. In the language of vector bundles, we could say $T(X) \cong r^*(T(X)|\{0\})$. Here $T(X)$ is the totality of all velocity vectors, the fiber $T(X)_x$ is $\mathbb{R}^3$, the restricted bundle $T(X)|\{0\}$ is precisely the space $\{0\} \times \mathbb{R}^3 \cong \mathbb{R}^3$ and the induced bundle $r^*(\{0\} \times \mathbb{R}^3)$ is the trivial bundle $X \times \mathbb{R}^3$. □

Let us next consider the case that X is a submanifold of a Euclidean space Y, say the 2-sphere in 3-space. Every velocity vector $\dot{c}(t)$ can be considered an element of the tangent space of Y at the point $c(t)$ and hence an element of the tangent space of Y at 0 (due to the translation described above), i.e., as n-tuple of derivatives of the coordinate functions of c. Already by reason of dimensions it is clear that, in general, the tangent space $T(X)_x$ of all velocity vectors at x of paths through x cannot be identified with the full tangent space $T(Y)_0$ but only with some subspace and it may be different for each x.

<u>Example 1</u>. $X = S^1$ and $Y = \mathbb{R}^2$

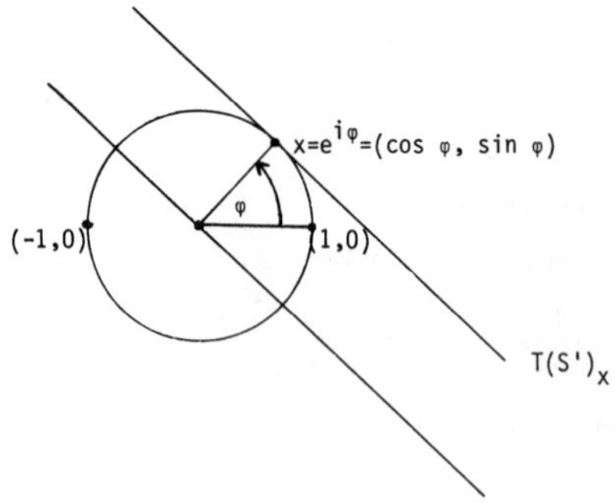

If we follow the translation with a rotation through the angle ϕ then we have identified all tangent spaces $T(S^1)_x$ with the same subspace of $T(\mathbb{R}^2)_0$ or with the tangent space $T(S^1)_{(1,0)}$. We write $T(S^1) = q^*(T(S^1)/\{(1,0)\})$ where $q\colon S^1 \to \{(1,0)\}$ is the retraction and $T(S^1)|\{(1,0)\}$ can be identified with $\{(1,0)\} \times \mathbb{R}$. Consequently $T(S^1) \cong S^1 \times \mathbb{R}$, i.e., the tangent bundle of S^1 is trivial.

In analytic terms this circumstance is expressed by saying that there exists a nowhere vanishing tangential vector field on S^1, i.e., at each point a non-vanishing velocity vector can be chosen and in a continuous fashion. For example choose at $x = (\cos\phi, \sin\phi)$ the unit velocity vector $\dot{c}(\phi/2\pi)$ where $c\colon I \to X$ is given by $c(t) = (\cos 2\pi t, \sin 2\pi t)$. Usually we write $\frac{d}{d\theta}(x)$ instead of $\dot{c}(\phi/2\pi)$. Then $\frac{d}{d\theta}$ is the nowhere vanishing vector field and it defines an isomorphism $T(S^1) \to S^1 \times \mathbb{R}$. The isomorphism is, for all $x \in S^1$, given by the map $T(S^1)_x \to \mathbb{R}$ which assigns to a velocity vector the value λ if it is equal to $\lambda \cdot \frac{d}{d\theta}(x)$. □

Example 2. $X = S^2$ and $Y = \mathbb{R}^3$.

Here the situation is different. There is no canonical way of identifying all tangent spaces. In other words - the tangent bundle $T(S^2)$ is non-trivial. If not there would have to exist two tangential vector fields on S^2 which are linearly independent at each point of S^2. But on S^2 there is not even a single vector field which vanishes nowhere (Exercise III.1.11).

Example 2 shows that it might be useful to distinguish the different tangent spaces at the different points. As one goes on to consider manifolds without an explicitly given imbedding in Euclidean spaces - so in physics with the theory of relativity - the notion of a bundle becomes indispensable. □

The modern concept of a bundle evolved from the topology and geometry of manifolds as practiced by Heinz Hopf and others since the 1920's. In the 1950's the notion was precisely formulated, and the classification of bundles and their systematic employment in deep problems of geometry and analysis started. (For the theory of "characteristic classes", see [Hirzebruch 1966, 49 f.].) Crudely expressed, the success of these methods derives from their utilizing given or manufactured "classical" structues (such as the tangent bundles or bundles of differential forms) on the manifolds under discussion to the greatest extent possible and thus shifting the plane of study from manifolds - which are conceptually simpler but harder to understand - to vector bundles which are more easily analyzed:

While for topological manifolds and other "triangulizable" spaces one depends at first on the combinatorial methods of the analysis of "cell decompositions", and while for differentiable manifolds only the group of diffeomorphisms is a priori available for investigation, vector bundles offer more opportunity for manipulations because of their richer structure. Principally, large parts of linear algebra can be used directly. (Incidentally, these play an important role also for the other method albeit under the surface.) For example (see above Exercise 3) one can perform linear constructions with vector bundles such as forming direct sums and quotients - which is impossible to do with manifolds. Also, the "clutching functions" which can be used to build complicated manifolds from simpler ones (see e.g., Exercise II.2.12), i.e., the diffeomorphisms, become linear only in the first derivative ("functional matrix" or "Jacobian"). In contrast, Theorem 1 shows how much (via linear clutching functions) the topology of vector bundles can be reduced to the geometry of the matrix spaces of linear algebra.

At the start of the 1960's linear algebra had "matured" enough with the Periodicity Theorem (for the (stable) homotopy groups of invertible matrices) discovered by Raoul Bott just before, and Michael Francis Atiyah and Friedrich Hirzebruch extracted from these methods an abstract formalism - K-theory - which they developed as a generalized cohomology theory using stability classes of vector bundles; see Chapter III.1. □

Literature

Standard Works

Adams, R. A.: Sobolev Spaces, Academic Press, New York, 1975.

Ahlfors, L.: Complex Analysis, McGraw-Hill, New York, 1953.

Bröcker, Th. & K. Jänich: Introduction to Differential Topology, Cambridge Univ. Press, New York, 1982 (transl. from German).

Courant, R. & D. Hilbert: Methods of Mathematical Physics I and II, Interscience Publishers, New York, 1953/62 (transl. from German).

Dugundji, J.: Topology, Allyn and Bacon, Boston, 1966.

Dym, H. & H. P. McKean: Fourier Series and Integrals, Academic Press, New York, 1972.

Eilenberg, S. & N. Steenrod: Foundations of Algebraic Topology, Princeton Univ. Press, Princeton, 1952.

Greenberg, M. J.: Lectures on Algebraic Topology, Benjamin, New York, 1967.

Guillemin, V. & A. Pollack: Differential Topology, Prentice-Hall, Englewood Cliffs, 1974.

Hörmander, L.: Linear Partial Differential Operators, Springer, Berlin-Heidelberg-New York, 1963/1969 (Russ. transl. Moscow 1965).

Karoubi, M.: K-Theory - An Introduction, Springer-Verlag, Berlin-Heidelberg-New York, 1978.

Kline, M.: Mathematical Thought from Ancient to Modern Times, Oxford Univ. Press, New York, 1972.

Kumano-Go, H.: Pseudo-differential Operators, MIT Press, Cambridge, Mass., 1981.

Seifert, H. and W. Threlfall: A Textbook of Topology, Academic Press, New York, 1980 (1st German ed. 1934).

Singer, I. M. & J. A. Thorpe: Lecture Notes on Elementary Topology and Geometry, Scott Foresmann, Glenview, Ill., 1967; reprinted, Springer, New York.

Steenrod, N.: The Topology of Fibre Bundles, Princeton Univ. Press, Princeton, 1951.

Taylor, M. E.: Pseudo-differential Operators, Princeton Univ. Press, Princeton, 1981.

Treves, F.: Introduction to Pseudo-Differential and Fourier Integral Operators I & II, Plenum Press, New York, 1980.

Triebel, H.: Interpolation Theory-Function Spaces, Differential Operators, North-Holland/Elsevier, New York, 1978 (transl. from German).

UNESCO (ed.): Mathematics Applied to Physics, with contributions from G. A. Deschamps and others. Edited by E. Roubine for UNESCO, Springer, Berlin-Heidelberg-New York, 1970.

Yosida, K.: Functional Analysis, Springer-Verlag, Berlin-Heidelberg-New York, 1974.

Evolution of the Index Formula

Atiyah, M. F.: Harmonic spinors and elliptic operators. Workshop lecture, Bonn, 1962.

_________: The Grothendieck ring in geometry and topology, Proc. Int. Congr. Math. (Stockholm 1962), Uppsala 1963a, 442-446.

_________: The index of elliptic operators on compact manifolds, Sém. Bourbaki, Exp. 253, Paris 1963b.

_________: The Euler characteristic theorem I. Arbeitstagung lectures, Vervielfältigt, Bonn, 1966a.

_________: K-Theory, Benjamin, New York, 1967a (Russ. transl. Moscow 1967).

_________: Algebraic topology and elliptic operators, Comm. Pure Appl. Math. 20 (1967b), 237-249.

_________: Bott periodicity and the index of elliptic operators, Quart. J. Math. Oxford (2), 19 (1968a), 113-140.

_________: Global aspects of the theory of elliptic differential operators, Reports of the Internat. Congress of Mathematics (Moscow 1966), Verlag Mir, Moscow, 1968b, 57-64.

_________: Algebraic topology and operators in Hilbert space. In: C.T. Taam (ed.), Lectures in modern analysis and applications I. Lecture Notes in Mathematics 103, Springer, Berlin (West) - Heidelberg - New York, 1969, 101-121.

_________: Topology of elliptic operators, Amer. Math. Soc. Symp. Pure Math., 16 (1970a), 101-119.

_________: Global theory of elliptic operators, Proc. Int. Conf. Functional Analysis, Tokyo, 1970b, 21-30.

_________: Vector fields on manifolds, Arbeitsgemeinschaft für Forschung des Landes Nordrhein-Westfalen 200 (1970c).

_________: Elliptic operators and singularities of vector fields, Actes, Congres Intern. Math. (Nice 1970), Tome 2, 1970d, 207-209.

_________: Riemann surfaces and spin structures, Ann. Sci. École Norm. Sup. (4) 4 (1971), 47-62.

_________: The role of algebraic topology in mathematics, J. London Math. Soc. 41 (1966b), 63-69.

_________: Elliptic operators and compact groups, Lecture Notes in Mathematics 401, Springer, Berlin (West) - Heidelberg-New York, 1974.

_________: Eigenvalues and Riemannian geometry. In: Proc. Internat. Conf. on Manifolds and Related Topics in Topology (Tokyo 1973), Univ. Tokyo Press, Tokyo, 1975a, 5-9.

_________: Classical groups and classical differential operators on manifolds, Centro Internazionale Matematico Estivo (Varenna 1975), Edizioni Cremonese, Rome, 1975b, 5-48.

_________: Elliptic operators, discrete groups and von Neumann algebras, Soc. Math. de France Astérisque 32/33 (1976a), 43-72.

_________: Trends in pure mathematics. In: Proc. of the 3rd Internat. Congress on Mathematical Education, Karlsruhe, 1976b, 61-74.

_______ & R. Bott: The index problem for manifolds with boundary, Coll. Differential Analysis, Tata Institute, Bombay, Oxford University Press, Oxford, 1964a, 175-186.

_____, _____: On the periodicity theorem for complex vector bundles, Acta Math. 112 (1964b), 229-247.

_____, _____: Notes on the Lefschetz fixed point theorem for elliptic complexes, Notes, Harvard University, 1965 (Russ. transl. Moscow, 1966).

_____, _____: A Lefschetz fixed point formula for elliptic differential operators, Bull. Amer. Math. Soc., 72 (1966), 245-250.

_____, _____: The Lefschetz fixed point theorem for elliptic complexes I. Ann. of Math. 86 (1967), 374-407.

_____, _____: The Lefschetz fixed point theorem for elliptic complexes II. Ann. Of Math. 88 (1968), 451-491.

_____, _____, & V. K. Patodi: On the heat equation and the index theorem, Invent. Math. 19 (1973), 279-330 (and Errata, 28 (1975), 277-280) (Russ. transl., Moscow, 1973).

_____ & J. L. Dupont: Vector fields with finite singularities, Acta Math. 128 (1972), 1-40.

_____ & F. Hirzebruch: Riemann-Roch theorems for differentiable manifolds, Bull. Amer. Math. Soc. 65 (1959), 276-281.

_____, _____: Vector bundles and homogeneous spaces, Amer. Math. Soc. Symp. Pure Math. 3 (1961), 7-38.

_____, _____: Analytic cycles on complex manifolds, Topology 1 (1962), 25-46.

_____, V. K. Patodi & I. M. Singer: Spectral asymmetry and Riemannian geometry, Bull. London Math. Soc. 5 (1973), 229-234.

_____, _____, _____: Spectral asymmetry and Riemannian geometry I/II/III. Math. Proc. Camb. Phil. Soc. 77 (1975a), 43-69/78 (1975b), 405-432/79 (1976), 71-99.

_____ & G. B. Segal: The index of elliptic operators II. Ann. of Math. 87 (1968), 531-545 (Russ. transl. in Yonexn Matem. Hayk 23 (1968), Nr. 6 (144), 135-149).

_____ & I. M. Singer: The index of elliptic operators on compact manifolds, Bull. Amer. Math. Soc. 69 (1963), 422-433.

_____, _____: The index of elliptic operators I. Ann. of Math. 87 (1968a), 484-530 (Russ. Transl. in Yonexn Matem. Hayk 23 (1968), Nr. 5 (143), 99-142).

_____, _____: The index of elliptic operators III. Ann. of Math. 87 (1968b), 546-604 (Russ. transl. in Uspehi Mat. Nauk 24 (1969), Nr. 1 (145), 127-182).

_____, _____: Index theory for skew-adjoint Fredholm operators, Publ. Math. Inst. Hautes Etudes Sci. 37 (1969), 305-325.

_____, _____: The index of elliptic operators IV. Ann. of Math. 93 (1971a), 119-138 (Russ. transl. in Uspehi Mat. Nauk 27 (1972), Nr. 4 (166), 161-178).

_____, _____: The index of elliptic operators V. Ann. of Math. 93 (1971b), 139-149 (Russ. transl. in Uspehi Mat. Nauk 27 (1972), Nr. 4 (166), 179-188).

Booss, B.: Elliptische Topologie von Transmissionsproblemen, Bonner Mathematische Schriften Nr. 58, Bonn 1972.

Bott, R.: Homogeneous differential operators. In: Differential and combinatorial topology (S. S. Cairns, ed.), Princeton Univ. Press, Princeton 1965, 167-186.

Boutet de Monvel, L.: Boundary problems for pseudo-differential operators, Acta Math. 126 (1971), 11-51.

Bojarski, B. W.: On the index problem for systems of singular integral equations, Bull. Acad. Polon. Sci. Sér. Sci. Math. Astronom. Phys. 11 (1963), 653-655.

Brieskorn, E. et al.: Die Atiyah-Singer-Indexformel. Seminarvorträge, Bonn, 1963.

Breuer, M.: Fredholm theories in von Neumann algebras I/II. Math. Ann. 178 (1968), 243-254, und 180 (1969), 313-325.

Cartan, H. & L. Schwartz et al.: Théorème d'Atiyah-Singer sur l'indice d'un opérateur différentiel elliptique. Séminaire Henri Cartan 16e année 1963/64, École Normale Supérieure, Paris 1965.

Calderon, A.P.: The Analytic calculation of the index of elliptic equations, Proc. Nat. Acad. Sci. 57 (1967), 1193-1194.

Coburn, L. A., R. G. Douglas, D. G. Schaeffer & I. M. Singer: C*-algebras of operators on a half space II - Index theory, Publ. Math. Inst. Hautes Études Sci. 40 (1971), 69-79.

Fedosov, B. W.: Analytic formulae for the index of elliptic operators. Trans. Moscow Math. Soc. 30 (1974), 159-240 (translation from 30 (1974), 159-241).

Gelfand, I. M.: On elliptic equations, Russ. Math. Surveys 15, No. 3 (1960), 113-123 (translation from Uspehi Mat. Nauk 15, 3 (1960), 121-132).

Gilkey, P. B.: The index theorem and the heat equation, Mathematics Lecture Series No. 4, Publish or Perish Inc., Boston, 1974.

__________: The boundary integrand in the formula for the signature and Euler characteristic of a Riemannian manifold with boundary, Advances in Math. 15 (1975), 334-360.

Hirzebruch, F.: Topological Methods in Algebraic Geometry, 3rd ed., Springer Berlin (West)-Göttingen-Heidelberg, 1966 (1st German ed. 1956).

__________: Elliptische Differentialoperatoren auf Mannigfaltigkeiten, Arbeitsgemeinschaft für Forschung des Landes Nordrhein-Westfalen 33 (1966), 583-608.

_____ & D. Zagier: The Atiyah-Singer theorem and elementary number theory, Mathematics Lecture Series 3, Publish or Perish Inc., Boston, 1974.

Hörmander, L.: On the index of pseudodifferential operators. In: G. Anger (ed.), Elliptische Differentialgleichungen, Band II, Akademie-Berlag, Berlin 1971, 127-146 (Russ. transl., Moscow, 1970).

Mayer, K. H.: Elliptische Differentialoperatoren und Ganzzahligkeitssätze für charakteristische Zahlen. Topology 4 (1965), 295-313.

McKean, H. P., Jr. & I. M. Singer: Curvature and the eigenvalues of the Laplacian, J. Differential Geometry 1 (1967), 43-69.

Palais, R. S. (ed.): Seminar on the Atiyah-Singer index theorem, Ann. of Math. Studies 57, Princeton Univ. Press, Princeton, 1965 (Russ. transl. Moscow, 1970).

Ray, D. & I. M. Singer: R-torsion and the Laplacian on Riemannian manifolds, Advances in Math. 7 (1971), 145-210.

_____, _____: Analytic torsion for complex manifolds, Ann. Of Math., 98 (1973), 154-177.

Schulze, B. W.: On the set of all elliptic boundary value problems for an elliptic pseudodifferential operator on a manifold, Math. Nachr. 75 (1976a), 271-282.

_________: Elliptic operators on manifolds with boundary, Proc. "Globale Analysis", Ludwigsfelde, Berlin, 1976b (notes).

Seeley, R. T.: Integro-differential operators on vector bundles, Trans. Am. Math. Soc. 117 (1965), 167-204.

_________: The resolvent of an elliptic boundary problem, Amer. J. Math. 91 (1969), 889-920.

Singer, I. M.: On the index of elliptic operators. In: Outlines Joint Soviet Amer. Sympos. Partial Diff. Equations (Nowosibirsk 1963), Moscow, 1963.

_________: Elliptic operators on manifolds. In: Pseudo-Differential Operators, Centro Internazionale Matematico Estivo Stresa 1968, Edizioni Cremonese, Rom, 1969, 333-375.

_________: Operator theory and K-theory. Lecture presented in the Arbeitsgemeinschaft für Forschung des Landes Nordrhein-Westfalen, 1970.

_________: Future extensions of index theory and elliptic operators. In: Prospects in Mathematics by F. Hirzebruch et al., Princeton University Press, Princeton 1971a.

_________: Operator theory and periodicity, Indiana Univ. Math. J. 20 (1971b), 949-951.

_________: Recent applications of index theory for elliptic operators, Amer. Math. Soc. Symp. Pure Math. 23 (1973), 11-31.

_________: Eigenvalues of the Laplacian and invariants of manifolds, Proc. Int. Congr. Math. (Vancouver 1974), Vancouver, 1975, 187-200.

Part I

Atkinson, F. W.: The normal solubility of linear equations in normed spaces, Mat. Sb. 28 (70) (1951), 3-14 (Russian).

Brieskorn, E.: Über die Dialektik in der Mathematik. In: M. Otte (ed.) Mathematiker über die Mathematik, Springer-Verlag, Berlin (West)-Heidelberg-New York, 1974, 221-285.

Brown, L. G., R. G. Douglas & P. A. Fillmore: Unitary equivalence modulo the compact operators and extensions of C*-algebras. In: P. A. Fillmore (ed.), Proceedings of a Conference on Operator Theory, Lecture Notes in Mathematics 345, Springer, Berlin (West)-Heidelberg-New York, 1973.

Calkin, J. W.: Two-sided ideals and congruence in the ring of bounded operators in Hilbert space, Ann. of Math. 42 (1941), 839-873.

Devinatz, A.: On Wiener-Hopf-operators. In: B. Gelbaum (ed.), Functional analysis, Thompson, Washington 1967, 81-118.

Douglas, R. G.: Banach algebra techniques in operator theory, Academic Press, New York, 1972.

Fillmore, P. A. (ed.): Proceedings of a conference on operator theory, Lecture Notes in Mathematics 345, Springer-Verlag, Berlin (West)-Heidelberg-New York, 1973.

Gelfand, I. M., D. A. Raikov & G. E. Shilov: Kommutative normierte Algebren VEB Deutscher Verlag der Wissenschaften, Berlin, 1964 (translation from Russian).

Gohberg, I. Z. & I. A. Feldman: Faltungsgleichungen und Projektionsverfahren zu ihrer Lösung, Akademie-Vlg., Berlin, 1974 (translation from Russian)

_____ & M. G. Krein: The basic propositions on defect numbers, root numbers, and indices of linear operators, Amer. Math. Soc. Transl. (2) 13 (1960), 185-264 (translation from Uspehi Mat. Nauk 12 (1957), Nr. 2, 44-118).

_____, _____: System of integral equations on the half-line with kernels depending on the difference of the arguments, Amer. Math. Soc. Transl. (2) 14 (1960), 217-287 (translation from Uspehi Mat. Nauk 13 (1958), Nr. 2, 3-72).

Hellinger, E. & O. Toeplitz: Integralgleichungen und Gleichungen mit unendlichvielen Unbekanten. In: Enzyklopädie der mathematischen Wissenschaften II, C 13, Teubner, Leipzig, 1927, 1335-1601.

Jörgens, K.: Lineare Integraloperatoren, Teubner, Stuttgart, 1970; Eng. transl. Pitman, San Francisco, 1982.

Illusie, L.: Contractibilité du groupe linéaire des espaces de Hilbert de dimension infinie, Séminaire Bourbaki 17 (1964/65), No. 284.

Ivachenko, A. G. & V. G. Lapa: Cybernetics and forecasting techniques, American Elsevier, New York, 1967 (translation from Russian).

Kolmogorov, A. N.: Interpolation and extrapolation of stationary stochastic sequences, Izv. Akad. Nauk SSSR Ser. Mat. 5 (1941), 3-14 (Russian).

Krein, M. G.: Integral equations on a half-line with kernel depending upon the difference of the arguments, Amer. Math. Soc. Transl. (2) 22 (1962), 163-288 (translation from Yonexn Matem. Hayk 13 (1958), Nr. 5, 3-120).

Kuiper, N. H.: The homotopy type of the unitary groups of Hilbert space, Topology 3 (1965), 19-30.

MacLane, S.: Homology, Springer-Verlag, Berlin (West)-Heidelberg-New York, 1963.

Noether, F.: Ober eine Klasse singulärer Integralgleichungen, Math. Ann. 82 (1921), 42-63.

Przevorska-Rolevicz, D. & S. Rolewicz: Equations in linear spaces, Panstvove Vydavnistvo Naukove, Warsaw, 1968.

Schechter, M.: Principles of functional analysis, Academic Press, New York, 1971.

Titchmarsh, E. C.: Introduction to the theory of Fourier integrals, Oxford University Press, Oxford, 1937, I/1948 II.

Wiener, N.: The Fourier integral and certain of its applications, Cambridge University Press, Cambridge, 1933.

________: Extrapolation, interpolation and smoothing of stationary time series. With engineering applications, M.I.T. Press, Cambridge, 1949.

________: Cybernetics - Control and communication in the animal and the Machine, 2nd ed., M.I.T. Press, New York, 1961.

Zypkin, J. S.: Adaption and Lernen in kybernetischen Systemen., Oldenbourg Blg., München - Wien, 1970 (translation from Russian).

Part II

Agranovich, M. S.: Elliptic singular integro-differential operators, Russian Math. Surveys, 20 (1965), Nr. 5/6, 1-121 (translation from Uspehi Mat. Nauk 20 (1965), Nr. 5 (125), 3-120).

Bers, L., F. John & M. Schechter: Partial differential equations, Interscience Publishers, New York-London-Sidney, 1964 (Russ. translation, Moscow, 1966).

Calderon, A. P.: Boundary value problems for elliptic equations. In: Outlines of the Joint Soviet-Amer. Symp. Partial Diff. Equations (Novosibirsk, 1963), Acad. Sci. USSR, Siberian Section, Moscow, 1963, 303-304.

_____ & A. Zygmund: On singular integrals, Amer. J. Math. 78 (1956), 289-309.

Dynkin, E. B. & A. A. Yushkevich: Sätze und Aufgaben über Markoffsche Prozesse. Heidelberger Taschenbücher 51, Springer, Berlin (West)-Heidelberg-New York, 1969 (translation from Russian).

Eskin, G. I.: Boundary value problems and the parametrix for systems of elliptic pseudodifferential equations, Trans. Moscow Math. Soc. 28 (1973), 74-115 (translation from Uspehi Mat. Nauk, 28 (1973), 75-116).

________: Boundary value problems for elliptic pseudo-differential equations, Amer. Math. Soc. Transl. 52, 1981 (translation from Russian).

Grubb, G. & G. Geymonat: The essential spectrum of elliptic systems of mixed order, Math. Ann. 227 (1977), 247-276.

Hartmann, P.: Ordinary differential equations, Wiley, New York, 1964.

Hermann, R.: Vector bundles in mathematical physics I/II, Benjamin, New York, 1970.

Hörmander, L.: Pseudo-differential operators and non-elliptic boundary problems, Ann. Math. 83 (1966a), 129-209 (Russ. transl., Moscow, 1967).

________: Pseudo-differential operators and hypoelliptic equations, Amer. Math. Soc. Symp. X on Singular Integral Operators, 1966b, 138-183 (Russ. transl., Moscow, 1967).

________: The calculus of Fourier integral operators. In: Prospects in Mathematics (F. Hirzebruch et al.), Princeton University Press, Princeton, N.J., 1971a, 33-57.

________: Fourier integral operators I, Acta Mathematica 127 (1971b), 79-183 (Russ. transl., Moscow, 1972).

________: On the existence and the regularity of solutions of linear pseudo-differential equations, Enseignement Math. 17 (1971c), 99-163 (Russ. transl. in Uspehi Mat. Nauk 28 (1973), Nr. 6 (174), 109-164).

Ize, J.: Bifurcation theory for Fredholm operators, Mem. Am. Math. Soc. 7/174 (Sept. 1976).

Karoui, N. & H. Reinhard: Processus de diffusion dans $\mathbb{R}^n$. In: Sém. Probabilités VII, Univ. de Strasbourg, Springer LN 321, Berlin (West)-Heidelberg-New York, 1973, 95ff.

Lions, J. L. & E. Magenes: Problèmes aux limites non homogènes et applications, I, Dunod, Paris, 1968.

Meyer, P. A.: Probabilités et potentiels, Hermann, Paris, 1966.

Morrey, C.: Multiple integrals and the calculus of variations, Springer, Berlin (West)-Heidelberg-New York, 1966.

Narasimhan, R.: Analysis on real and complex manifolds, Masson, Paris, 1973.

Nemitsky, V. V. & V. V. Stepanov: Qualitative theory of differential equations, Princeton Univ. Press, Princeton, N.J., 1960 (transl. from Russian).

Nirenberg, L.: Pseudo-differential operators, Amer. Math. Soc. Symp. Pure Math. 16 (1970), 149-167.

Pontryagin, L. S.: Ordinary Differential equations, Pergamon Press, London, 1962 (transl. from Russian).

Prössdorf, D.: Über eine Algebra von Pseudodifferentialoperatoren im Halbraum, Math. Nachr. 52 (1972), 113-139.

________: Einige Klassen singulärer Gleichungen, Academie-Verlag, Berlin, 1974 (Engl. translation North-Holland/Elsevier, New York, 1978).

Schwartz, L.: Théorie des distributions I, II, Hermann, Paris, 1950/51.

Taylor, M.: Pseudo differential operators, Lecture Notes in Mathematics 416, Springer, Berlin (West)-Heidelberg-New York, 1974.

Vaynberg, M. M. & V. A. Trenogyn: Theorie der Lösungsverzweigungen bei nichtlinearen Gleichungen, Akademie-Vlg., Berlin, 1973 (transl. from Russian).

Vishik, M. I. & G. I. Eskin: Normally solvable problems for elliptic systems of equations in convolutions, Math. USSR-Sbornik 3 (1967), 303-332 (transl. from Matem. Co. 74 (1967), Nr. 3, 326-356).

Wallace, A. H.: Differential topology - first steps, Benjamin, New York, 1968.

Part III

A. A. Albert: Introduction to Algebraic Theories, Univ. of Chicago Press, Chicago, 1941.

Alexandroff, P. & H. Hopf: Topologie I, Springer, Berlin, 1935.

Bellman, R. & K. L. Cooke: Modern elementary differential equations, Addison-Wesley, Reading, 1971.

______ & R. Kalaba: Quasilinearization and nonlinear boundary value problems, American Elsevier, New York, 1965.

Benedetto, J. J.: Spectral synthesis, Teubner, Stuttgart, 1975.

Brieskorn, E.: The development of geometry and topology, Notes of introductory lectures given at the University of La Habana in 1973. M. z. Berufspraxis Math. Heft 17, Bielefeld 1976, 109-203.

Brill, A. & M. Noether: Die Entwicklung der Theorie der algebraischen Funktionen in älterer und neuerer Zeit. Jber. Deutsch. Math.-Verein. 3 (1892/93), 107-566.

Dieudonné, J.: Cours de géometrie algébrique, Vol. 1: Apercu historique sur le développement de la géometrie algebrique, Presses Univ. de France, Paris, 1974.

Hirsch, M. W.: Differential topology, Springer-Verlag, Berlin (West)-Heidelberg-New York, 1976.

Hirzebruch, F.: Hilbert modular surfaces, Enseignement Math. 19 (1973), 183-281.

Husemoller, D.: Fibre bundles, McGraw-Hill, New York, 1966 (Russ. transl., Moscow, 1970); 2nd Edition, Springer-Verlag, New York, 1975.

Kodaira, K.: On the structure of compact complex analytic surfaces I, Am. J. Math. 84 (1964), 751-798.

Koschorke, U.: Frame fields and nondegenerate singularities, Bull. Amer. Math. Soc. 81 (1975), 157-160.

Milnor, J. W.: Morse theory, Princeton Univ. Press, Princeton, N. J., 1963.

______ & J. D. Stasheff: Characteristique classes, Ann. of Math. Studies 76, Princeton Univ. Press, Princeton, 1974.

Mumford, D.: Introduction to algebraic geometry, Vol. 1, Springer, Berlin (West)-Heidelberg-New York, 1976.

Schwarzenberger, R. L. E.: Appendix One. Zu: F. Hirzebruch, Topological methods in algebraic geometry, Springer, Berlin (West)-Heidelberg-New York, 1966, 159-201.

Seeley, R. T.: Complex powers of an elliptic operator, Amer. Math. Soc. Proc. Sympos. Pure Math. 10 (1967), 288-307 (Russ. transl., Moscow, 1968).

Segal, G.: Equivariant K-theory, Publ. Math. Inst. Hautes Études Sci. 34 (1968), 105-151.

Siegel, C. L.: Topics in complex function theory I-III, Wiley-Interscience, New York, 1969/1971/1973.

Thomas, E.: Vector fields on manifolds, Bull. Amer. Math. Soc. 75 (1969), 643-683.

Toledo, D. & L. L. Tong: A parametrix for $\overline{\partial}$ and Riemann-Roch in Čech theory, Topology 15 (1976), 273-301.

Ulam, S. M.: A collection of mathematical problems, Interscience, New York-London, 1960.

Vekua, I. N.: Systeme von Differentialgleichungen erster Ordnung vom elliptischen Typus und Randwertaufgaben, VEB Deutscher Vlg. der Wissenschaften, Berlin, 1956 (transl. from Russian).

________: Generalized analytic functions, Pergamon Press, Oxford, 1962 (transl. from Russian).

Wallach, N. R.: Harmonic analysis on homogeneous spaces, Decker, New York, 1973.

Weyl, H.: The concept of a Riemann surface, 3rd ed., Addison-Wesley, Reading, Mass., 1964 (1st German ed., 1913).

Part IV

Atiyah, M. F., N. J. Hitchin, V. G. Drinfeld & Y. I. Manin: Construction of instantons, Phys. Lett. 65A (1978), 185-187.

Atiyah, M. F., N. J. Hitchin & I. M. Singer: Deformations of instantons, Proc. Nat. Acad. Sci. 74 (1977), 2662-2663.

________: Self-duality in four-dimensional Riemannian geometry, Proc. Roy. Soc. London A362 (1978), 425-461.

Atiyah, M. F. & R. S. Ward: Instantons and algebraic geometry, Commun. Math. Phys. 55 (1977), 117-124.

Belavin, A. A., A. M. Polyakov, A. S. Schwarz & Y. S. Tyupkin: Pseudo-particle solutions of the Yang-Mills equations, Phys. Lett. 59B (1975), 85-87.

Bernard, C. W., N. H. Christ, A. H. Guth & E. J. Weinberg: Instanton parameters for arbitrary gauge groups, Phys. Rev. D16 (1977), 2967-2977.

Bleecker, D. D.: Gauge Theory and Variational Principles, Addison-Wesley, Reading, Mass., 1981.

Cantor, M.: Elliptic operators and the decomposition of tensor fields, Bull. Amer. Math. Soc. 5 (1981), 235-262.

Chern, S. S. & J. Simons: Characteristic forms and geometric invariants, Ann. Math. 99 (1974), 48-69.

Donaldson, S. K.: An application of gauge theory to four-dimensional topology (Preprint), The Mathematical Institute, Oxford, 1983.

Drechsler, W. & M. E. Mayer: Fiber Bundle Techniques in Gauge Theories, Lecture Notes in Physics, Vol. 67, Springer-Verlag, New York, 1977.

Eguchi, T., P. B. Gilkey & A. J. Hanson: Gravitation, gauge theories, and differential geometry, Phys. Reports 66 (1980), 213-393.

Eisenhart, L. P.: Riemannian Geometry, Princeton University Press, Princeton (1966).

Freed, D. & K. Uhlenbeck: Instantons and four-manifolds, (MSRIP 1), Springer-Verlag, New York, 1984.

Freedman, M. H.: The topology of four-dimensional manifolds, J. Diff. Geom. 17 (1982), 357-453.

Freedman, M. H. & A. Kirby: Geometric proof of Rohlin's theorem, Proc. Symp. Pure Math., Vol. 32, Amer. Math. Soc., Providence, R.I. (1978), 85-97.

Jackiw, R., C. Nohl & C. Rebbi: Conformal properties of pseudo-particle configurations, Phys. Rev. D15 (1977), 1642-1646.

Kaluza, T.: Sitzber. Preuss Akad. Wiss. (1921), 966.

Klein, O.: Quantentheorie und fünfdimensionale Relativitätstheorie, Z. Phys. 37 (1926), 895.

________: On the theory of charged fields (in "New Theories in Physics", Proc. of a Conf. held in Warsaw, 1938), Institute for Intellectual Collaboration, Paris.

Kobayashi, S. & K. Nomizu: Foundations of Differettial Geometry, Vol. 1, Wiley, New York, 1963.

________: Foundations of Differential Geometry, Vol. 2, Wiley, New York, 1969.

Lang, S.: Differentiable Manifolds, Addison-Wesley, Reading, Mass., 1972.

Lawson, H. B.: The theory of gauge fields in four dimensions, Lectures from the CBMS Regional Conference held at the University of California at Santa Barbara, August 1983.

Madore, J.: Geometric methods in classical field theory, Phys. Reports 75 (1981), 125-204.

Milnor, J. & D. Husemoller: Symmetric Bilinear Forms, Springer-Verlag, New York, 1973.

Milnor, J. & J. Stasheff: Characteristic Classes, Princeton Univ. Press, Princeton, 1974.

Mitter, P. K. and C. M. Viallet: On the bundle of connections and gauge orbit manifold in Yang-Mills theory, Commun. Math. Phys. 79 (1981), 457-472.

Palais, R. S.: Foundations of Global Non-linear Analysis, Benjamin, New York, 1968.

Parker, T. H.: Gauge theories on four dimensional Riemannian manifolds, Commun. Math. Phys. 85 (1982), 563-602.

Schiff, L. I.: Quantum Mechanics, McGraw-Hill, New York, 1968.

Schwarz, A. S.: Instantons and fermions in the field of an instanton, Commun. Math. Phys. 64 (1979), 233-268.

Spanier, E. H.: Algebraic Topology, McGraw-Hill, New York, 1966.

Stong, R. E.: Notes on Cobordism Theory, Princeton Univ. Press, Princeton, 1958.

Taubes, C. H.: Self-dual Yang-Mills connections on non-self-dual 4-manifolds, J. Diff. Geom. 17 (1982), 139-170.

Trautman, A.: Fiber Bundles, Gauge Fields, and Gravitation, in General Relativity and Gravitation I, edited by A. Held, Plenum Press, New York, 1980, 287-307.

Uhlenbeck, K. K.: Removable singularities in Yang-Mills fields, Commun. Math. Phys. 83 (1982), 11-29.

________: Connections with L^p bounds on curvature, Commun. Math. Phys. 83 (1982), 31-42.

Weyl, H.: The Classical Groups, Princeton Univ. Press, Princeton, 1946.

Index of Notation for Parts I, II, and III

α	Bott isomorphism $K(\mathbb{R}^2 \times X) \cong K(X)$	240
$\lvert\alpha\rvert$	degree of multiindex α	85, 104
b	Bott class	240
B^n	n-dimensional ball	61
B_H	closed unit ball of Hilbert space H	17
$B(X)$	covariant ball bundle of Riemannian manifold X	199
$\mathcal{B}(H)$	Banach algebra of bounded linear operators on a Hilbert space H	3
$\mathcal{B}^x$	group of units of $\mathcal{B}$	35
β^+	boundary isomorphism for elliptic boundary-value problem	192f, 196
$\dot{c}(0)$	directional derivative	131
$c_i(E)$	i-th Chern class of vector bundle E	273
ch(E)	Chern character	273
Coker(T)	cokernel of operator T	3
C^r	r-times differentiable, $r \geq 0$	30ff
$C^0(X)$	complex-valued, continuous functions on X	128
C^∞	infinitely differentiable, smooth	128
$C^\infty(X)$	complex-valued C^∞ functions on the C^∞ manifold X	128
C^ω	analytic functions	(not used)
$C^r(E)$	C^r sections of the vector bundle E, $r = 0,1,2,\ldots,\infty,\omega$	136, 404
C_0^r	C^r functions with compact support	148
$C(X)$	semi-group of equivalence classes of complexes	249
$\mathbb{C}$	field of complex numbers	4
$\mathbb{C}_+$	half-plane of $\mathbb{C}$	89
$\mathbb{C}^N$	complex N-space	4
$\mathbb{C}_X^N$	trivial product bundle with fiber $\mathbb{C}^N$ over X	136
$\chi(E)$	Euler class of vector bundle E	279
d	exterior derivative	117, 413
deg	degree, mapping degree	224, 245
deg(P)	local index of an elliptic operator P	204
det	determinant	
$d\phi\rvert_x$	differential of ϕ at point x	134

dim(E)	fiber dimension of vector bundle E	404
dim(V)	dimension of vector space V	3
dim(X)	dimension of manifold X	128
div	divergence (of a vector field)	140
D^+	signature operator	279
D_E^+	signature operator with coefficients in E	261
D^α	partial derivative of multiindex α	85, 104
D	$-i \times$ ordinary derivative operator	82
$\overline{D}_V$	Riemann-Roch operator	291
$\mathrm{Diff}_k(E,F)$	set of differential operators of order $\leq k$	136
δ	Dirac distribution	100
$\mathcal{D}$	divisor	290
Δ	Laplace operator	110, 176, 279
∇	C^∞ connection	176
∂X	boundary of manifold X	141
$\frac{\partial}{\partial \bar{z}}$, $\bar{\partial}$	Cauchy-Riemann operator	125, 288
e(X)	Euler number of manifold X	8, 278
$E\|Y$	restriction of a bundle	403
[E]	isomorphism class of E in K(X)	63, 232
$E \oplus F$	direct sum of vector bundles	406
$E \otimes F$	tensor product of vector bundles	406
E*	dual bundle	406
E_x	fiber of bundle E above point x	402
$E^\cdot$	complex of vector bundles	249
$\mathrm{Ell}_k(E,F)$	space of elliptic operators of order k between vector bundles E and F	184
$\mathrm{Ell}_c(X)$	elliptic operators of order 0 which equal the identity at infinity on the paracompact space X	247
f_*	differential ("Jacobian matrix") of map f	132
f^*	pull-back of differential forms via map f	134
f*E	pull-back of bundle E via map f	403
$f \sim g$	map f homotopic to map g	47, 315
$\mathcal{F}(H)$	space of Fredholm operators on the Hilbert space H	3
g(X)	genus of Riemann surface X	224, 278, 284ff
$GL(N,\mathbb{C})$	general linear group	48
H_V	Hopf bundle over $\mathbb{P}V$	404f
H_k, H^k	homology, cohomology	8

$H_0(S^1)$	Hardy space of periodic functions	92
$H_0(\mathbb{R})$	Hardy space	99
index F	index of operator F	3
index T	index bundle of Fredholm family T	64
I	unit interval	47
I(X)	group of stable equivalence classes of bundles	233
Id	identity	10
Im(T)	image of operator T	3
Im(z)	imaginary part of the complex number z	120
$\mathrm{Iso}_{SX}(E,F)$	symbol space	187
$\mathrm{Iso}_{SX}^{\infty}(E,F)$	C^{∞} symbol space	185
K_{ϕ}	convolution operator on half-line	98
K(X)	ring of virtual vector bundles	64, 230-236
Ker(T)	kernel of the operator T	3
$\mathcal{K}$	ideal of compact operators	18
κ	Cayley transformation	99
L(f)	Lefschetz number	300
$L^1(X)$	Banach space of complex-valued absolutely-integrable functions on X	79ff
$L^2(X)$	Hilbert space of complex-valued square-integrable functions on X	4, 79ff
$L^2(E)$	Hilbert space of norm-square integrable sections of vector bundle E	172
Λ	canonical Riesz operator	174
M^{α}	partial multiplication operator	83
M_f	multiplication by f	92
$\mathcal{M}_{\eta}^{\pm}$	characteristic vector space for boundary-value problem	191f
NX	normal bundle of Riemannian submanifold X	134
$\mathcal{N}$	normal operators in a *-algebra	77
OP_k	operators of order k	182
Ω	space of complex-valued differential forms	260
$\Omega^{\pm}$	domain of signature operators	200
$\Omega^k(X)$	differential forms of degree k on X	413
P^*	formal adjoint of operator P	139, 172
P#Q	tensor product of P and Q over product manifold	187
$\mathbb{P}V$	projective space of vector space V	404
$\pi^*(E)$	pull-back of E to T'X via $\pi: T'X \to X$	136
$\pi_1(X,x_0)$	fundamental group	223

$\pi_n(G)$	n-th homotopy group of G	225
R^α	Riesz operator of multiindex α	150
Re(z)	real part of complex number z	120
Res(T)	resolvant set of the operator T	40
$\mathbb{R}$	field of real numbers	17
$\mathbb{R}_+$	non-negative real numbers	98
$\mathbb{R}^n$	real n-space	17
$\text{shift}^\pm$	shift operators	4
sign(X)	signature of a 4q-dimensional manifold X	279
S^1	circle (1-sphere)	50
S^n	n-sphere	54
SX	covariant sphere bundle	185
$\text{Smbl}_k(X)$	k-symbols of trivial line bundle over X	164
$\text{Symb}_k(E,F)$	k-symbols from E to F	137
Spec(T)	spectrum of T	39f
$\text{Spec}_c(T)$	continuous spectrum	40
$\text{Spec}_e(T)$	essential spectrum	39f
$\text{Spec}_p(T)$	point spectrum	39f
$\text{Spec}_r(T)$	residual spectrum	39f
$\text{supp}(\phi)$	support of function	105
σ_k	k-symbol	162f
σ	symbol (highest order)	137
T_f	discrete Wiener-Hopf operator of f	92
$\|\|T\|\|$	operator norm of T	3
T^n	n-dimensional torus	178
T*	operator adjoint to T	16
TX	tangent bundle of manifold X	132
T'X	covariant (dual tangent) bundle of X without zero section	136
T*X	covariant bundle of X	133f
$(TX)_x$	tangent space of X at x	131
$\mathcal{T}$	Toeplitz algebra of discrete Wiener-Hopf operators	102
U(E)	total orientation class of oriented vector bundle E	272
U(m)	unitary group	49
$\mathcal{U}(H)$	unitary group of Hilbert space H	49
V_X	trivial bundle over X with vector space V as fiber	403
Vect(X)	semigroup of isomorphism classes of complex vector bundles over X	407
$\text{Vect}_N(X)$	isomorphism classes of complex N-dimensional continuous vector bundles over X	407

$W(f,0)$	winding number of $f : S^1 \to \mathbb{R}^2 \setminus 0$ about 0	219-223
W_f	continuous Wiener-Hopf operator	100
$W^s(\mathbb{R}^n)$	Sobolev space of order s on $\mathbb{R}^n$	174
$W^m(\mathbb{R}^n_+)$	Sobolev space for a half-space	177
$X_1 \cup_f X_2$	clutching of manifolds via f	142
$X_1 \times X_2$	product manifold	187
$[X,Y]$	homotopy set	48
$\mathbb{Z}$	ring of integers	4
$\mathbb{Z}_+$	non-negative integers	4
$\langle ..,.. \rangle$	scalar product	3
$\langle ..,.. \rangle_x$	Riemannian metric at point x	134
$\langle ..,.. \rangle_0$	scalar inner product in L^2-spaces	173
$\hat{}$	Fourier transform	81f, 82
$\wedge$	wedge product for differential forms	412
$*$	convolution	80
$\lvert \cdot\cdot \rvert_0$	L^2-norm	173
$\lVert \cdot\cdot \rVert_s$	norm in Sobolev space	174
$\otimes$	tensor product	74, 96, 187
$\boxtimes$	outer tensor product	236
$\oplus$	direct sum	5
$\perp$	orthogonal	14

Index of Notation for Part IV

G	Lie group of matrices	332
$\pi: P \to M$	principal G-bundle	332
$\mathcal{G}$	Lie algebra of G	333
ω	connection 1-form on P	333
X^H, X^V	horizontal, vertical part of $X \in TP$	333-4
D^ω	covariant derivative rel. ω	334
Ω^ω	curvature of connection ω	334
$[\omega,\omega]$	bracket of $\mathcal{G}$-valued 1-form	334
$P \times_G W$	associated vector bundle	334
$C(P,W)$	equivariant W-valued functions on P	335
$\overline{\Omega}^k(P,W)$	equivariant W-valued k-forms on P	335
$\langle\cdot,\cdot\rangle$	$C^\infty(M)$-valued pairing on $C^\infty(P \times_G W)$	336
$(\cdot,\cdot)$	L^2-inner product on $C^\infty(P \times_G W)$	336
ν	volume element on M	336
$\|\|\cdot\|\|$	L^2-norm on $C^\infty(P \times_G W)$	337
$*$	(Hodge) star operator	337
Ω_-, Ω_+	spaces of (anti-) self-dual 2-forms	337
δ^ω	covariant codifferential	338
$C(P)$	space of C^∞ connections on P	339
$GA(P)$	group of C^∞ gauge transformations of P	339
Ad	adjoint action of G on G	340
$\mathcal{Ad}$	adjoint action of G on $\mathcal{G}$	340
$C(P,G)$	G-valued equivariant functions on P	340
$\Phi: C(P,G) \to GA(P)$	isomorphic identification	340
$\mathrm{Exp}: C(P,\mathcal{G}) \to C(P,G)$	exponential map	340
$F\cdot\alpha$	action of $GA(P)$ on $C(P)$ or $\overline{\Omega}^k(P,W)$	340
$\mathfrak{s}\cdot\alpha$	inf. action of $C(P,\mathcal{G})$ on $C(P)$ or $\overline{\Omega}^k(P,W)$	341-2
$L(M) \to M$	bundle of linear frames	342
ϕ	fundamental 1-form on $L(M)$	343
$F(M) \to M$	orthonormal frame bundle	343
θ	Levi-Civita connection	343
$R(X,Y,Z,W)$	curvature tensor	344
$\tilde{\Omega}^\theta$	curvature tensor as in $\Omega^2(M,\Lambda^2(M))$	344
$Ricc(X,Y)$	Ricci tensor	344

S	scalar curvature	344
$\mathcal{R}$	vector space of "curvature tensors"	344
S^2	symmetric bilinear forms	344
$r: \mathcal{R} \to S^2$	Ricci map	344
$s: \mathcal{R} \to \mathbb{R}$	scalar map	344
$v: S^2 \times S^2 \to \mathcal{R}$	"vee product"	344
$\mathcal{R}_1, \mathcal{R}_2, \mathcal{R}_3$	O(n)-irred. decomp. of $\mathcal{R}$	345
C_i, $i=1,\ldots,4$	SO(4)-irred. decomp. of $\mathcal{R}$ for $n = 4$	347
$\mathcal{W}^-, \mathcal{W}^+$	(anti-) self-dual part of $\mathcal{R}_3$	348
$P \bullet F(M)$	fibered-product	349
$T^{r,s}(W)$	$P \times_G$ W-valued tensors	350
$\Omega^k(W)$	$P \times_G$ W-valued k-forms	350
$\nabla^{\omega\oplus\theta}$	covariant derivative on $T^{r,s}(W)$	350
$D^{\omega\oplus\theta}$ $(= D^{\omega})$	exterior derivative on $\Omega^k(W)$	350
$t^{j_1\ldots j_s}_{i_1\ldots i_r \mid k}$	components of $\nabla^{\omega\oplus\theta}(t)$	351
Δ^{ω}	covariant Laplacian on $\Omega^k(W)$	352
$\mathcal{O}_-, \mathcal{O}_+$	(anti-)self-dual parts of $\mathcal{O}(4)$	355
$s_k(\tilde{\Omega})$	Chern forms	356
$c_k(E)$	Chern classes	356
$\chi(M)$	Euler characteristic	357
$P_0 \to M$	holonomy bundle	358
G_0	holonomy group	358
I_{ω}	gauge isotropy subgroup at ω	358
ad	derivative of Ad at identity	360
$YM(\omega)$	Yang-Mills functional	360
k	Killing form	360
k_c	complexified Killing form	361
$\mathcal{G}_c$	$\mathcal{G} \otimes \mathbb{C}$	361
$\eta_i, \bar{\eta}_i$	't Hooft matrices	364
σ_i	Pauli matrices	364
θ_+, θ_-	decomp. of Levi-Civita conn.	364
A^+, A^-	pull-backs of θ_+ & θ_- to $\mathbb{R}^4$	364
F^+, F^-	field strengths of A^+, A^-	364
θ^h_i	local Levi-Civita conn. forms	366
$\lvert\ \rvert_h$	norm rel. $h = f^2 ds^2$	367

$\| \|_e$	norm rel. ds^2	367
λ	size of instanton	368
M	space of moduli of connections	373
S_ω	slice through ω	374
E	$P \times_G G_c$ and/or all-purpose bundle	375
T	operator, $\Omega^1(E) \to \Omega^0(E) \oplus \Omega^2_-(E)$	375
$C(P)^+$	space of wkly-irred. self-dual conns.	380
$L^p(E)$	L^p sections of $E \to M$	380
$W^{p,k}(E)$	L^p Sobolev space of sections	380
$\| \|_{p,k}$	Sobolev norm	380
$C(P)^{p,k}$	Sobolev space of connections	382
$C(P)^+_m$	mildly-irred. self-dual connections	388
M^+	moduli of $C(P)^+_m$	388
s_M	"intersection form" on $H^2(M, \mathbb{Z})$	390
$M_1 \# M_2$	connected sum	391
$\sigma(s_M)$	signature of s_M	393
r(b)	rank of bilinear form b	393
n(b)	no. of solution pairs of $b(x,x) = 1$	393
V_ε	space of instantons of size $< \varepsilon$	395

Index of Names/Authors

Cursive numbers refer to the literature index.

Subject Index

Cursive page numbers indicate the introduction and/or definition of the entry.

9087